TL
507
.P75

Progres~
astronaut
and
aeronauti
an Americ

WEST GEORGIA TECH LIBRARY

MW01388190

Teleoperation and Robotics in Space

Teleoperation and Robotics in Space

Edited by
Steven B. Skaar
University of Notre Dame
Notre Dame, Indiana

Carl F. Ruoff
Jet Propulsion Laboratory
California Institute of Technology
Pasadena, California

Volume 161
PROGRESS IN
ASTRONAUTICS AND AERONAUTICS

A. Richard Seebass, Editor-in-Chief
University of Colorado at Boulder
Boulder, Colorado

Published by the American Institute of Aeronautics and Astronautics, Inc.
370 L'Enfant Promenade, SW, Washington, DC, 20024-2518

Copyright © 1994 by the American Institute of Aeronautics and Astronautics, Inc. Printed in the United States of America. All rights reserved. Reproduction or translation of any part of this work beyond that permitted by Sections 107 and 108 of the U.S. Copyright Law without the permission of the copyright owner is unlawful. The code following this statement indicates the copyright owner's consent that copies of articles in this volume may be made for personal or internal use, on condition that the copier pay the per-copy fee ($2.00) plus the per-page fee ($0.50) through the Copyright Clearance Center, Inc., 222 Rosewood Drive, Danvers, Massachusetts 01923. This consent does not extend to other kinds of copying, for which permission requests should be addressed to the publisher. Users should employ the following code when reporting copying from this volume to the Copyright Clearance Center:

1-56347-095-0/94 $2.00 + .50

Data and information appearing in this book are for informational purposes only. AIAA is not responsible for any injury or damage resulting from use or reliance, nor does AIAA warrant that use or reliance will be free from privately owned rights.

ISSN 0079-6050

Progress in Astronautics and Aeronautics

Editor-in-Chief
A. Richard Seebass
University of Colorado at Boulder

Editorial Board

Richard G. Bradley
Lockheed Fort Worth Company

John L. Junkins
Texas A&M University

Allen E. Fuhs
Carmel, California

Martin Summerfield
*Princeton Combustion Research
Laboratories, Inc.*

Jeanne Godette
Director
Book Publications
AIAA

Table of Contents

Preface

Part 1. Introduction

Part 2.   Human-Machine Interface

Chapter 3.   Human Enhancement and Limitation in
Thomas B. Sheridan, *Massachusetts Institute of Technology, Cambridge, Massachusetts*

Chapter 4.   Ground Experiments Toward Space Teleoperation

Blake Hannaford, *University of Washington, Seattle, Washington*

Part 3.   Planning and Perception

Chapter 7.   Automatic Planning in Robotic Applications 161

Chapter 8.   Techniques for Collision Prevention, Impact Stability, and Force Control by Space Manipulators 175

Richard Volpe, *Jet Propulsion Laboratory, California Institute of Technology, Pasadena, California*

Chapter 9.   Versatile and Precise Vision-Based Manipulation 213

Steven B. Skaar and Emilio Gonzalez-Galvan, *University of Notre Dame, Notre Dame, Indiana*

Part 4. Dynamics and Control

Chapter 10. Tutorial Overview of the Dynamics and Control of Satellite-Mounted Robots 237
Richard W. Longman, *Columbia University, New York, New York*

Chapter 11. Reorientation of Free-Flying Multibody Structure Using Appendage Movement 259
Ranjan Mukherjee, *Naval Postgraduate School, Monterey, California,* and
   Yoshihiko Nakamura, *University of Tokyo, Tokyo, Japan*

Chapter 12. Transfer Functions of Flexible Beams and Implications of Flexibility on Controller Performance 291

Sabri Cetinkunt, *University of Illinois at Chicago, Chicago, Illinois*, and
Wayne J. Book, *Georgia Institute of Technology, Atlanta, Georgia*

Chapter 13. Stability and Control of Robotic Space Manipulators 315

John L. Junkins, *Texas A&M University, College Station, Texas*, and Youdan
Kim, *Seoul National University, Seoul, Korea*

Part 5.   Telerobot System Design and Applications

Chapter 14.   Teleoperation: From the Space Shuttle to the

Phung K. Nguyen, *Spar Aerospace Ltd., Brampton, Ontario, Canada*, and
Peter C. Hughes, *University of Toronto, North York, Ontario, Canada*

Chapter 15.   Overview of International Robot Design for Space Station Freedom 411

W. Brimley, D. Brown, and B. Cox, *Spar Aerospace Limited, Brampton, Ontario, Canada*

Chapter 16.   Use of Manipulators in Assembly of Space Station Freedom 443

Patrick L. Swaim, Jeffrey J. Arend, Pat J. Bevill, Roy J. Decker, John C. Dunn, David A. Read, Robert E. Reiher, Brian J. Richard, Ken J. Ruta, and Scott Teplitz, *McDonnell Douglas Space Systems, Houston, Texas*

Chapter 17.   Space Station Robotics Task Validation and Training 475

Donald Woods, Mike Kearney, Dave Crosse, and Mike Massimino, *McDonnell*
   Douglas Aerospace, Houston, Texas

Preface

There is a great deal of interest in expanding the scope of commercial and scientific activity in space, but because of increasingly difficult economic constraints, governments are demanding higher productivity from their space programs and are reluctant to spend large sums on space projects that do not have clear benefits. Manned space operations entail a great deal of risk and are extremely costly due to associated demands for room, life support, and high reliability. As a result, expanding the scope of space activity will require the extensive use of robots and other forms of automation. Versatile, reliable space robots are a clear and attractive alternative to direct manual efforts by astronauts. They will be used for a variety of operations including maintenance, assembly, and repair.

Terrestrial robots, which have undergone significant development, have a great deal in common with space robots, but there are important differences arising from microgravity, vacuum, the thermal environment, and the need to minimize mass. In addition, requirements for versatility, reliability and good work throughput, coupled with the fact that the technology for building intelligent robots does not yet exist, imply that space robots must be supervised or controlled, sometimes at a detailed level, by human operators. Even with human supervision or control, however, space robots must have some onboard intelligence, since the operators may be ground-based, communicating with the robots through channels that have sharply limited data rates and significant transmission delays.

This volume is intended to provide a unified source that deals with some of the principal technical issues associated with space robotics and the justification, design, and operation of existing and anticipated space robot systems. It is intended for a broad audience, including engineers, scientists, managers, policymakers, students, and anyone with interests in human-supervised space robots. It addresses a range of current topics of interest to both the general technical reader and the specialist. These include the human-machine interface, planning, perception, dynamics, control, and space robot system design and applications. Dynamics, control, and human-machine considerations, many of which are unique to the space environment, are discussed in detail. The overview chapter is provided to establish the overall context.

We hope the volume proves to be a useful contribution.

Acknowledgments

The editors wish to thank for their indispensable help and support: Evelyn Addington, Christine Kalmin, Thomas Mueller, John-David Yoder, Wenzong Chen, Eric Baumgartner, and Umesh Korde. The support of the United States Office of Naval Research, Grant No. N00014-91-J-1054, is also gratefully acknowledged.

Steven Skaar
Carl Ruoff
August 1994

Part 1.   Introduction

Chapter 1

Overview of Space Telerobotics

Carl F. Ruoff[*]

*Jet Propulsion Laboratory, California Institute of Technology,
Pasadena, California 91109*

Introduction

T HIS chapter discusses the need for space teleoperators and robots, here generically called telerobots, and describes the basic elements of such systems as
well as their current capabilities, limitations, and needed technical improvements.
Rovers, which are intended for in-situ lunar and planetary exploration, as well as
reconnaissance,[1-3] involve many of the same basic technologies but are discussed
only briefly here.

In the years since 1957, when the launching of Sputnik thrust the world into the
Space Age, vast resources have been invested in developing space systems. This
investment has been enormously successful. Earth-orbiting satellites have revo-
lutionized communications, intelligence gathering, weather prediction, resource
management and navigation, and scientific satellites and spacecraft have provided
a wealth of data that have dramatically improved our scientific understanding of
the Earth, the solar system, and the universe. Apollo astronauts have visited the
moon, and spacecraft have flown by every planet in the solar system save Pluto
(a Pluto mission, the Pluto Fast Flyby, which uses a microspacecraft, is currently
under study by NASA). Automated orbiters and landers, including Mariner, Magellan, Surveyor, Viking, and the Russian Veneras, have explored the Moon, Mars,
and Venus, and the Space Shuttle and expendable launch vehicles now provide
reliable means for placing large payloads in orbit.

Space-based capabilities and the improved scientific knowledge resulting from
the investment in space are having significant practical effects. It is now possible,
for example, using communications satellites in geosynchronous orbit, to broadcast television to billions of people in once-inaccessible areas. It is also possible
to search for minerals and monitor environmental effects, such as ozone depletion

Copyright © 1994 by the American Institute of Aeronautics and Astronautics, Inc.
The U.S. Government has a royalty-free license to exercise all rights under the copyright
claimed herein for Governmental purposes. All other rights are reserved by the copyright
owner.
*Manager, Robotic Systems and Advanced Computer Technology Section.

and the shrinkage of forested areas, from space and, using the global positioning system (GPS), to determine positions on or near the surface of the Earth to within 100 m using inexpensive hand-held receivers[4] (military receivers are accurate to within about five meters). With more elaborate scientific receivers, accuracies of about 2 cm can be achieved. (Scientists used accurate GPS position determinations to great advantage in the recent Northridge earthquake in Los Angeles. Using fixed receivers, they were able to determine the movements of points in Los Angeles county nearly instantaneously.)

Despite its successes, the international space program is in a state of flux due to economic pressures and the redefinition of priorities caused by the end of the cold war. The need for a space program and its focus are subject to intense debate, as illustrated by the battles in the U.S. Congress over funding for the Space Station. The economic and political expense of large missions is so high that international cooperation is a necessity, and space agencies are actively seeking ways to streamline operations and reduce mission costs. The costs of assembly or emplacement, inspection, servicing, and maintenance of large space assets will be significant. Large space assets cannot be launched already assembled. They must be launched in sections and assembled in space or emplaced on planetary surfaces after sites are prepared. Afterward, on an ongoing basis, they must be inspected, serviced, and maintained.

Need for Space Teleoperators and Robots

The Apollo astronauts repeatedly demonstrated their ability to function on the Moon, and the brilliantly successful refurbishment of the Hubble Space Telescope in December, 1993[6] demonstrated that astronauts can perform assembly, maintenance, and repair operations in space. The use of astronauts on a large scale for such operations, however, is far too costly and entails significant safety risk.[7] Accordingly, space teleoperators and robots, generically called space telerobots, are being developed in the United States, Europe, Canada, Russia, and Japan for use on the Space Station and in planetary and lunar scientific missions.[8] Space telerobots can extend astronaut capabilities and performance, thereby increasing mission performance and reducing costs. The Space Shuttle arm has been extremely useful in this regard.

The Space Station, which now includes the Russians as partners and is again being redesigned to reduce its scope and cost, is currently the only space facility under development that will make use of telerobots for construction, servicing and repair. (Rover missions are being implemented, however. The Russian Mars '96 and United States Mars Pathfinder missions plan to put rovers on Mars in 1996 to make scientific observations and perform rover technology experiments.) Many mission concepts employing telerobots exist, however, and a telerobot flight experiment is being implemented under NASA sponsorship (Professor David Akin of the University of Maryland is working on a telerobot flight experiment called Ranger). The Space Station is being designed to employ several telerobots, including the Canadian Space Station Remote Manipulator System (SSRMS) and the Special Purpose Dexterous Manipulator (SPDM) (see later chapters in this book) as well as a pair of Japanese telerobots, part of the Japanese Experiment Module (JEM), to perform various operations including assembly, inspection, repair, experiment tending, and so on. Other mission concepts currently under study that will require space telerobots (including lunar observatories and bases, science satellites, and ground support operations) are described in the following sections.

Lunar Observatories and Bases

Details depend on whether manned bases or unmanned observatories are being considered. In either case, telerobots would be useful for excavation, construction, assembly, maintenance, inspection, calibration, and repair, and would be used in nuclear reactor and observatory site preparation and assembly, solar panel emplacement, and habitation construction.[9]

Science Satellites

Large astrophysical observatories, such as infrared telescopes, might be placed in high orbits, which are largely inaccessible to astronauts. Telerobots could be used to replace cryogens, replenish attitude control system propellant, and perform module changeout, providing that the observatories are designed for ease of servicing. Telerobots could also be used to service Earth observing satellites, if they, again, are designed to permit it.

Ground Support Operations

Ground support operations are not missions in the usual sense, but support many individual missions by automating aspects of the ground operations necessary for fabrication, testing, and launch. Examples of ground support telerobot systems now under development include the satellite test assistant robot (STAR), which is being developed to automate operations inside the large environmental test chamber at the Jet Propulsion Laboratory (STAR has already been used to test Cassini components), HFIR, which automates the HEPA air filter inspection at the Kennedy Space Center in Florida, and the Tile Inspection and Maintenance Robot, which automates the waterproofing and inspection of the heat-resistant tiles on the Space Shuttle.

Whether or not telerobots are used in the preceding missions will depend upon economic tradeoffs. Placing telerobots and the necessary support equipment in orbit or on the moon will be expensive, and it must be cheaper to use them than it is to use astronauts and astronaut-operated equipment for construction and maintenance. It must also be cheaper to service and maintain space assets than simply to discard them. Telerobots could be used, for example, to service and maintain constellations of communication satellites like those in the Calling system.[10, 11] Calling satellites, however, are projected to cost less than ten million dollars each, and will be launched several at a time. Space servicing would have to be inexpensive to be justifiable. Economic issues are addressed in Chapter 2.

Autonomy, Intelligence, and Performance

Autonomy and intelligence are independent concepts. In robotics, an autonomous system is generally taken to be capable of achieving an externally specified goal without further external inputs. Intelligent systems are able to cope with complex situations, drawing conclusions and making control decisions appropriate to achieving their goals. A simple system can operate autonomously in simple situations, but may not be successful at operating autonomously in situations with significant degrees of complexity and variability. Heat-seeking missiles are a good example: a heat-seeking missile with a relatively simple infrared sensor can function autonomously when a single unambiguous infrared object, such as a hot exhaust pipe, is present. Its control objective is simply to keep the centroid of the infrared intensity distribution centered in its field of view. If the sensor and control system can control the missile with sufficient bandwidth, it will fly up the exhaust pipe, destroying the target. When countermeasures, such as magnesium flare decoys are deployed, aiming the missile toward the centroid of the infrared distribution will fail. To destroy the target, the missile must have a more intelligent sensing and control system that can determine which of the multiple infrared objects is the appropriate one. Developing a sensing and control system that will work in the presence of countermeasures (which are designed to increase variability and complexity) is a challenging problem.

Temporal response requirements and problem complexity together dictate computational power requirements. Dealing with high-performance targets requires making control decisions in hard real time to achieve sufficient control bandwidth.

Space Telerobots

Telerobots are machines that perform physical tasks. The motivation for using them in space is to accomplish, more efficiently and safely, tasks that would otherwise have to be done by humans, and to perform tasks that humans are incapable of doing. We therefore want telerobots to perform tasks like inspection, maintenance, repair, module changeout, cleanup, tending science experiments, performing repetitive operations, capturing and despinning satellites, and so on.

Telerobot operations involve bringing objects, including telerobots themselves, into prescribed mechanical states with respect to other objects. This means the telerobot must be able to locate and fetch relevant objects in space, move them to the appropriate locations, apply the correct forces when mating them, recognize error conditions, control mobility and attitude, etc. Telerobots may also need to cooperate with astronauts, other telerobots, and various types of equipment.[12]

A typical space telerobot system includes a control station and a telerobot as shown schematically, along with command and data flow, in Fig. 1. The control station, also called the *local* site, includes the interface which the operator uses both to comprehend the remote task and control the telerobot. The interface displays a graphical or video representation of the telerobot's worksite (in stereo, perhaps) along with information describing the state of the telerobot, the task, etc. It also includes command menus and the devices, like joysticks, hand controllers, and mice, which the operator uses to interact with, and command, the telerobot. The control station, which may be supported by powerful simulation and planning computers, communicates with the telerobot through a data link.

The telerobot, or *remote* site, physically performs tasks under the control of the operator. A telerobot comprises one or more substantially anthropomorphic manipulators, each with several degrees of freedom, mounted on a platform which

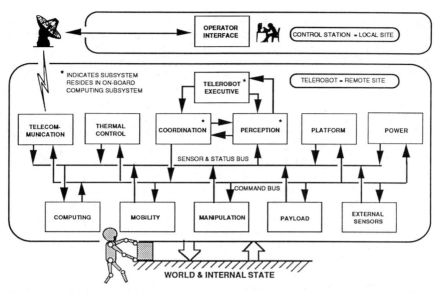

Fig. 1   Space telerobot system showing command and data flow; this diagram shows the major elements of a space telerobot system along with command and data pathways. The control station, which may be located in space or on the ground, is the local site; the telerobot is the remote site.

might be mobile (or free flying). Illustrations apear in Chapters 14–17. A space telerobot also has a sensor suite, usually including arm, platform, and mobility state sensors, force sensors, cameras, and necessary computation and support systems as shown in Fig. 1. The cameras may also be mounted on multi-degree-of-freedom platforms. A telerobot's major subsystems are computing, coordination, external sensors, manipulation, mobility, payload, perception, platform, power, telecommunication, thermal control, and the telerobot executive. These are described further in the appendix.

If a telerobot has a high degree of autonomy, that is, if it can perform operations with no human assistance while they are being performed, it is operating robotically. If it more or less mimics motions input by a human operator using a hand controller or joystick, or if human operators must issue detailed commands, again while the operations are being performed, it is operating as a teleoperator. Typical telerobots merge both teleoperator and robotic capabilities.

Space Telerobots Versus Industrial Robots

Robots have limited intelligence and ability to perceive. To compensate, applications have historically relied either on human presence in the control loop or on the imposing of significant order on tasks so they could be programmed as relatively fixed motion sequences. In industrial robotics, imposing order has been achieved mechanically through fixturing coupled with robot repeatability and positioning precision (machine vision systems and improved controllers are now permitting the relaxation of fixturing and positioning precision requirements). Achieving good control performance and positioning precision with current control technology requires stiff manipulators, which tend to be massive. Relying on fixturing and positioning precision is expensive, and is justifiable only when production runs are long enough to recover setup and tear-down costs. Space telerobot operations have high added value, so runs need not be long, but massive telerobots and fixtures are unacceptable because mass is the strongest driver of launch costs. The need to minimize launch mass leads to robot and support structures that are lightweight, hence flexible, difficult to model (resulting in positioning uncertainty) and difficult to control. Thermal effects can cause ringing (this was initially a problem with the Hubble Space Telescope) which also degrades positioning precision and creates control problems. In addition, space telerobotic tasks are typically more varied than those of industrial robots and do not involve nearly as much slavish repetition. Finally, handling the diversity of tasks with a population of special-purpose robots is not tenable, due again to launch mass constraints as well as development costs. Space telerobots, therefore, cannot rely upon mechanically imposed order. They must be versatile and adaptable, and must be capable of performing many different types of tasks with minimal setup and reconfiguration unless that setup and reconfiguration can be done easily.

Finally, the space environment itself is more difficult. Thermal and radiation effects, the lack of atmosphere, degradation due to atomic oxygen, and so on, all create engineering problems. Furthermore, the lack of gravity requires that objects must be positively retained. They cannot be simply set down or released, unless this is done with the utmost control, because they might escape and become high-velocity life- and mission-threatening projectiles.

Telerobot Control System Capabilities

As described in the previous section, models of a space telerobot, its operating environment, and the objects it manipulates will have significant uncertainty. Operating effectively and safely in space, then, requires that a telerobot *system,* which, along with the remote telerobot, includes the human operators and ground support, be capable of accommodating uncertainties. The system must be able to perform tasks which cannot initially be precisely defined, and which may have unplanned side effects, while avoiding unintended damage to itself and objects around it. To do this, the system must be able to determine the state of the task and relevant objects, iteratively determine the actions to take, predict their effects, and coordinate the subsystems to perform the actions while monitoring their effects to make sure they are consistent with the predictions. If the predicted and observed effects are not consistent, there is a potential problem. Finally, the control system must be able to monitor and maintain system health.

These capabilities, in turn, require that the system acquire and maintain representations of the workspace, the telerobot state, and the task, that it be able to recognize error conditions and success, that it be able to generate plans for error recovery, and so on. (Automatic plan generation is discussed in Chapter 7.) It is also useful for the system to be able to improve its performance by representing and organizing past experience. Examples include learning the positions of objects, adjusting control gains to improve positioning performance, and on-line hand-eye calibration. The capabilities are generic, arising, in part from the subtle interplay of action and sensing (for example, perceiving the compliance of a structural element involves determining the response of the element to applied forces).

We have considered the overall system including human operators. If timely human control and assistance cannot be provided because of operational constraints, the capabilities, which endow the system with a measure of autonomy, must be present in the telerobot's onboard control system. The requisite system sophistication depends on task complexity.

Spacecraft like the Galileo Jupiter orbiter and probe have historically been termed robotic, but they differ in important ways from planetary rovers and the telerobots being considered here. They are more like preprogrammed machine tools than robots. Satellites and spacecraft have been restricted to operating in simple environments (empty space) and with control objectives that can be relatively easily characterized. Such systems do not have, or need, the ability to sense and classify complex external situations and make quick decisions onboard. They operate in an open-loop manner for long periods, and most of the control decisions are made on the ground. In addition, they are not usually required, other than landing, to interact mechanically with the environment (the Viking lander's arm interacted mechanically with Martian soil during sampling operations). Rovers and space telerobots, on the other hand, must be capable of performing mechanical operations at reasonable rates in uncertain natural or man-made environments. (This is driven by the need to complete missions in a reasonable time, which is driven, in turn, by system survivability and mission operations costs. If one could operate extremely slowly in a static environment, pure remote control would be acceptable even with significant data rate limitations and communication delays.) If they are being used where data rates are limited or communication delays are appreciable, they must have onboard sensing, decision-making, and control capabilities that can deal with greater uncertainty and complexity, especially if the task state can change unexpectedly. (For example, moving along a trajectory in free space toward a planet whose position can be predicted to extremely high precision is a computationally simpler task than moving about and operating in an imprecisely known cluttered space, without suffering or causing damage.)

Practical Approach to Space Telerobot Control

Developing intelligent robots that have great dexterity and motor skill, that can reason deeply about tasks and operate independently with great versatility in complex environments, is an extremely difficult unsolved problem. While scientists generally agree that conscious, intelligent behavior arises from some sort of computational activity, after years of effort we still do not know how to characterize realistic environments in ways suitable for robotics, to describe the procedures for such apparently effortless activities (for humans and animals) as recognizing objects, or to build and program computing hardware to perform the computations. Human judgment is still essential for difficult situations.

While we cannot yet automate intelligent behavior, lower level behaviors like moderately complex sensor data interpretation, motion and force control are easily automated. Furthermore, robots perform well in environments that are relatively predictable and on tasks that have reasonable dexterity demands. Once an (accessible) object is pointed out to a robot system, for example, it is fairly straightforward for the robot to acquire it automatically if it is easy to grasp and manipulate. Even complex tasks are comprised of simpler elemental subtasks, so a powerful approach to building robotic systems is to automate the lower level functions and detail management while relying on humans to provide overall guidance and handle difficult situations. Thus humans, who become bored during repetitive tasks, perform the high-level intelligent functions at which they excel, while telerobots physically perform the tasks. Where a telerobot is capable, it operates autonomously. When it needs assistance, the operator provides it, issuing detailed commands when necessary. That is the essence of telerobotics.

To summarize, a practical approach to developing useful telerobot systems[12] is to 1) automate lower level functions by developing reliable control algorithms that adapt on the basis of sensory information, 2) rely on human operators for providing overall task guidance and supervision, and for handling special situations (repair of damage is an example), and 3) develop advanced interfaces and tools that aid in planning and managing telerobot tasks and permit the operator to communicate easily with the system at multiple levels of detail.

Currently we must give telerobots detailed commands, either in the form of macros or in the form of motion commands generated by joysticks and hand controllers, and human operators must be prepared to assist them in locating and identifying objects. As autonomous system technology advances, we will be able to delegate higher levels of decision making to telerobots, reducing the load on human operators, ground control, and telecommunication systems while improving telerobot performance. This is represented schematically in Fig. 2.

Ground Control

The use of telerobots, like the Shuttle Remote Manipulator System (SRMS), controlled by astronauts in space, can significantly increase the productivity of space operations, as been demonstrated on numerous Shuttle flights. Space-controlled telerobots allow astronauts, working in the comfortable shirtsleeve environment inside their spacecraft (intravehicular activity, or IVA), rather than outside in space suits (extravehicular activity, or EVA), to acquire and manipulate payloads such as satellites, Space Station modules, and orbital replacement units (ORUs). IVA reduces both preparation time, which is significant, and risk over EVA (according to the Fisher-Price study, the time spent in preparation is at least five times the time spent doing EVA work). Because a telerobot control system can scale human-generated command inputs, astronauts can also use (smaller) telerobots to perform small operations like calibrating instruments and tending laboratory experiments. (Different robots would be used, but the control stations could be identical. It should also be noted that teleoperator-based devices are now being applied in noninvasive surgery. By scaling and filtering, the effective resolution of a surgeon's motions can be improved and the ever-present tremor can be attenuated.)

Space-controlled telerobots are important for the Space Station because, as the Fisher-Price study[7] has shown, there is far less astronaut EVA time available than is needed to perform the necessary inspection and maintenance operations. The study also showed that the performance of IVA astronauts using simple telerobots

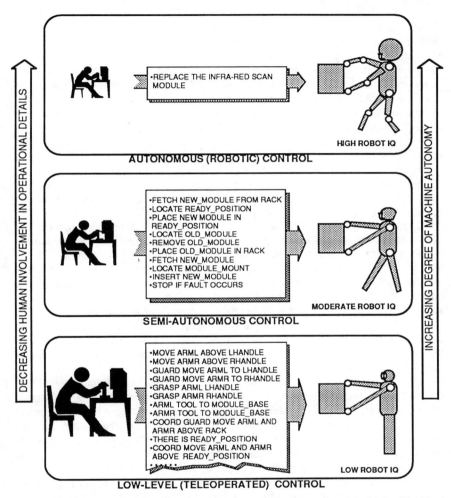

Fig. 2 Telerobot control modes. In low-level, or teleoperated, control, the telerobot has little autonomy, forcing the operator to issue detailed commands; this is tedious, and requires high data rate communication channels for the operator's visual and state displays. As telerobot autonomy improves, less detailed commands are required; this eases the burden on the operator and reduces data rate requirements because less state data is needed and visual displays need not be updated as often. Missions with limited data rate channels and significant communication delays require a high degree of autonomy to achieve good performance.

to perform tasks was equal to or better than the performance of EVA astronauts performing the same tasks directly.

Controlling telerobots from the ground (ground control) is potentially much more effective than controlling them from space. The total hourly cost of a ground-based operator is orders of magnitude less than that of an astronaut in space, and ground-based operators are more efficient because they work in a normal office environment. They are also at much less risk. In addition, ground control stations can have greater computing resources available, can be configured to support many operators, and are easier to maintain. The Fisher-Price study[7] recommended ground control for the Space Station to reduce crew workload. Other studies have recognized the advantages of ground telerobot control, and have recommended it for both Space Station and lunar operations.[9,13] Ground control would also enable telescience, allowing terrestrial scientists to perform (remotely) experiments in space.

Ground control of space telerobots is attractive, but it places significant demands on the control system. To avoid damage and be capable of performing difficult tasks like acquiring tumbling satellites, a telerobot, whether it employs space or ground control, must be able to respond quickly to rapidly evolving events (like accidents and motion perturbations) that cannot be predicted. In addition, telerobot movements must always be stable and predictable. Unanticipated (by either human controllers or automatic planning systems) motions cannot be tolerated (reflexes may be an exception, but they are planned in a sense, in that the system must enable or anticipate them and must control their scope). All this requires high-rate, possibly data-rich control loops with low latency. Because of data rate limitations and communication delays inherent in space communication systems (up to several seconds even for low Earth orbit) and resulting potential stability problems, it is impossible to close such loops on the ground (for example, downlinking real-time video from orbit for ground processing and control signal calculation and uplinking the results).

Ground control naturally partitions the overall control system into ground (local) and space (remote) components. Because of the latency and limited data rates of communications with the ground, the space-based component must be capable of operating with some degree of autonomy. That is, given intermittent high-level terrestrial instructions and assistance, it must be capable of gathering and interpreting data, making decisions, and taking appropriate actions in response to the local situation and task goals. As in the example of the heat-seeking missile described earlier, a simple control system will suffice if the environment is simple and control requirements are not demanding. As the environment becomes more complex and control requirements escalate, greater onboard autonomy and intelligence are required. This may be true even if astronauts are on site and all control loops are closed in space, because data traffic on the Space Station itself may involve appreciable latency.

State of the Art in Space Telerobotics

As we have seen, telerobot systems incorporate both teleoperator and robotic elements. Some current telerobot systems are nearly pure teleoperators, while others have significant robotic capabilities.

The operator interfaces of both pure teleoperator and telerobot systems include visual displays, panels or menus to select operating modes and control lights, cameras, and auxiliary devices. In addition, they usually include one or more position controllers or joysticks for generating commands for the remote site (see Fig. 2). In many laboratory systems the visual displays are implemented in powerful graphics computers that can display color stereo pairs which can be viewed with special active glasses that are synchronized with the frame rates of the displays. Advanced telerobot systems under development also include tools for task management, collision avoidance, visual recognition and tracking, planning, modeling, calibration, and so on, which may be implemented at both the local and remote sites. See Chapters 5–9.

Joysticks provide rate information. That is, the deflection of the joystick represents the instantaneous velocity of the remote device according to the convention currently in effect. It may represent commanded axis velocity or the commanded velocity of a coordinate frame embedded in the telerobot. The position of a position controller actually represents the commanded position of the remote device in the current (remote) coordinate frame. If they are suitably equipped, position controllers can be backdriven by contact or error signals from the remote site (or simulated remote site) to give a sense of contact to the operator. This is referred to as force reflection. A joystick or position controller can also be used much like a mouse to move a cursor around the visual display, pull down menus, etc.

Teleoperators

Teleoperators are often master-slave systems. The master is a (perhaps scale) replica of the slave. The operator, watching a visual representation of the worksite, performs a task by moving the master as if it were performing the task. The master is thus a position controller because the operator is issuing position commands. The time-varying positions of the master arm joints are uplinked to the remote site as time-varying position control inputs for the slave servos. The slave therefore executes a scaled copy of the master's motion. When the slave contacts the environment, position error signals are generated. These are downlinked to the master and are used to backdrive its joints, giving a sense of contact with the remote task. A slave's axes are coordinated implicitly because it is geometricaly similar to the master.

In nonreplica master-slave systems the master and slave are not geometrically similar. Axis coordination is handled by a computer in the control loop that continually maps the present position of the master handle into the (scaled) position of the slave hand. The computer uses master kinematics to calculate the Cartesian position of the master handle in space and slave kinematics to transform this position into position commands for the slave axes, thus making the slave hand perform the same (scaled) Cartesian motion as the master handle. Again, error and contact signals are used to backdrive the master, giving a sense of contact with the remote environment.

When joysticks rather than position controllers are used, a computer again maps the velocities commanded by the joystick into instantaneous velocities of a coordinate frame embedded in the hand, using slave kinematics to calculate the corresponding slave joint velocities.

The master-slave exoskeleton being developed at the Jet Propulsion Laboratory (JPL), which has five-fingered hands and replicates much of a human's arm and finger motion, reflects forces to the operator. (The exoskeleton is described briefly in Chapter 5.) A so-called *human-equivalent* system, it can be used in situations that demand dexterity and fine motor control. As in all teleoperator systems, the operator views the worksite, or an image of it, and moves the arm and hands to complete the task. Because forces are reflected to the joints, the operator's sense of presence at the remote site, *telepresence,* is enhanced.

If the teleoperator control system implements *shared* control, some task degrees of freedom can be controlled by the control system while others are controlled by the operator. In cleaning a window, for example, the operator could control the x-y position of the cleaning head while the control system controls the normal force and makes sure the head remains normal to the window's surface.

Consider inserting a threaded fastener into its hole prior to tightening. The fastener is already being held in a power driver that is being grasped, in turn, by the slave arm. To complete the task in teleoperator mode with low-latency high-data rate communications using a position controller, the operator would note the position of the hole and obstacles in the visual display of the remote workspace. The operator would then decide how to move the fastener to the hole and, watching the fastener in the display, would move the position controller to bring the fastener to the hole while avoiding collisions, checking visually for alignment, and watching for the position controller excursions and stiffness changes that indicate contact or collisions. When the fastener entered the hole or contacted the surface near the hole, the operator would mentally observe the stiffness behavior of the position controller in response to small motions about the hole. The stiffness behavior has a different character depending on whether the fastener is entering the hole, in the hole, bottomed, on the edge, or on the surface near the hole. Once the fastener was in the hole the operator would actuate the driver while exerting a small axial thrust force and nulling the radial forces and moments. If the system had shared control, it could restrict velocity and control normal forces as the operator moved to contact. Then, as the driver was actuated, it could maintain axial alignment as the operator sensed and adjusted the normal force during tightening.

If there is a significant communication delay, both force reflection and shared control can have stability problems. In addition, for delays greater than about half a second, force reflection and visual delay can become confusing to the operator. Hence, in cases with significant communication delay and/or limited data rates, normal teleoperation cannot be performed. In those cases it is necessary for the operator to interact with computer-aided design (CAD) models of the slave and the task that may be overlaid on visual imagery. The operator uses the position controller or joystick to drive the CAD model of the slave. Virtual contact force feedback to the operator can be used as well, based on simulating the mechanical interaction of the CAD model of the slave, driven by the operator, with the CAD model of the task. The resulting simulated motions are previewed on the display. If they are suitable, the commands are formatted and uplinked to the slave, which will respond after some delay. This is similar to the off-line programming of industrial robots. It requires that the slave have a more sophisticated control

system, with greater autonomy, than that required for simple master-slave control.[†]
It is necessary, for example, for the remote site to know the operator's *intent* in
issuing particular motion commands, as it must correct small anomalies itself.
Intent, which can be associated with an operating mode, can be uplinked to the
slave as part of the command stream. Module insertions and other tasks have been
demonstrated in the laboratory using this approach as is described in Chapters 3–6.

Using CAD models requires calibration. The CAD model must be accurately
aligned with the actual imagery, if it exists. In any case, the master and slave
must be aligned (calibrated) with the remote worksite. This problem also exists
for telerobots, as will be described more fully below. When dealing with limited
data rates and communication delays, the distinction between teleoperator and
telerobotic systems becomes blurred.

The Shuttle manipulator (SRMS) is essentially a teleoperator that is controlled
with a pair of rate joysticks, each with three degrees of freedom. The operator
views the manipulator and task directly, or through various cameras, and moves
the two joysticks to control arm velocity (speed and direction). One joystick
controls arm position; the other controls hand orientation. The SRMS has been
invaluable to Shuttle operations, but using it is tedious because little is offered
in the way of automation tools for reducing workload. For example, waiting for
bending transients to die out between motions and checking for collisions, current
operational requirements, are time consuming. (Collision prevention, stability,
flexibility, and control are discussed in Chapters 9–13.)

Telerobots

Systems with robotic capabilities are able to use predefined CAD models of
the telerobot, the workspace, and workpieces, along with macros or routines for
specific tasks, to generate small task plans and collision-free paths. Moving the
hand to contact is an example of a specific task that might be defined by a routine.
The operator supervises the task, resolving difficult situations and determining
what routines or macros to use. He can specify a task and then relinquish control
to the telerobot, which returns control when either the task is complete or an
impasse is reached. The operator can also seize control at any time. Thus control
is *traded* back and forth between the operator and the telerobot.

The operator specifies a task through a combination of moving the telerobot with
the joysticks or position controllers to record positions along with forces, perhaps,
and other data, selecting macros or task names from a menu or by entering text
(giving symbolic names to important task elements like coordinate frames and part
features), and designating objects and/or locations in the display in response to

---

[†]Long communication delay and/or a low data rate means that local autonomy is critical
to achieve throughput. It can take a significant amount of time to get imagery back from the
remote site. From Mars, for example, it can take several days to return a panorama. Sending
back parameters that characterize the environment is a potential simplifying approach, but
that requires the on-board capability for characterizing the remote environment. For remote
missions, the number of control actions per day may be severely limited. In the 1996 Mars
pathfinder mission, for example, there will be only about one control interaction with the
microrover per day. In Earth orbit, the situation is not so difficult.

prompts from the system. Parameters are either stored in the macros and routines or are supplied by the operator at the system's request.

In performing the fastener-insertion task described earlier with such a telerobot, the required CAD models and sensor-based macros or routines for inserting fasteners would have been defined in advance or created as necessary by the operator on line. Using a keyboard or a menu, the operator would specify that the system was to insert the fastener. If the system did not know the location of the hole, it would prompt the operator to designate its location in the display. If the system did know the hole location it would highlight the location in the visual display for the operator to verify. If the location in the visual display did not correspond with the highlighted location, it would indicate calibration errors which would be addressed as described later. It might also request the fastener identity, if it were unknown, so appropriate control parameters could be recalled. If it did not know the path, the system would prompt the operator to designate points along the path. If it did know, or could plan, the path, it would display the points and prompt the operator to verify them. The system would then display a preview of the insertion. If acceptable, the operator would allow the system to proceed.

If visual calibration were a problem, the system would move the fastener, still in the driver, to a location visible in the visual display and show an overlay indicating where it thought the fastener was located (based on arm kinematics), prompting the operator to designate, in the display, where the fastener was actually located. The system could thus calculate the local calibration correction, assuming arm kinematics were well calibrated.

On-line local calibration is a significant problem that has a representational component and a kinematic component. The representational component deals with ensuring that the internal representation of the worksite is metrically equivalent to the external environment. It is important because the processes for planning, collision avoidance, and deciding where to look for features necessarily operate on an internal representation of the world. The kinematic component deals with ensuring that the commands for positioning an appendage at a physical location or aiming the vision system at a physical point actually do so. Calibration has been locally corrected by designating corresponding features on the visual displays and their CAD model overlays.[14] It can also be corrected using image processing and computer vision techniques.

Telerobots with features similar to those just described exist in a number of laboratories. Satellite tracking and capture along with autonomous servicing[15] have been demonstrated at JPL. Supervised autonomy has been demonstrated many times (see Chapters 3–6). The German robotics technology experiment (ROTEX) demonstrated autonomous vision-based free-floating object capture and telerobotic assembly operations.[16, 17]

Flight systems incorporating many of the same features are being developed for the Space Station as described in Chapters 14–17. The new systems represent a significant improvement over the Shuttle manipulator, but advances are still needed to increase productivity and safety and improve the ability of telerobots to perform difficult tasks, such as repair, autonomously. Repair operations can involve structural damage, which distorts geometry in unknown ways and therefore cannot be modeled in advance. Repair can also involve cutting and forming operations, which are more difficult than insertion and fastening operations, and may require enhanced dexterity. The operator must be intimately involved in repair tasks to comprehend the situation, construct models and decide how to proceed. As a class, servicing tasks are much simpler for telerobots than repair tasks if space systems are rationally designed.

Needed Telerobot System Improvements

Needed telerbot system improvements include 1) automatic worksite modeling for collision avoidance with rapid, interactive on-line model building for recognizing, and keeping track of, objects in the workspace; 2) automatic/interactive calibration of CAD models with the physical worksite and the identification of anomalies; 3) on-line mechanical calibration to compensate for thermal effects and drift; 4) ability to lock on and track objects, based upon their geometrical structure, without requiring labels or targets; 5) improved dexterity and contact motion control; 6) integration of flexible structure control to suppress bending modes; and 7) systematic integration of perception, reasoning, planning, control, and interface capabilities so it is easy for operators to define new tasks and convey intent, and so the number of uplinked operator commands can be reduced.

Supporting Technologies

Supporting technologies needed to realize the preceding telerobot improvements include 1) improved telerobot architectures that can naturally fuse information from different sensory modalities, can easily store and retrieve relevant memories, and can easily deal with the dynamic control reconfiguration involved in identifying, controlling, and coordinating the telerobot and its entire set of sensors and actuators as it moves about its environment performing tasks; 2) more capable, low-power, low-mass, error-tolerant flight computing, including alternative forms like special purpose processors and hardwired neural networks, so the demanding computational requirements of autonomous robots can be satisfied; 3) computing and interface standards to make the system programming problem tractable; 4) integration of sensors and structures (smarter structures) so robust task and telerobot state estimates can be made; and 5) efficient, low-mass sensors and actuators.

Conclusions and Summary

Space telerobots, which merge teleoperator and robot characteristics, are needed to improve the productivity of space missions, and are needed for missions such as lunar base construction and Space Station maintenance because of the high cost and low productivity of EVA astronauts. Telerobots might also be used for other tasks, but the decision to do so will be based on economic tradeoffs. Controlling space telerobots from the ground could make them extremely attractive, but ground control places stringent demands on their control systems because of communication delays, data rate limitations, and task uncertainties. To compensate while preserving performance, telerobot systems will need greater intelligence and autonomy, which are independent concepts related to environmental and task complexity that have a profound impact on telerobot system performance. Extremely intelligent robots are far beyond the state of the art, but useful systems can be built around the concept of humans supervising telerobots. This approach permits robots to perform low-level operations automatically while freeing human operators to concentrate on higher level task elements.

The Shuttle manipulator, the only operational space telerobot, is useful but has little autonomy. State-of-the-art laboratory telerobots have advanced features, such as automated collision avoidance and interactive task execution using macros, that could notably improve space operations and are being incorporated in future flight systems. Problems that should be addressed by telerobot research and development programs, such as the need to wait for bending transients to die out, the need for on-line calibration, and the need for more automatic model generation, remain, however. Developing these capabilities will require advances in a number of supporting technologies including system architectures, flight computing, sensors, and actuators.

Advancing space telerobot technology will improve industrial robots, helping to make them more adaptable and easier to set up and use, and will therefore improve economic competitiveness. Because space telerobots need greater autonomy, space telerobotics will have a beneficial effect on service robots and aids for the disabled, including smart manipulators, legged wheelchairs, active braces for stabilizing limbs, and perception systems that can sense hazards for the blind. Medical robotics also stands to gain. Telesurgical operations on animal tissue from the United States to Italy have already been demonstrated, for example. Likewise, advances in these other areas will benefit space telerobotics.

Appendix: Telerobot Systems

This appendix briefly describes the telerobot subsystems that do not have obvious descriptions. These include the computing, coordination, external sensors, manipulation, mobility, perception, payload, platform, telecommunication, and telerobot executive subsystems. The subsystems are shown schematically in Fig. 1, which also shows the control station along with command and data flow.

The *computing* subsystem, which is under the control of the telerobot executive, is the aggregate of computational devices aboard a telerobot, including general and special purpose computers, low level controllers, sensor preprocessors, and other dedicated electronics. All control resides in the computing subsystem. The telerobot executive, perception, and coordination subsystems reside in the computing subsystem as well.

The *coordination* subsystem, which performs kinematic and dynamic computations as necessary and coordinates the behavior of the various subsystems and devices that are under the control of the executive resides in the computing subsystem as hardware and/or software. For example, it coordinates the manipulator, mobility, and platform actuators during cooperative moves, issuing commands to the actuator controllers. It receives state information from telerobot actuators and from external sensors (interpreted by the perception subsystem) and can, in advanced systems, send predicted state information to the perception system so task evolution consistency can be determined, that is, so the system can determine if the task is proceeding as predicted. The coordination subsystem, which also controls the payload subsystem, is analogous to the motor control systems of animals.

The *external sensor* subsystem senses the state of the external world. Sensors may include (possibly stereo) cameras and other noncontact devices such as laser and multispectral scanners and thermometers. Sensor pointing and deploying devices are considered part of the platform; their motions are coordinated with other devices by the coordination subsystem. Sensor information is interpreted by the perception subsystem.

The *manipulation* subsystem is a telerobot's arms and hands, including arm state sensors and actuators. Manipulator control resides in the computing subsystem.

The *mobility* subsystem includes the devices, with the necessary state sensors, that move a telerobot about. Mobility system types include those with wheels, thrusters, legs, and rails. Telerobots operating in zero *g* would probably employ thrusters, legs (with grasping feet), tracks, rails, or another manipulator (the Canadian SPDM is slated to be positioned by the SSRMS). Those operating on planetary surfaces might use legs, wheels, or tracks, the choice depending upon the predominant operating environment. As with the manipulation system, mobility system control resides in the computing system. Legged mobility and manipulator systems are quite similar. Both employ limbs, specialized for particular roles, that can be used to manipulate objects or move the telerobot.

The *perception* subsystem receives input from the various state and external sensors as well as the telecommunication system. In sophisticated telerobots it computes a summary of the world, telerobot, and task states, which is used by the telerobot executive and the coordination subsystem. In simple systems, the perception subsystem may just perform transformations on sensory data.

The *payload* subsystem is the collection of elements, not part of a telerobot itself, that a telerobot uses and (perhaps) controls to perform its tasks. Tools like power fastener drivers and cutters are examples. If a telerobot were being used for scientific purposes, its payload might include various scientific instruments which the telerobot would position and actuate to make scientific measurements.

The *platform* subsystem comprises the actuators and devices, like camera and antenna pointing systems and body joints, that make up the body of a telerobot.

The *telecommunication* subsystem, which receives information from the control station, provides a bit stream of commands and data that is decoded by the perception subsystem. The telecommunication system also encodes bit streams from telerobot sensors and the executive, transmitting them to the control station, which has its own receiver/transmitter (not shown). The telecommunication system does not include any required antenna pointing systems. They are considered part of the platform.

The *telerobot executive,* which resides in the computing system, schedules and controls the overall high-level behavior of a telerobot's subsystems, except for automatic fault protection and reflexes. It receives operator comands and instructions as well as world, telerobot, and task state information from the perception subsystem. In advanced telerobots the executive includes planning, reasoning, behavior prediction, and fault diagnosis tools. The executive can issue commands to both the perception and coordination subsystems. The perception subsystem might, for example, be commanded to look for a particular pattern as the coordination subsystem scans the environment with a camera system.

Acknowledgment

Jacob Matijevic of JPL helped define the overall approach to this chapter. His contribution is gratefully acknowledged. This chapter was prepared at the Jet Propulsion Laboratory, California Institute of Technology, under contract with the National Aeronautics and Space Administration. The views and conclusions expressed in this document are those of the author and should not be interpreted as representing the official policies, either expressed or implied, of NASA, the U.S. Government, or the Jet Propulsion Laboratory, California Institute of Technology.

References

[1]Reynolds, K., "Rocky IV, Literally Out of This World," *Road and Track Magazine,* Vol. 44, No. 8, April 1993, pp. 92–97.

[2]Pivirotto, D., and Dias, W., "United States Planetary Rover Status-1989," Jet Propulsion Lab., JPL Pub. 90-6, Pasadena, CA, June 1990.

[3]Ruoff, C., Bowyer, J., Brooks, T., Hanson, J., Holmes, K., and Wilcox, B., "Autonomous Ground Vehicles: Control System Technology Development," Jet Propulsion Lab., Final Rept., JPL Task No. 80–2021, Pasadena, CA; Rept. ETL-075, U.S. Army Engineer Topographic Labs., Ft. Belvoir, VA, Oct. 1984.

[4]Bayer, W., "Interview With Randy Hoffman, President and CEO, Magellan Systems Corp.," *Space News,* Vol. 5, No. 2, Jan. 10–16, 1994, p. 22.

[5]Hotz, R. L., "Where in the World Are You," *Los Angeles Times,* Vol. 113, Feb. 7, 1994, pp. A1, A18.

[6]Begley, S., and Annin, P., "We Slam-Dunked It," *Newsweek Magazine,* Vol. 122, No. 25, Dec. 20, 1993, pp. 100–102.

[7]Fisher, W., and Price, C., "Space Station Freedom External Maintenance Task Team, Final Report," NASA Lyndon B. Johnson Space Center, Houston, TX, July 1990.

[8]Whittaker, W., Kanade, T., Allen, P., Bejczy, A., Lowrie, J., McCain, H., Moutemerlo, M., and Sheridan, T., "Space Robotics in Japan," Japanese Technology Evaluation Center, JTEC Panel Report, Loyola College, Baltimore MD, Jan. 1991, NTIS Rept. #PB91–100041.

[9]Dias, W., Matijevic, J., Venkataraman, S., Smith, J., Lindemann, R., and Levin, R., "Directions for Lunar Construction: A Derivation of Requirements from Construction Scenario Analysis," *Engineering, Construction, and Operations in Space III,* Proc Space '92, edited by W. Sadeh, S. Sture, and R. Miller, American Society of Civil Engineers, New York, 1992, pp. 357–367.

[10]Tuck, E., Patterson, D., Stuart, J. R., and Lawrence, M., *The Calling^{SM} Network: A Global Telephone Utility,* Teledesic Corporation, Kirkland, WA, 1993.

[11]Seitz, P., "New LEO Satellite Venture Proposes Ambitious Plans," *Space News,* June 28–July 11, 1993, p. 16.

[12]Varsi, G., Ruoff, C., Burdick, J., and Culick, F., "Projected Automation Technology Requirements 1992–2011," Jet Propulsion Lab., JPL Pub. JPL D-9306, Pasadena, CA, May 1992.

[13]Swanson, P., "Space Station/OACT Robotics Technology Study," Vols. I and II, McDonnell-Douglas Corp., and Oceaneering Space Systems, Oct. 1993.

[14]Bon, B., Wilcox, B., Litwin, T., and Gennery, D., "Operator-Coached Machine Vision for Space Telerobotics," SPIE Symposium on Advances in Intelligent Systems, Conference on Cooperative Intelligent Robots in Space (Boston, MA), SPIE, Bellingham, WA, Nov. 1990.

[15]Schenker, P., et. al., "NASA Telerobot Testbed Development and Core Technology Demonstration," *Proceedings of the SPIE Conference on Space Station Automation (IV)*, Vol. 1006 (Cambridge, MA).

[16]Sanchez, V., "Robust Robot Vision Techniques for Grasping Floating Objects under μ-Gravity," Proceedings of the 3rd European In-Orbit Operations Technology Symposium, European Space and Research Technology Center, Noordwijk, The Netherlands, June 22–24, 1993.

[17]Hirzinger, G., Landzellel, K., and Dietrich, J., "Sensorbased Space Robotics—ROTEX and its Telerobotic Features," Proceedings of the 44th Congress of the International Astronautical Federation, Paper No. IAF-93-U.6.591, Graz, Austria, Oct. 16–22, 1993.

Chapter 2

Economics of Automation in Space: Implications of Automated Versus Manned Operations

Wayne F. Zimmerman*

*Jet Propulsion Laboratory, California Institute of Technology,
Pasadena, California 91109*

Nomenclature

A = subset of viable man-machine alternatives composed of a_{kt}
a_{kt} = identified man-machine alternative t for a given subsystem k
B_{kt} = net benefit from automating subsystem k with alternative t, made up of the benefit from manhours saved M_{kt} and other incremental net dollar benefits c_{kt} [i.e., B_{kt} is $(M_{kt} + c_{kt})$]
C_{kt} = net subsystem k cost not considering the benefit of automating alternative t
C_t = cost target for the total system
H_{kt} = incremental hazard exposure time reduction due to automating subsystem k with alternative t
P_{kt} = incremental power impact of automating subsystem k with alternative t
P_t = system-level power constraint
W_{kt} = incremental weight impact of automating subsystem k with alternative t
W_t = system-level weight constraint

Introduction

T HE continued high cost and risk of placing astronauts in space have placed a considerable burden on NASA to cut costs and consider other means of achieving mission goals both effectively and safely. Additionally, future science missions that might place a tremendous burden on Shuttle availability, or require extended vehicle duty cycles on the lunar and Mars surfaces, might preclude

Copyright © 1994 by the American Institute of Aeronautics and Astronautics, Inc. The U.S. Government has a royalty-free license to exercise all rights under the copyright claimed herein for Governmental purposes. All other rights are reserved by the copyright owner.

*Technical Task Manager, Robotic Systems and Advanced Computer Technology Section.

the presence of astronauts altogether. The solution to these dilemmas involves recognizing the essential mission and cost parameters that must be controlled, as well as recognizing that determining which functions to leave to astronauts and which functions to automate is actually a complex resource allocation problem. A four-tiered mathematical model has been designed, built, and tested on Space Station subsystems to assist in the decision process associated with making human-machine tradeoffs. The model is composed of 1) a task decomposition step (involving both task time and safety elements as part of selecting the likely automation candidates), 2) development of conceptual designs and costs for the candidates, 3) cost and benefits assessment, and 4) optimization of automation candidates relative to a set of constraints. Although simple in concept, the actual process must consider many nonquantitative variables that must be merged with the quantitative factors in formulating the solution to the competing objectives and resources problem. Analytical advancements in the area of assessing human-machine tradeoffs are discussed in this chapter. Recent conclusions addressing likely near-term targets for robotics and automated systems on Space Station, based on human-machine tradeoff studies, are also described.

Background

Classical system resource allocation problems have employed time and motion studies, project evaluation and review techniques (PERT), linear regression analysis, and dynamic programming.[1,2] These techniques are effective under two conditions, 1) when the resource variables are well defined and an empirical database exists to quantify the variables, and 2) when the relationships between variables are approximately linear. But these techniques have limitations because not all resource allocation problems have well-defined variables, nor can they be uniquely configured with clean linear relationships between variables. These limitations became particularly apparent when the Jet Propulsion Laboratory (JPL) began developing new analytical techniques for making human-machine tradeoffs while injecting new, advanced automation technology into application designs.

The JPL of the California Institute of Technology has historically been one of the leading NASA centers for unmanned missions. Deep space probes such as Voyager, and more recently, Magellan and Galileo, are semi-autonomous vehicles. As such, JPL has been one of the leaders in autonomous system design and control for some time. In the 1970s, JPL was involved in transferring expertise and technology for automated systems to the private sector through the Energy and Technology Applications Program. The first exposure to having to explore new ways of managing limitations of techniques such as those mentioned earlier came through an advanced underground automated mining project sponsored by the Department of Energy, Bureau of Mines, in the late 1970s. Although, on the surface, the insertion of automation into the mining process appeared to be a classical production optimization problem, upon further study it turned out to be a complex interplay of several factors. A means of assessing production impact considering limited workforce and dollar resources, variations in coal properties, mining techniques, and machine operator ability had to be developed. Not only production, but health, safety, system complexity, reliability, and technology variables had to be assessed in the final selection of the best human-machine combination. Although the coal mining industry had an extensive historical productivity and task workhour database, except for picking out obvious production stumbling blocks, the database was useless because the automated system performance required the model to project time, productivity, safety, and cost impacts. Lastly, new technology was needed to replace old technology, and a new machine configuration had to be conceptually generated to facilitate cost and benefit projections. Given all of the preceding analytical factors, it was clear that a new technique had to be developed to combine the various quantitative and qualitative variables, and provide a solid means for choosing human-machine options. It was also clear that in this new realm of competing objectives and resource allocation, the idea of the optimal solution had to yield to a best or reasonable solution.

The problem was ultimately solved by building a model consisting of 1) a task network analysis (PERT) to drive out the major production barriers, 2) an interactive industry survey of potential quantitative impacts of automation on the major production stumbling blocks, 3) a technology assessment and conceptual design of the primary automation candidates, and 4) a cost and benefits analysis which included health and safety impacts analysis. The final results, which identified the highest payback design option, were presented as a multivariable envelope consisting of cost, production impact, projected reduction in average yearly injuries, and reduction in exposure to coal dust. The modeling tool was eventually employed by the Utah Power and Light Company to automate one of their high-seam longwall mining operations.[3]

The experience gained from the preceding initial modeling effort set the stage for the improved version(s) developed for the Space Station. The newer versions of the human-machine tradeoff model, and associated supporting analytical processes, are discussed in the following sections.

Problem of Limited On-Orbit Resources

One of the recurring problems which is constantly addressed by the NASA manned spacecraft community is the availability of crew workhours. The Shuttle Mission Operations Directorate (MOD) at the NASA Johnson Space Center (JSC) spends months meticulously planning each crewmember's daily and hourly schedule. Typically, after subtracting out housekeeping activities, equipment monitoring, and anomaly/failure troubleshooting, an average on-orbit day only has 3–5 h of productive time available for payload tending and science.[4] It is clear, based on the Shuttle workload history, that crew availability is one of the most important resources to be managed. The recent Shuttle mission calling for retrieval and repair of the International Telecommunications Satellite Organization (INTELSAT) communication satellite (STS-49, May 14, 1992) is a good example of the importance of crew availability for contingency. What was originally planned for a two extravehicular activity (EVA) crew and one intravehicular activity (IVA) crewmember, eventually required a three EVA crew and one IVA crewmember after two unsuccessful attempts to retrieve the satellite.

Similarly, current workload studies suggest that the Space Station crew is going to be under equal pressure to maintain an extremely aggressive day-to-day work schedule with little time available for contingency.[5,6] Most importantly, the combined crew workload relative to EVA and IVA tasks must not jeopardize astronaut, and overall system, safety. Concurrently, every effort must be exerted to meet mission objectives. This operational dilemma has historically placed significant strain on manned spacecraft systems. Whereas at one time astronauts felt strongly about exercising complete control over all spacecraft functions, the newer breed of astronaut (particularly mission specialists) recognizes the potential workload-mission conflict and welcomes assistance through robotic and automated systems.[6]

Determination of Major Resource Variables

The key resource variables, which must be consistently balanced to solve the allocation puzzle, were determined by talking directly to the MOD at the JSC responsible for making crewsystem allocation decisions.[7] The primary variables include 1) system cost (initial investment and life cycle), 2) weight impact, 3) power consumption, 4) crew workhour impact, 5) reliability (impact on system mean time between failures), and 6) safety (reduction in hazard exposure hours). The secondary variables are associated with minimizing the technology developmental risk (i.e., making sure the necessary technology is developed at a timely rate which allows it be incorporated into the overall system by launch). These variables include 1) technology availability (both immediate and far-term), 2) retrofit amenability, and 3) technology importance to mission(s) completion. These last three variables do not normally have quantitative cost values, but they have major cost and schedule implications and are primarily used as branching or bounding decision factors in making human-machine design tradeoffs.

A branching or bounding decision factor means that the variable is used as a limiting or cutoff point. For example, a particular telerobotic design option might appear functionally sound, but might require technology which will not be available for several years past the intended launch date. Given a hard launch constraint, the only recourse is to branch to another design option. In this example, technology availability is used as a decision point to further limit the subsets of viable technology/design options.

Summary of Analytical Approaches

Several different types of modeling approaches have been used to develop a solid design tradeoff foundation for the manned spacecraft problem. Some of the approaches are based on well established operations research techniques, and others are extensions or refinements of more recent decisionmaking structures. It is not the purpose of this section to revisit and derive the host of analytical equations and examples leading up to the final human-machine tradeoff model. This information can be found in the several references cited in this chapter. Rather, this section provides the reader with insight into how the Space Station community is using techniques like this to select reasonable solutions to an extremely complex tradeoff problem. Subsequently, a summary of the analytical techniques is provided here along with supporting references for the interested reader.

Guidelines for Automation

As a preface to the discussion of analytical approaches, it is important to recognize the basic human-factor principles that provide a high-level filter for the initial selection of task candidates that represent good targets for automation. This high-level filter feeds the first step in the analytical process—the task network analysis.

Accepted human engineering standards suggest that the following criteria be used for determining which functions to automate in complex manned systems. Designers should automate, in prioritized order, tasks which 1) endanger crew safety, 2) cause perceptual saturation, 3) must be completed on compressed timelines, 4) exceed operator bandwidth limitations, 5) are complex, or require quickened response, 6) represent complex mathematical/logic problems, and 7) are time consuming, repetitive, sequential, boring, or require extensive memorization.[8]

Criterion 1 is obvious because tasks that directly expose astronauts to hazards should be automated. The reader should note that there is an implied acceptable level of risk associated with this item. Clearly, sending astronauts into space is hazardous by itself. But given that NASA accepts this risk, the implication here is that once on orbit, we should look for ways to reduce exposure to hazards which affect both astronaut and mission safety.

Criteria 2–6 are all equally weighted in that they all address the problem of errors. Errors caused by having to process considerable information and react quickly can not only result in component failures, but can have safety impacts as well. The last criterion is related to the preceding five in that the long-term effects of tedious tasks can result in operational errors.

When the given criteria were applied to the Space Station system design and projected operations, the following functions became obvious robotic or automated system candidates[9]: 1) EVA (assembly and long-term servicing of the station/payloads), 2) subsystem monitoring, 3) subsystem state verification/calibration, 4) mission/operations planning, 5) subsystem state assessment/change.

The candidate of greatest concern to NASA right now is, based on current EVA/IVA workload studies (see the section on the problem of limited resources), the EVA function. Recent decisions on how NASA will proceed with the phased solution to this problem will be discussed after the summary of the supporting analytical techniques.

Solution Structure

As stated earlier in the Introduction, the model developed for the Space Station assessment was a four-tiered system which enabled a logical analysis, filtering, and grouping of data into the final cost-benefit and design tradeoff algorithms. That logical structure starts with a detailed task network analysis. This analysis identifies the obvious high-workload, and potentially hazardous, task elements. The resulting functional and time (task and hazard exposure) data are stored for use in the conceptual design, cost-benefit, and tradeoff modules. The next component of the problem structure is the determination of how the functional candidate will be automated. This requires both a design and technology assessment. Cost estimates are developed for the conceptual design using a bottom-up component-by-component (hardware and software) approach. This step provides a solution to the cost side of the problem. The benefits side considers the task time, safety, and other potential benefits (e.g., reduction in number of launches) obtained by reducing the human involvement. Using constraints, or bounds, on the key resource variables (given earlier), the various robotic or automation design options are interatively assessed until the safest, lowest cost option(s) surface.

The preceding discussions and sections provided a foundation for the reader to understand the primary tradeoff, or resource variables, the near-term automation candidates, and the solution structure. The next section provides a summary of the various solution techniques developed around the given foundation.

Analytical Techniques

Task Network Analysis

The first component, task network analysis, is used uniformly in all of the analytical models which have evolved. The network analysis is a standard PERT-type approach, exactly the same as that presented in classical operations research textbooks. The two major departures from classical PERT analysis are derived from the recognition that human resources in space cannot be looked at in the same way as they are in the private sector. The idea that time is money is not exactly correct. Although an approximate value of a workhour in space can be determined based on the cost of putting a human on orbit, what is most important is how time is put to good use for science (the concept of quality time). Therefore, the workhour time variable is analyzed by superimposing either EVA/IVA limits (based on overall schedule demands and life support constraints), and exposure time limits to hazards [based on predetermined historical constraints (e.g., radiation, EVA life support system constraints)]. When the schedule and hazard constraint template is placed over the projected task timelines, the high task time drivers exceeding the constraint template then represent the primary targets for telerobotic or automation augmentation.[7,10]

The second departure was more a matter of foundation than process. It was recognized early on in modeling stages that the task analysis was tedious and obviously could not draw on Earth-based industry experience. The PERT steps of the analysis, the process component, was straightforward. However, a detailed list of historical tasks in space had to be compiled to enable task sequence construction. This process was initiated by studies such as the Massachusetts Institute of Technology "Automation, Robotics, and Machine Intelligence Systems" effort,[11] and the McDonnell Douglas "Human Role in Space" activity.[12] Drawing on these data, the bank of Shuttle task data, and more recent contractor projections for workload on the Space Station, JPL developed a relational database which allows one to assemble any desired task sequence based on historical and projected task/time data using the telerobot joint analysis system.[13,14] This tool supports the tasking analysis front end to the human-machine tradeoff process.

Technology Assessment and Conceptual Design Generation

The technology assessment and conceptual design element of the human-machine tradeoff process has also been used uniformly throughout the evolution of the modeling effort. Similar to the earlier automated mining problem, the spacecraft problem requires the analyst to develop concepts which augment, or replace, the historical human function highlighted by the preceding task constraint template. This was done by pooling the results of NASA, national laboratory, industry, and university projections on technology needs/availability. The assembly of a conceptual design followed standard automated system design practice (see Fig. 1), and included the following criteria[7, 10]:

1) The telerobotic or automated system must have, at the lowest level, sensors to provide intelligence to the control system.

2) The system must have low-level dedicated processors which provide reflex/safing control (e.g., shutting off fuel flow in the event of a leak).

3) Subsystem level processors must be present to manage incoming/outgoing command and status information and provide limited diagnostics/fault management for functions requiring rapid response.

4) System level processors must be included to provide an interface to the operator or fault manager, and contain sufficient memory and on-line analytical tools to enable supervision of control operations and full fault recovery.

The network communication and actuator components exist whether the system is manually controlled or semi-autonomously controlled and, therefore, are not included as part of the system design. The two technology and conceptual design outputs are used 1) to provide another screening variable relative to whether or not a needed piece of technology will be available on time, and 2) to feed the cost-benefit assessment.

Cost and Benefit Projections

Two primary cost-benefit approaches were developed as part of the human-machine tradeoff process, namely, 1) incremental cost reduction, and 2) net cost savings.[7, 10] Both of these approaches are summarized here. The costing approach for both techniques is provided first, followed by a separate section on assessing tradeoffs utilizing both techniques.

Incremental cost reduction. The incremental cost reduction technique provides a means of successively paring down automation options to eliminate the

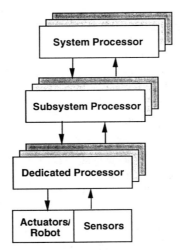

Fig. 1 Standard automated system design.

low performers before finally attempting to separate out the best set of automation candidates. Additionally, the approach utilizes a variable which measures the fractional increase/decrease in a particular cost variable as a result of automating a particular function(s). The process starts by taking the results of the task and conceptual design analyses and establishes an efficient frontier of viable options. Only the major variables of cost and productivity are used at this time. The cost of each telerobotic or automation design option is estimated using a bottom-up approach. The design is broken down into its hardware and software components (i.e., at the level of sensors, servos, actuators, processors, software module) and, using state-of-the-art industry historical costs, summed to obtain a implementation/delivery cost estimate. Spares are included and software is estimated using current versions of the COCOMO software costing model.[15] If a component is not commercially available, current development costs and technology availability projections are used to project time-of-availability costs. The new technology cost projection can be computed by combining a historical percentage breakdown on first flight unit cost, as a function of total flight development cost.

The efficient frontier is established by plotting each candidate's relative projected impact on productivity as a function of relative cost. The candidates exhibiting the best productivity impact with the least cost are the winners. Equation (1) provides the mathematical representation of the decision process.

$$\text{Max}_{x} \sum_{i=1}^{n} P_i x_i \tag{1a}$$

subject to

$$\sum_{i=1}^{n} c_i(x) x_i \leq c \tag{1b}$$

where P_i are incremental crew hours saved if function i is automated; x_i is the automation decision variable, equal to 1 if automated, and equal to 0 if not; $c_i(x)$ is the net incremental cost for automating function i; and c is the cost of most expensive man-machine alternative.

The next step in the paring-down process is to take the preceding subset of viable candidates and determine the projected life cycle cost impact of inserting the technology into each respective subsystem. The mathematical representation of this next step is shown in Eq. (2).

$$\sum_{i=1}^{n} (C_{LC_i} C_{LCB_i})_{PV} = \sum_{i=1}^{n} \left[\sum_{j=1}^{m} (1 \pm b_{LCVE_j}) C_{LCVE_j} \right]_{iPV} \tag{2}$$

where C_{LC_i} is the life cycle cost of subsystem i, C_{LCB_i} is the life cycle cost benefit of automating subsystem i, b_{LCVE_j} is the fractional change in life cycle variable element j resulting from automation, C_{LCVE_j} is the life cycle cost variable element j (e.g., development, operations, maintenance, training, manpower), and PV is the present value taken.

The fractional increase/decrease in a particular cost variable was determined through three primary sources of historical data, including 1) NASA/Department of Defense/aerospace industry data on automated spacecraft/aircraft/ship based systems, 2) the automobile industry (U.S., Japan), and 3) other industries (IBM, Honeywell, FMC Corporation). Where life cycle cost projections for Space Station components existed, they were used. Unknown cost variables were determined by using a historical total percentage breakdown for space-based systems (e.g., the Shuttle). If one, or more, major cost component(s) (e.g., development/first unit cost) is known, each of the other components can be represented as a function of the known variables. Each of the unknown variables are then projected as a fraction or multiple of the known cost components. Upon completion, a major piece of the tradeoff puzzle is available (an example can be seen in Ref. 16).

The third step in the paring process considers not only the newly derived quantitative tradeoff factors (productivity, cost/savings), but incorporates other less quantitative factors as well. This total consideration of quantitative and nonquantitative factors is the list of resource variables described earlier in this chapter. This third step employs a well-established decision analysis technique called multiattribute decision analysis (MADA) to rank order the automation options across all tradeoff factors, or attributes.[17,18] The objective/attribute hierarchy is shown in Fig. 2. The technique basically calls for the analyst to establish measures for each attribute [i.e., cost (dollars), productivity (hours saved), hazard exposure (hours exposed), reliability (mean downtime or time between failures), and technology availability (years to introduction)]. Approximate value ranges for each attribute are assembled based on the earlier analysis, historical data, and industry design projections. The analyst then carefully selects a group of experts in the field of automation/telerobotics and proceeds to interactively question them on their preferences for the different automation options within the range of the projected state values for each attribute. The expert is asked to globally examine the full range of state values, and, through a structured question and answer process, provide his assessment of candidate performance within the expected range of state values. The expert's projection of performance is equated to a utility value; and, once the utility values are known, the utility curve can be approximated. A utility value is selected within the range of 0–1; with 0 representing low performance, 1 representing high performance, and 0.5 representing average (or the point of indifference) performance. An overall utility value for a particular option is calculated by combining the various subutility values across all attributes. The outcome is a weighted product of the experts' responses.

The analyst now has a ranking of the best-to-worst telerobotic or automation options considering the host of tradeoff variables. The solution is not considered optimal, but reasonable. The last step, the system tradeoff and best-mix analysis, can now be performed using a much smaller subset of viable automation candidates. It should be noted that the simplifying assumption that only a given subsystem is affected by a particular automation option is conservative. Indeed, automation has a cross-correlation effect across subsystems (i.e., subsystems can share computing and diagnostic resources to meet redundancy needs).

Net cost savings. The net cost savings approach is more straightforward than the aforementioned incremental cost reduction technique, and was derived as an evolution of that model. All of the same sources of data/techniques are employed to determine the various development, flight unit, operating, and maintenance cost parameters. The difference revolves around the basic cost savings equation.

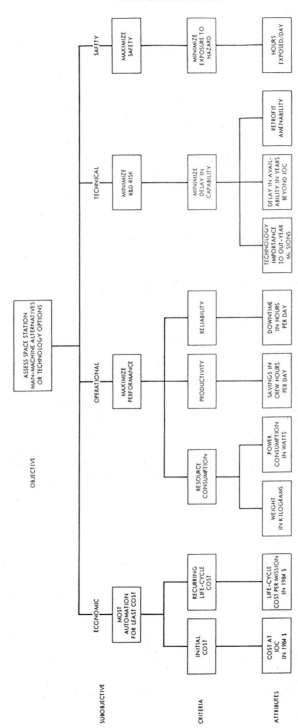

Fig. 2   Hierarchy of objectives, criteria, and attributes.

Whereas the incremental approach uses a automation cost reduction factor, coupled with multiattribute decision analysis for the harder to quantify design/cost attributes, the cost savings technique specifically concentrates on the cost-based variables and proceeds to calculate the net cost savings between the system without automation and the system with automation. The cost savings relationship is shown in Eqs. (3).

$$\text{TotO\&M}_{\text{PV}}^{\text{EVA+IVA}} - \text{TotO\&M}_{\text{PV}}^{\text{EVA+IVA+Auto}} - CI_{\text{PV}} \geq 0 \tag{3a}$$

given

$$CI_{\text{PV}} = \sum_{t=0}^{n} (\text{DDTE}_t + \text{FLT}_t + L_c)_{\text{PV}} \tag{3b}$$

where $\text{TotO\&M}^{\text{EVA+IVA}}$ is the total operations and maintenance cost without automation; $\text{TotO\&M}^{\text{EVA+IVA+Auto}}$ is the total operations and maintenance cost with automation; and CI are the design, development, test, and evaluation (DDTE), flight unit (FLT), and launch costs (L_c).

Upon completion of this step, the analyst has established a matrix with option cost along the horizontal, and cost savings along the vertical. Each of the automation candidates occupies its respective position within the matrix. As with the incremental cost reduction approach, the last step, the system tradeoff analysis, can now be performed using any one of the candidates in the matrix.

Design Tradeoff Analysis

The last analytical procedure calls for the analyst to iteratively sort through the telerobotic and automation design options and find the best overall system mix/design which meets the budget, productivity, and other attribute constraints. This last subsection completes the preceding cost-benefit discussion by addressing the two primary techniques for picking the best telerobotic or automation option(s).

Incremental cost reduction optimization. It is now possible to impose some form of system optimization, having pared down the array of automation options to a subset which represents a reasonable solution pool. A standard linear regression technique was employed for the final tradeoff step (note that it is not necessary to assume linearity, but linearity appears to provide a reasonable solution given the uncertainty associated with the various attribute data). The solution of the problem calls for picking the best system-wide subset of automation options, across all subsystems, which satisfy all major constraints. Mathematically, the problem is stated as shown in Eqs. (4).

$$\max_{A,a} \sum_{k=1}^{n} \sum_{t=1}^{m} B_{kt}(a_{kt}) \tag{4a}$$

subject to

$$\sum_{k=1}^{n} \sum_{t=1}^{m} (C_{kt} - B_{kt})(a_{kt}) \leq C_t \tag{4b}$$

$$\sum_{k=1}^{n}\sum_{t=1}^{m} W_{kt}(a_{kt}) \leq W_t \qquad (4c)$$

$$\sum_{k=1}^{n}\sum_{t=1}^{m} P_{kt}(a_{kt}) \leq P_t \qquad (4d)$$

$$\sum_{k=1}^{n}\sum_{t=1}^{m} H_{kt}(a_{kt}) = 0 \qquad (4e)$$

where

$$\sum_{t=1}^{n} a_{kt} = 1, k = 1\ldots n, t = 1\ldots n, a_{kt} \geq 0$$

for all k, all m. (See also the Nomenclature at the beginning of this chapter.)

The branch and bound technique is the preferred method of solution and is a standard operations research optimization process. The technique begins by obtaining a bound on the objective function (i.e., maximize productivity hours/value) by suppressing the dependencies at the system level. This is done by fixing the values in the preceding equations at their upper values and solving the resultant integer linear program. Next, the branch and bound method requires that the set of all feasible solutions (i.e., the telerobotic and automation candidates selected, by subsystem, using the previous paring down process) first be partitioned into subsets. This means that we start by picking one candidate from each subsystem and form a single system solution subset. This process continues until all candidates are partitioned into subsets. Because the objective function requires that we maximize available on-orbit workhours and net benefits, without exceeding a cost ceiling, any solution subset which surpasses the cost ceiling is excluded. The remaining subsets are partitioned further and examined in the same manner. This process is repeated until a feasible solution is found which satisfies the objective function. It is possible that no solution subset meets the objective function. In this event, the next step is to select the subsets that minimize the difference between the cost target and upper bound on each of the solution subsets.

Net cost savings optimization. The net cost savings optimization process aims at establishing an envelope of acceptable telerobotic or automation solutions. In the process of performing the cost savings calculation, the high-cost operations and maintenance drivers become apparent. If a telerobotic/automated system is going to be cost effective, it must offset these high-cost drivers. The ability of a given design option to effect these high costs is determined by plotting cost savings as a function investment cost, and then parametrically varying the major operating cost drivers. By doing this, the analyst can see how much, and how quickly, changes in the major cost drivers affect the net cost savings. As with the incremental cost reduction approach, the final best design may only offer a reasonable payback.

The application domain for this technique was the Space Station Flight Telerobot Servicer (FTS).[10] The problem was to find an FTS configuration which could provide enough functionality to reduce EVA hours sufficiently to allow the station to be assembled in the prescribed number of flights (i.e., early workload studies showed that insufficient EVA astronaut hours were available to assemble/ maintain the station in the first several years of operation). Figure 3 displays a graph of required EVA assembly hours, vs the EVA budget (driven by constraints of the EVA life support system and fatigue). As can be seen, the addition of the FTS allowed the budget to be met. When the cost savings analysis was done it was discovered that even with the highest estimated dollar value of an EVA hour, the real cost driver was the flight manifesting and launch costs. This variable was critical because if assembly or maintenance on-orbit is not completed on schedule, then additional Shuttle flights must be scheduled and payloads must be remanifested to meet both weight and workload constraints. This remanifesting effort is done at great expense. Therefore, by varying launch costs one could see the allowable FTS investment cost range that would still allow the system to break even or save money. A set of graphical plots showing an envelope bounded by cost savings and an FTS cost investment range were generated by parametrically varying Shuttle launch costs. An example of one of these plots is shown in Fig. 4. Other cost saving envelopes were generated using other drivers such as EVA hours, discount rate, and combinations (e.g., EVA hours and Shuttle manifesting) to obtain a clear picture of the FTS design configuration offering the best impact on meeting the EVA hour budget, while still offering a savings given current/projected Shuttle launch costs.

Implications of Telerobotics and Automation for the Space Station

The preceding discussion provided the reader with an understanding of some of the current modeling techniques being employed to weigh the human vs machine partitioning problem. With the opportunity to perform more unmanned and manned missions in space, coupled with the growing availability of advanced telerobotic and automation technology, it is clear that for safety, efficiency, and

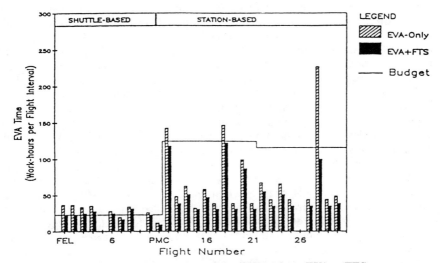

Fig. 3   Low-range EVA estimates for EVA-only vs EVA + FTS cases.

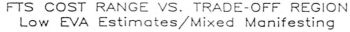

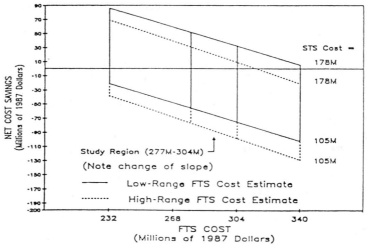

Fig. 4   High-range vs low-range FTS cost.

cost reasons we must develop a clearer understanding of how to resolve this extremely complex problem. We cannot take a standard human capital approach
to this tradeoff problem (i.e., replace human beings with machines when the net
cost/profit of the machine exceeds the benefits offered by humans). In the space
environment, there are many functions better performed by astronauts than by
machines. For example, the tending of science payloads often require the performance of off-normal anomaly troubleshooting tasks. Data collection and reduction
often requires complex on-line analysis. Finding the right mix and match of hu-
man/machine skills requires an understanding of non-cost-related variables such
as these. This last section explores NASA's current thoughts and plans relative to
introducing telerobotic and automation technology into the manned space arena.

There are many factors to consider in injecting more telerobotic and automation
technology into an environment classically dominated by human operators. Earlier
in this chapter reference was made to the realization of the new wave of astronauts for greater augmentation by automated systems. Several risk factors have
affected the development and injection process. The Challenger disaster placed
NASA's leadership role in cutting-edge science and technology development at
risk. Numerous questions have been raised about the true need to continue to
place humans at risk in the space environment at great cost. At the same time, the
research and development budgets within NASA have been tightly constrained
over the last several years by a squeezed national economy. This condition has
required several station redesigns and made it difficult for the technology to keep
pace with flight program requirements and demands. Therefore, NASA currently
is taking a very cautious approach to having astronauts and machines concurrently
controlling functions or telerobotic devices.

In the previous section the primary automation candidates were listed. Although the EVA function was part of that list, distinct tasks were not included. A more complete list of planned automation and telerobotic functions which resulted from the tradeoff studies is provided here: 1) subsystem state monitoring, 2) subsystem state verification/calibration, 3) subsystem state assessment/change, 4) fault diagnostics/recovery, 5) mission operations/planning, 6) station module placement on orbit, 7) module assembly, 8) inspection (including preassembly, postassembly, work cell preparation, and scheduled/unscheduled maintenance), 9) work cell setup/teardown, 10) station/payload servicing [orbital replacement unit (ORU) level], and 11) logistics support to EVA (including pallet handling, EVA tool/component holding, and holding/jigging components being worked on by EVA astronauts).

Items 1–5 are more automation oriented, whereas items 6–11 are telerobotic functions. In the automation area, expert systems are being built to monitor sensed subsystem state data, simulate subsystem performance, and determine whether a state change must be initiated. The target subsystems include almost all of the station subsystems, such as power, life support, thermal, navigation/control, and data management. Much of this work is being done within the NASA centers for eventual transfer to the primary ground operations control center, JSC. Before any of the expert system software will be allowed to migrate to on-orbit systems, it will be used in the ground operations monitoring facility. The automated control software will actually operate simultaneously with the more manual on-orbit subsystem, but in the simulation mode. Anomalies and faults will be simulated and diagnosed in the ground version using the actual on-orbit sensing data. The ground version will initially provide support to the ground operations crew. As the ground-based automated systems get debugged and confidence is established, the expert systems will be moved on orbit. Several of the ground-based automated tools for mission planning, state assessment/change, and fault diagnostics are already in use for unmanned spacecraft such as Voyager, Magellan, and Galileo.

The telerobotic functions will primarily be performed using teleoperation from the Shuttle and/or Space Station operator control stations (i.e., basically real time). Current autonomous control augmentations to manual teleoperation will initially take the form of proximity position accommodation, joint limit monitoring, manipulator pose monitoring, or force threshold monitoring and control. As with the automated systems, all telerobotic technology will be exercised in a ground-based task verification environment before being migrated on orbit. The ground-based environment will operate primarily in a support mode for the on-orbit IVA teleoperator. As the technology establishes a confidence base, more functions will be offloaded to the autonomous control system. This process will be necessary in the early phases of operation of the Space Station because of onboard computational constraints. On the ground, task sequences will be built, partitioned between operator and autonomous control, and simulated/tested before being telemetered up to the station operator control station.[?]

In addition to the migration of technology on orbit, NASA is also developing the technology base necessary to allow ground-remote operation of telerobotic systems. When implemented, the technology will initially be used for inspection (noncontact) type tasks until a strong confidence base is built. Again, each inspection task will be simulated and verified on the ground before a command sequence is telemetered to the robot controller(s).[16]

Conclusions

In conclusion, the author has endeavored to expose the reader to some of the key issues and variables that drive the human-machine partitioning of functions in space, along with a summary of some of the analytical techniques used in the tradeoff process, as well as provide an indication of the direction in which NASA is moving in getting the applications and technologies offering the most immediate cost and functionality payback implemented. This tradeoff process is being executed in a rapidly changing design and economic environment which has affected the rate at which needed technologies are being developed. This environment may impact the readiness of the various emerging technologies. The reader will be exposed to many of these emerging technologies and their use in subsequent chapters of this book.

References

[1]Ackoff, R., and Sasieni, M., *Fundamentals of Operations Research,* John Wiley and Sons, Inc., New York, 1968.

[2]Miller, D., and Starr, M., *Executive Decisions and Operations Research,* Prentice Hall, Inc., Englewood Cliffs, NJ, 1969.

[3]Zimmerman, W., Aster, R., Harris, J., and High, J., "Automation of the Longwall Mining System," Jet Propulsion Lab., JPL Pub. 82-99, Pasadena, CA, Nov. 1982.

[4]Freeman, D., *STS Work Day Handbook,* NASA, Johnson Space Center, Operations Planning Section, Feb. 1980.

[5]Fisher, W., and Price, C., "Space Station Freedom External Maintenance Task Team Final Report," Vol. 1 and 2, NASA, Johnson Space Center, July 1990.

[6]Weeks, D., Zimmerman, W., and Swietek, G., "Space Station Freedom Automation and Robotics: An Assessment of the Potential for Increased Productivity," NASA/MITRE Corp., Houston, TX, Dec. 1989.

[7]Zimmerman, W., Bard, J., and Feinberg, A., "Space Station Man-Machine Automation Trade-off Analysis," Jet Propulsion Lab., JPL Pub. 85-13, Pasadena, CA, Feb. 1985.

[8]Von-Tiesenhaussen, G., "An Approach Toward Function Allocation Between Humans and Machines in Space Station Activities," NASA, Marshall Space Flight Center, NASA TM-82510, Nov. 1982.

[9] Anon., Advanced Technology Advisory Committee (ATAC), "Advancing Automation and Robotics Technology for the Space Station and for the U.S. Economy," NASA TM-87566, Vol. II, March 1985.

[10]Smith, J., Gyamfi, M., Volkmer, K., and Zimmerman, W., "The Space Station Assembly Phase: Flight Telerobotic Servicer Feasibility, Volume 2—Methodology and Case Study," Jet Propulsion Lab., JPL Pub. 87-42, Pasadena, CA, Sept. 1987.

[11]Miller, R., Minsky, M., Smith, D., and Akin, D., "Space Applications of Automation, Robotics, and Machine Intelligence Systems (ARAMIS)," Massachusetts Inst. of Technology, Cambridge, MA, and NASA, Marshall Space Flight Center, Aug. 1982.

[12] Anon., McDonnell Douglas Corp., "The Human Role in Space (THURIS)," sponsored by NASA, Marshall Space Flight Center, 1984.

[13] Smith, J., and Drews, M., "Generic Extravehicular (EVA) and Telerobot Task Primitives for Analysis, Design, and Integration," Jet Propulsion Lab., JPL Pub. 90-10, Pasadena, CA, March 1990.

[14] Drews, M., "Telerobotics/EVA Joint Analysis System (TEJAS), Version 1.0 User's Guide," Jet Propulsion Lab., JPL Pub. 90-11, Pasadena, CA, March 1990.

[15] Boehm, B., *Software Engineering Economics,* Prentice Hall Inc., Englewood Cliffs, NJ, 1981.

[16] Zimmerman, W., "Closeout of FY90 Space Exploration Initiative Surface Transport Vehicles/Human, Automation, Robotic Tradeoff Deliverables," Jet Propulsion Lab., JPL IOM 3475-90-66, Pasadena, CA, Nov. 7, 1990.

[17] Kenny, R., and Raiffa, H., *Decisions with Multiple Objectives: Preferences and Value Trade-offs,* Wiley, New York, 1976.

[18] Tarvainen, K., and Haimes, Y., "Coordination of Hierarchical Multi-objective Systems: Theory and Methodology," *IEEE Transaction Systems, Man, and Cybernetics,* Vol. SMC-12, No. 6, 1982.

[19] Zimmerman, W., "Supervisory Autonomous Local-Remote Control System Design: Near-Term and Far-Term Applications," NASA/U.S. Air Force, SOAR, Aug. 1992.

Part 2.   Human-Machine Interface

Chapter 3

Human Enhancement and Limitation in Teleoperation

Thomas B. Sheridan*

Massachusetts Institute of Technology, Cambridge, Massachusetts 02139

Introduction

T HIS chapter offers a perspective on the question: "Why perform sensing and manipulation in space by using a combination of a human and a machine, as compared to either by itself?" In other words, why and under what circumstances should required tasks be performed through teleoperation rather than by an astronaut or robot? A second salient question is: "To what extent can the human be designed into a space telemanipulation system in a rational, objective, and predictable way, as normally practiced (or aspired to) by engineers; or to what extent is dealing with the human an art, or even worse, a matter of chance?"

Toward answering these questions, the chapter first states what the author believes is the undeniable trend in space, as well as other sophisticated high-technology systems, namely, the rapid evolution from direct manual control to a flexible mix of human and automatic control called supervisory control. The chapter then reviews some basics of what we know about modeling human performance characteristics in sensing and control tasks, characteristics which are unchanging for the time scale of interest. Next are discussed some salient lessons learned about integrating humans into otherwise automatic and remotely controlled sytems in other fields, such as aviation and nuclear power. The chapter concludes with reviews of our experimentally derived understanding of human enhancement and limitation in several specific aspects of teleoperation, among them command language, vision and force feeback, coping with time delay, and telepresence. These same topics are picked up by other authors in the chapters which follow.

Human Versus Robot in Space: A False Dilemma

The media characterize robots as intelligent, often malevolent, machines which willfully seek to better mankind. The human is mostly portrayed as the robot's

Copyright © 1994 by Thomas B. Sheridan. Published by the American Institute of Aeronautics and Astronautics, Inc., with permission. Released to AIAA to publish in all forms.

*Professor of Engineering and Applied Psychology, Department of Mechanical Engineering, and Professor of Aeronautics and Astronautics.

victim, not its friend and benefactor. NASA and the space bureaucracy have been very conservative about robotics, have inclined toward protection of the astronaut's role, and have continuously emphasized the need for direct human observation and manipulation of experiments in space by a person bodily present. For reasons not altogether clear, ground-to-Space Shuttle communications were not designed to accommodate teleoperation from the ground. The Canadian Remote Manipulator System (RMS) served well on many missions as an astronaut-operated device for very limited tasks, and not until 1993 did the German Rotex experiment demonstrate something closer to current teleopreration capability from the ground. Hence, it is no wonder that the public (and the U.S. Congress) are confused about the possibilities for human-robot cooperation, and have come to see the relation as more of an adversarial than a cooperative one.

Naturally there are those tasks which are easily and less expensively fully automated, and there surely are some which we do not yet know how to automate. However, to those who have thought about the problem of human-machine cooperation there is very little question but that the combination of a human and a machine provides much more capability than either by itself for many tasks one might wish to do in space. The question remains: what forms of combination, and what degrees of automation, are best for what tasks?

Fitts List and the Automation Scale

In 1951 psychologist Paul Fitts published a list (Table 1) of functions in which people are better than machines, and those in which machines are better than people. More recently the author published a 10-point scale (Table 2) from no automation to full automation, the purpose of which was to provoke discussion of what level of automation would be best for what tasks. Both scales have now seen many variants. Both lists, being qualitiative, do not give answers immediately usable by engineers; they merely pose the questions.

Synthesizing the Human and Machine for Teleoperation

In rough mechanical form, the allocation of functions between human and machine has been somewhat obvious. The teleoperator can endure the space

Table 1   Fitts (1951) MABA-MABA list (abbreviated)

Men are better at	Machines are better at
Detecting small amounts of visual, auditory, or chemical energy	Responding quickly to control signals
Perceiving patterns of light or sound	Applying great force smoothly and precisely
Improvising and using flexible procedures	Storing information briefly, erasing it completely
Storing information for long periods of time, and recalling appropriate parts	Reasoning deductively
Reasoning inductively	Doing many complex operations at once
Exercising judgment	

Table 2   Scale of degrees of automation

1. The computer offers no assistance, human must do it all.
2. The computer offers a complete set of actiofi alternatives, and
3.   narrows the selection down to a few, or
4.   suggests one, and
5.   executes that suggestion if the human approves, or
6.   allows the human a restricted time to veto before automatic execution, or
7.   executes automatically, then necessarily informs the human, or
8.   informs him after execution only if he asks, or
9.   informs him after execution if it, the computer, decides to.
10. The computer decides everything and acts autonomously, ignoring the human.

environment, and can perform long reaches which the human cannot. The human must provide the intelligence to perform one-of-a-kind, sometimes unpredictable real-time sensing and manipulation tasks. This is in sharp contrast to industrial robots, which, because they perform the same task over a large number of repetitions, can justify extensive programming efforts. The environmental (visual or haptic) pattern recognition required of the human typically goes well beyond that which computers are currently capable of. The control required of the human must be adaptive and multidimensional, but high speed and large forces are not required. Relatively simple fully automated satellites have visited distant planets, but the more complex experiments conducted on the Space Shuttle have required astronauts, sometimes suited for extravehicular activity (EVA), sometimes operating the Remote Manipulator in the Shuttle bay as a direct arm extension.

From Manual to Supervisory Control

Except for some very crude telemanipulations from the ground to the lunar rover Surveyor in the late 1960s and the very recent demonstrations by the Germans mentioned earlier, telesensing and telemanipulation in space have thus far consisted of direct manual (joystick rate control and master-slave position control) with the operator peering out the window or at a video close-up view. The tasks were kept simple, and when they could not be, EVA astronauts supported at the end of the RMS arm did the job. There was no time delay or communication channel bandwidth constraint and essentially no automatic control.

This situation should be changing soon. Many researchers and NASA laboratories have for many years been proving out various forms of automation, which can be added to the otherwise purely manual teleoperation, resulting in what has come to be called telerobotics (where etymologically robot implies at least some degree of autonomy). The role of the human, for reasons now to be given, best assumes the role of supervisor of the telerobot.

Basic Idea of Supervisory Control

The term supervisory control is derived from the close analogy between the supervisor's interaction with subordinate staff members in a human organization, and a human operator's interaction with an "intelligent" automated subsystem. A supervisor of humans gives directives that are understood and translated into detailed actions by staff subordinates. In turn, subordinates collect detailed information about results and present it in summary form to the supervisor, who then must infer the state of the system and make decisions for further action. The intelligence of the subordinates determines how involved their supervisor becomes in the process. Automation and semi-intelligent subsystems permit the same sort of interaction to occur between a human supervisor and the computer-mediated process.[1]

Strictly speaking, supervisory control means that one or more human operators are intermittently programming and continually receiving information from a computer that closes an autonomous control loop itself through artificial effectors and sensors to the controlled process or task environment. In a less strict usage, supervisory control means that one or more human operators are continually programming and receiving information from a computer that interconnects through artificial effectors and sensors to the controlled process or task environment—with no necessary autonomous control through the computer that excludes the human. In both definitions the computer transforms information from human to controlled process and from controlled to human process, but only under the strict definition is the computer an autonomous controller for some variables at least some of the time.

In the strict forms of supervisory control, the human supervisor programs by specifying to the computer goals, objective tradeoffs, physical constraints, models, plans, "if-then-else" procedures, and at the very minimum, set points for simple control loops. This specification is usually and most conveniently put in high-level natural language—in terms of desired relative changes in the controlled process, rather than in terms of control signals. Once the supervisor turns control over to the computer, the computer executes its stored program and acts on new information from its sensors independently of the human, at least for short periods of time. The human may remain as a supervisor, or may from time to time assume direct control (this is called traded control), or may act as supervisor with respect to control of some variables and direct controller with respect to other variables (shared control).

Figure 1 illustrates the concept of supervisory control in detail. At the bottom are multiple tasks, each interacting with a task-interactive computer (TIC), in the normal sense of automatic control. The tasks could be sensing tasks, manipulation tasks, or something else. At the top are shown a number of functions performed by the human and/or the human-interactive computer (HIC) which will be detailed hereafter. Human-to-system communication is largely in terms of symbolic commands (concatenations of typed symbols or specialized key presses). However, some fraction of the commands may be analogic (hand-control movements isomorphic to the space-time-force continuum of the physical task) to point to objects or otherwise demonstrate to the computer relationships that are difficult for the operator to put into symbols. The HIC must be human-friendly, able to indicate that it understands the message, and able to point out that a specification is incomplete. In this way it should help the operator to edit the message correctly. It also needs to interpret signals from the distant telerobot, storing and processing them to generate meaningful integrated graphic displays. Finally, the HIC should contain a knowledge base and a model of the controlled process and task environment and be able to answer queries put to it by the operator.

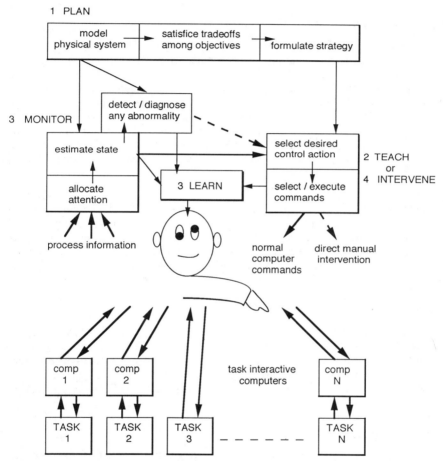

1 PLAN

3   MONITOR

2 TEACH
or
4 INTERVENE

Fig. 1   Generic supervisory control framework: at the bottom are subtasks, any or all of which may be automated for short periods through a task-interactive micro-computer; at the top are the supervisory functions (plan, teach, monitor, intervene) subdivided into elements, all of which must be attended to by the human supervisor, but any of which may be aided by a computer-based expert system or on-line deci-sion aid. In addition, off-line and not shown is the supervisory function learn. The supervisor must allocate his or her attention among all the boxes. From Ref. 1.

Meanwhile, each subordinate TIC that accompanies some remote controlled process must receive commands, translate them into executable strings of code, and perform the execution, closing each control loop though the appropriate actuators and sensors.

Five Generic Supervisory Functions

The human supervisor's functions are 1) planning what task to do and how to do it; 2) teaching (or programming) the computer what was planned; 3) monitoring the automatic action to make sure all is going as planned and to detect failures; 4) intervening (which means that the supervisor supplements ongoing automatic control activities, takes over control entirely after the desired goal state has been reached satisfactorily, or interrupts the automatic control in emergencies to specify a new goal state and reprogram a new procedure); and 5) learning from experience so as to do better in the future. These are usually time-sequential steps. Most of the steps are represented by separate boxes in Fig. 1, except for learning.

Planning

This is the hardest function to model. Formally it means 1) gaining experience and understanding of the physical process to be controlled, including the constraints set by nature and circumstances surrounding the job; 2) setting goals that are attainable, or objectives along with tradeoffs, that the computer can understand sufficiently well to give proper advice or make control decisions; and 3) formulating a strategy for going from the initial state to the goal state.

Teaching the Computer

The supervisor must translate goals and strategy into detailed instructions to the computer such that it can perform at least some part of the task automatically, at least until the instructions are updated or changed or the human takes over by manual control. This includes knowing the requisite command language sufficiently well that goals and instructions can be communicated to the computer in a correct and timely fashion.

Monitoring Automatic Control

Once the goals and instructions are properly communicated to the computer for automatic execution of that part of the task, the supervisor must observe this performance to ensure that it is done properly, using direct viewing or whatever remote sensing instruments are available. Essential parts of the supervisor's job include prompt detection of the presence and location of failures, prompt detection of conflicts between actions and goals, and the anticipation that either of these is about to occur.

Intervening to Update Instructions or Assume Direct Control

If the computer signals that it has accomplished its assigned part task, or if it has apparently run into trouble along the way, the human supervisor must step in to update instructions to the computer or to take over control in direct manual fashion, or some combination of the two. Because the controlled process is an ongoing dynamic system, not a machine that can be arbitrarily stopped and started again like a computer, the takeover itself must be smooth so as not to cause instability. Similarly, reverting to the automation must be smooth.

Learning from Experience

The supervisor must ensure that appropriate data are recorded and computer-based models are updated so as to characterize current conditions with the most accurate information. Historical data must continuously be analyzed for trends or contingencies leading to abnormalities. All such information must be in a form usable in the future in the four preceding steps.

Modeling the Human Black Box: Epistemological Pitfalls

The engineer seeks to model, or to use models developed by others, to predict the performance of any system design. Modeling humans and human-machine systems, at least in terms which are usable to engineers, is a relatively new activity compared to mechanics and other technolgical fields, having a history of only several decades. This section reviews what salient characteristics of humans as system operators we can generalize and closes with some modeling difficulties which we face as we look forward to automation, the solutions to which currently evade us.

Dynamic Range and Discrimination: Decibels of Light and Force

Experimental psychology of the nineteenth century clearly established the fact that perceived magnitude of sensation scales logarithmically to the physical intensity of the stumulus. The bel and later the decibel (one tenth of a bel) were first established as metrics of the psychological metric loudness (as a function of sound pressure level). The piano keyboard was laid out in proportion to human sensation of pitch (logarithmic to frequency). Similarly, the sensation of brightness is logarithmic to physical radiation intensity. In cases where the human observer may closely anchor on the physical measure itself (e.g, in observing the length of something, or the weight) there is a problem of judging the physical metric rather than the true perceived magnitude of sensation (this is called the stimulus error).

Similarly, sensation discriminability is well established; the just noticeable difference (JND) in physical units is proportional to the reference magnitude in physical units. This empirical fact (Weber's law) characterizes all the senses, more or less.[2]

Human Senses of Vision, Hearing, Reaction Time, Etc.

There is a rich body of experimental literature on vision, hearing, reaction time and other topics which are relatively easy to measure in highly controlled laboratory conditions.[3] One might say these are the characteristics of the ports of the multi-input-multi-output block box, one port at a time, with most of its normal functioning disabled. The literature then becomes a set of observations within each of a large number of pigeon holes of experimental conditions in isolation. Unfortunately, real-life tasks involve many transformations on many variables all interacting simultaneously in a dynamic world setting. So-called integrative models which combine larger clusters of behavior are very few and not particularly quantitative.

Quasilinearity and Adaptation in Manual Control

An example of one semi-integraive behavior is for traget tracking, where a human operator serves as control element in a simple manual feedback control loop. One example (and indeed the one that supported much research) was flying an airplane straight and level under random air turbulence. It was argued that aircraft control system design could not be accomplished until reasonably good equations for the pilot were developed. Since roughly 1950 there has been much effort devoted to trying to apply conventional linear control theory to model simple manual control systems (Fig. 2), where the human operator is the sole in-the-loop control element and the controlled process can be represented by linear differential equations. Initially an independent model of the pilot was sought, so that the pilot model could be combined with whatever the controlled process was. However, this was soon found to be impractical. The characteristics of the human operator proved to be totally dependent on the controlled process, in fact adapting so as to compensate for the controlled process and thereby stabilizing the closed-loop system and providing satisfactory transient response.

McRuer and Jex[4] modeled the human operator and the controlled process as a single forward loop element, to combine the two upper blocks of Fig. 2. This meant there would be only minor variation of the combined human and controlled process from application to application. This approach proved very successful. The result is the simple crossover model which has the form

$$x/e = Ke^{-j\omega T}/j\omega$$

a combined pure time delay and integrator, where ω is frequency. The values of parameters K and T are established in the literature and vary only slightly with the disturbance bandwidth and the order of the controlled process. The main idea in this model is that the combined dynamics of human operator plus controlled process is what is invariant, not the human operator per se.

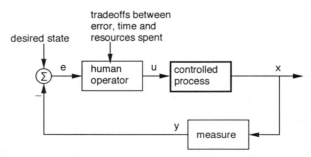

Fig. 2   Simple compensatory manual control paradigm; action u causes simple nulling of error e between current measured state x and desired state (goal).

Perception and Information Processing

Shannon's theory of information, first published in the 1940s, was adopted by a number of experimental psychologists in the early 1950s who saw in it an opportunity to characterize human performance characteristics of many kinds. Two examples of application are cited, examples which have at times been used to characterize aspects of teleoperation.

Hick[5] proposed a model for the time it takes to choose which of several alternative movements i to make when the choice is based on an immediately displayed signal calling for that move and the move time itself is brief and constant

$$H_{\text{choice}} = \Sigma_i p_i \log_2(1/p_i), \quad T_{\text{choice}} = a + b H_{\text{choice}}$$

where H_{choice} is information in bits, p_i is the probability of signal i, T_{choice} is the time required to choose, and a and b are scaling constants dependent on task conditions. Here, a includes at minimum the base reaction time for making the slightest hand movement in response to a visual stimulus.

Fitts[6] also used the information measure for his model of the time required for making a discrete arm movement

$$H_{\text{move}} = \log_2(2A/B), \quad T_{\text{move}} = a + b H_{\text{move}}$$

where (Fig. 3) H_{move} is information in bits (sometimes also called index of difficulty), A is the distance moved, B is the tolerance to within which the move must be made (for Fitts' experiment a tap between two lines), T_{move} is task completion time, and a and b are again scaling constants, different for different conditions.

The Fitts model has found wide application in telemanipulation experiments. When applied to simple one-dimensional movements to within tolerances, the model has withstood the test of time and been robust over a wide range of values of A and B and other task conditions such as barehanded vs master-slave manipulator. It was successfully fitted to experimental data in a number of the studies described earlier. However, like so many elegant models for human behavior, Fitts' model breaks down for more complex manipulations.

Thompson,[7] in his studies of manipulation, showed that the time required to mate one part to another was a function of the degrees of constraint (the number of positions and orientations that simultaneously have to correspond before the final mating could take place). Figure 4 illustrates the idea of degrees of constraint.

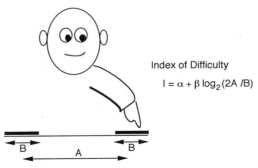

Index of Difficulty

$$I = \alpha + \beta \log_2(2A/B)$$

Fig. 3   Experimental determination of Fitts' index of difficulty.

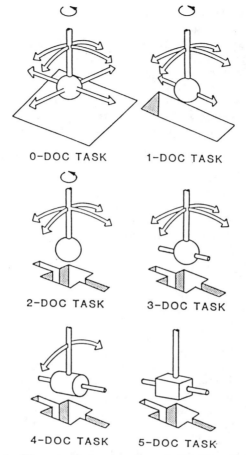

0-DOC TASK   1-DOC TASK

2-DOC TASK   3-DOC TASK

4-DOC TASK   5-DOC TASK

Fig. 4   Thompson's degrees of constraint. From Ref. 7.

Data based on both the Fitts and the Thompson measures will be shown later when time-delayed manipulation is discussed.

Information Value

Let $V(u_j \mid x_i)$ be any measure of the gain or reward for taking action u_j when an event x is in state i. If x_i is known exactly, then a rational decision maker adjusts u_j (selects j) to maximize V for each occurrence of x_i, in each instance yielding $\max_j[V(u_j \mid x_i)]$. In this case the average reward over a set of x_i is

$$V_{\text{avg}} = \sum_i p(x_i)\{\max_j[V(u_j \mid x_i)]\}$$

If x_i is known only as a probability density $p(x_i)$, then the best a rational decision maker can do is to adjust u_j once, to be the best in consideration of the whole density function $p(x_i)$. In this case the average reward over a set of x_i is

$$V'_{avg} = \max_j \left[\sum_i p(x_i) V(u_j \mid x_i) \right]$$

Information value, then, is the difference between the gain in taking the best action given each specific x_i as it occurs, and the gain in taking the best action in ignorance of each specific x_i, i.e., knowing only $p(x_i)$. This difference is

$$V^*_{avg} = V_{avg} - V'_{avg}$$

and is what the specific information is really worth in the given control situation.

Subjective Expected Utility of a Consequence

As with human judgment of the magnitude of sensation, the judged magnitude of the relative worth of some physical event or thing does not scale linearly with the physical measure of that event or thing. Thus, it is encumbent on decision theorists and experimentalists (and others concerned with human tradeoff judgments) to measure how humans scale relative worth, called utility. The most accepted means was that proposed by Von Neumann and Morganstern.[8]

Formally, utility is determined by finding for what consequence, were it to occur with certainty, a judge is indifferent as compared to a lottery of two consequences which are defined to have given utility. For example, a judge might be indifferent between a sure gift of $300 (call it C) and a 50-50 chance of $1000 (call it A) or nothing (call it B). If $1000 is the maximum consequence with an assumed utility of one, and 0 is the minimum consequence with an assumed utility of zero, the basic Von Neumann axiom states that the utility of C is

$$U(C) = 0.5 U(A) + 0.5 U(B) = 0.5$$

By repeating this same procedure, either changing the probabilities for A and B or using A and B consequences whose utilities have already been determined, one may find the utility of any other consequence. This procedure usually produces a concave-downward function which is interpreted as risk aversion or conservatism, i.e., people generally prefer a sure reward of less than half over a 50% chance of a full reward. Similar procedures can be used to determine the utility for losses and serve to explain why people are willing to buy insurance. And they can be extended to multiattribute or multidimensional consequences.

Von Neumann utility is mathematically sound and operationally tractable. However it has been a disappointment in terms of experimental results. Evidence suggests that subjects have difficulty judging consequences with which they are not familiar and must, in any case, be sophisticated in terms of understanding the implications of probability. Von Neumann utility, however, is currently the best game in town for judging the relative importance of alternative consequences and making tradeoffs, including those related to accidents and safety.

Attention Allocation, Workload, Learning and Fatigue

Typically in complex human-machine systems such as space teleoperation tasks there are many more things for the human to attend to than there is time for. The question is whether the human allocates his attention among those signals which convey the most important information in terms both of information "surprisal" and in terms of importance when they do occur, in other words information value as was previously described. Experiments by Senders et al.[9] indicate that for relatively sophisticated subjects sampling among displays is allocated in proportion to bandwidth, which is rational, but with a slight tendency to hedge their bets, i.e., to undersample high-bandwidth signals and oversample low-bandwidth signals.

Allocation is not the only consideration, however, because human attention deteriorates (gets bored, goes to sleep) if the operator is not given enough to do. If mental workload is excessive, it is a precusor to a sudden and catastrophic decrement in human performance, though by measuring performance directly one cannot usually anticipate that performance is about to drop off. Mental workload has been measured[10] for airline pilots and nuclear power plant operators, and it is appropriate to measure it for human operators of teleoperators.

Two other effects are at work, both cumulative, which makes them differ from the momentary effects described earlier. Operators learn, and they fatigue. Maximum feasible learning is desirable, and of course extensive training for any space teleoperation task would be expected. Fatigue is the cumulative effect of mental and physical effort, and science has not easily been able to distinguish the two. There has been plenty of evidence of physical fatigue in teleoperation, by astronauts in space as well as by terrestrial operators of telemanipulators who perform repetitive tasks in nuclear laboratories and facilities.

Memory and Chunking

Miller, in his famous paper "The Magical Number Seven, Plus or Minus Two"[11] makes use of information ideas to characteize human immediate memory for simple stimuli. Reviewing a number of experiments he showed that, after a period of training on the symbolic identity of each of a set of stimuli (this one is A, this one is B, etc.) along a one-dimensional sensory continuum (different points along a line, different brightnesses of a light, etc.) one can only identify without error up to seven or eight (roughly) different stimuli, or approximately three bits worth. When the stimulus dimensionality is doubled (points in a square, lights which have both different brightness and color) the information theoretically adds, which would mean six bits or 64 items, and so on for higher dimensions. In fact, with each added dimension there is some melt, but nevertheless humans are able to distinguish reliably a seeming infinite number of items of high dimensionality (e.g., faces, voices).

One way we apparently cope with large numbers of items or features is that we recode them into clusters, where each cluster is a bite that we can handle. A simple example is to ask someone to remember the bit string 101110100010100101001. The clever person will make it an easy task by recoding (chunking) each subsequence of three bits into octal, transforming the bit string into 5642451. The bite in this case is a byte!

Mental Models

A mental model is a hypothetical mental simulation or representation of events and their interrelation which is capable of being run dynamically, much as a computer program: e.g., if the apple is dropped, here is what would happen. Verbal elicitations of mental models (e.g., of peoples' understanding of how certain physical things work) seem to be couched in some combination of graphic images and verbal rules.

Clearly, mental state is not directly and physically measurable, and historically some psychologists (traditionally the behaviorists) have asserted that even attempts to infer mental state are fruitless. In spite of this seemingly overwhelming constraint, there is much current interest in mental state, motivated largely by computer analogies to behavior as well as modern approaches to studying the relation of language to thinking and problem solving. Cognitive science is an energetic and growing field.

Affect and Alienation

Engineers, even those concerned with human-machine systems, try their best to avoid consideration of affect, the feelings of people towards objects and events (and each other). It is not a matter easily reduced to equations. Nevertheless, some scholars claim that the effects of affect on performance of even mechanical tasks in industry, in piloting aircaft, and surely in operating space teleoperators is quite significant.

Some Further Modeling Challenges, Especially with a Change to Supervisory Control

Modeling Systems Which Include Creativity and Free Will

One reason we put people into systems is because they are able to exercise ingenuity, be creative, and supposedly exercise their free will to cope with unexpected events and compensate for deficiencies of the machinery. By definition creativity and free will are not predictable, and therefore, it would seem, not modelable.

Modeling the Human Versus Modeling the Human-Machine Combination

It was mentioned earlier that in simple manual control systems the human tends to adopt the inverse dynamics of the controlled process, such as to keep the open-loop characteristic of human plus controlled process constant.[4] There is every reason to believe that humans attempt to do the same for more complex systems, yet our understanding of this seeming invariant property of human-machine systems remains poorly tested and understood beyond the most simple linear control loops.

Modeling Aided Human Control

When the human controller is using some form of decision aiding, which suggests or tells him what to do, the question arises as to whether the human is slavishly following the decision aid, or whether the human is thinking and deciding independently of the decision aid. Of course, if it is the human plus computer in tandem that one is modeling (as per the earlier discussion), then a model may not have to discriminate between human and computer.

Lessons from Aviation and Other Human-Machine Systems

Many problems of dealing with humans are common across a variety of human-machine systems, particularly as automation and human supervisory control become more prevalent. This section mentions a few caveats drawn from experience in other such contexts, namely, aviation and nuclear power plants.

Monitoring and Failure Detection

There is much evidence that people are not good monitors of automation. If human operators are taken out of control but expected to continuously monitor for incipient failures, they do not do a good job. They become inattentive to the automation and are easily distracted to tasks in which they are obliged to play a more active role, or they become bored and fall asleep if there are no demanding tasks. This phenomenon is well established from observation of factory and process plant operators responsible for monitoring automatic systems, aircraft pilots responsible for monitoring autopilots, and others.

Human Error

Ever since Alexander Pope's 18th century phrase "to err is human" people have been attributing system failures to human error, and in recent years there has been a resurgence of interest in human error.[12] Automation in high-technology systems such as aircraft and nuclear plants has become quite reliable, and most system failures are probably correctly attributed to human error rather than random unpredictable machine failures. Further, attribution of errors to human operators alone is probably overdone, because the other humans who design, program, install, maintain, and manage systems also play significant roles, and their missteps can also precipitate failures.

Why do we continue to keep people in systems? It is not because people do not err. Probably it is because, unlike machines, people are good at catching and correcting their own errors, and their erratic, but sometimes ingenious means of perceiving problems and fixing things is complimentary to the way machines do it, thus often (curiously) rendering the human-machine system more reliable than either by itself.

Automation Mode Confusion

It has been suggested that automation does not mean a reduction of human factor concerns, and that aircraft automatics have become so reliable that human errors now account for most of the accidents. Sometimes the human errors occur because the pilots assume more of the autopilots than they should, and sometimes they simply forget what they had asked the automation to do—what mode they have set it in. The Airbus A320 is a case in point. It was designed to appeal to developing nations, where pilot training is not so sophisticated, and therefore it seemed that more automation would be better. Unfortunately the A320 has the worst accident record of any of the newer aircraft, even though the automation itself has not been faulted. Human error has almost always been the charge.

Designers of intelligent systems essentially have three alternative ways of assisting the human operator:

1) Make the control automatic, so that the driver is relieved of control, except as a backup if the automation fails. In an aircraft this characterizes the autopilot. Once set, the autopilot holds the airplane at a particular altitude, heading, rate of descent, or whatever.

2) Let the computer figure out the best control action to take in each situation and give the driver specific guidance to take that action. In an aircraft such a display is the flight director, which presents a pitch, bank, and heading bug as a command to be matched by pilot control of the actual aircraft orientation indicator.

3) Present the facts (for example about the location of other air traffic with which there may be a collision threat, or the weather condition in a region to which one is headed) and let the human operator decide on the appropriate response.

Modern aircraft embody all three types, in some cases offering the pilot a choice. Having different levels of automation for different functions and in some cases choices among the levels for the same function has not simplified piloting. "Killing us with kindness" is one expression pilots have used to characterize the benevolence of modern technology.

Mental Models in Relation to Other Internal Models in Teleoperator Systems

Figure 5 shows four loci of models in a telerobot system: 1) mental models (presumably resident in the supervisor's head), 2) software-based models in the computer, 3) representations of the telerobot task in the configuration of the human's hand controls, and 4) representations of the telerobot task in the graphics-text presentation on the supervisor's displays (including computer-generated, video,

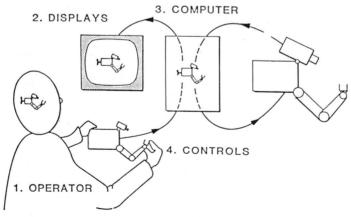

Fig. 5   Four loci of models in the supervisory control system: 1) mental model, 2) display representation, 3) control configuration, 4) computer model. From Ref. 1.

and other-than-visual displays). The latter two are not typically considered models, but they truly are. The arrangement of what the operator sees and what he does with his hands and how this corresponds with what he thinks is critical. Miscorrespondence (inability or difficulty in mapping from eye or other sensor to mind, and then to hand) is likely to cause delays and errors. The computer's internal models will run at their own speed and in terms of whatever the programmer decided, but these must bear correspondence to the mental and display-control models. All, of course, must correspond to the external reality. Designing in this correspondence is not a trivial matter. In the parlance of conventional human factors, notions of display-control compatibility, user expectation stereotypes, naturalness, and transparency are often used.

Command Language

This and the chapters to follow discuss the most important and rapidly evolving aspects of automation and supervisory control. These topics are discussed by other authors in more detail in chapters to follow.

Manipulating Objects and Their Representations

As mankind has evolved there are two ways in which the human body has affected its environment: by manipulation and by communication. Both can be called tool using. In manipulation, humans have used their hands, or other parts of the body or mechanical tools in the hands, to apply forces and displacements to physically modify the environment. In communication, the vocal cords or other parts of the body have been used as tools to make representations for others to observe and understand. The move toward computer mediation in performing tasks means humans using their bodies more for communication than for direct manipulation.

There is a correspondence between human verbal language and digital computer language which is widely appreciated. Both are represented as alphanumeric strings which have meaning as a function of their syntax and parsability. Not so widely appreciated is the close correspondence between human verbal language and human manipulation of objects in the physical environment. This disparity in appreciation is all the more surprising because the subject-verb-object sentence (with appropriate modifiers on each) corresponds rather directly to the hand-action-tool sentence, or more generally the logic of hand/tool action upon external object. It has taken the artifact of the teleoperator to make us aware of this close relation. However, surprisingly, we still do not have an acceptable representational language for manipulation.

Symbolic and Analogic Command Languages

Depending on the command hardware and software available, the teaching or programming of any manipulation or mobility task, including specification of a goal state and a procedure for achieving it, and including necessary constraints and criteria, can be formidable or quite easy. By command hardware we mean the way in which human motor action—hand, foot, or voice—is converted to physical signals to the computer. As was mentioned briefly earlier, command hardware can be either analogic or symbolic.[1]

Analogic command means that there is an isomorphism in space and time between human physical action, semantic meaning, and/or feedback display. For example, moving a control up to increase the magnitude of a variable, which causes a display indicator to move up, would be analogic correspondence. Pointing to an object to identify it is another example of analogic language. Manual force or position tracking amounts to continuous analogic command.

Symbolic command, by contrast, is accomplished by depressing one or a series of keys (as when typing words on a typewriter), or uttering one or a series of sounds (as in speaking a sentence), each of which has a particular meaning. For symbolic commands, one series or concatenation of such responses has a different meaning from another concatenation. In this case spatial or temporal correspondence to the meaning or the desired result is not requisite. Sometimes analogic and symbolic can be combined, e.g., where up-down keys are both labeled and positioned accordingly.

Humans naturally intermix analogic and symbolic commands or even use them simultaneously. This happens, for example, when one talks and points at the same time, or plays the piano and conducts a choir with one's head. Alphanumeric symbols might be used to teach the telerobot to make a movement, and they might also serve to communicate heuristics for avoiding obstacles or making tactile discriminations (which cannot be done so easily by analogic demonstration). Symbols might be better than analog demonstration for communicating to the telerobot how to draw given geometric figures like squares or triangles.

Some Early Telerobot Command Language

Early industrial robots were often programmed by means of teach pendants (small hand-held switch boxes) by which the teacher, standing adjacent to the robot, could command its joint movements one degree of freedom at a time, thus programming simple routines which could be refined and later replayed many times for assembly line operations. Gradually, symbolic macros were employed. Ernst's MH-1,[13] Unimation's PAL and HAL,[14] and Barber's MANTRAN[15] were examples. Paul[16] reviews the computational aspects of robot command language in Ref. 16.

A high-level command system called SUPERMAN was developed by Brooks.[17] The supervisor could teach the computer by performing a manipulation with a master arm and simultaneously coding the objects and the movements using a symbolic keyboard. Later, when the human operator required a particular already-trained manipulation, the operator could initialize a new coordinate system relative to the old one by moving the teleoperator hand to the starting point of the task (e.g., grasping a particular nut or valve handle) and signaling for execution on this object. The computer automatically retransformed the old coordinates to a new coordinate system and performed the desired task, possibly also following commands to terminate the execution at a previously identified location or object. Brooks' supervisory programs could, upon certain touch conditions' becoming true, branch into other programs. For example, the telerobot hand could grasp a nut, unscrew it half a turn, pull back to test whether it was off, and, if it was, place it in a (previously located by this demonstration procedure) bucket, or if it was not, repeat the operation.

Brooks demonstrated his command language on six manipulation tasks: retrieval of a tool from a rack, returning a tool to its rack, taking a nut off, grasping an object and placing it in a container, opening or closing a valve, and digging sand and putting it in a container. In addition, four manual control modes were employed: switch (on-off reverse, or fixed rate), joystick (variable rate), master-slave position control, and master-slave position control with force feedback. For all these combinations of conditions both direct manual teleoperator and supervisory control were compared. In all cases error rates for supervisory control were lower than those for direct manual teleoperation. Even with no time delay, supervisory control was found to be more effective (as determined from the task completion times and manipulation errors) than switch rate control, joystick rate control, and master-slave position control. Bilateral force-reflecting master-slave control was found to be slightly faster than supervisory control but more prone to errors. Since the experiments were performed under ideal conditions, it could be predicted in 1979 that supervisory control would show an even greater advantage when used with degraded sensor or control loops (time delays, limited bandwidth, etc.).

Yoerger[18] extended Brooks' work, developing a more extensive and robust supervisory command system that enabled a variety of arm-hand motions to be defined, called upon, and combined under other commands. Part of Yoerger's system was an on-line computer simulation and display that allowed motions of the manipulator to be simulated in all six DOF to test programs before they were actually executed. In one set of experiments Yoerger compared three different procedures for teaching a robot arm to perform a continuous seam weld along a complex curved workpiece.

Continuous Trajectory Analogic Demonstration

The human teacher first moved the master (with slave following in master-slave correspondence) relative to the workpiece in the desired trajectory. The computer would memorize the trajectory, and then cause the slave end effector to repeat the trajectory exactly.

Discrete Point Analogic Specification with Machine Interpolation

The human teacher moved the master (and the slave) to each of a series of positions, pressing a key to identify each. The human supervisor would then key in additional information specifying the parameters of a curve to be fitted through these points and the speed at which it was to be executed, and the computer would then be called on for execution.

Analogic Specification of Reference and Symbolic Goal Specification Relative to that Reference

In this mode the supervisor used the master-slave manipulator to contact and trace along the workpiece, to provide the computer with knowledge of the location and orientation of the surfaces to be welded. Then, using the keyboard, the supervisor would specify the positions and orientations of the end effector relative to the workpiece (e.g., to move along exactly 1 in. away from the designated surface at a given speed and a given angle). The computer could then execute the task instructions relative to the geometric references given.

Measures were made of both position error and orientation error in system performance after teaching in each of the three modes. Results showed consistently that the best performance was for identifying the geometry of the workpiece and analogically and then giving symbolic instructions about what to do relative to the workpiece. It was further shown that the graphic interface decreased the operator's dependence on visual feedback.

Part of Yoerger's system was an on-line computer simulation and display that allowed motions of the manipulator to be simulated in all six DOF to test programs before they were actually executed. The operator could view the simulation from any angle, could translate or zoom the display, could run various simulations faster than in real time, and could then call for an actual execution of some already-stored trajectory at some specified new location.

Cannon[19] conducted experiments in analogic task teaching in unstructured environments wherein the supervisor's role was limited to task conception and pointing, and the telerobot did the rest. (The phrase "put that there" was originally popularized by Bolt[20] as a graphics interface technique.) Using a mobile robot with a six-DOF arm and two pan-tilt charge-coupled device (CCD) video cameras mounted on a post, the human operator aimed camera reticles at the crucial objects and destinations to accomplish a real-world task. From camera angle triangulations, in concert with combinations of meaningful subphrases such as put that...and that...there, the telerobot interactively built arm and mobile base trajectories for the immediate or delayed execution of tasks. The tasks involved objects and locations about which the robot had essentially no foreknowledge, such as putting tools in a tool box, items of trash in a wastebasket, and blocks on a pallet; all the items had been strewn randomly in a workspace. Cannon demonstrated that his technique reduced human supervisory control time and sped up task-execution time.

In cases involving significant time delays, the point-and-select approach eliminates the need for the move-and-wait stepping actions of telemanipulation. This is because destinations, rather than incremental motions, are prescribed so that the robot can move quickly and continuously between locations of importance. Task-specific criteria, such as proper welding speed and offset requirements, could be incorporated such that the phrase "weld from there to there" creates a routine that installs a welding rod and then, using proximity sensors, follows the contours of any curved surface between the two points while keeping the angle and the spacing of the rod correct for welding on that surface. The command "inspect all but that" could become meaningful with advanced natural language systems. In all such cases, the role of the human remains at the highest level of supervisory control commensurate with defining tasks involving objects and destinations about which the robot has no specific foreknowledge.

More Recent Developments in Command Language

Hirzinger and Heindl[21] and Hirzinger and Lanzettel[22] proposed and experimentally implemented a technique by which the human operator, either on or off line, could continuously specify a position trajectory to a telerobot in space, using an isometric force joystick or sensor ball to control rate in six axes. Especially in cases where there was a time delay, a local computer simulation or duplicate hardware teleoperator could be used to produce immediate and easily observable feedback. When the telerobot sensed contact with the environment, or alternatively when the operator chose, a force control mode could be switched on where the forces applied to the six axes of the sensor ball served as reference signals to a force control loop closed locally at the remote teleoperator.

The former (rate) mode presumably would be used in free space and the latter (force) mode when contact with external objects threatens instability and/or the operator needs force feedback (e.g., in fitting a peg in a hole). Note that Hirzinger's system was not true force feedback. The spring restoring forces from the six-axis joystick were what was felt by the operator, and these were also giving reference signals to the local force control loops. Ranging sensors attached to the end effector could be made to act as pseudoforce sensors in contributing to the balance of end-effector forces to commanded forces.

Asada and Yang[23] demonstrated a system by which they could capture the deburring skill of an experienced machinist and transfer that skill to a robot. The machinist taught the computer by repeated demonstrations, and at the same time a variety of sensor signals were recorded, such as forces, positions, grinding wheel speed and torque, and sounds. Analysis of these data by means of discriminant functions determined an average mapping or control law from process state variables (inputs to the human) into control actions (human outputs). Asada and Liu[24] proposed similar teaching by means of neural net conditioning.

Funda et al.[25] developed a system wherein the operator programs by kinesthetic as well as visual interactions with a (virtual) computer simulation. The instructions to be communicated to the actual telerobot are generated automatically in a more compact form than record and playback of analog signals. Several free-space motions and several contact, sliding and pivoting motions, which constitute the terms of the language, are generated by automatic parsing and interpreting of kinesthetic command strings relative to the model. These are then sent on as instruction packets to the remote slave. The Funda et al. technique also provides for error handling. When errors in execution are detected at the slave site (e.g., because of operator error, discrepancies between the model and real situation and/or the coarseness of command reticulation), information is sent back to help update the simulation. This is to represent the error condition to the operator and allow him to more easily see and feel what to do to correct the situation.

Brooks[26] reports on experiments with five progressively graded levels of human supervisory control of a telerobotic vehicle, which he calls action, direction, agreement, negation, and delegation. By action he means either real-time direct and continuous control by the operator or continuous record and then playback. In either case the operator must do each and every step of the task. Direction means that the operator specifies each in a series of small incremental goals, and the computer interprets and executes these one step at a time. Agreement means that the computer selects an action but waits for the operator to agree to it; if he does not, another action can be selected, and so on. Negation means that the computer seeks to carry out a task autonomously but the human operator may override it. Delegation means that the human operator specifies overall goals, then turns over part or all of a task to the computer to perform as it sees fit. The computer in this case has no responsibility to inform the operator what it decides.

Other Issues in Command Language

Computer-Graphic Aids for Avoiding Obstacles

Various computer-graphic control aids have been developed by which a supervisor can plan and try out an anticipated telerobot arm movement in a simulation, before committing to the actual move. For example, one developed by Park[27] assumed that for some obstacles the positions and orientations are already known and represented in a computer model. The user commanded a series of straight-line moves to a subgoal point in threespace by designating a point on the floor or the lowest horizontal surface (such as a table top) by moving a cursor to that point (say in Fig. 6) and clicking. Then lifting the cursor by an amount corresponding to the desired height of the subgoal point (say A) above that floor point, he observed on the graphic model a blue vertical line being generated from the floor point to the subgoal point in space. This process was repeated for successive subgoal points (say B and C). The user could view the resulting trajectory model from any desired perspective (though the real environment could be viewed only from the perspective provided by the video camera's location). Either of two collision-avoidance algorithms could be invoked: a detection algorithm which indicated where on some object a collision occurred as the arm was moved from one point to another, or an automatic avoidance algorithm which found (and drew on the computer screen) a minimum-length no-collision trajectory from the starting point to the new subgoal point. Experiments with this technique showed that it was easy to use and that it improved safety greatly.

Chiruvolu[28] experimented with the idea of superposing on the video model-based images allowing the operator to see through the teleoperator arm and hand.

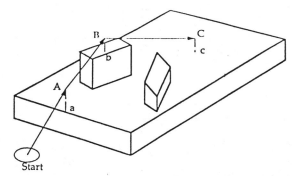

Fig. 6   Park's display of computer aid for obstacle avoidance: human specification of subgoal points on a graphic model.

This allows for peg-in-hole or other assembly operations which otherwise are not possible with limited camera views.

Intermediate Feedback

Ideally, communication with a machine requires only giving action commands and getting back indications of state. However, reflection on how humans communicate with one another reveals that something more is going on: intermediate feedback for both giving commands and getting state information, as shown in Fig. 7. A human worker (or an intelligent machine) might verify an instruction by saying to the supervisor, "This is what I understand you have asked me to do; is that correct?", leaving the supervisor the opportunity to confirm or clarify. Similarly, when the supervisor gets a report on the state of objects or events, he is likely to nod or otherwise indicate what he understands, or to ask the subordinate for clarification. Machines need to enable such communication.

Voice Command

Voice-recognition systems (for voice command by human operators) are now widely available at reasonable prices. Naturally there are tradeoffs involving size of vocabulary, speaker training, speaking speed and style, number of different speakers to be accommodated by the same algorithm (one speaker is always preferred), recognition reliability, and cost. If speech consists of disconnected words, recognition is much more straightforward than if words are connected as in natural speech. Further, a single speaker will modify the sound of his or her voice

FUNCTION OF HUMAN   FORM OF   FUNCTION OF COMPUTER
                          COMMUNICATION
                          (analogic or symbolic)

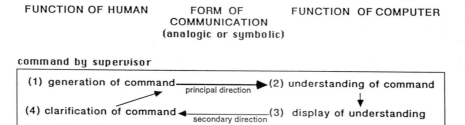

command by supervisor

(1) generation of command ——principal direction——▶ (2) understanding of command

(4) clarification of command ◀——secondary direction—— (3) display of understanding

feedback of state

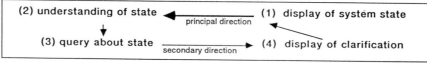

(2) understanding of state ◀——principal direction—— (1) display of system state

(3) query about state ——secondary direction——▶ (4) display of clarification

Fig. 7   Intermediate feedback in command and display, heavy arrows indicate the conventional understanding of functions, light arrows indicate critical additional functions which tend to be neglected.

as a function of stress, fatigue, and attention. Voice command for teleoperation has been experimented with for at least a decade,[29] but thus far its acceptance has been limited.

Vision, Force, and Motion Feedback in Teleoperation

This section reviews the operator interfaces for vision, muscle and tendon force sensing, cutaneous touch, and the motion and position senses.

Video

Vision is clearly the most important sensory communiction channel for remote manipulation. For teleoperation tasks which are not too distant direct viewing is preferred over video. Whether or not this is wise depends on how close the video camera can be brought to the objects of interest, whether the camera has pan, tilt, or translational capability, what its contrast and illumination ranges and frame rate are, how well it reproduces color, and how important any of these factors is for the job to be done.

Massimino and Sheridan[30] compared telemanipulation capability for direct vision vs video in simple block-insertion tasks. They found that mean task-completion times decreased dramatically as the subtended angle of the critical objects in the visual field increased beyond 1 deg and the frame rate increased beyond 3 frames per second (Fig. 8). However, for broadcast standard resolution there was no significant difference between direct viewing and video when the total visual field of objects to be manipulated was the same.

The importance of color is obviously task dependent, and may be overplayed. In this regard it is interesting to note that Murphy et al.[31] found that experienced dermatologists could diagnose skin lesions as well over black-and-white video as over color video.

Gaining a sense of depth is the most difficult visual problem when viewing through closed circuit video. To recreate the stereo sense of depth obtained when viewing a real object with two eyes, different two-dimensional images must be obtained from two horizontally separated viewpoints, then presented to the corresponding left and right retinas. In direct viewing, other strong three-dimensional cues are accommodation, shadows, prior knowledge of relative size and of what object is behind what other object, and motion parallax (i.e., the ability to move the head from side to side and gain a different viewpoint). None of the latter cues requires two eyes. In televiewing accommodation is not available, and motion parallax is available only with head-mounted or other head-position-measuring display techniques.

For teleoperation the images can be obtained by two separate video cameras, or by a single camera outfitted with two optical paths sharing the video field in time or space, or by a geometric model run in a computer. Presentation of the images can be by means of two separate optical paths (one to each eye) or by a single display which provides two images in rapid alternation. The latter can be separated for each eye by color filtering (wearing red and green glasses) or by temporal shuttering (alternate presentation to each eye of each corresponding

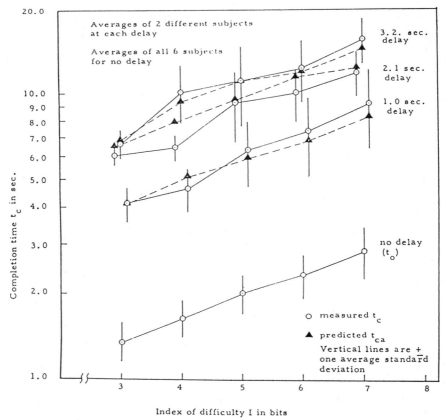

Fig. 8 Ferrell's results for time-display in telemanipulation, experiments were performed in simple two-DOF grasp-and-place tasks with various accuracy requirements (Fitts' index of difficulty) and pure time delays. From Ref. 48.

image). Image transmission must maintain proper size, shape, brightness, and color (if color is displayed as such and not used for binocular channel separation). Many studies confirm a significant reduction in task performance time by use of stereovision.[32, 33]

Other artificial means of providing depth cues on a video or computer screen are: multiple views (not always possible because of camera location restrictions), artifical shadows,[34] superposing in the video display some computer-generated perspective grid lines with equidepth reference lines drawn from the reference grid to important objects,[35] and superposing other graphics onto the video picture, much as the aircraft pilot's head-up display is superposed onto the windscreen.

Force Reflection to Muscles, Joints, and Tendons

What the human body's joint, muscle, and tendon receptors feel as the net reaction force and torque acting on the hand, i.e., the vector resultant of all of the component forces and torques of the hand acting on the environment, may be called resolved force feedback. (This is in contrast to cutaneous touch feedback, which is a differential pattern of forces distributed in time and space.) The limbs can measure such net forces over a wide dynamic range and with a just-noticeable difference of 6–8% in the 2–10 N range. In force-reflecting master-slave systems such resolved forces are measured at the slave end either by strain-gauge bridges in the wrist (so-called wrist-force sensors), by position sensors in both master and slave (which, when compared, indicate the relative deflection in six DOF, which in the static case corresponds to force), or by electrical motor current or hydraulic actuator pressure differentials.

Display of feedback to the operator can be straightforward in principle; in force-reflecting master-slave systems the measured force signals drive motors on the master arm which push back on the hand of the operator with the same forces and torques with which the slave pushes on the environment. This might work perfectly in an ideal world where such slave-back-to-master force tracking is perfect, and the master and slave arms impose no mass, compliance, viscosity, or static friction characteristics of their own. But not only does reality not conform to this dream, it can also be said that we hardly understand what the deleterious effects of these spurious mechanical characteristics are in masking the sensory information that is sought by the telemanipulation operator, or how to minimize these effects. Corker and Bejczy[36] showed that master and slave need not have the same kinematics if force reflection is to be used; a computer can transform coordinates and still produce proper force feedback. It has also been shown that force reflection can be applied to a rate-control joystick.[37]

Force Feedback Masking and Teleoperation Performance

There are several factors in master-slave teleoperation which contribute to insensitivity to contact or other forces. These can result in instability because the operator may not feel the forces imposed on the slave by the environment and will keep moving the master when force feedback should signal him to stop or to reverse direction.

There is effective masking of forces felt by the operator because the mechanism of the force-reflecting hand controller may have significant coulomb friction (stiction) force F_c, viscous friction force F_v (F_c and F_v would not occur at the same time), inertial force F_i, and gravity force F_g between the force feedback actuators/sensors and the handgrip, all of which can cancel or, because they might be larger than feedback forces from the slave, confuse the operator as to what their source is. These masking forces add to the operator's own sensory threshold F_s for force detection. This effect is multiplied by whatever ratio R obtains for force feedback transferred to the master relative to forces applied to the slave by the environment. Combining these factors results in a net force threshold F_T:

$$F_T = R(F_c + F_v + F_i + F_g + F_s)$$

F_c and F_s are usually the major culprits.

Numerous studies have been performed over the years to evaluate whether, and under what circumstances, force feedback helps performance.[17,38–40] Hannaford et al.[41] found that force feedback made a consistently significant difference.

Impedance Control, and What Impedance is Best for Telemanipulation

Impedance is normally defined as the relation between applied force F and velocity V. For a linear system impedance Z is commonly defined as

$$Z = F/V = Ms + B + K/s$$

where M is mass, B is viscous damping, K is stiffness, and s is the Laplace argument or time-derivative equivalent in the time domain. This implies a common null point for all terms.

The compliance null point can be moved in position, whereas the compliance and/or viscosity parameters (mass is usually neglected) can be adjusted independently. Thus a relatively constant force may be applied to an object in spite of small arm motions relative to it by commanding the end point compliance to be soft and adding a large equivalent position bias in the desired force direction. A relatively constant position may be imposed on an object in spite of disturbing forces by making the end point act like a stiff spring (what we usually think of as position control). In fact, subject to constraints of stability and actuator limits, any desired end point impedance to motion (or admittance of forces) may thus be programmed to mimic the desired compliance, viscosity, and mass parameters of the end point. These parameters may even be different in different directions, or change with time, which seems to be what we do with our own limbs in catching balls, threading needles, and other ordinary manipulation tasks.[42]

There is a diversity of opinion about what constitutes the best impedance for a master-slave teleoperator. One argument is that an ideal teleoperator is one that is transparent, i.e., the equivalent of an infinitely stiff and weightless mechanism between the end effector of the master arm and the operator's hand assembly of the master arm. Vertut and Coiffet[39] have suggested instead that operators get tired when holding their arms in fixed and awkward positions and/or applying constant forces (as master-slave systems often require), and the author's experience confirms this. Raju et al.[43] and Bejczy and Handylykken[40] report that there seem to be different best combinations of force-feedback gain (from slave to master) and feedforward gain (from master to slave) for different tasks.

Providing for the operator to adjust the impedance of the master and/or the slave may be a promising way of making a master-slave teleoperator more versatile than if the compliance-viscosity-inertance parameters remained fixed. A carpenter may carry and use within one task several different hammers, and a golfer many clubs, because each different tool provides an impedance characteristic appropriate for particular task conditions which are expected. Carrying many teleoperators into space, for example, may be avoided by making the impedance adjustable between slave and task and/or between human and master.

Skin Senses of Touch

Touch refers to the sense of differential forces (or, equivalently, displacements) on the skin in time and in space, both normal and tangential to the skin surface. (The skin can sense other stimuli, of course, such as heat, cold, pain, etc.) The skin is a poor sensor of absolute magnitude of force, and it adapts quickly.

Five types of nerve fibers mediate touch[44,45]: 1) Merkel cells, 1–100 Hz, are acutely responsive to edges and regions of curvature. 2) Meissner corpuscles, 2–200 Hz, are less spatially and temporally acute than type 1. 3) Ruffini structures have a low-pass temporal response (0–10 Hz), and are primarily responsive to horizontal skin stretch. 4) Pacinian corpuscles have a temporal bandpass of 20–1000 Hz with peak sensitivity at 300–400 Hz and are extremely sensitive (100–1000 Å peak to peak) to skin amplitude vibrations generally. 5) Hair follicle receptors should also be added, which respond to light axial or bending forces, with spatial discrimination from 0.1–3 cm.

Touch sensing and perception may be considered in three different sensorimotor contexts: 1) forces imposed on the skin by the environment without any overt intentional movements (passive touch), 2) voluntarily movements to explore or identify some portion of the mechanical environment (pure active or haptic touch), and 3) touch sensing as part of active manipulation. Context 1 may seem to be the equivalent of visual and electromagnetic image recognition and understanding. Unfortunately, cutaneous patterns do not seem to be perceived with enough resolution and memory to make much of this available theory applicable. Context 2 is now seeing some research in telerobotics, which may also offer a way into context 3, which poses difficulty in research because the modes of touching in the precess of doing are so many and varied. An excellent review is that provided by Loomis and Lederman,[44] which includes determination of absolute and differential thresholds as a function of force magnitude and direction relative to the skin, time, frequency, body locus, two-point separation, stimulus size and shape (including texture), recognition among previously learned patterns, and the effects of masking on all of these.

Touch Sensing and Display Devices

In the last few years devices for artificial teletouch sensing have become available. They variosly use magnetic, resistive, capacitive, or optical sensing elements.

The most difficult problem for teletouch is not sensing but display. How should artificially sensed pressure patterns be displayed to the human operator? One would like to display such information to the skin on the same hand that is operating the joystick or master arm which guides the remote manipulator. This has not been achieved successfully, the major reason being that the skin receptors are masked by the forces of gripping the handle as well as the reaction forces of inertia, friction, and spring centering (if any) of the master. An option is to display to the skin at some other location than the handle-gripping surfaces. Much of the early research in tactile displays was directed toward aiding the blind. Included have been arrays of electromagnet vibrators, or piezoelectric bimorphs (up to 64 × 64 such bimorph vibrators have been packaged into a 7 × 7-in. array). More recently alloys such as TiNi which change their length when heated have shown promise.

Most of the success in teletouch has been achieved by displaying remote tactile information to the eyes using a computer-graphic display. The preceding problems for tactile teleoperation occur in spite of the fact that without vision one can easily track a randomly moving tactile stimulus almost as well as a visual one.[46,47]

Kinesthesis and Proprioception

Kinesthesis is literally sense of motion. Proprioception is literally awareness of self, but is taken to mean position awareness. Kinesthesis and proprioception are terms often used together by psychologists, at least in part because the same receptors in the human body's muscles and tendons mediate both. Gravity also provides a strong vestibular head position cue to the otolith organs, and head movement is sensed by the semicircular canals. Muscle reflexes are driven automatically to maintain posture and body-position awareness. With a teleoperator these cues are normally missing, or at least there is a severe problem of establishing anything approaching the tight coupling between kinesthetic and proprioceptive sensors and the brain. Experience has shown it to be easy for the operator to lose track of the relative position and orientation of the remote arms and hands and how fast they are moving in any given direction. This is particularly aggravated by one's having to observe the remote manipulation through video without peripheral vision or very good depth perception, or by not having master-slave position correspondence, i.e., when a joystick is used. Potential remedies are multiple views, a wide field of view from a vantage point which includes the arm base, and computer-generated images of various kinds (the latter will be discussed further hereafter). Providing a better sense of depth is critical to telemanipulation anywhere.

Coping with Time Delay

Continuous teleoperation in Earth orbit or deep space by human operators on the Earth's surface is seriously impeded by signal transmission delays imposed by limits on the speed of light (radio transmission) and computer processing at sending and receiving stations and satellite relay stations. For vehicles in low Earth orbit, round-trip delay (the time from sending a discrete signal until any receipt of any feedback pertaining to the signal) is minimally 0.4 s; for vehicles on or near the moon these delays are typically 3 s. Usually the loop delays are much greater, approaching 6 s in the case of the Earth-orbiting Space Shuttle because of multiple up-down links (Earth to satellite or the reverse) and the signal buffering delays which occur at each device interface.

Instability

Continuous closed-loop control over a finite time delay is not possible, because any energy entering the loop at such a frequency that half a cycle is equal to the time delay will result in positive feedback rather than negative, so that if the loop gain exceeds unity at this frequency (which it normally would at low frequency to achieve good control) there is an inherent instability. Of course in the case of supervisory control, wherein commands are sent by the human operator through the time delay to a computer, the computer then implements the commands by closing loops local to itself, reporting back to the supervisor when the task is completed. The computer's local loop closure has no delay in it and therefore causes no instability. Also, because of the intermittent nature of the supervisor's control, the delay in the command-feedback loop will not cause instability.

Early Experiments with Time Delay

Communication delays in a continuous telemanipulation loop, it has been shown, make the time for a human operator to accomplish even simple manipulation tasks increase manifold over the no delay case, depending of course on the length of the time delay and the complexity of the task. This is because the human operator, to avoid instability, must adapt a move-and-wait strategy, wherein he commits to a small incremental motion of the remote hand or vehicle, stops while waiting (the round-trip delay time) for feedback, then commits to another small motion, and so on.

Ferrell[48] first demonstrated experimentally the predictability of teleoperation task performance as a function of the delay, the ratio of movement distance to required accuracy, and other aspects of delayed feedback in teleoperation. Ferrell's results (Fig. 8) are for simple two-axis-plus-grasp manipulations on a table. Black[49] performed similar experiments with a conventional six-axis-plus-grasp master-slave manipulator. Thompson[7] showed how task-completion time was affected not only by time delay but also by degrees of constraint (see Fig. 4). Thompson's experimental results are shown in Fig. 9. Held and Durlach[50] showed that sensory-motor adaptation is essentially impossible for delays as small as 0.3 s, and that experimental subjects dissociate the teleoperator hand movements from those of their own hand at these delays.

This problem has discouraged continuous control of space vehicle systems from the ground. However, as more and more devices are put in space, the requirements increase for humans to perform remote manipulation and control, and if this can be done entirely from Earth even at the cost of time, there are great savings to be gained in dollars and risk to life.

Predictor Displays

A predictor display presents a computer-derived estimate of future system state. There are two types of predictors. The first is based on current state and time derivatives—i.e., Taylor-series extrapolation. The second involves inputting into a model the current state and time derivatives, as well as expected near-future control signals.[51] Such displays have been employed in gunsights, on ships and submarines, and as head-up optical landing aids for aircraft pilots. When there is significant transmission delay (say more than 0.5 s) and a slow frame rate (say less than one frame per 4 s), a predictor display can be very useful.

Sheridan and Verplank[52] implemented an experimental predictor of the second type for a simulated planetary rover. A computer model of the vehicle was repetitively set to the present state of the actual system, including the present control input, then allowed to run at roughly 100 times real time for a few seconds before it was updated with new initial conditions. During each fast-time run, its response was traced out in a display as a prediction of what would happen over the next time interval (say several minutes) "if I keep doing what I'm doing now." Such techniques are adequate for continuous control of single-entity or rigid body vehicles, but not for telemanipulation, where it is necessary to predict, relative to the environment, the simultaneous positions of a number of parts—i.e., a spatial configuration in multiple DOF, not just a single point.

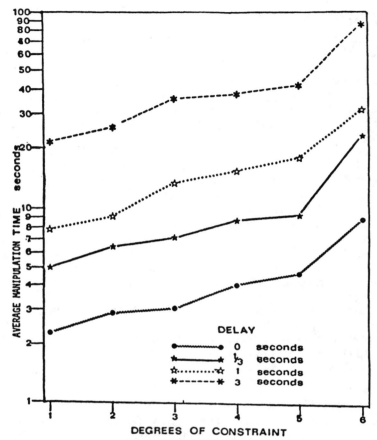

Fig. 9 Thompson's results for time delay and degrees of constraint, averaged times include transport of peg to hole, positioning, and inserting. From Ref. 7.

The first predictor display for telemanipulation was built by Noyes[53] and Noyes and Sheridan.[54] A newly available computer technology for superposing artificially generated graphics onto a regular video picture was used. The video picture was a time-delayed picture from the remote location, generated as a coherent frame (snapshot) so that all picture elements in a single scan were equally delayed. (Otherwise the part of the screen refreshed last would be delayed more than the part refreshed first.) As shown in Fig. 10, the predictor display was a line drawing of the present configuration of the manipulator arm or vehicle or other device. The latter was generated by using the same control signals that were sent to the remote manipulator to drive a kinematic model of it. The computer model could be drawn instantaneously on the video display in the same location where it would be seen to be on the video after one round-trip time delay. Since the graphics were generated in perspective and scaled relative to the video picture, if one waited at least one

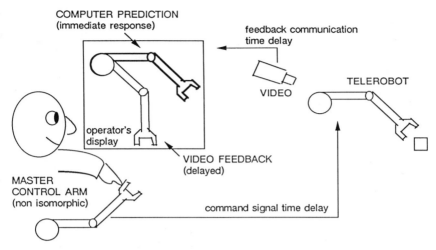

Fig. 10   Noyes' telemanipulation predictor display: diagram of experimental setup.

round-trip delay without moving the control, both the graphics model and video picture of the manipulator (device) could be seen to coincide. The effectiveness of these techniques has been demonstrated for simple models of the manipulator arm and simple tasks. [55-57] With such a display, operators could lead the actual feedback and take larger steps with confidence, reducing task performance time by 50% (Fig. 11).

Recent Predictor Instruments

When the motion of vehicles or other objects not under the operator's control can be predicted, e.g., by the operator indicating on each of several successive frames where certain reference points are and then extrapolating, these objects can be added to the predictor display. With any of these planning and prediction aids, the display can be presented from any point of view relative to the manipulator or vehicle, which is not possible with the actual video camera.

A prediction architecture proposed by Hirzinger et al.[58] includes this notion (Fig. 12) as well as dynamic prediction. The stick-figure overlay on the delayed video is driven by a dynamic model (whereas Noyes and Sheridan[54] used a kinematic model). In the figure this is constituted by the sum of the A and/or B feedback coefficients operating on correspondingly delayed commands. In the middle of the diagram is the implementation of the canonical first-order $x(k + 1) = Ax(k) + Bu(k)$, where k corresponds to what is going on instantaneously with the space telerobot. The $x(k + 1)$ estimate is corrected in the usual way by Kalman gain-multiplied discrepancy between the estimated $y(k - nd)$ and the corresponding actual downlink signal. The delay line on the right side is required to estimate $y(k - nd)$. By estimating $x(k)$—i.e., what is happening in space—activities such as rendezvous and docking can be coordinated with clock-determined events which are not under the control of this human operator.

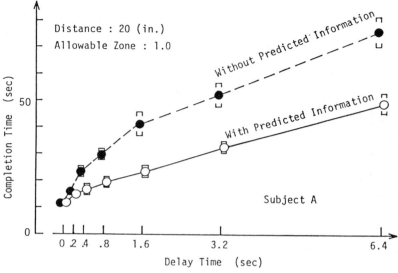

Fig. 11   Hashimoto results for predictor display evaluation in simple task of repositioning a block 20 in. to within a 1-in. tolerance, data shown are for one subject, brackets are standard error of the mean for repeated trials. From Ref. 55.

Time-Delayed Force Feedback

Force feedback with time delay is a different problem from that with visual feedback. Ferrell[59] showed that it is unacceptable to feed resolved force continuously back to the same hand that is operating the control. This is because the delayed feedback imposes an unexpected disturbance on the hand which the operator cannot ignore and which, in turn, forces an instability on the process. With visual delay the operator can ignore the disturbance and can avoid instability by a move-and-wait strategy or by supervisory control.[60]

Since those early experiments by Ferrell there have been various proposals: 1) display force feedback in visual form on a computer display; 2) display force feedback to the hand that is not on the master hand or joystick; 3) feedback forces greater than a certain magnitude to the controlling hand for a brief period, at the same time cutting off or reducing the loop gain to below unity, and subsequently reposition the master to where it was at the start of the event; 4) predict the force feedback at that time which compensates for the delay, and feed the predicted force but not the real-time force back to the operator's hand.

Buzan[61] and Buzan and Sheridan[62] evaluated the latter approach experimentally, using an open-loop model-based prediction to drive both a visual predicted-position display and a force exerted back on the operator through a master positioning arm. The experiments used a one-DOF teleoperator system, a 3-s time delay, and two challenging computer-simulated tasks. The first task was to extend the arm to make contact with (and avoid accelerating) a floating mass, then grasp it with a discrete action (an additional half-DOF) before it got away. The second task was to push an object into a stiff slot with enough force to get it in and have

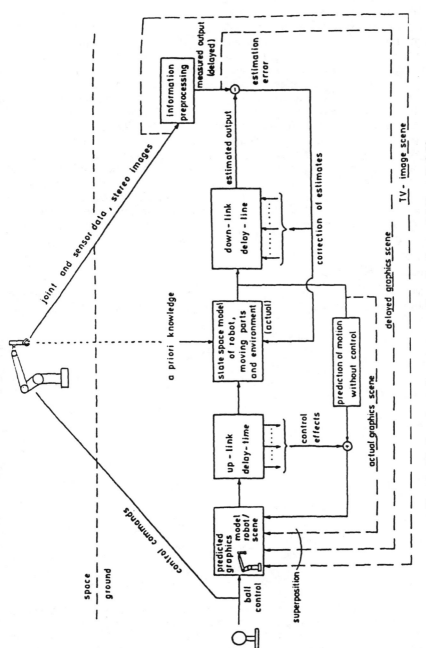

Fig. 12   Hirzinger's predictor incorporating adaptive model. From Ref. 58.

static friction hold it there, but not so much force that the object would go right out the other side. Figure 13 illustrates the second task.

Buzan and Sheridan tried three force-feedback-display techniques. In one, which he called direct force feedback, he simply presented the predicted force (but not the delayed real force) to the active hand, the hand commanding the teleoperator position. In a second method, which he called dual force feedback, he presented the delayed force to an inactive hand and the predicted force to the active hand. In the third display technique, which he called complimentary force feedback, he presented to the active hand the sum of a low-pass-filtered delayed force feedback and a high-pass-filtered predicted force feedback.

Buzan's results showed, among other things, that end-point impedance made a big difference in these tasks. The contact-and-grasp task was easiest with a soft end point compliance, whereas the slot task favored a stiff end point. Buzan also found that the complementary force feedback proved difficult to use. When the visual predictor was used and was perfect, the predicted force feedback had a negligible effect on performance. When telemanipulation was blind, both the direct and the dual force feedback worked quite well, enabling the operator to do the tasks where he otherwise could not.

Sensory Substitution

Massimino[63] tested concept 2, feedback to the nonactive hand, for the teleop-eration task of inserting a four-sided peg into a four-sided hole. Instead of force

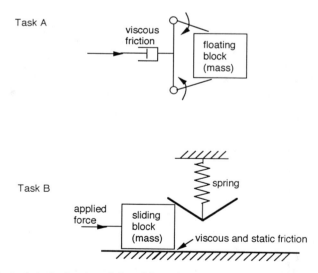

Fig. 13   Buzan's tasks for time-delayed force feedback experiments: in task A, human subject was to reach out and grasp block (which was freely floating in space) without inadverently accelerating it out of reach before grasp could be achieved; in task B, human was to push block to center of spring clamp until light static friction held it there and not let it pop out the other side. Models used to generate predictor displays were simplifications of (simulated) real-time tasks (e.g., no static friction in task B). From Ref. 61.

he used vibrators on the skin (four vibrators in a square, the amplitue of vibration corresponding to the interactive force between the peg and the corresponding side of the hole), and in another case auditory tones (high and low pitch for contact with top ond bottom of hole, left and right ears for contact with left and right sides of the hole). Results showed clear advantages of sensory substitution in either auditory or vibratory form when time delay otherwise would cause instability.

Desynchronization in Time and Space for Telemanipulation Planning

Conway et al.[64] extended the predictor idea of Noyes and Sheridan[54] and combined it with a planning model in what they call disengaging time-control synchrony using a time clutch and disengaging space control synchrony using a position clutch. In their scheme, the time clutch allows the operator to disengage synchrony with real time, to speed up making inputs and getting back simulator responses for easy maneuvers and to slow down the pace of such commands and simulator responses for hard maneuvers where more sample points are needed. The computer buffers the command samples and later feeds them to the actual control system at the real-time pace, interpolating between sampled points as necessary. (This is not unlike a driver speeding up on the straightaways and slowing down on the curves and, in fact, is what anyone would do to make the best use of planning time.) The only requirement is that the progression of planned actions keep ahead of what must be delivered right now for real-time control (and also take into account any time delay).

Disengaging the position clutch allows one to move the simulator in space without committing to later playback, this for the purpose of trying alternative commands to see what they will do. Disengaging the position clutch necessarily disengages the time clutch and creates a gap in the buffer of command data. Re-engaging the position clutch may require path interpolation from the previous position by the actual telerobot controller.

Conway et al. tested these ideas experimentally using a Puma robot arm, a joystick hand controller, and a simple two-dimensional positioning task. They compared teleoperation under three conditions: without any predictor display, with predictor display, and with predictor display plus time clutch. Plots of task-completion time as a function of task difficulty ratio (distance moved divided by diameter of target) yielded results for the first two conditions which confirmed the Hashimoto et al.[55] results that the predictor by itself made significant improvement (they found up to 50% shorter completion times for some subjects). They also found that adding the time clutch could make further improvement (of up to 40%) if the slewing speed of the robot arm was constrained to be very slow and if the operators used finesse and were careful not to overdrive the system. Various other researchers have adopted versions of the time clutch.

Editing of Commands for Prerecorded Manipulation Command

At the extreme of time desynchronization is recording and editing a whole task on a simulator, then sending it to the telerobot for reproduction. This might be workable when one is confident that the simulation matches the reality of the telerobot and its environment, or when small differences would not matter (e.g., in programming telerobots for entertainment). Doing this would certainly make it possible to edit the robot's maneuvers until one was satisfied before committing them to the actual operation. Machida et al.[65] demonstrated such a technique by which commands from a master-slave manipulator could be edited much as one edits material on a video tape recorder or a word processor. Once a continuous sequence of movements had been recorded, it could be played back either forward or in reverse at any time rate. It could be interrupted for overwrite or insert operations. Their experimental system also incorporated computer-based checks for mechanical interference between the robot arm and the environment.

Other Ways of Coping with Delay

Placing energy dissipating (damping) elements (as shown in Fig. 14) guarantees stability, in spite of the time delay and independent of its time constant.[66] (It is common experience that gripping a slightly unstable master arm tightly, or adding friction at the slave end makes the oscillations go away.) Use of this technique may reduce the pace of the task to an unacceptable level.

The reader is reminded that time delay in a supervisory contol loop would not normally cause instability. In such a situation commands are sent by the human operator through the time delay to a computer, and the computer then implements the commands by closing a loop local to itself, reporting back to the supervisor through the time delay intermittently or when the task is completed. The computer's local loop normally has either no or only an insignificant delay in it and therefore causes no instability.

For the most part, space operations need not be done in a hurry. In current space teleoperation on the Shuttle very gentle movements are performed mostly by joystick rate control without force feedback. Movements are carefully planned and rehearsed on the ground, and great care is taken before end-effector contact is made with environmental objects. Natural or haptic feeling around with the telemanipulator is seldom if ever done. For this kind of manipulation, the gains can be effectively reduced to less than one at the frequencies of interest and/or control can be open loop for long periods. Considering the time it takes to get astronauts into space, and their limited working hours when there, urgency does not seem to be a major criterion.

For movements in free space, as noted earlier, predictor displays work well, and ameliorate the effects of time delay. For movements involving contact and assembly, predictor displays do not help much, but accommodation by soft compliance or impedance seems to be in order. This change from stiff position control for free positioning to soft compliance can be automatic on close approach to contact,

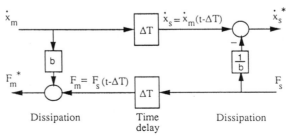

Fig. 14   Damping to stabilize a telecommunictions process. From Ref. 66.

as suggested in several papers cited earlier, or it can be switched on manually as needed. It is described further in other chapters of this book.

Using new techniques for transformation of control signals to wave variable propagation within the communications system, Niemeyer and Slotine[66] offer interesting possibilities for preventing instability. These techniques bring with them the undesirable side effects of position mismatch and continuous drift. Amelioration is possible by adding termination impedances, but more research remains to be done to determine under what circumstances this is preferable to either slow, damped movements, move-and-wait, or supervisory control.

Telepresence and Virtual Presence

Telepresence, the experience of feeling present or being there, is commonly claimed to be important for direct manual telemanipulation. It has yet to be shown how important it is to performance as compared to simply having high resolution, a wide field of view, and other attributes of good sensory feedback.

If a computer-generated picture is substituted for the video picture and similarly referenced to the head oreientation, the viewer can be made to feel present (virtual presence, virtual environment, artificial reality, or virtual reality) within a computer-generated artificial world. In addition to displays this can include controls (which one actuates by moving one's hand in body-referenced space).

Telepresence Research

For the most part telepresence refers to experiencing the visual environment by wearing a head-mounted display the position and orientation of which drives a video camera in the remote environment to a corresponding position and orientation. Telepresence is usually identified with direct manual teleoperation, though it may be just as important to be able to feel present when supervising a semiautonomous telerobot.

Auditory telepresence (binaural localization and spectral correspondence to sounds experienced in a real world terrestrial environment) is now technologically available. It may be useful for space teleoperation as an artificial cue. Resolved force (muscle force) telepresence, tactile (skin sense) telepresence, and vestibular telepresence (achievable through a motion platform driven by the same disturbances the operator would be subjected to were he at the remote, or virtual, location) are not yet developed to a credible level.

Many mechanical-force or other disturbances which might contribute to one's sense of telepresence are the very things one seeks to avoid. For example, one often seeks to avoid vibration or sudden unexpected mechanical forces which interfere with visual-motor skills. One seeks to avoid extremes of temperature and pressure, explosions, and other hazards which might contribute further to telepresence but which may be the very reasons for teleoperation in the first place. Therefore, full telepresence is a questionable goal in many situations.

Tachi and Arai[67] developed and evaluated the hardware components to implement teleoperator telepresence with a head-mounted, binocular, color liquid-crystal (320×220 pixel) display. Head position was measured to 2.5 mm in translation and 0.5 deg in rotation within a 1.5-m^3 workspace. This drove a 7-DOF telemanipulator.

A virtual window is another technique to achieve geometrically correct visual telepresence. Schwartz[68] describes a fixed high-resolution stereo-video system with head tracking, corresponding camera positioning, and image reproduction to each eye to correspond to what the viewer would see were he looking through a fixed window. Merrit[69] reports ongoing research to utilize both of these techniques in a sophisticated telepresence viewing system.

Virtual Presence Research

Ellis[32] provides a sampler of recent research on pictorial displays for both virtual and tele environments. Fisher et al.[70] and Furness[71] were among the first to demonstrate these techniques, using a head-mounted display (HMD) as described earlier. By now there have been many more HMDs developed, and they have been well publicized.

Virtual acoustic displays may play an important role in virtual presence. This is largely due to people's ability to resolve and identify meaningful sound patterns spatially even though their signal strength is but a fraction of the total sound energy entering the ear. The head and pinna structures as well as room configuration and damping characteristics modify power spectral transfer functions for sounds reaching the eardrum from sources at different external locations. An electronic device called the Convolvotron can use this principle to produce in earphones a realistic experience of multiple sound sources as a function of head position and orientation.[72]

Patrick[73] added a tactile buzzer to a VPL data glove and programmed it so that the wearer can reach out and touch something—in this case, when the index finger or the thumb or both are correctly positioned they feel a vibrotactile stimulus. Such a tactile display, however, gives the impression of touching a tuning fork, not an inert object.

Psychophysical Determinants of the Sense of Presence

It is natural to seek an objective measure or criterion that can be used to say that telepresence or virtual presence have been achieved. However, telepresence (or virtual presence) is a subjective sensation, much like mental workload—it is not so amenable to objective physiological definition and measurement. Some might assert that a subjective report from the person having the experience is the only measure. An objective criterion might be a test analogous to the Turing test for computer intelligence: if the observer cannot reliably tell the difference between telepresence (or virtual presence) and direct presence, then the telepresence (virtual presence) has been fully achieved. A practical criterion of telepresence proposed by Held and Durlach[74] is the degree to which the observer responds in a natural way to unexpected stimuli—e.g., by blinking his eyes or ducking his head when he sees that an object is about to hit him. We are far from meeting this strict criterion in most applications.

Sheridan[75] proposed that there are three principal and independent determinants of the sense of presence: extent of sensory information (the transmitted bits of information concerning a salient variable to appropriate sensors of the observer), control of relation of sensors to environment (e.g., ability of the observer to modify his viewpoint for visual parallax or visual field, or to reposition his head to modify binaural hearing, or ability to perform haptic search), and ability to modify the physical environment (e.g., the extent of motor control to actually change objects in the environment or their relation to one another). These determinants may be represented as three orthogonal axes (see Fig. 15), because the three can be varied independently in an experiment.

Design Criteria for Telepresence and Virtual Presence

High-quality visual telepresence or virtual presence requires that the viewed image follow the head motion with no apparent lag or jitter (this is a servo-control problem and has been hard to achieve in existing systems), that an object in the display subtend the same retinal angle as it would in direct vision, and that motion parallax and other head-motion cues also correspond to direct viewing. Other problems are in achieving sufficient field of view (it should be at least 60 deg), depth of field, correct focal length, image separation for stereoscopic fusion and luminance, resolution, color, and other image-quality factors, particularly at the fovea. When the image is computer generated, additional problems lie in achieving sufficient image-generation speed and frame rate, grayscale, and variable accommodation (in contrast to fixed focus at infinity). As one might expect, there are also serious problems of cost, size, and weight.

Criteria for haptic telepresence have been suggested by Jex,[76] based on much experience with aircraft and automobile simulators. The practical issue is to achieve realistic feel of hand controls (control sticks in aircraft, steering wheels in automobiles, etc.):

1) When all other simulated forces and the mass or inertia of the simulated hand control are set to zero, it should feel like a stick of balsa wood (have negligible

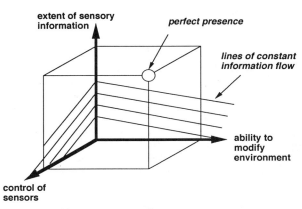

Fig. 15   Determinants of the sense of presence: a) three components of presence, hypothetical lines of constant information suggest that for purposes of providing a sense of presence, information channels are better used for control of sensors and modification of the environment than for higher resolution displays, b) salient independent and dependent variables, the relations of which are yet to be understood.

lag, jitter, friction or other forces) up to the highest frequency that a finger grip can impose, or about 7 Hz.

2) When pushed against simulated hard stops, the hand control should stop abruptly, with no sponginess, and it should not creep as force continues to be applied.

3) When set for pure Coulomb friction (i.e., within a noncentering hysteresis loop), the hand control should remain in place, without creep, sponginess or jitter, even when repeatedly tapped.

4) When set to simulate a mechanical centering detent and moved rapidly across the detent, the force reversal should be crisp and give a realistic clunk with no perceptible lag or sponginess.

5) For simulations in which operator steering of a vehicle is involved, to keep mental workload and performance within acceptable bounds, any simulation delay artifact must be less than about one-fourth of the effective operator response delay.

Conclusions

This chapter asserts that pure automation and pure human control for space tasks are extremes of a continuum, and that a combination of a human and a computer usually offers more promise. In particular, it is proposed to have the human act as supervisor to one or many automated processes, where the human plans, teaches the computer, monitors its automatic execution of the programmed actions, intervenes as necessary, and learns from experience. Because putting the human into this supervisory role poses deep questions about how to model the human and apply human talents in conjunction with automation, approaches to behavioral modeling are then reviewed, along with lessons learned from human-automation integration in other fields. The chapter concludes by examining the problems and what we know in four categories of human interaction with teleoperators: command language; vision, force, and motion sensory feedback; coping with time delay; and telepresence and virtual reality.

References

[1]Sheridan, T. B., *Telerobotics, Automation, and Human Supervisory Control,* MIT Press, Massachusetts Inst. of Technology, Cambridge, MA, 1992.

[2]Atkinson, R. C., Herrnstein, R. J., Lindzey, G., and Luce, D. R., *Stevens' Handbook of Experimental Psychology,* 2nd ed., Wiley, New York, 1988.

[3]Boff, K., Kaufman, L., and Thomas, J. P., (eds.), *Handbook of Perception and Human Performance,* Vol. 2, Wiley, New York, 1986.

[4]McRuer, D. T., and Jex, H. R., "A Review of Quasi-Linear Pilot Models," *IEEE Transactions Human Factors in Electronics,* Vol. HFE-8, No. 3, 1967, pp. 231–249.

[5]Hick, W. E., "On the Rate of Gain of Information," *Quarterly Journal of Experimental Psychology,* Vol. 4, 1952, pp. 11–26.

[6]Fitts, P. M., "The Information Capacity of the Human Motor System in Controlling the Amplitude of Movement," *Journal of Experimental Psychology,* Vol. 47, 1954, pp. 381–391.

[7] Thompson, D. A., "The Development of a Six Degree-of-Freedom Robot Evaluation Test," *Proceedings of 13th Annual Conference on Manual Control,* Massachusetts Inst. of Technology, Cambridge, MA, 1977.

[8]Von Neumann, J., and Morganstern, O., *Theory of Games and Economic Behavior,* Princeton Univ. Press, Princeton, NJ, 1944.

[9]Senders, J. W., Elkind, J. I., Grignetti, M. C., and Smallwood, R. P., "An Investigation of the Visual Sampling Behavior of Human Observers," NASA CR-434. Bolt, Beranek and Newman, Cambridge, MA.

[10]Moray, N. (ed.), *Mental Workload: Its Theory and Measurement,* Plenum Press, New York, 1979.

[11]Miller, G. A., "The Magical Number Seven, Plus or Minus Two: Some Limits on our Ability to Process Information," *Psychology Review,* Vol. 63, 1956, pp. 81–97.

[12]Reason, J., *Human Error,* Cambridge Univ. Press, 1990; also Senders, J. W. and Moray, N. P. *Human Error: Cause, Prediction and Reduction,* Erlbaum, Hillsdale, NJ, 1991.

[13] Ernst, H., "MH-1, a Computer-Operated Mechanical Hand," Sc.D. Thesis, Massachusetts Inst. of Technology, Cambridge, MA, 1961.

[14] Nof, S. Y. (ed.), *Handbook of Industrial Robotics,* Wiley, New York, 1985.

[15] Barber, D., "MANTRAN, a Symbolic Language for Supervisory Control of an Intelligent Manipulator," S.M. Thesis, Massachusetts Inst. of Technology, Cambridge, MA, 1967.

[16]Paul, R. P., *Robot Manipulators: Programming and Control,* MIT Press, Massachusetts Inst. of Technology, Cambridge, MA, 1981.

[17]Brooks, T. L., "SUPERMAN: a System for Supervisory Manipulation and the Study of Human-Computer Interactions," S.M. Thesis, Massachusetts Inst. of Technology, Cambridge, MA, 1979.

[18]Yoerger, D. R., "Supervisory Control of Underwater Telemanipulators: Design and Experiment," Ph.D. Thesis, Massachusetts Inst. of Technology, Cambridge, MA, 1982.

[19]Cannon, D., "Point-and-Direct Telerobotics: Interactive Supervisory Control at the Object Level in Unstructured Human-Machine System Environments," Ph.D. Thesis, Stanford Univ., Standford, CA, 1992.

[20]Bolt, R., "'Put that there': Voice and Gesture at the Graphics Interface," Proceedings of SIGGRAPH 80," 1980; *Computer Graphics,* Vol. 14, No. 3, 1980 pp. 262–270.

[21]Hirzinger, G., and Heindl, J., "Sensor Programming: a New Way for Teaching a Robot Paths and Force-Torques Simultaneously," *Proceedings of 3rd International Conference on Robot Vision and Sensory Controls,* Cambridge, MA, Nov. 7–10, 1983.

[22]Hirzinger, G., and Lanzettel, K., "Sensory Feedback Structures for Robots with Supervised Training," IEEE International Conference on Robotics and Automation, St. Louis, MO, March 1985.

[23]Asada, H., and Yang, B-H., "Skill Acquisistion from Human Experts Through Pattern Processing of Teaching Data," *Proceedings of 1989 IEEE International Conference on Robotics and Automation,* Scottsdale, AZ, IEEE, New York, 1989, pp. 1302–1307.

[24]Asada, H., and Liu, S., "Acquisition of Task Performance Skills from a Human Expert for Teaching a Machining Robot," *Proceedings of 1990 American Control Conference,* 1990, pp. 2827–2832.

[25]Funda, F., Lindsay, T. S., and Paul, R. P., "Teleprogramming: Toward Delay-Invariant Remote Manipulation," *Presence: Teleoperators and Virtual Environments,* Vol. 1, No. 1, Winter, 1992, pp. 29–44.

[26]Brooks, T. L., "Supervisory Control of Multiple Vehicles," *Proceedings of 15th Annual Symposium Association for Unmanned Vehicles,* San Diego, CA, June 6–8, 1988.

[27]Park, J. H., "Supervisory Control of Robot Manipulators for Gross Motions," Ph.D. Thesis, Massachusetts Inst. of Technology, Cambridge, MA, 1991.

[28]Chiruvolu, R. K., "Virtual Display Aids for Teleoperation," S.M. Thesis, Massachusetts Inst. of Technology, Cambridge, MA, 1991.

[29]Bejczy, A. K., Dotson, R. S., and Mathur, F. P., "Man-Machine Speech Interaction in a Teleoperator Environment," *Proceedings of Symposium on Voice Interactive Systems,* DOD Human Factors Group, Dallas, TX, May 11–13, 1980.

[30]Massimino, M., and Sheridan, T. B., "Variable Force and Visual Feedback Effects on Teleoperator Man-Machine Performance," *Proceedings of the NASA Conference on Space Telerobotics,* Pasadena, CA, Jan.31–Feb.2, 1989.

[31]Murphy, R. L. H., Fitzpatrick, T. B., Haynes, H. A., Bird, K. T., and Sheridan, T. B., "Accuracy of Dermatalogical Diagnosis by Television," *Archives of Dermatology,* Vol. 105, June 1974, New York, pp. 833–835.

[32]Ellis, S. R. (ed.), *Pictorial Communication in Virtual and Real Environments,* Taylor and Francis, 1991.

[33]Smith, D. C., Cole, R. E., Merritt, J. O., and Pepper, R. L., "Remote Operator Performance Comparing Mono and Stereo TV Displays: the Effects of Visibility, Learning and Task Factors," Naval Ocean Systems Center, TR-380, San Diego, CA, 1979.

[34]Winey, C. M., "Computer Simulated Visual and Tactile Feedback as an Aid to Manipulator and Vehicle Control," S.M. Thesis, Massachusetts Inst. of Technology, Cambridge, MA, 1981.

[35]Kim, W., Ellis, S. R., Tyler, M., and Stark, L., "Visual Enhancements for Telerobotics," *Proceedings of 1985 IEEE International Conference on Systems, Man and Cybernetics,* Tuscon, AZ, 1985.

[36]Corker, K., and Bejczy, A. K., "Recent Advances in Telepresence Technology development," *Proceedings of 22nd Space Congress,* Kennedy Space Center, FL, April 22–25, 1985.

[37]Lynch, P. M., "Rate Control of Remote Manipulators with Force Feedback," S.M. Thesis, Massachusetts Inst. of Technology, Cambridge, MA, 1972.

[38]Hill, J., "Two Measures of Performance in a Peg-in-Hole Manipulation Task with Force Feedback," *Proceedings of 13th Annual Conference on Manual Control,* Massachusetts Inst. of Technology, Cambridge, MA, 1977.

[39]Vertut, J., and Coiffet, P., *Robot Technology, Volume 3A: Teleoperation and Robotics: Evolution and Development.* Prentice-Hall, Englewood Cliffs, NJ, 1986. *Volume 3B: Teleoperation and Robotics: Applications and Technology,* Prentice-Hall, Englewood Cliffs, NJ, 1986.

[40] Bejczy, A. K., and Handylykken, M., "Experimental Results with Six-Degree-of-Freedom Force-Reflecting Hand Controller," *Proceedings of 7th Annual Conference on Manual Control,* (Los Angeles, CA), 1981, pp. 465–477; also Bejczy A. K., "Sensor, Controls and Man-Machine Interface for Advanced Teleoperation," *Science,* Vol. 208, No. 4450, 1980, pp. 1327–1335.

[41]Hannaford, B., Wood, L., Guggisberg, B., McAffee, D., and Zak, H., "Performance Evaluation of a Six-Axis Generalized Force-Reflecting Teleoperator," Jet Propulsion Lab. Pub. 89–18, California Inst. of Technology, Jet Propulsion Lab., Pasadena, CA, 1989.

[42]Hogan, N., "Impedance Control: an Approach to Manipulation, Part 1: Theory, Part 2, Implementation, Part 3: Applications," *ASME Journal of Dynamic Systems, Measurement, and Control,* 1985.

[43]Raju, G. J., Verghese, G., and Sheridan, T. B., "Design Issues in 2-Port Network Models of Bilateral Remote Manipulation," *Proceedings of IEEE International Conference on Robotics and Automation* (Scottsdale, AZ), 1989, pp. 1316–1321.

[44]Loomis, J. M., and Lederman, S. J., "Tactual Perception," *Handbook of Perception and Human Performance,* edited by K. Boff, L. Kaufman and J. P. Thomas, Vol. 2, Wiley, New York, 1986, Chap. 31, pp. 1–31.

[45]Sherrick, C. E., and Cholewiak, R. W., "Cutaneous Sensing," *Handbook of Perception and Human Performance,* edited by K. Boff, L. Kaufman and J. P. Thomas, Vol 1, Wiley, New York, 1986, Chap. 11.

[46]Weissenberger, S., and Sheridan, T. B., "Dynamics of Human Operator Control Systems Using Tactile Feedback," *Journal of Basic Engineering,* June, 1962.

[47]Jagacincki, R. J., Miller, D. P., and Gilson, R. D., "A Comparison of Kinesthetic, Tactual and Visual Displays in a Critical Tracking Task," *Human Factors,* Vol. 21, 1983, pp. 79–86.

[48]Ferrell, W. R., "Remote Manipulation with Transmission Delay," *IEEE Trans. Human Factors in Electronics,* Vol. HFE-6, No. 1, 1965.

[49]Black, J. H., "Factorial Study of Remote Manipulation with Transmission Time Delay," S.M. Thesis, Massachusetts Inst. of Technology, Cambridge, MA, 1971.

[50]Held, R., and Durlach, N., " Telepresence, Time Delay and Adaptation," NASA Conf. Publ. 10032, Chap. 28, 1987.

[51]Sheridan, T. B., and Ferrell, W. R., *Man-Machine Systems,* MIT Press, Massachusetts Inst. of Technology, Cambridge, MA, 1974.

[52]Sheridan, T. B., and Verplank, W. L., "Human and Computer Control of Undersea Teleoperators," MIT Man-Machine Systems Lab. Rept., Massachusetts Inst. of Technology, Cambridge, MA, 1978.

[53]Noyes, M. V., "Superposition of Graphics on Low Bit-rate Video as an Aid to Teleoperation,," S.M. Thesis, Massachusetts Inst. of Technology, Cambridge, MA, 1984.

[54]Noyes, M. V., and Sheridan, T. B., "A Novel Predictor for Telemanipulation Through a Time Delay," *Proceedings of Annual Conference on Manual Control,* Field, CA, NASA Ames Research Center, Moffett Field, CA, 1984.

[55]Hashimoto, T., Sheridan. T. B., and Noyes, M. V., "Effects of Predicted Information in Teleoperation Through a Time Delay," *Japanese Journal of Ergonomics,* 22, No. 2, 1986.

[56]Mar, L. E., "Human Control Performance in Operation of a Time-Delayed Master-slave Manipulator," S.B. Thesis, Massachusetts Inst. of Technology, Cambridge, MA, 1985.

[57]Cheng, C.-C., "Predictor Displays: Theory, Development and Application to Towed Submersibles," Sc.D. Thesis, Massachusetts Inst. of Technology, Cambridge, MA, 1991.

[58]Hirzinger, G., Heindl, J., and Lanzettel, K., "Predictor and Knowledge-Based Telerobotic Control Concepts," *Proceedings of IEEE International Conference on Robotics and Automation* (Scottsdale, AZ), IEEE, New York, 1989, pp. 1768–1777.

[59]Ferrell, W. R., "Delayed Force Feedback," *Human Factors,* Oct. 1966, pp. 449–455.

[60]Ferrell, W. R., and Sheridan, T. B., "Supervisory Control of Remote Manipulation," *IEEE Spectrum,* Vol. 4, No. 10, pp. 81–88.

[61]Buzan, F., "Control of Telemanipulators with Time Delay," Sc.D. Thesis, Massachusetts Inst. of Technology, Cambridge, MA, 1989.

[62]Buzan, F., and Sheridan. T. B., "A Model-Based Predictive Operator Aid for Telemanipulators with Time Delay," *Proceedings of 1989 IEEE International Conference on Systems, Man and Cybernetics,* Cambridge, MA, Nov. 1989, pp. 14–17.

[63]Massimino, M. J., "Sensory Substitution for Force Feedback in Space Teleoperation," Ph.D. Thesis, Massachusetts Inst. of Technology, Cambridge, MA, 1992.

[64]Conway, L., Volz, R., and Walker, M., "Teleautonomous Systems: Methods and Architectures for Intermingling Autonomous and Telerobotic Technology," *Proceedings of the 1987 IEEE International Conference on Robotics and Automation* (Raleigh, NC), 1987, IEEE, New York, pp. 1121–1130.

[65]Machida, K., Toda, Y., Iwata, T., Kawachi, M., and Nakamura, T., "Development of a Graphic Simulator Augmented Teleoperator System for Space Applications," *Proceedings of the 1988 AIAA Conference on Guidance, Navigation, and Control, Pt. I,* AIAA, Washington, DC, 1988, pp. 358–364.

[66]Sheridan, T., "Space Teleoperation Through Time Delay: Review and Prognosis," *IEEE Transactions on Robotics and Automation,* Vol. 9, No.5, Oct. 1993, pp. 592–606.

[67]Tachi, S., and Arai, H., "Study on Tele-Existence II: Three-Dimensional Color Display with Sensation of Presence," *Proceedings of the 1985 International Conference on Advanced Robotics,* Tokyo, Japan, Sept. 9–10, 1985.

[68]Schwartz, A., "Head Tracking Stereoscopic Display," *Proceedings of the Society for Information Display,* 27, No. 2, 1986, pp. 133–137.

[69]Merritt, J. O., "Visual-Motor Realism in 3-D Teleoperator Display Systems," *Proceedings of SPIE, True Imaging Techniques and Display Technologies,* Vol. 761, 1987.

[70]Fisher, S. S., McGreevy, M., Humphries, J., and Robinett, W., "Virtual Interface Environment for Telepresence Applications," *Proceedings of the ANS International Topical Meeting on Remote Systems and Robotics in Hostile Environments,* edited by J. D. Berger, 1987.

[71]Furness, T., "The Super Cockpit and its Human Factors Challenges," *Proceedings of the 1986 Annual Meeting of the Human Factors Society,* Vol. 1, 1986, pp. 48–52.

[72]Wenzel, E. M., Wightman, F. I., and Foster, S. H., "A Virtual Display System for Conveying Three-Dimensional Acoustic Information," *Human Factors,* Vol. 32, 1988, pp. 86–90.

[73]Patrick, N., "Design, Construction and Testing of a Fingertip Tactile Display for Interaction with Virtual and Remote Environments," S.M. Thesis, Massachusetts Inst. of Technology, Cambridge, MA, 1990.

[74]Held R., and Durlach, N., "Telepresence," *Presence: Teleoperators and Virtual Environments,* Vol. 1, No. 1, 1992, pp. 109–112.

[75]Sheridan, T. B., "Musings on Presence and Telepresence," *Presence: Teleoperators and Virtual Environments,* Vol. 1, No. 1, 1992, pp. 120–125.

[76]Jex, H., "Four Critical Tests for Control Feel Simulators," *Proceedings of 1988 Annual Conference on Manual Control,* Massachusetts Inst. of Technology, Cambridge, MA, 1988.

Chapter 4

Ground Experiments Toward Space Teleoperation with Time Delay

Blake Hannaford[*]

University of Washington, Seattle, Washington 98195

I. Introduction

A N important component of research into advanced telemanipulation systems is performance evaluation. Advances in computation, mechanization, and control must be calibrated in terms of measurable improvements in manipulation performance. Performance evaluation of telemanipulators is a difficult task which of necessity involves many test operators, training sessions, and well-defined evaluation tasks. Literature on many studies of this type performed over the past 20 years is surveyed in Hannaford et al.[1,2]

In all teleoperation systems, some time delay will be present in the communication between master and slave subsystems. Because of the large distances involved, this delay is especially prominent in contemplated applications in space. Early studies such as Sheridan and Ferrell[3] looked at the effects of time delay on the control of a remote manipulator without force feedback. A study by Ferrell[4] found force feedback to be useful with time delay, but revealed a degradation of performance and the potential for unstable operation. More recently, studies have simulated time delay with digital memory buffers and studied its effect on teleoperation with force reflection in single axis[5] and multiaxis systems.[6]

Since early teleoperators were first remotized electronically, reflection of force information to the operator was recognized as a key to higher performance remote manipulation. Many force reflecting teleoperation systems are implemented by sending a position or velocity signal from master to slave and a force signal from slave back to master. In this scheme, the communication delay appears twice in a larger system involving human operator, hand controller, communication link, slave robot, and environment. It has been widely observed that this delay can cause instability of the force reflecting control system.

Copyright © 1994 by Blake Hannaford. Published by the American Institute of Aeronautics and Astronautics, Inc., with permission. Released to AIAA to publish in all forms.

*Associate Professor, Department of Electrical Engineering.

This chapter will review several studies which have attempted to quantify the performance of teleoperation with force reflection. Because it is relatively easy to simulate the main delay characteristics of space teleoperation, several studies have measured the decline in task performance as a function of simulated delay. Finally, control methods which use the force information locally at the task site, can eliminate some performance problems induced by communications delays.

II. Performance Measures

Teleoperation performance cannot be quantified without defining measures of the quality with which a task is accomplished. The most commonly used measure, completion time, is taken directly from comparable studies of human task performance. However, because teleoperation technology is still relatively imperfect, some aspects of performance which are not an issue in direct human operation, such as contact force control, must be quantified with new measures.

A. Completion Time

The simplest and most long-standing task performance measure, completion time, can be determined from the length of the data file containing the force-torque data. To eliminate possible bias or noise in this measure, our analysis software identified the starting and ending contact-force transients and used them as the basis of completion time.

B. Force Performance

The magnitude of the forces used to accomplish a task is assumed to be related to the risk of equipment damage and robot maintenance costs. Thus, an important element in the evaluation of performance must be a measure of the applied forces. Unfortunately, there is no agreed upon measure of force performance. Two measures that have been used are sum of squared force (SOSF) and root mean squared (RMS) force.

SOSF is computed by taking a nondecreasing sum of the square of the force or torque values:

$$\text{SOSF} = \sum_{i=1}^{N} f_i^2 \, dt$$

where N is the number of data samples, f_i the ith sample of force or torque, and dt is the sampling interval.

SOSF is better than a normalized measure such as RMS force because it is additive between different experimental runs with different completion times. For example, it is simple to add results from 10 repetitions and have them equally weighted. The RMS measure cannot be averaged because each sample is in general weighted by a different N. SOSF weights larger forces more than small ones (because of the squaring operation). This is good because even very brief applications of high force can cause damage. SOSF might be roughly considered as a measure of mechanical impulse interaction between the robot hand and the environment.

Although peak forces could be important in system safety analysis, they are hard to measure because of the limited dynamic range of available force/torque sensors. In most laboratory systems, transients can easily saturate the force sensors if they are designed with sufficient sensitivity for dexterous manipulation.

C. Error Rate

A third way in which performance can be measured is through an observer's notations of the quality with which a task is performed. In our experiments, a set of errors was defined and explained to the test operators and experimenters. An experimenter watched each repetition of the experiment and counted occurrences of each error.

Individual differences among error scorers might be a source of bias in this measure. To detect this possible bias, four of the experimenters scored the same experimental run. Their error counts were substantially in agreement. Error rate will not be discussed in this review. For some experimental results, see Hannaford et al.[1,2]

D. Task Segmentation Analysis

The application tasks that are planned for advanced telemanipulation systems are characterized by complex sequences of steps. Each step will generally involve different capabilities such as precision motion control, sensor guided motion, compliant motion, and interaction with unusual materials such as thermal insulation. Each phase of the task, therefore, will be affected differently by the independent variables of the experiment such as control mode or time delay. Task segmentation analysis attempts to look inside the data recording to create submeasures of performance for individual parts of the task. For example, in a task involving contact and noncontact phases, completion time and SOSF should be computed for the two phases separately. Software has been developed[7] to automatically identify the transitions between task phases and accumulate performance measures for each individual phase. To reliably detect transitions between task phases in spite of variability and noise in sensor traces requires a model of the task performance. Hidden Markov Models have been used[8,9] to guide this automatic data segmentation.

III. Delay Sources

In the work reviewed here, time delays were tested over a wide range (2–4096 ms) because contemplated applications impose a wide range of delays. Application examples include sources such as multiple satellite links to low Earth orbit (LEO)(2–8 s), geosynchronous operation (0.5 s), spacecraft local area networks (50 ms), and general purpose computer-based short-distance designs (1–50 ms).

There are two main sources of time delays, computation, and communication delays. Although communication delays can be estimated by using the distance divided by the speed of light, few actual communication systems involve line of site communication. For example, communication with the Space Shuttle involves the TDRSS system—an elaborate network of geosynchronous satellites and computers which imposes delays of 2–8 s. Although few teleoperation missions have been proposed which do not use the Shuttle, significant opportunities exist for ground-based teleoperation applications in LEO or geosynchronous orbit. Geosynchronous orbit imposes delays of about 0.25 s on each up and down link, but the time delay is constant and line of sight communication from the control station to telerobot is quite practical. Although there are difficulties in deploying a reusable telerobot into geosynchronous orbit, there is significant commercial potential. LEO offers delays as low as a few milliseconds and low launch costs. The disadvantage of LEO from the point of view of ground-based teleoperation is the limited opportunity for line of sight communication from the ground. However, the option of intermittent operation during the times that a LEO telerobot is within the view of a ground station deserves study and economic analysis.

Time delays are easily simulated in laboratory telemanipulation systems using digital memory buffers. A curious fact about teleoperation systems is a marked asymmetry between the data rates required to and from the remote site. This is chiefly due to video signals required for operator orientation. Fortunately, realistic simulations can be obtained by lumping all delay into the uplink. The data rate from control station to slave robot is on the order of kilobits per second in foreseeable systems.

IV.   Teleoperation Modes

There are vast numbers of possible modes in which a multiaxis teleoperation system can be controlled. For example, a recently studied six-axis system[2] had 10 possible control modes for each of the six axes—a total of 10^6 modes! (For a prototype of some software with which an operator can manage this complexity, see Ref. 10.) There is no theory available with which to choose the best mode for a given task so that any selection of modes for experimentation is somewhat arbitrary. A reasonable way to constrain this huge search is to perform experiments on systems with all axes set to the same control mode in a given experiment.

Some paradigm modes which have been extensively studied, are basic teleoperation, kinesthetic force feedback (KFF), and shared compliant control (SCC).

Basic teleoperation involves a position command being generated by the operator through a joystick which is sent to the slave robot. No contact force or joint torque information is sent back to the operator. With KFF, contact force measured between the slave and the environment is applied back to the operator through a mechanism in the joystick. The idea is to reproduce the kinesthetic sensations of actual task performance. SSC refers to a recently tested mode in which force information is used locally at the slave side, as well as being fed back to the operator. The local feedback loop does not have significant time delays and can comply to applied forces with a higher bandwidth.

V.   Experimental Design

A significant issue in teleoperator performance evaluation is experiment design. The wide number of task variables, control mode possibilities, and variations between test operators quickly generate large demands on the available experiment time. The basic performance evaluation experiment is to vary one or more independent variables and try to test the hypothesis that this variation is related to a change in performance. Because a telemanipulation system involves so many controls and information displays, there are usually a huge number of potentially testable hypotheses. For example, each control mode potentially generates a test condition for performance evaluation. Performance also varies with the task, so each task generates a dimension of the testing matrix. Finally, performance will depend on the quality of all sensory information available to the operator so that visual display parameters are also potential independent variables. A significant constraint on the experiment design is that the laboratory time required grows geometrically with the number of independent variables.

Human test operators introduce additional complexities. For example, during the performance of a long experiment, the operators will acquire experience which may improve their performance on later trials. Alternatively, skills learned with one task or control mode may interfere with performance on a subsequently tested mode or task. There are two ways to minimize this transfer.

First, training time should be provided before data is taken in each test condition. Of course, the amount of training time is difficult to determine. One commonly used method is to monitor task performance and stop training when it is asymptotic, i.e., when the relative improvement from one trial to the next drops below a threshold. Unfortunately, especially for difficult tasks, performance is usually nonmonotonic. A practical compromise can be reached by a combined criterion of improvement rate, absolute performance threshold, and a maximum number of training trials. A limit to the number of training trials is necessary because, especially in very difficult circumstances, performance may be highly variable and not show evidence of convergence. A typical training rule could be, "Test operator will perform training trials until one of the following is true: completion time is below 30 seconds; the improvement from one training trial to the next is less than 3 seconds; or 10 training trials have been performed."

Second, the test conditions should be administered to test operators in a random order. When results are averaged over the test operator population, randomization of the test sequence will largely eliminate the transfer effects. The number of test operators which must be used is a difficult question which deserves further study.[1] Past work has shown significant variabilities in the performance among operators.[1]

The tasks used for teleoperator evaluation fall naturally into two classes: generic tasks and application tasks. Generic tasks are idealized, simplified tasks that are designed to test specific telemanipulation capabilities. Application tasks are designed as much as possible to mimic real-world uses for teleoperation. Evaluations based on generic tasks illustrate the capabilities of existing telerobotic technology, while application tasks guide the technology in the direction of greatest payoff.

VI. Models and Passivity

Modeling is an essential part of the study of advanced teleoperation systems which will not be reviewed in detail here. Models of the dynamical details of kinesthetic force reflecting systems have coalesced around the two-port network model from circuit theory. Forms of this model include the Z matrix,[11] and the H matrix[12] models which are locally linearized versions of the system dynamics around a selected operating point. Higher level models exist including factors such as the cognitive activities of the human operator[7] which are well reviewed in Ref. 13.

In the late 1980s it was recognized that it was not impossible to have stable force reflecting teleoperation with time delay.[14] Using the analogy between electrical and mechanical dynamic systems, it was realized that electrical transmission line systems exhibit properties analogous to force reflection, and are stable in spite of any time delay. The key idea was to control the communication link between master and slave so that it would emulate a passive transmission line and hence remain stable under all time delays.[14] If the master and slave control systems are designed to be passive, and if suitable assumptions about the human operator and environment are made, system stability is assured. Later work has optimized this idea to include impedance matching between the communication link and the master and slave control systems.[15]

Considerable excitement was generated by the passivity theory result. However, in spite of experimental implementation of these algorithms, little performance evaluation of passivity-based control has been previously attempted.

Two approaches can be followed to produce a passive communication law between master and slave. The first developed is that of Anderson and Spong[14] using scattering theory. A scattering operator S can be defined for a two-port network by the relationship between force and velocity.

$$F - V = S(f + v) \tag{1}$$

Where S is a matrix in the frequency domain, and F, V and f, v represent the master and slave forces and velocities. Any communication law can be tested for stability using a theorem stating that a two-port network is passive if and only if the norm of its scattering operator is less than or equal to one.

Anderson and Spong manipulated the transmission line equations to obtain control laws for passive behavior of the communications block

$$V(t) = v(t - T) - F(t) + f(t - T) \tag{2}$$

$$f(t) = F(t - T) + v(t) - V(t - T) \tag{3}$$

Where T is the communication time delay. These communication laws are passive (stable) for all time delays. Under steady-state conditions, the forces and velocities of master and slave are identical.

Assuming the human operator, master, slave, and environment can all be represented as passive systems; the passive communication block guarantees stable operation in spite of any time delay. However, Eqs. (2) and (3) may affect the overall performance of the system. Anderson and Spong,[14] and later Niemeyer and Slotine,[15] showed that the stiffness between master and slave (h_{22}^{-1} in the H parameter notation of Hannaford[12]) was effectively reduced by the passive communication law. The effective stiffness of Anderson and Spong's teleoperation system can be inferred from the experimental results given in their paper which show the effective stiffness of the teleoperator getting lower as the time delay increases. For a time delay of 200 ms, for example, the stiffness of their system was reduced by a factor of approximately one third.

The second approach, developed by Niemeyer and Slotine,[15] uses an energy-based formulation. The total power flow into the teleoperator network is given by

$$P = F \cdot V - f \cdot v \tag{4}$$

The power flows can also be formulated with wave variables. Wave variables are motivated by the physical concept of waves with an input and output wave at each port of a network. In this manner, the total power flow can be written as

$$P = \tfrac{1}{2}u_m^2 - \tfrac{1}{2}v_m^2 + \tfrac{1}{2}u_s^2 - \tfrac{1}{2}v_m^2 \tag{5}$$

Where u and v are the input and output wave variables. Equating Eqs. (4) and (5) leads to a set of transformation equations between power variables and wave variables.

VII. Experimental Results

The following data will illustrate some of the types of results which can be obtained in teleoperation performance evaluation studies.

A. Example Recordings

In early comprehensive experiments with a six-axis system[1,2] (See Chapter 5 for details on this system), a high-precision peg-in-hole task was performed by five test operators in three control modes. Before the computation of performance measures, it is often useful to examine raw sensor information. For example, we can consider a single axis of contact force in a direction parallel to the peg (Fig. 1). In this task, the operator starts by tapping the grasped peg against the task board and moving it to the hole. A large positive x force indicates the struggle to insert the peg, which is complete when the jaws release it. The jaws reclose on empty space to tap the back of the inserted peg, the peg is regrasped, and withdrawal begins almost immediately. After the peg is completely removed and contact forces cease, there is a short translation to the mark and a final tap. During translation through space, there is no contact force.

When selected experimental records from the several control modes are compared, the x-axis force trace tells most of the story for the peg-in-hole task because of the alignment between the task axis and the force-torque sensor's x axis. Comparison of performance in the several control modes shows the reduced completion times and force levels achieved as more capability is added to the system.

In the visual feedback condition, the operator used position control without kinesthetic force feedback, but received contact force-torque information by means of a graphic display. With only the visual force-torque information (Fig. 1d), the task was completed in about 100 s, and peak forces of up to 10 lb and -9.5 lb were observed in the insertion and extraction phases, respectively. When kinesthetic force feedback was used (Fig. 1c), completion time dropped to 82 s, and peak forces dropped to approximately ± 5 lb.

Sensed force and torque information can be used locally by the manipulator control system to implement an effective compliance or, in general, mechanical

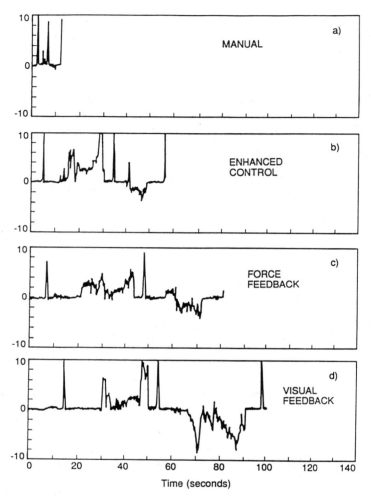

Fig. 1   Illustration of force record changes with different control modes for four repetitions of the peg-in-hole task; individual records of X axis force: a) manual control, b) enhanced control—hybrid force accommodation, c) force feedback, and d) visual force feedback.

impedance of the manipulator. Performance with one example of this type of shared control is shown in Fig. 1b. In this mode, the orientation axes were controlled automatically to move in the direction of sensed torques. Position axes were controlled by the operator. Completion time was reduced to 60 s. Forces were substantially reduced only for the extraction phase, in which the peak force was −4 lb.

Finally, when the task was performed manually (Fig. 1a), the task-related forces were practically invisible compared to the taps. Completion time was about 15 s, and peak forces reached +3 lb and −0.5 lb.

B. Kinesthetic Force Feedback Versus Task

When performance measures are computed, it is possible to compare control modes and to make statistical conclusions. For example, in Hannaford et al.,[2] performance was measured for several examples of generic and application tasks. Five test operators performed multiple repetitions of tasks including high-precision peg-in-hole insertion as described earlier, attachment/detachment of Velcro® patches, and mating/unmating of standard electronic connectors such as DB-25 and 1/4-in. phone plugs.

The effect of force feedback was not the same for all of the tasks (Figs. 2 and 3). For the peg-in-hole task, completion time followed the expected course, dropping by almost a factor of two from 105 to 59 s as force reflection was added. For the velcro-blocks task, completion time increased from 72 s to 83 s. Both of these changes were statistically significant. For the electrical connectors, when force feedback was added, only a slight change from the completion time of 70 s was observed which was not statistically significant. Of course, all of the tasks were completed much faster by the barehanded operator. The average time in this case was about 15 s.

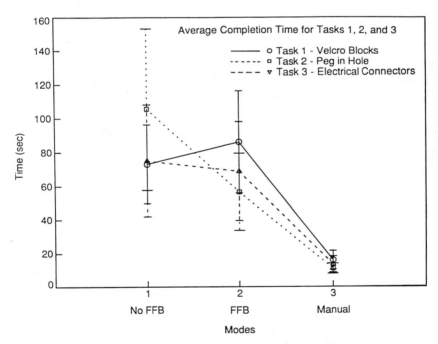

Fig. 2   Completion time as a function of control mode broken down by the individual task type: velcro blocks (solid line, circles), peg-in-hole (dotted line, squares) and electrical connectors (dashed line, triangles). Force feedback affects each task differently; the changes are all statistically significant except the slight drop in CT for electrical connectors with force feedback. Note that force feedback significantly increased CT for the velcro task.

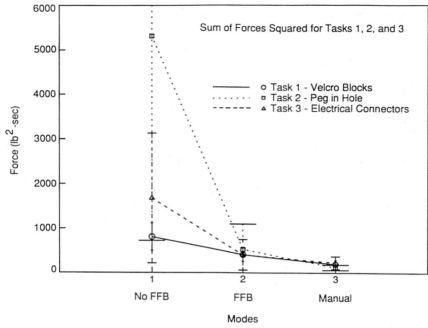

Fig. 3 **SOSF measure for the three task types; key same as for Fig. 2. SOSF dropped significantly for all of the tasks when force feedback was used; levels with force feedback were still approximately double those recorded from the barehanded operator.**

The SOSF data (Fig. 3) tell a different story. As with completion time for the peg-in-hole task there was a dramatic drop in SOSF (from 5400 to 500 lb^2-s) as force reflection was added. The increase in completion time seen for the velcro task was accompanied by a significant decrease in SOSF (from 800 to 400 lb^2-s). For the electrical connectors, the SOSF measure declined significantly despite the unchanged completion time.

C. Performance with Delay

As time delays are introduced between master and slave, KFF control systems can become unstable and the usefulness of force information to the human operator declines. Before evaluating manipulation tasks under conditions of simulated delay, stable control of contact force can be evaluated using a simpler task in which the operator presses against a surface to generate a constant contact force.

This task was to make contact with a flat panel in the task board surface and control the slave to exert a constant contact force of 10 lb normal to the surface using KFF. Operators were instructed to make occasional use of a graphical display of contact force for numerical accuracy of the force magnitude. The delays tested in this experiment were 2, 4, 8, 16, 32, 64, 128, 256, 512, and 1024 ms. Robot motion was disabled in all but the Cartesian axis normal to the task board. Earlier work has described the effect of operator mechanical impedance on a single-axis model of this system.[5] For this study, in the six-axis system, we repeated the experiment under two instructions to the subjects: "maintain a rigid-as-possible grasp on the hand controller," and "maintain a loose grasp on the hand controller." After the experiments, the force signal during each contact maintenance task was analyzed with the fast Fourier transform (FFT) to see if there was an oscillation frequency at which a significant peak amplitude occurred.

Four operators performed the contact task for 30 s at each of the specified values of time delay. The dominant frequency of oscillation (if any) determined by the FFT analysis was plotted against time delay (Fig. 4) for each subject in both the rigid and loose instructions. Also plotted is a dotted-dashed line (marked with × s) illustrating the frequency at which the added time delay is equivalent to a phase of lag of 180 deg. With the rigid grasp instruction, oscillation was nonexistent or less than 0.5 Hz for all subjects and delays (Fig. 4) This indicates that the subjects were able to eliminate oscillations under all tested delay conditions. With the loose grasp instruction (Fig 4b), as delay increased, there was a delay at which each operator's force level began to oscillate. The delay at which the transition from stable to oscillatory control occurred varied among the operators from 64 to 512 ms. The frequency of oscillation was in each case less than $1/2T$, with most operators' oscillation frequency converging to the $1/2T$ line as the imposed delay T became large compared to the system's intrinsic delays.

A peg-in-hole task with 3.6-ml clearance and no chamfer was used. The hole diameter was 1.0016 in. The peg was 4.75 in. in length and 0.998 in. in diameter. The operator made momentary contacts with a designated 1×1-in. square near the hole to indicate the beginning, middle, and end of the task. The resulting force spikes in data records provided well-defined benchmarks (referred to as taps) for measurement and interpretation of the progress of the task. Each operator was asked to perform the peg-in-hole task sequences described earlier. The two control modes evaluated with the peg-in-hole task were KFF and SCC.

In the KFF condition, the human operator received kinesthetic force feedback through a force-reflecting hand controller. Four time delays of 2, 16, 256, and 512 ms were tested. Longer delays were very difficult to control and were not tested due to significant risk of equipment damage.

In the SCC condition, the human operator did not receive kinesthetic force feedback. However, a compliant control algorithm[6] allowed the slave to deflect in response to applied forces and torques. Seven time delays of 2, 16, 256, 512, 1024, 2048, and 4096 ms were tested. Unlike the experience with the KFF system, long time delays did not cause instability to the SCC system. The compliance and damping parameter values used in the experiment were 0.43 in./lb and 3.3 lb s/in. respectively (bandwidth = 0.11 Hz). For simplicity, identical parameter values were used for all six Cartesian position/orientation axes, and no serious attempt was made to find the optimal parameter values.

Typical traces of the force along the peg axis for KFF (Fig. 5a), and SCC (Fig. 5b) under a 1024-ms time delay illustrate that task performance with SCC is superior to the performance with KFF at this time delay. The KFF trace shows spurious contacts with the task board ($t = 30, 40$ s) and large periodic oscillations resulting in forces saturating the sensor during the insertion phase ($t = 50–100$ s). The overall completion time is almost 160 s. In the SCC trace, the taps (part of the instructed task; $t = 5, 48, 90$ s) are of appropriately brief duration, and the

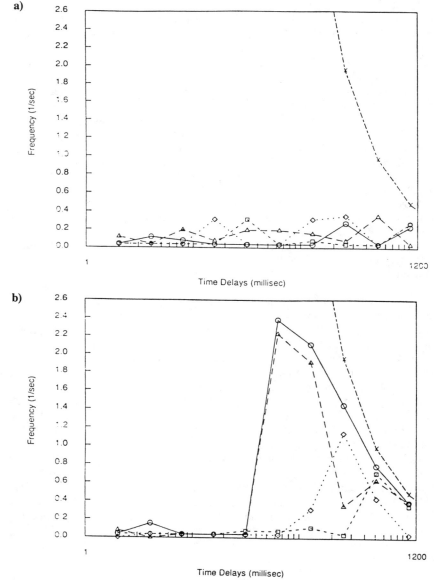

Fig. 4 Oscillation under time delay conditions in a constant force maintenance task with advanced teleoperator system employing a kinesthetic force feedback. The peak frequency of oscillation (computed with FFT) is plotted against applied time delay; the predicted frequency of oscillation based purely on the applied delay, 1/2T, is plotted as well (double-dashed-lines, × s). Data were taken for four test operators with different instructions: a) "maintain constant force with a rigid grasp," and b) "maintain constant force with a relaxed (loose) grasp."

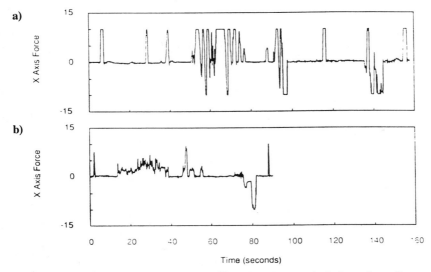

Fig. 5 Sample records of contact force during the peg-in-hole task performance under 1024-ms time delay a) with kinesthetic force feedback and b) with shared compliant control; note lack of oscillation and saturation, reduced completion time, and reduced levels of force when shared compliance control was used under the same delay condition (b vs a).

magnitude and duration of the forces in the insertion ($t = 15$–40 s) and extraction ($t = 75$–85 s) phases are greatly reduced compared to the KFF performance.

These effects can be quantified by comparison of the completion time (CT) and the sum of square forces (SOSF) measures as a function of the time delay in the peg-in-hole task. Averages over all subjects were computed and plotted to directly compare compliant and force feedback performance (Fig. 6). The superiority of SCC over KFF in the time-delay operation is clearly shown. Only SCC enabled the operator to complete the task at time delays above 1 s. The completion time rose at a rate of 80 times the imposed delay with KFF (Fig. 6a, dashed line), whereas it rose at a much lower rate of 33 times the introduced delay with SCC (Fig. 6a, solid line). This lower rising rate amounted to large reductions in total completion time at delays above 125 ms (e.g., 67 vs 127 s at 1-s time delay). When no significant time delay (a delay of 2 ms) was introduced, performance with the two control modes was approximately equal: the completion time for the task was about 38 s, slightly better but comparable to the times reported for a similar task in earlier KFF experiments (Fig. 2, Ref. 2). The SOSF measure of applied force (Fig. 6b) showed even more dramatic differences. Again, the performance started out approximately the same for the two modes at negligible delay (700 lb^2 s, also in agreement with earlier results, Fig. 3) but diverged at substantially different slopes as delay was introduced. With KFF, the SOSF increased at a rate of 3900 lb^2 whereas for the SCC mode, the measure rose at only 138 lb^2.

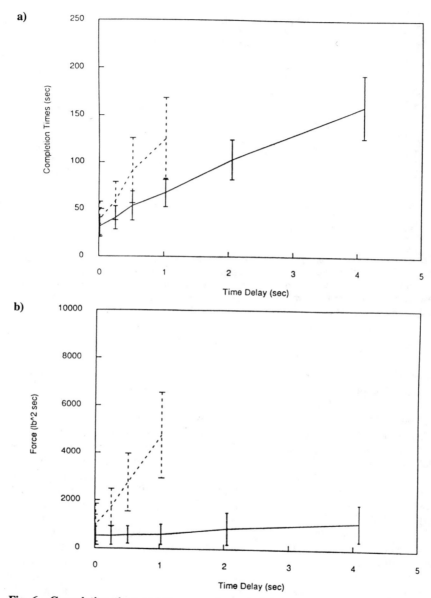

Fig. 6 Completion time and the sum of squared forces for the five operators performing the peg-in-hole task with kinesthetic force feedback (dashed line) and shared compliant control (solid line); note that shared compliant control enabled task performance for time delays above 1 s.

D.   Performance with Passive Communication

Recently, we have begun to study the performance of the *passive* communication laws under time delays in a single-axis system.[16, 17] We implemented six representative control laws at a 1000-Hz sample rate with a minimum delay of 2 ms. (For details of the algorithm, see Ref. 17.) The tasks we studied were 1) pure position response without contact, 2) constant force maintenance against a rigid obstacle, and 3) a nonlinear impedance similar to the detent tasks used by Raju.[11]

Completion time for the nonlinear task (Fig. 7) increased approximately linearly with delay for all algorithms except those with the passive communication law. The slope for the nonpassive algorithms is about 20 times the imposed delay.

The algorithms with passive communication laws showed markedly higher completion times at all values of delay. For example, at even 20 ms of added delay, the average completion time for the two passivity algorithms was 18.5 s vs about 10 s for the others. As time delay increased, the completion times for the passivity-based algorithms increased rapidly without a clear linear trend. However, a slope of approximately 50 can be fit to these data using linear regression.

The hard contact task (Fig. 8) exercised the system in what is usually considered the extreme case for the control of force: contact with a rigid object. This task is inherently difficult because of its discontinuous nature and the extremely high stiffness (impedance) of the block. Completion time is a somewhat nonlinear measure of performance on this task because nominal performance of the task requires a prespecified time. In this case, the completion time should not be

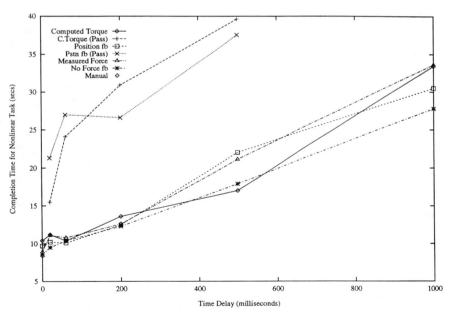

Fig. 7   Completion time for a one-axis nonlinear task time vs delay; results are presented for six control modes and direct maximal teleoperation.

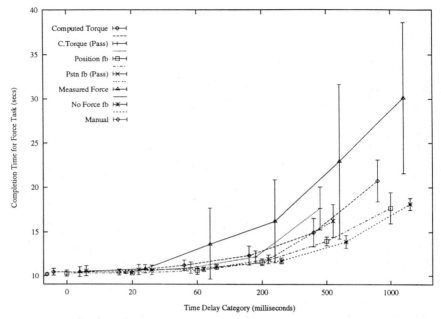

Fig. 8   Force task completion time by category; the data points are spread around the time delay category for clarity, actual time delays were identical for all modes.

different from about 10 s unless the force trajectory is variable enough to cause the force limits (± about 6 N) to be exceeded. This would cause a timer to restart and thus change the desired completion time. With zero or 20 ms of delay, all algorithms gave completion times tightly clustered around 10 s. At 60 ms of delay, the mean completion time of the measured force algorithm increased to about 12 s, an indication of the beginning of problems with instability. At increasing values of time delay, completion time increased slowly for all algorithms, with the measured force algorithm increasing significantly faster.

Total completion time for all tasks is shown in Fig. 9. For small time delays, all algorithms except those using passive communications laws, performed similarly. Segmentation of the data into subtasks[7, 16] showed that the increased total completion time for the passive communication law was dominated by the difficulty of attaining the peak position (i.e., the unstable equilibrium point) in the nonlinear task. For longer time delays, the complete absence of force feedback gave the best completion times.

VIII.   Discussion

These preliminary ground experiments serve two roles in relation to flight experiments. First, they define performance benchmarks. Recent flight experiments, such as the German Robotic Technology Experiment[18] (ROTEX) (flight date May 1993), employed some experimental tasks similar to the ones described here. If similar performance measures are used, detailed studies of performance can

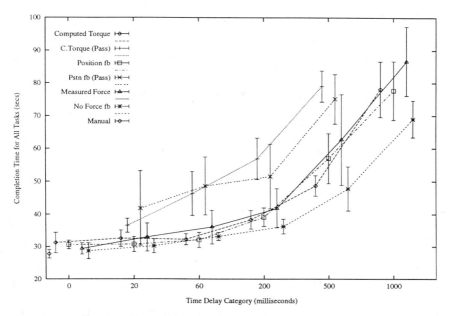

Fig. 9   Total completion time for all tasks by category.

compare the space and ground environments. Secondly, an understanding of how the time delay affects performance should impact the planning of future experiments and missions. In particular, the performance issues associated with time delays will affect the allocation of tasks to ground controlled robots. For example, as described earlier, the initial results from experiments like ROTEX may be disappointing in terms of the capabilities of ground operation because of the large delays (2–4 s) imposed by the Shuttle communications channels. However, future unmanned systems featuring line of sight communication will have the opportunities for substantially higher performance. A system in geosynchronous orbit, for instance, will have the comparatively benign delay of 0.5 s imposed from a large region of the Earth. A system in low Earth orbit may even be able to do practical force reflection to the ground at the price of availability limited to the time the system is in sight overhead.

Most of the data presented here are consistent with a linear increase in completion time with the imposed delay. Apparent exceptions in Figs. 7–9 are due to time structures imposed on the task such as minimum durations or timeouts. One possible interpretation of this linear behavior is that the task consists of discrete actions which take a fixed duration without delay and which are each delayed by the additional time delay. Mathematically, where N is the number of discrete movements, t_0 is the elemental movement time, and T is the delay

$$CT = N(t_0 + T) \qquad (6)$$

For the nonpassive control modes of Fig. 4, for example,

$$CT = 20(0.5 + T) \qquad (7)$$

The elemental task duration, 0.5 s in this example, comes from the time axis intercept (i.e., the completion time with no delay) divided by the slope of the performance/delay trend. When experimental conditions are varied, the slope varies. For example, in Fig. 6a, the use of KFF changed the delay slope from 33 to 80. In the preceding model, this would suggest that under adverse conditions, such as the use of a difficult control mode, the task is broken up by the operator into smaller chunks—each of which can be accomplished with a reasonably high probability. In the preceding example, the intercept was 38 s. The two performance trends are thus modeled

$$CT = 80(2.11 + T) \tag{8}$$

for the KFF case and

$$CT = 33(0.868 + T) \tag{9}$$

for the shared compliant control case. The higher elemental task durations in Eqs. (8) and (9) compared to that in Eq. (7) reflect the greater difficulty of a high-precision multi-axis task.

In their pioneering study, Sheridan and Ferrell[3] measured the completion time for a noncontact positioning task of varying difficulty under time delays from 0 to 3.2 s. They used Fitts' information measure to quantify the task difficulty which was 3–7 bits for their tasks. Completion time was not a linear function of time delay for most of their data. However, the data became more linear for the lower indices of difficulty. Fitting a linear increase to their data at an index of difficulty of three bits, we get

$$CT = 1.667(1.2 + T) \tag{10}$$

This much lower slope is consistent with the interpretation given earlier and the definition of their experiment in which task completion consisted of a single move. A possible corrective move accounts for the additional 0.667. The tasks reported in this chapter are complex multistep operations and so it is reasonable to assume that they might be composed of many more elemental operations.

Sheridan and Ferrell derived a model of completion time as a function of time delay and the index of difficulty. In our experiments, the difficulty is not as easy to quantify (for a modified Fitt's measure for the peg-in-hole task, see Hannaford et al.[2]). However, holding the difficulty constant, and lumping the operator's reaction time into the elemental move time, their model reduces to the form of Eq. (6).

In his early experiments with delayed force feedback from tasks involving contact, Ferrell[4] also found linear increases in completion time with imposed delay. The slope depended somewhat on the strategy employed by the operators varying from 3.6 to 5.3. Here t_0 varied from 0.28 to 0.49.

Of course, an alternate explanation is that the number of elemental tasks can stay the same, but the time per task can increase. This is more clearly expressed if Eq. (6), for example, is written

$$CT = T_0 + NT \tag{11}$$

One of these two theories might be able to be confirmed with future experiments or further analysis of existing data. One method might be to apply Fourier analysis to the sensor traces and look for periodicities which might reveal elemental durations giving an independent measure of N.

Acknowledgments

Most of the experiments reported here were performed at the Jet Propulsion Laboratory of the California Institute of Technology under contract to NASA. The passivity experiments were performed at the University of Washington with the support of the National Science Foundation.

References

[1]Hannaford, B., Wood, L., Guggisberg, B., McAffee, D., and Zak, H., "Performance Evaluation of a Six-Axis Generalized Force-Reflecting Teleoperator," Jet Propulsion Lab., JPL Pub. 89-18, Pasadena, CA, June 15, 1989.

[2]Hannaford, B., Wood, L., McAffee, D., and Zak, H., "Performance Evaluation of a Six Axis Generalized Force Reflecting Teleoperator," *IEEE Transactions on Systems, Man, and Cybernetics,* Vol. 21, No. 3, 1991, pp. 620–633.

[3]Sheridan, T. B., and Ferrell, W. R., "Remote Manipulative Control with Transmission Delay," *IEEE Transactions on Human Factors in Electronics,* Vol. HFE-4, 1963, pp. 25–29.

[4]Ferrell, W. R., "Delayed Force Feedback," *Human Factors,* Oct. 1966, pp. 449–455.

[5]Hannaford, B., and Anderson, R., "Experimental and Simulation Studies of Hard Contact in Force Reflecting Teleoperation," *Proceedings of the IEEE Conference on Robotics and Automation,* IEEE Press, Piscataway, NJ, April, 1988, pp. 584–589.

[6]Kim, W. S., Hannaford, B., and Bejczy, A. K., "Force-Reflection and Shared Compliant Control in Operating Telemanipulators with Time Delay," *IEEE Transactions on Robotics & Automation,* Vol. 8, No. 2, 1992, pp. 176–185.

[7]Hannaford, B., and Lee, P., "Hidden Markov Model of Force Torque Information in Telemanipulation," *International Journal of Robotics Research,* Vol. 10, No. 5, 1991, pp. 528–539.

[8]Hannaford, B., and Lee, P., "Multi-Dimensional Hidden Markov Model of Telemanipulation Tasks with Varying Outcomes," Proceedings of the IEEE International Conference on Systems, Man, and Cybernetics, Los Angeles, CA, Nov. 1990.

[9]Fiorini, P., and Giancaspro, A., "A Procedure for the Frequency Analysis of Telerobotic Tasks Data," *Proceedings of the IEEE/RSJ International Conference on Intelligent Robots and Systems* (Raleigh, NC), Vol. 2, 1992, pp. 873–880.

[10]Lee, P., Bejczy, A., Schenker, P., and Hannaford, B., "Telerobot Configuration Editor," Proceedings of the IEEE International Conference on Systems, Man, and Cybernetics, Los Angeles, CA, Nov. 1990.

[11]Raju, G. A., "Operator Adjustable Impedance in Bilateral Remote Manipulation," Ph.D. Dissertation, Massachusetts Inst. of Technology, Cambridge, MA, 1988.

[12]Hannaford, B., "A Design Framework for Teleoperators with Kinesthetic Feedback," *IEEE Transactions on Robotics and Automation,* Vol. 5, No. 4, 1989, pp. 426–434.

[13]Sheridan, T. B., "Telerobotics, Automation, and Human Supervisory Control," MIT Press, Cambridge, MA, 1992.

[14]Anderson, R. J., and Spong, M. W., "Bilateral Control of Teleoperators with Time Delay," *Proceedings of the IEEE International Conference on Systems, Man, and Cybernetics* (Beijing, China), IEEE Press, Piscataway, NJ, Vol. 1, 1988, pp. 131–138.

[15]Niemeyer, G., and Slotine, J. J., "Stable Adaptive Teleoperation," *IEEE Journal of Oceanic Engineering,* Vol. 16, Jan 1991, pp. 152–162.

[16]Lawn, C. A., "Design and Construction of a Single-Axis Master-Slave Teleoperation System and the Performance Evaluation of Control Algorithms and Passive Communication Laws with Time Delay," MSME Thesis, Univ. of Washington, Seattle, WA, 1992.

[17]Lawn, C. A., and Hannaford, B., "Performance Testing of Passive Communication and Control in Teleoperation with Time Delay," *Proceedings of the IEEE International Conference on Robotics and Automation,* Vol. 3, May 1993, pp. 776–781.

[18]Hirzinger, G., Brunner, B., Dietrich, J., and Heindl, J., "Sensors-Based Robotics— ROTEX and its Telerobotic Features," *IEEE Transactions on Robotics and Automation,* Vol. 9, No. 5, Oct. 1993, pp. 649–663.

Chapter 5

Toward Advanced Teleoperation in Space

Antal K. Bejczy*

*Jet Propulsion Laboratory, California Institute of Technology,
Pasadena, California 91109*

Introduction

T HE term "teleoperation" refers to the use of manipulators (mechanical arms) and mobility devices, equipped with some sensing capabilities, and remotely controlled by a human operator. Historically, mechanical arms are the most important teleoperated devices.

In general, teleoperation implies *continuous* perceptive and cognitive human operator involvement in the control of remote manipulators. Typically, the human control is a *manual* one, and the basic information feedback is through *visual images*. Continuous human operator involvement in teleoperation has both advantages and disadvantages. The disadvantages become quite dramatic when there is an observable, two-way communication time delay between the operator and the remotely controlled equipment. Modern development trends in teleoperator control technology are aimed at amplifying the advantages and alleviating the disadvantages of the human element in teleoperation through the development and use of various nonvisual sensing capabilities, intelligent or task-level computer controls, computer graphics or virtual reality displays, and new computer-based human-machine interface devices and techniques in the information and control channels between the operator and the remotely controlled manipulators. These development trends are typically summarized under the popular titles of telepresence and supervisory control technologies. In this chapter, those two titles are lumped under the term advanced teleoperation.

Several notes should be added to the objective description of telepresence and supervisory control technologies. First, none of them eliminates the human operator from the control operation, but both change the operator's function assignments and employ human capabilities in new ways. Second, both technologies promise

Copyright © 1994 by the American Institute of Aeronautics and Astronautics, Inc. The U.S. Government has a royalty-free license to exercise all rights under the copyright claimed herein for Governmental purposes. All other rights are reserved by the copyright owner.

*Senior Research Scientist, Robotic Systems and Advanced Computer Technology Section.

the performance of more complex tasks with better results; but, in doing so, both technologies also make a close reference to human capabilities of operators who will use evolving new devices and techniques in the control station. Third, both telepresence and supervisory control make reference to evolving capabilities of other technologies like sensing, high-performance computer graphics, computerized electromechanical devices, algorithm-based flexible automation, expert systems for planning and error recovery, and so on. Thus, the progress in both technologies are tied to rich multidisciplinary activities. Fourth, both technologies need the evaluation and validation of their results relative to application environments.

This chapter focuses on the description and some practical evaluation of an experimental advanced teleoperation system, developed at the Jet Propulsion Laboratory (JPL) during the past seven to eight years. First we describe the JPL Advanced Teleoperator (ATOP) system and its control station where a variety of operator interface devices and techniques are integrated into a functional setting, accommodating a primary operator and secondary operators. Then we will summarize the results of some generic and application task experiments. In the third part of the chapter, we will highlight the lessons learned so far. The chapter will conclude with a brief description of an ongoing work on an anthropomorphic (human-like) advanced telemanipulation system.

Jet Propulsion Laboratory Advanced Teleoperator System

The basic underlying idea of the JPL ATOP system setting is to provide a dual arm robot system together with the necessary operator interfaces to extend the two-handed manipulation capabilities of a human operator to remote places. The system setting intends to include all perceptive components that are necessary to perform sensitive remote manipulation efficiently, including nonrepetitive and unexpected tasks. The general goal is to elevate teleoperation to a new level of task performance capabilities through enhanced visual and nonvisual sensing, computer-aided remote control, and computer-aided human-machine interface devices and techniques. The overall system is divided into two major parts: the remote (robot) work site and the local (control station) site, with electronic data and TV communication links between the two sites.

The remote site is a workcell. It comprises: 1) two redundant eight-DOF arms (produced by AAI Company, Inc.) in a fixed base setting, each covering a hemispheric work volume, and each equipped with the latest JPL-developed model C smart hands that contain three-dimensional force-moment sensors at the hands' base and grasp force sensing at the base of the hand claws; 2) a JPL-developed control electronics and distributed computing system for the two arms and smart hands; and 3) a computer controllable multi-TV gantry robot system with controllable illumination. This gantry robot currently accommodates three color TV cameras, one on the ceiling plane, one on the rear plane, and one on the right side plane of the workcell. Each camera can be position controlled in two translational DOF in the respective plane, and in two orientation directions (pan and tilt) relative to the respective moving base. Zoom, focus, and iris of each TV camera can also be computer controlled. A stereo TV camera system is also available which can be mounted on any of the two side camera bases. The total size of the rectangular remote work site is about 5 m in width, about 4 m in depth, and about 2.5 m in height. See Fig. 1 for the ATOP remote workcell.

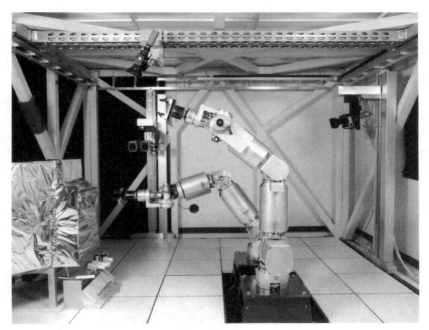

Fig. 1   JPL ATOP dual arm workcell with Gantry TV frame.

The control station site organization follows the idea of accommodating the human operator in all levels of human-machine interaction, and in all forms of human-machine interfaces. Presently, it comprises 1) two general purpose force-reflecting hand controllers (FRHC), 2) three TV monitors, 3) TV camera/monitor switchboards, 4) a manual input device for TV control, and 5) three graphics displays. One of these graphics displays is connected to the primary graphics workstation (IRIS 4D/310 VGX) which is used for preview/predictive displays and for various graphical user interfaces (GUIs) in four-quadrant format. The second is connected to an IRIS 4D/70 GT workstation and is solely used for sensor data display. The third one is connected to a SUN workstation (SparcStation 10) and is used as a control configuration editor (CCE), which is an operator interface to the manipulators' control software based on an X-window environment. See Fig. 2 for the ATOP local control station.

Advanced Teleoperator Hand Controllers

The human arm and hand are functionally both powerful mechanical tools and delicate sensory organs through which information is received from and transmitted to the world. Therefore, the human arm-hand system (thereafter simply called hand here) is a key communication medium in teleoperator control. With hand actions, complex position, rate, or force commands can be formulated and very physically written to the controller of a remote robot arm system in all workspace directions. At the same time, the human hand also can receive force, torque, and

Fig. 2   JPL ATOP control station.

touch information from the remote robot arm-hand system. Furthermore, the human fingers offer additional capabilities to convey new commands to a remote robot controller from a suitable hand controller. Hand controller technology is, therefore, an important technology in the development of advanced teleoperation. Its importance is particularly underlined when one considers computer control which connects the hand controller to the remote arm system. The direct and continuous (scaled or unscaled) relation of operator hand motion to the remote robot arm's motion behavior in real time through a hand controller is in sharp contrast to the computer keyboard type commands which, by their very nature, are symbolic, abstract, and discrete (noncontinuous), and require the specification of some set of parameters within the context of a desired motion.

In contrast to the standard force-reflecting, replica master-slave systems, a new form of bilateral, force-reflecting manual control of robot arms has been implemented at the JPL ATOP project. The hand controller is a backdrivable six-DOF isotonic joystick. It is dissimilar to the controlled robot arm both kinematically and dynamically. But, through computer transformations, it can control the motion of any robot arm in six task space coordinates (in three position and three orientation coordinates). Forces and moments sensed at the base of the robot hand can backdrive the hand controller through proper computer transformations so that the operator feels the forces and moments acting at the robot hand while he controls the position and orientation of it. This hand controller can read the position and orientation of the hand grip within a 30-cm cube in all orientations, and can apply arbitrary force and moment vectors up to 20 N and 1.0 Nm, respectively, at the hand grip.

The overall schematic of the six-DOF FRHC is shown in Fig. 3. [The mechanism of the hand controller was designed by J. K. Salisbury Jr., now at the Massachusetts Institute of Technology (MIT), Cambridge, Massachusetts.] The kinematics and the command axes of the FRHC are shown in Fig. 4. The hand grip is supported by a gimbal with three intersecting axes of rotation ($\beta_4, \beta_5, \beta_6$). A translation axis ($R_3$) connects the hand gimbal to the shoulder gimbal which has two more intersecting axes (β_1, β_2). The motors for the three hand gimbal and translation axes are mounted on a stationary drive unit at the end of the hand controller's main tube. This stationary drive unit forms a part of the shoulder gimbal's counterbalance system. The moving part of the counterbalance system is connected to the R_3. It serves 1) to maintain the hand controller's center of gravity at a fixed point and 2) to maintain the tension in the hand gimbal's drive cables as the hand gimbal changes its distance from the stationary drive unit. The actuator motors for the two shoulder joints are mounted to the shoulder gimbal frame and to the base frame of the hand controller, respectively.

The self-balance system renders the hand controller neutral against gravity. Thus, the hand controller can be mounted both horizontally or vertically, and the calculation of motor torques to backdrive the hand controller does not require gravity compensation. In general, the mechanical design of the hand controller provides a dynamically transparent input/output device for the operator. This is accomplished by low backlash, low friction and low effective inertia at the hand

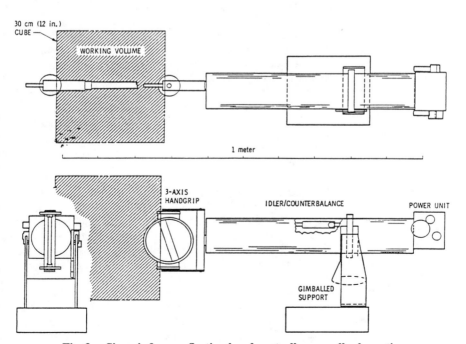

Fig. 3   Six-axis force-reflecting hand controller overall schematic.

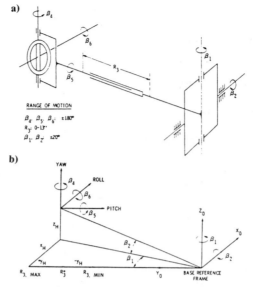

Fig. 4   Hand controller kinematics and command axes.

grip. More details of the mechanical design of this hand controller and on hand controller technology in general can be found in Refs. 1 and 2. A computer-based control system establishes the appropriate kinematic and dynamic control relations between the FRHC and the robot arm. The FRHC can control any robot arm and can receive force/torque feedback from any robot arm equipped with a three-dimensional force-moment sensor at the base of the robot hand.

The computer-based control system supports four modes of manual control: position, rate, force-reflecting, and compliant control in task space (Cartesian space) coordinates. The operator, through an on-screen menu, can designate the control mode for each task space axis independently. The *position control mode* servos the slave position and orientation to match the master's. The *indexing function* allows slave excursions larger or smaller than the 30-cm cube hand controller work volume. In the *force-reflecting mode*, the hand controller is backdriven based on force-moment data generated by the robot and sensed during the robot hand's interaction with objects and environment. The *rate control mode* sets the slave endpoint velocity in task space based on the displacement of the hand controller. This is implemented through a software spring in the control computer of the hand controller. Through this software spring, the operator has a sensation of the commanded rate, and the software spring also provides a zero-referenced restoring force. The rate mode is useful for tasks requiring large translations. The *compliant control mode* is implemented through a low-pass software filter acting on the robot hand's force-torque sensor data in the hybrid position-force loop. This permits the operator to control a springy or less stiff robot. Active compliance with damping can be varied by changing the filter parameters in the software menu. Setting the spring parameter to zero in the low-pass filter will reduce it to a pure damper which results in a high stiffness hybrid position-force control loop.

The present FRHC has a simple hand grip equipped with a deadman switch and with three function switches. To better utilize the operator's finger input capabilities, an exploratory project recently evaluated a design concept that would place computer keyboard features attached to the hand grip of the FRHC. To accomplish this, three DATAHAND^{TM} (Ref. 3) switch modules were integrated with the hand grip as shown in Fig. 5. Each switch module at a finger tip contains five switches as indicated in Fig. 6. Thus, the three switch modules at the FRHC hand grip can contain fifteen function keys which can directly communicate with a computer terminal. This eliminates the need for the operator to move his/her hand from the FRHC hand grip to a separate keyboard to input messages and commands to the computer. A recent test and evaluation, using a mock-up system and ten test subjects, indicated the viability of the finger-tip switch modules as part of a new hand grip unit for the FRHC as a practical step towards a more integrated operator interface device in the ATOP system. More on this concept and evaluation can be found in Ref. 4.

Advanced Teleoperator Control System

The overall ATOP control organization permits a spectrum of operations between full manual, shared manual and automatic, and full automatic (call traded) control, and the control can be operated with variable active compliance referenced to force-moment sensor data. More on the overall ATOP control system can be found in Refs. 5–8. Only the salient features of the original ATOP control system are summarized here. The overall control/information data flow diagram (for a single arm) is shown in Fig. 7 . It is noted that the computing architecture of this original ATOP system is a fully synchronized pipeline, where the local servo loops at both the control station and the remote manipulator nodes can operate at a 1000-Hz rate. The end-to-end bilateral (i.e., force-reflecting) control loop can operate at a 200-Hz rate. More on the computational system critical path functions and performance can be found in Ref. 9.

The actual data flow depends on the control mode chosen. The different selectable control modes are the following: freeze mode, neutral mode, current mode, joint mode, task mode. In the *freeze mode* the brakes of joints are locked, the motors are turned off, and some joints are servoed to maintain their last positions. This mode is primarily used when the robot is not needed for a short period of time but turning it off is not desired. In the *neutral mode* all position gains are set to zero, gravity compensation is active to prevent the robot from falling down. In this mode the user can manually move the robot to any position and it will stay there. In the *current mode* the six motor currents are directly commanded by the data coming in from the communication link. This mode exists for debugging only. In the *joint mode* the hand controller axes control individual motors of the robot. In the *task mode* the inverse kinematic transformation is performed on the incoming data, and the hand controller controls the end effector tip along the three Cartesian and pitch, yaw, and roll axes. This mode is the most frequently

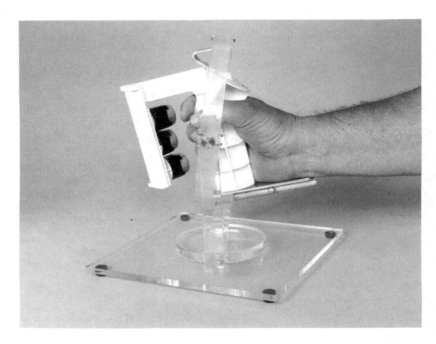

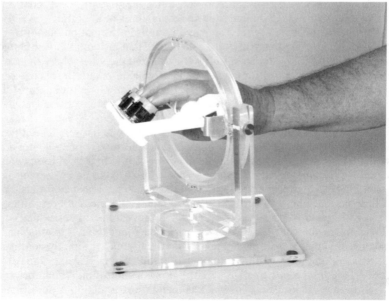

Fig. 5   DATAHAND™ (Ref. 3) switch modules integrated FRHC hand grip.

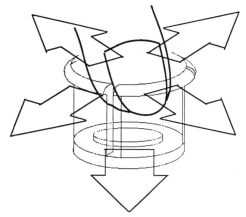

1. Each module contains five switches.

2. Switches can give tactile and audio feedback.

3. Switches require low strike force.

4. Switches surround finger creating differential feedback regarding key that has been struck.

Fig. 6   Five key-equivalent switches at a DATAHAND™ fingertip switch module.

used for task execution or experiments, and this is the one shown explicitly in Fig. 7.

The control system on the remote site is designed to prevent sudden robot motions. The motion commands received are incremental and are added to the current parameter under control. Sudden large motions are also prevented in case of mode changes. This necessitates proper initialization of the inverse kinematics software at the time of the mode transition. This is done by inputing the current Cartesian coordinates from the forward kinematics into the inverse kinematics.

The data flow diagram shown in Fig. 7 illustrates the organization of several servo loops in the system. The innermost loop is the position control servo at the robot site. This servo uses a PD control algorithm, where the damping is purely a function of the robot joint velocities. The incoming data to this servo is the desired robot trajectory described as a sequence of points at 1 ms intervals. This joint servo is augmented by a gravity compensation routine to prevent the weight of the robot from causing a joint positioning error. Because this servo is a first-order servo, there will be a constant position error that is proportional to the joint velocity.

In the basic Cartesian control mode the data from the hand controller are added to the previous desired Cartesian position. From this the inverse kinematics generate the desired joint positions. The joint servo moves the robot to this position. From the actual joint position the forward kinematics compute the actual Cartesian positions. The force-torque sensor data and the actual positions are fed back to the hand controller side to provide force feedback.

This basic mode can be augmented by the addition of compliance control, Cartesian servo, and sticktion/friction compensation. Figure 8 shows the compliance control and the Cartesian servo augmentations. There are two forms of compliance, an integrating and a spring type (see Fig. 9). In integrating compliance the velocity of the robot end effector is proportional to the force felt in the corresponding direction. To eliminate drift a deadband is used. The zero velocity band does not have to be a zero force, a force offset may be used. Such a force offset is used if, for example, we want to push against the task board at some given force

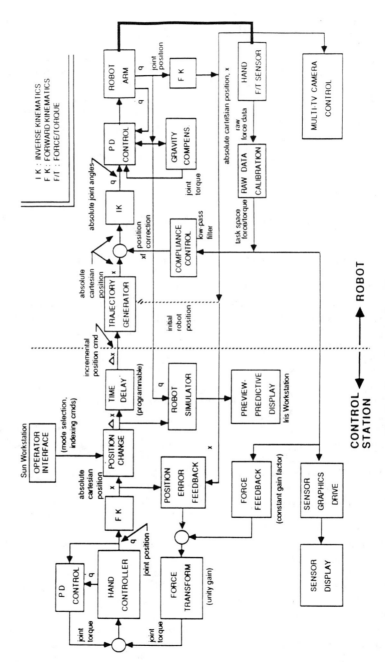

Fig. 7   Control system flow diagram.

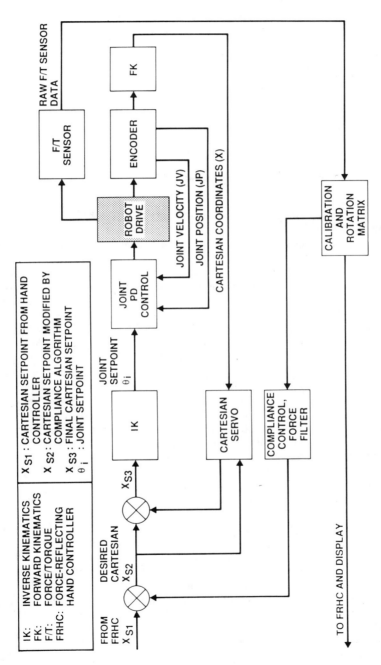

Fig. 8   Control schemes: joint servo, Cartesian servo, compliance control.

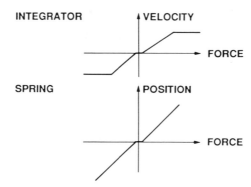

Fig. 9   Compliance components and interpretations.

while moving along other axes. Any form of compliance can be selected along any axis independently. In the case of the spring-type compliance the robot position is proportional to the sensed force. This is similar to a spring centering action. The velocity of the robot motion is limited in both the integrating and spring cases.

There is a wide discrepancy between the robot response bandwidth and the force readings. The forces are read at a 1000-Hz sampling rate. The robot motion command has an output response at a 5-Hz bandwidth. To generate smooth compliance response, the force readings go through two subsequent filters. The first one is a simple averaging of ten force readings. This average is called 100-Hz force and is computed at a 100-Hz rate. From this 100-Hz force a 5-Hz force reading is computed by a first-order low-pass filter. This 5-Hz force reading is also computed at a 100-Hz rate. The 5-Hz force is used for compliance computations.

As is shown in Fig. 8, the Cartesian servo acts on task space (X, Y, Z, pitch, yaw, roll) errors directly. These errors are the difference between desired and actual task space values. The actual task space values are computed from the forward kinematic transformation of the actual joint positions. This error is then added to the new desired task space values before the inverse kinematic transformation determines the new joint position commands from the new task space commands.

A trajectory generator algorithm was formulated based on observations of profiles of task space trajectories generated by the operators manually through the FRHC. Three important features were observed in hand-generated task space trajectory profiles:

1) The operators always generated trajectories as a function of the relative distance between start point and goal point in the task space or, in general, as a function of the present position state relative to the desired position state of the end effector in the task space. In other words, the operators did not manually generate trajectories based on time (on clock signals).

2) The velocity-position phase diagrams of motion typically resembled a harmonic (sine) function.

3) Between the start and completion phases, the operator-generated trajectories typically attained a constant velocity profile.

Based on these observations, we formulated a harmonic motion generator (HMG) with a sinusoidal velocity-position phase function profile as shown in

Fig. 10. The motion is parameterized by the total distance traveled, the maximum velocity, and the distance used for acceleration and deceleration. Both the accelerating and decelerating segments are quarter sine waves, with a constant velocity segment connecting them. This scheme still has a problem, the velocity being 0 before the motion starts. This problem is corrected by adding a small constant to the velocity function.

It is noted that the HMG discussed here is quite different from the typical trajectory generator algorithms employed in robotics which use a polynomial position-time function. Our algorithm generates the motion as a trigonometric (harmonic) velocity vs position function. The position vs time and the corresponding velocity vs time functions generated by the HMG are shown in Fig. 11. More on performance results generated by HMG, Cartesian servo, and force-torque sensor data filtering in compliance control can be found in Refs. 6 and 10. Illustrative examples are shown in Fig. 12 and Fig. 13.

Advanced Teleoperation Computer Graphics

Task visualization is a key problem in teleoperation, because most of the operator's control decisions are based on visual or visually conveyed information. For this reason, computer graphics play an increasingly important role in advanced teleoperation. This role includes 1) planning actions, 2) previewing motions, 3) predicting motions in real time under communication time delay, 4) helping operator training, 5) enabling visual perception of nonvisible events like forces and moments, and 6) serving as a flexible operator interface to the computerized control system.

The capability of task planning aided by computer graphics offers flexibility, visual quality, and a quantitative design base to the planning process. The capability of graphically previewing motions enhances the quality of teleoperation by reducing trial-and-error strategies in the hardware control and by increasing the operator's confidence in control decision making during task execution. Predicting consequences of motion commands in real time under communication

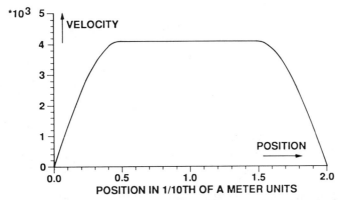

Fig. 10   Harmonic motion generator velocity-position function.

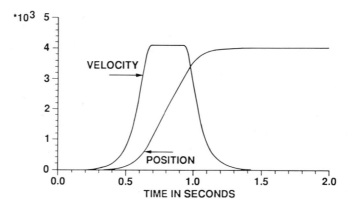

Fig. 11   Harmonic motion generator position and velocity time functions.

time delay permits longer action segmentations as opposed to the move-and-wait control strategy normally employed when no predictive display is available, increases operation safety, and reduces total operation time. Operator training through a computer graphics display system is a convenient tool for familiarizing the operator with the teleoperated system without turning the hardware system on. Visualization of nonvisible effects (like contact forces) enables visual perception of different nonvisual sensor data, and helps manage system redundancy by providing some suitable geometric image of a multidimensional system state. Last, but not least, computer graphics as a flexible operator interface to the control systems; it replaces complex switchboard and analog display hardware in a control station.

The actual utility of computer graphics in teleoperation depends to a high degree on the fidelity of graphics models that represent the teleoperated system, the task, and the task environment. In the past few years the JPL ATOP project developed high-fidelity calibration of graphics images to actual TV images of task scenes. This development has four major ingredients: first, the creation of high-fidelity three-dimensional graphics models of robot arms and objects of interest for robot arm tasks; second, the high-fidelity calibration of the three-dimensional graphics models relative to given TV camera two-dimensional image frames which cover the sight of both the robot arm and the objects of interest; third, the high-fidelity overlay of the calibrated graphics models over the actual robot arm and object images in a given TV camera image frame on a monitor screen; fourth, the high-fidelity motion control of robot arm graphics image by using the same control software that drives the real robot.

The high-fidelity fused virtual and actual reality image displays became very useful tools for planning, previewing, and predicting robot arm motions without commanding and moving the robot hardware. The operator can generate visual effects of robot motion by commanding and controlling the motion of the robot's graphics image superimposed over TV pictures of the live scene. Thus, the operator can see the consequences of motion commands in real time, before sending the commands to the remotely located robot. The calibrated virtual reality display system can also provide high-fidelity synthetic or artificial TV camera views to

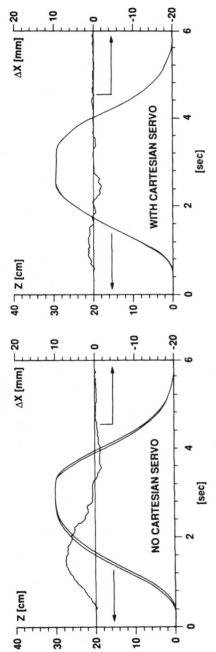

Fig. 12   Vertical (Z) straight line trajectory from manual control and ΔX error.

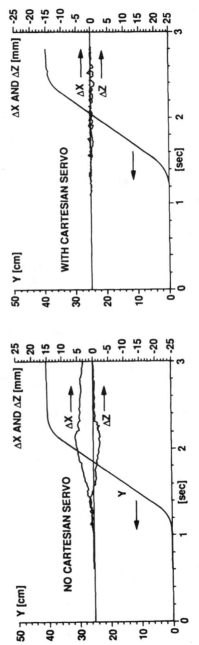

Fig. 13   Horizontal (Y) straight line trajectory from harmonic motion generator and ΔX and ΔZ errors.

the operator. These synthetic views can make critical motion events visible that are otherwise hidden from the operator in a given TV camera view or for which no TV camera view is available. More on the graphics system in the ATOP control station can be found in Refs. 11–15.

High-Fidelity Graphics Calibration

A high-fidelity overlay of graphics and TV images of work scenes requires a high-fidelity TV camera calibration and object localization relative to the displayed TV camera view. Theoretically, this can be accomplished in several ways. For the purpose of simplicity and operator-controllable reliability, an operator-interactive camera calibration and object localization technique has been developed, using the robot arm itself as a calibration fixture, and using a nonlinear least-squares algorithm combined with a linear algorithm as a new approach to compute accurate calibration and localization parameters.

The current method uses a point-to-point mapping procedure, and the computation of camera parameters is based on the ideal pinhole model of image formation by the camera. In the camera calibration procedure, the operator first enters the correspondence information between the three-dimensional graphics model points and the two-dimensional camera image points of the robot arm to the computer. This is performed by repeatedly clicking with a mouse a graphics model point and its corresponding TV image point for each corresponding pair of points on a monitor screen which, in a four-quadrant window arrangement, shows both the graphics model and the actual TV camera image (see Fig. 14). To improve calibration accuracy, several poses of the manipulator within the same TV camera view can be used to enter corrresponding graphics model and TV image points to the computer. Then the computer computes the camera calibration parameters. Because of the ideal pinhole model assumption, the computed output is a single linear 4×3 calibration matrix for a linear perspective projection.

Object localization is performed after camera calibration by entering corresponding object model and TV image points to the computer for different TV camera views of the object. Again, the computational output is a single linear 4×3 calibration matrix for a linear perspective projection.

The actual camera calibration and object localization computations are carried out by a combination of linear and nonlinear least-squares algorithms. The linear algorithm, in general, does not guarantee the orthonormality of the rotation matrix, providing only an approximate solution. The nonlinear algorithm provides the least-squares solution that satisfies the orthonormality of the rotation matrix, but requires a good initial guess for a convergent solution without entering into a very time-consuming random search. When a reasonable approximate solution is known, one can start with the nonlinear algorithm directly. When an approximate solution is not known, the linear algorithm can be used to find one, and then one can proceed with the nonlinear algorithm. More on the calibration and object localization technique can be found in Refs. 16 and 17.

After completion of camera calibration and object localization, the graphics models of both the robot arm and the object of interest can be overlaid with high fidelity on the corresponding actual images of a given TV camera view. The overlays can be in wire-frame or solid-shaded polygonal rendering with varying

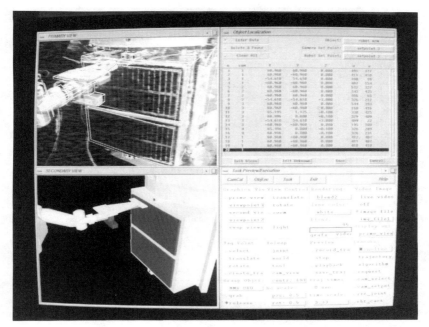

Fig. 14   Graphics user interface for calibrating virtual (graphics) images to TV images.

levels of transparency, providing different task details. In the wire-frame format, the hidden lines can be removed or retained by the operator, depending on the information needs in a given task.

Advanced Teleoperator Graphics Operator Interface

The first development of a graphic system as an advanced operator interface was aimed at parameter acquisition, and was handled as and called a teleoperation configuration editor (TCE).[18] This interface used the concepts of windows, icons, menus, and a pointing device to allow the operator to interact, select, and update single parameters as well as groups of parameters. TCE utilizes the direct manipulation concept, with the central idea of having visible objects such as buttons, sliders, and icons that can be manipulated directly, i.e., moved and selected using the mouse, to perform any operation. A graphic interface of this type has several advantages over a traditional panel of physical buttons, switches, and knobs: the layout can be easily modified and its implementation cycle, i.e., design and validation, is significantly shorter than hardware changes.

The TCE, Fig. 15, was developed to incorporate all the configuration parameters of an early single-arm version of the ATOP system. It was organized in a single menu divided into several areas dedicated to the parameters of a specific function. Dependencies among different graphical objects are embedded in the interface so that, when an object is activated, the TCE checks for parameter congruency. A

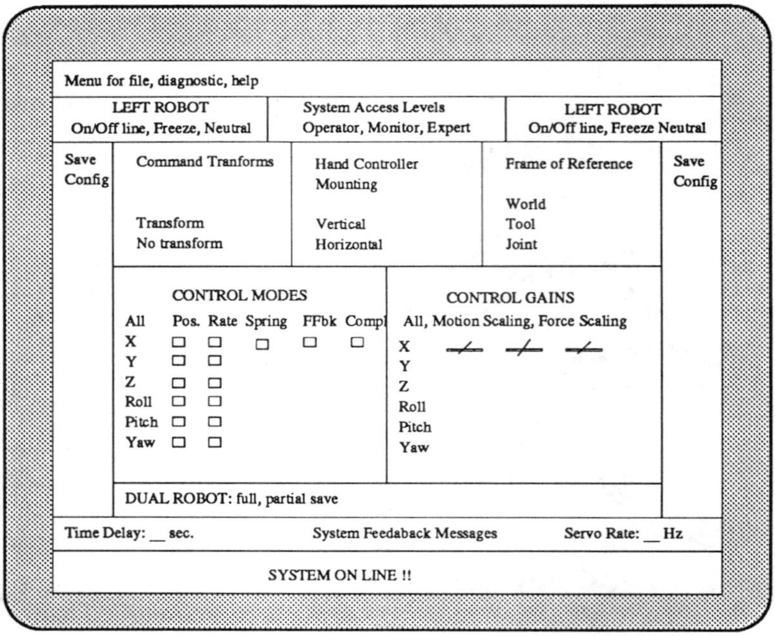

Menu for file, diagnostic, help		
LEFT ROBOT On/Off line, Freeze, Neutral	System Access Levels Operator, Monitor, Expert	LEFT ROBOT On/Off line, Freeze Neutral

Save Config | Command Tranforms | Hand Controller Mounting | Frame of Reference | Save Config

Transform / No transform — Vertical / Horizontal — World Tool Joint

CONTROL MODES
All   Pos. Rate Spring FFbk Compl
X   ☐ ☐   ☐   ☐   ☐
Y   ☐ ☐
Z   ☐ ☐
Roll ☐ ☐
Pitch ☐ ☐
Yaw ☐ ☐

CONTROL GAINS
All, Motion Scaling, Force Scaling
X   ⟋ ⟋ ⟋
Y
Z
Roll
Pitch
Yaw

DUAL ROBOT: full, partial save

Time Delay: __ sec.   System Feedaback Messages   Servo Rate: __ Hz

SYSTEM ON LINE !!

Fig. 15   Schematic layout of the TCE interface.

significant feature of this implementation is its capability to store and retrieve sets of parameters via macro buttons. When a macro command is invoked, it saves the current system configuration and stores it in a function button which can later restore it. The peg-in-hole task, for instance, requires mostly translational motions but when holes have a tight clearance, a compliance is necessary. An appropriate macro configuration is one that enables x, y, and z axes, with position control in the approach direction and automatic compliance on the other two axes. This configuration can be assigned to a macro button and then recalled during a task containing a peg-in-hole segment.

The continuing work on a graphic system as an advanced operator interface is aimed at the data presentation structure of the interface problem, and, for that purpose, uses a hierarchical architecture.[15] This hierarchical data interface looks like a menu tree with only the last menu of the chain (the leaf) displaying data. All the ancestors of the leaf are visible to clearly indicate the nature of the data displayed. The content of the leaf includes data or pictures and quickly conveys the various choices available to the operator. A schematic figure of this layout is shown in Fig. 16. Parameters have been organized in four large groups that follow the sequence of steps in a teleoperation protocol. These groups are 1) layout, 2) configuration, 3) tools, and 4) execution. Each group is further subdivided into specific functions. The layout menu tree contains the parameters defining the physical task structure, such as the relative position of the robots and of the FRHC, servo rates, etc. The configuration menu tree contains the parameters

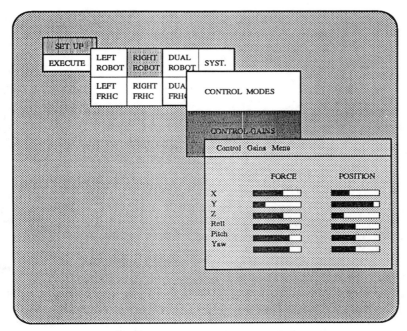

Fig. 16   Schematic layout of the hierarchical data interface.

necessary to define task phases, such as control mode and control gains. The tools tree contains parameters and commands for the off-line support to the operator, such as planning, redundancy resolution, and software development. Finally, the execution tree contains commands and parameters necessary while teleoperating the manipulators, such as data acquisition, monitoring of robots, hand controllers and smart hands, retrieval of stored configurations, and camera commands.

This hierarchical data interface helps solve the problem of displaying the large amount of data needed during a teleoperation task, but it does not address the issue of data entry by the operator. Traditional interfaces require serveral operators, using a different input device for each controlled function. A single operator using this interface needs to use different input devices: joysticks to move the manipulators, buttons for camera and video control, keyboard and mouse for parameters entry. Operating these devices is time consuming and distracts the operator from the task.

An alternative approach would be using the FRHC as the only input device for the operator. This scheme is similar to a virtual reality interface, where operators can accesss commands and data by simply moving their arms and hands. In our implementation concept, the operator would use the FRHC as a pointing device on the interface menus and would use the FRHC trigger to click the selected button. The discrimination between commands for the robot and those for the data interface could be done by the deadman switch on the FRHC handle. When not pressed, this switch inhibits the generation of motion commands for the robots, but FRHC motion data are still available and can be used to move the cursor. This scheme can also be combined with the DATAHAND™ switch module integrated with the FRHC hand grip, discussed previously in this chapter.

Advanced Teleoperator Control Experiments

To evaluate advanced teleoperation capabilities, two types of experiments were designed and conducted: experiments with generic tasks and experiments with application tasks. Generic tasks are idealized, simplified tasks and serve the purpose of evaluating some specific advanced teleoperation features. Application tasks simulate some real-world use of advanced teleoperation.

Generic Task Experiments

In these experiments, described in detail in Ref. 19, four tasks were used: attach and detach velcro, peg insertion and extraction, manipulation of three electrical connectors, and manipulation of a bayonet connector. Each task was broken down into subtasks. The test operators were chosen from a population with some technical background but not with an in-depth knowledge of robotics and teleoperation. Each test subject received 2–4 h of training on the control station equipment. The practice of individuals consisted of four–eight 30-min sessions.

As is pointed out in Ref. 19, performance variation among the nine subjects was surprisingly slight. Their backgrounds were similar (engineering students or recent graduates) except for one who was a physical education major with training in gymnastics and coaching. This subject showed the best overall performance by each of the measures. This apparent correlation between performance and prior background might suggest that potential operators be grouped into classes based on interest and aptitudes.

The generic task experiments were focused at the evaluation of kinesthetic force feedback vs no force feedback, using the specific force feedback implementation techniques of the JPL ATOP project. A typical generic experiment is shown in Fig. 17. The evaluation of the experimental data supports the idea that multiple measures of performance must be used to characterize human performance in sensing and computer-aided teleoperation. For instance, in most cases kinesthetic force feedback significantly reduced task completion time. In some specific cases, however, it did not, but it did sharply reduce extraneous forces. More information on the results can be found in Refs. 19 and 20. Also see Chapter 4 of this book "Ground Experiments Toward Space Teleoperation with Time Delay" by B. Hannaford.

Application Task Experiments

Two major experiments were performed: one without communication time delay and one with communication time delay.

The experiments without communication time delay were grouped around a simulated satellite repair task. The particular repair task was the duplication of the solar maximum satellite repair (SMSR) mission, which was performed by two astronauts in Earth orbit in the Space Shuttle Bay in 1984. Thus, it offered a realistic performance reference data base. This repair is a very challenging task, because this satellite was not designed for repair. Very specific auxiliary subtasks must be performed (e.g., a hinge attachment) to accomplish the basic

a)   b)

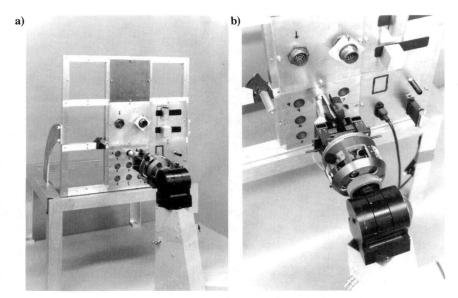

Fig. 17   Two views of the task board and the robot hand performing the peg-in-hole task; the modular task board design has nine 7- × 7-in. openings for task modules; shown here are the four task modules used in the experiments: a) bayonet connector and peg-in-hole tasks (upper and lower left modules), b) velcro and electrical connector tasks (upper and lower right modules). Task modules could be removed and mounted on a separate force-torque sensor for measurement of force and torques in hand operation.

repair which, in our simulation, is the replacement of the main electric box (MEB) of the satellite. The total repair, as performed by two astronauts in Earth orbit, lasted for about 3 h, and comprised the following set of substasks: thermal blanket removal, hinge attachment for MEB opening, opening of the MEB, removal of electrical connectors, replacement of MEB, securing parts and cables, replug of electrical connectors, closing of MEB, and reinstating thermal blanket. It is noted that the two astronauts were trained for this repair on the ground for about a year.

The SMSR simulation by ATOP capabilities was organized so that each repair scenario had its own technical justification and performance evaluation objective. For instance, in the first subtask-scenario performance experiments, alternative control modes, alternative visual settings, operator skills vs training, and evaluation measures themselves were evaluated.[21,22] The first subtask-scenario performance experiments involved thermal blanket cutting and reinstating, and unscrewing MEB bolts. That is, both subtasks implied the use of tools. Figure 18 illustrates these experiments.

Several important observations were made during the aforementioned subtask-scenario performance experiments. The two most important observations are that:

1) The remote control problem in any teleoperation mode and using any advanced component or technique is at least 50% a visual perception problem to the

Fig. 18   SMSR repair subtask simulation, reinstating the satellite's thermal blanket.

operator, influenced greatly by view angle, illumination, and contrasts in color or in shading.

2) The training or, more specifically, the training cycle has a dramatic effect upon operator performance. It was found that the first cycle should be regarded as a familiarization with the system and with the task. For a novice operator, this familiarization cycle should be repeated at least twice. The real training for performance evaluation can only start after completion of a familiarization cycle. The familiarization can be considered complete when the trainee understands the system I/O details, the system response to commands, and the task sequence details. During the second cycle of training, performance measurements should be made so that the operator understands the content of measures against which the performance will be evaluated. Note, that it is necessary to separate each cycle

and repetitions within cycles by at least one day. Once a personal skill has been formed by the operator as a consequence of the second training cycle, the real performance evaluation experiments can start. A useful criterion for determining the sufficient level of training can be, for instance, that of computing the ratio of standard deviation of completion time to mean completion time (that is, computing the coefficient of variation). If the coefficient of variation of the last five trials of a subtask performance is less than 20%, then a sufficient level of training can be declared. In the subtask-scenario experiments quoted here, the real training, on the average, required one week per subject. More details on application task experiments can be found in Refs. 21 and 22.

The practical purpose of training is, in essence, to help the operator develop a mental model of the system and of the task. During task execution, the operator acts through the aid of this mental model. It is, therefore, critical that the operator understands very well the response characteristics of the sensing and computer-aided ATOP system which has a variety of selectable control modes, adjustable control gains, and scale factors. (More on the human operator aspects of automation in teleoperation can be found in the chapter of this book by T. Sheridan, "Human Enhancement and Limitation in Teleoperation.")

The procedure of operator training and the expected behavior of a skilled operator following an activity protocol offers the possibility of providing the operator with performance feedback messages on the operator interface graphics, derived from a stored model of the task execution. A key element for such an advanced performance feedback tool to the operator is a program that can follow the evolution of a teleoperated task by segmenting the sensory data stream into appropriate phases.

A task segmentation program of this type has been implemented by means of a neural network architecture[23] and it is able to identify the segments of a peg-in-hole task. With this architecture, the temporal sequence of sensory data generated by the wrist sensor on the manipulators are turned into spatial patterns and a window of sensor observations which is related to the current task phase. A partially recurrent network algorithm was employed in the computation. Partially recurrent networks represent well the temporal evolution of a task, as they include in the input layer a set of nodes connected to the output units to create a context memory. These units represent the task phase already executed—the previous state. Several experiments of the peg-in-hole task have been carried out and the results have been encouraging, with a percentage of correct segmentations approximately equal to 65%. More on these experiments can be found in Refs. 23 and 24.

The performance experiments with communication time delay, conducted on a large laboratory scale in early 1993, utilized a simulated life-size satellite servicing task which was set up at the Goddard Space Flight Center (GSFC) and controlled 4000 km away from the JPL ATOP control station. Three fixed TV camera settings were used at the GSFC worksite, and TV images were sent to the JPL control station over the NASA-Select Satellite TV channel at video rate. Command and control data from JPL to GSFC and status and sensor data from GSFC to JPL were sent through the Internet computer communication network. The roundtrip command/information time delay varied between 4–8 s between the GSFC worksite and the JPL control station, dependent on the data communication protocol.

The task involved the exchange of a satellite module. This required inserting a 45-cm-long power screwdriver, attached to the robot arm, through a 45-cm-long hole to reach the module's latching mechanism at the module's backplane, unlatching the module from the satellite, connecting the module rigidly to the robot arm, and removing the module from the satellite. The placement of a new module back to the satellite's frame followed the reverse sequence of actions.

Four camera views were calibrated for this experiment, entering 15–20 correspondence points in total from three to four arm poses for each view. The calibration and object localization errors at the critical tool insertion task amounted to about 0.2 cm each, well within the allowed insertion error tolerance. This 0.2 cm error is referenced to the zoom-in view (fovy= 8 deg) from the overhead (front view) camera which was about 1 m away from the tool tip. For this zoom-in view, the average error on the image plane was typically 1.2–1.6% (3.2–3.4% maximum error); a 1.4% average error is equivalent to a 0.2-cm displacement error on the plane 1 m in front of the camera.

The idea behind placing the high-fidelity graphics image over a real TV image is that the operator can interact with it visually in real time on a monitor within one perceptive frame when generating motion commands manually or by a computer algorithm. Thus, this method compensates in real time for the operator's visual absence from reality due to the time-delayed image. Typically, the geometric dimensions of a monitor and the geometric dimensions of the real work scene shown on the monitor are quite different. For instance, an 8-in.-long trajectory on a monitor can correspond to a 24-in.-long trajectory in the actual work space, that is, three times longer than the apparent trajectory on the monitor screen. Therefore, to preserve fidelity between a previewed graphics arm image and actual arm motions, all previewed actions on the monitor were scaled down very closely to the expected real motion rate of the arm hardware. The manually generated trajectories were also previewed before sending the motion commands to the GSFC control system to verify that all motion data were properly recorded. Preview displays contribute to operational safety. To eliminate the problem associated with the varying time delay in data transfer, the robot motion trajectory command is not executed at the GSFC control system until all the data blocks for the trajectory are received. An element of fidelity between the graphics arm image and actual arm motion was given by the requirement that the motion of the graphics image of the arm on the monitor screen be controlled by the same software that controls the motion of the actual arm hardware. This is required to implement the GSFC control software in the JPL graphics computer.

A few seconds after the motion commands were transmitted to GSFC from JPL, the JPL operator could view the motion of the real arm on the same screen where the graphics arm image motion was previewed. If everything went well, the image of the real arm followed the same trajectory on the screen that the previewed graphics arm image motion previously described, and the real arm image motion on the screen stopped at the same position where the graphics arm image motion stopped earlier. After completion of robot arm motion, the graphics images on the screen were updated with the actual final robot joint angle values. This update eliminates accumulation of motion execution errors from the graphics image of the robot arm, and retains the robot arm graphics image position fidelity on the screen even after the completion of a force sensor referenced compliance control action.

The actual contact events (moving the tool within the hole and moving the module out from or into the satellite's frame) were automatically controlled by an appropriate compliance control algorithm referenced to data from a force-moment sensor at the end of the robot arm, implemented by the cooperating GSFC team and invoked by the JPL operator when needed.

The experiments have been performed successfully, showing the practical utility of high-fidelity predictive-preview display techniques, combined with sensor-referenced automatic compliance control, for a demanding telerobotic servicing task under communication time delay. More on these experiments and on the related error analysis can be found in Refs. 16 and 17. Figures 19a and 19b illustrate a few typical overlay views.

A few notes are in order here regarding the use of calibrated graphics overlays for time-delayed remote control.

1) There is a wealth of computation activities that the operator has to exercise. This requires very careful design considerations for an easy and user friendly operator interface to this computation activity.

2) The selection of the matching graphics and TV image points by the operator has an impact on the calibration results. First, the operator has to select significant points. This requires some rule-based knowledge about what is a significant point in a given view. Second, the operator has to use good visual acuity to click the selected significant points by the mouse.

Lessons Learned

The following general conclusions emerged so far from the development and experimental evaluation of the JPL ATOP:

1) The sensing, computer- and graphics-aided advanced teleoperation system truly provides new and improved technical features. To transform these features into new and improved task performance capabilities, the operators of the system have to be transformed from naive to *skilled operators*. This transformation is primarily an undertaking of education and training.

2) To carry out an actual task requires that the operator follow a clear procedure or protocol which has to be worked out off line, tested, modified, and finalized. It is this procedure or protocol following habit that finally will help develop the experience and skill of an operator.

3) The final skill of an operator can be tested and graded by the ability to successfully improve to recover from unexpected errors and complete a task.

4) The variety of I/O activities in the ATOP control station requires workload distribution between two operators. The primary operator controls the sensing and computer-aided robot arm system, while the secondary operator controls the TV camera and monitor system and assures protocol following. Thus, the *coordinated training of two cooperating operators* is essential to successful use of the ATOP system for performing realistic tasks. It is not yet known what a single operator could do and how. To configure and integrate the current ATOP control station for successful use by a *single operator* is challenging research and development work.

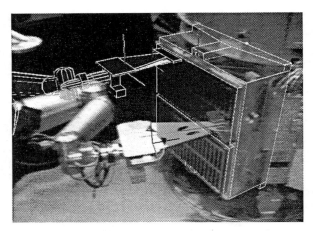

Fig. 19a   Predictive/preview display of end point motion.

Fig. 19b   Status of predicted end point after motion execution, from a different camera view for the same motion shown in Fig. 19a.

5) The problem of ATOP system development is not only to find ways to improve technical components and to create new subsystems. The final challenge is to integrate the improved or new technical features with the natural capabilities of the operator through appropriate human-machine interface devices and techniques to produce an improved overall system performance capability in which *the operator is part of the system* in some new way.

Perspectives of Anthropomorhic Telemanipulation

The robot arms employed in the JPL ATOP project are of the industrial type with industrial type parallel claw end effectors. This sets definite limits for the arms' task performance capabilities as dexterity in manipulation resides in the mechanical and sensing capabilities of the hands (or end effectors). The use of industrial-type arms and end effectors in space would essentially require researchers to design space manipulation tasks matching the capabilities of industrial-type arms and end effectors. Contrary to that, existing space manipulation tasks (except the handling of large space cargos) are designed for astronauts, including the tools used by astronauts. There are well over two hundred tools that are available today and certified for use by Extra Vehicular Activity (EVA) astronauts in space. Motivated by these facts, an effort parallel to the ATOP project was initiated at JPL to develop and evaluate human-equivalent or human-rated dexterous telemanipulation capabilities for potential applications in space because all manipulation related tools used by EVA astronauts are human rated.

The general technical approach adopted in this anthropomorphic telemanipulation project is the development and evaluation of an anthropomorphic (humanlike) exoskeleton master-slave force-reflecting arm-hand system. This technical approach implies the following: 1) the master arm is a replica of the slave arm, and each arm has seven DOF, 2) the master arm is solidly attached to the operator's arm, 3) forces acting on the slave arm can backdrive the master arm so that the operator can feel the forces/moments acting on the slave arm, 4) the slave hand is a humanlike fingered hand with a replica glove-like master controller attached solidly to the operator's hand, and 5) forces acting on the slave fingers can backdrive the fingers of the master glove so that the operator can feel the forces acting on the slave fingers. The ability of the operator to feel forces acting at the remote slave site provides kinesthetic telepresence to the operator. This enables the operator to perform sensitive, force-compliant manipulation tasks with or without tools.

The actual design and laboratory prototype development included the following specific technical features: 1) the system is fully electrically driven; 2) the hand and glove have four fingers (little finger is omitted) and each finger has four DOF; 3) the base of the slave fingers follow the curvature of the human fingers' base on the hand; 4) the slave hand and wrist form a mechanically integrated closed subsystem, that is, the hand cannot be used without its wrist; 5) the lower slave arm which connects to the wrist houses the full electromechanical drive system for the hand and wrist (altogether 19 DOF), including control electronics and microprocessors; and 6) the slave drive system electromechanically emulates the dual function of human muscles, namely, position and force control. This implies a novel and unique implementation of active compliance. All of the specific technical features taken together make this exoskeleton unique among the few similar systems. No other previous or ongoing developments have all the aforementioned technical features in one integrated system, and some of the specific technical features are not represented in any other similar systems at all. More on this system can be found in Ref. 25.

Currently, the JPL anthropomorphic telemanipulation system is assembled and tested in a terminus control configuration. In this configuration the master glove is integrated with our previously developed nonanthropomorphic six-DOF forcereflecting hand controller (FRHC), and the mechanical hand and forearm are mounted to an industrial robot (PUMA 560), replacing its standard forearm. The notion of terminus control mode refers to the fact that only the terminus devices (glove and robot hand) are of anthropomorphic nature, and the master and slave arms are nonanthropomorphic. The system is controlled by a high-performance distributed computer controller. Control electronics and computing architecture were custom developed for this telemanipulation system. The system is currently being evaluated, focusing on tool handling and astronaut equivalent task executions. The evaluation revealed the system's potential for tool handling but it also became evident that EVA tool handling operations in space require a dexterous, human-equivalent *dual* arm robot.

The anthropmorphic telemanipulation system in terminus control configuration is shown in Fig. 20. The master arm/glove and the slave arm/hand have 22 active joints each. The manipulator lower arm has five additional drives to control finger and wrist compliance. This active electromechanical compliance (AEC) system provides the muscle equivalent dual function of position as well as stiffness control. A cable links the forearm to an overhead gravity balance suspension system, relieving the PUMA upper arm of this additional weight. The forearm has two sections, one rectangular and one cylindrical. The cylindrical section, extending beyond the elbow joint, contains the wrist actuation system. The rectangular cross section houses the finger drive actuators, all sensors, and the local control and computational electronics. The wrist has three DOF with angular displacements similar to the human wrist. The wrist is linked to an AEC system that controls the wrist's stiffness. It is noted again that the slave hand, wrist, and forearm form a mechanically closed system, that is, the hand cannot be used without its wrist. A glove-type device is worn by the operator. Its force sensors enable hybrid position/force control and compliance control of the

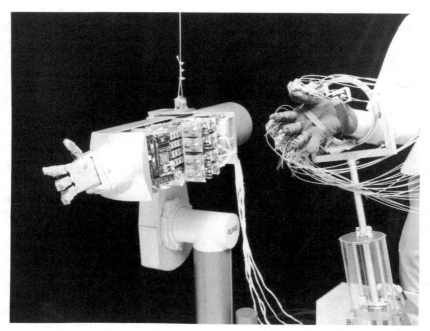

Fig. 20   Master glove controller and anthropomorphic hand.

mechanical hand. Four fingers are instrumented, each having four DOF. Position feedback from the mechanical hand provides position control for each of the 16 glove joints. The glove's feedback actuators are remotely located and linked to the glove through flex cables. One-to-one kinematic mapping exists between the master glove and slave hand joints, thus reducing the computational efforts and control complexity of the terminus subsystem. The exceptions to the direct mapping are the two thumb base joints which need kinematic transformations.

The present control electronics architecture for the master glove and the anthropomorphic hand/wrist is shown in Fig. 21. It is comprised of PC-based computational engines, using TMS320C40 (C40) processors and two custom designed intelligent controllers. The interface to the FRHC and the PUMA upper arm joints is provided by two separate universal motor controllers (UMC). The UMC has been described previously in Ref. 9. The C40s communicate with each other via a single duplex communication channel. The intelligent controllers are based on the Texas Instrument TMS320C30 (C30). The C30 was selected for this task because of its low cost and high performance (33 MFLOPS). The C30 is very similar to the C40 except that it lacks the six high-speed communication ports. The two intelligent controllers are placed near the systems's sensors, one is near the master glove, the other is near the anthropomorphic hand and wrist. The function of the controllers is to provide a sampling of analog signals, filter these signals, provide digital calibration of strain gages, model the actuator voltage-velocity curve, generate pulse-width modulated (PWM) signals, and communicate with the PC-based computational engine. All programs are written in the C language, using the SPOX Real-Time Operating System (Spectrum Microsystems) to facilitate the development of multipurpose programs. More on this system can be found in Ref. 26.

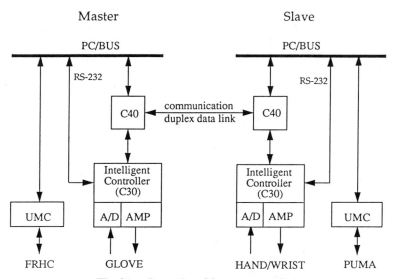

Fig. 21   Control architecture overview.

Testing and evaluation of this system is still in progress. It became clear during the tests, however, that tool handling EVA tasks require a dual-arm fingered hand system with at least four fingers and with 7-DOF compliant arms.

Remarks

It was not possible to include all aspects of the JPL ATOP work in this Chapter. Interested readers can find useful contributions to advanced teleoperator technology in the area of kinesthetic force feedback in microgravity, hand controllers, end effectors, stereo vision and parallel computation in telerobotics listed in Refs. 27–34.

Acknowledgments

This work was carried out at the Jet Propulsion Laboratory, California Institute of Technology under a contract by NASA, and supported by the NASA Code RC and later by the NASA Code CD telerobotics program. Several individuals contributed to various aspects of this work during the past eight or so years. In alphabetical order they are Ed Barlow, Eva Bokor, Thurston Brooks, Kevin Corker, Dan Diner, Hari Das, Ron Dotson, Amir Fijany, Paolo Fiorini, Blake Hannaford, Bruno Jau, Won-Soo Kim, Paul Lee, Sukhan Lee, Anthony Lewis, Doug McAffee, Tim Ohm, Eric Paljug, Paul Schenker, Zoltan Szakaly, Steve Venema, Lori Wood, and Haya Zak.

References

[1]Bejczy, A. K., and Salisbury, J. K., "Controlling Remote Manipulators Through Kinesthetic Coupling," *Computers in Mechanical Engineering,* Vol. 1, No. 1, 1983; also, "Kinesthetic Coupling Between Operator and Remote Manipulator," *Proceedings of the ASME International Computer Technology Conference Vol. 1* (San Francisco, CA), ASME, New York, 1980, pp. 197–211.

[2]Bejczy, A. K., "Teleoperation: The Language of the Human Hand," *Proceedings of the IEEE Workshop on Robot and Human Communication,* Tokyo, Japan, Sept. 1–3, 1992.

[3]Knight, L. W., and Retter, D., "Datahand™: Design, Potential Performance, and Improvements in the Computer Keyboard and Mouse," National Human Factors Society Conference, Denver, CO, Nov. 1989.

[4]Knight, L. W., "Single Operator Environment: Experimental Hand-Grip Controller for ATOP," Jet Propulsion Lab., DOE Summer Faculty Fellow (SFF) Reps., Pasadena, CA, July 31, 1992, and July 30, 1993.

[5]Bejczy, A. K., Szakaly, Z., and Kim, W. S., "A Laboratory Breadboard System for Dual Arm Teleoperation," *Third Annual Workshop on Space Operations, Automation and Robotics,* NASA Conf. Pub. 3059, Johnson Space Center, Houston, TX, July 1989, pp. 649–660.

[6]Bejczy, A. K., and Szakaly, Z., "Performance Capabilities of a JPL Dual-Arm Advanced Teleoperation System," Space Operations, Applications, and Research Symposium (SOAR '90) Proceedings, Albuquerque, NM, June 26, 1990, pp. 168–179.

[7]Bejczy, A. K., Szakaly, Z., and Ohm, T., "Impact of End Effector Technology on

Telemanipulation Performance," *Third Annual Workshop on Space Operations, Automation and Robotics,* NASA Conf. Pub. 3059, Johnson Space Center, Houston, TX, 1989, pp. 429–440.

[8]Bejczy, A. K., and Szakaly, Z., "An 8-D.O.F. Dual Arm System for Advanced Teleoperation Performance Experiments," Space Operations, Applications, and Research Symposium (SOAR '91), NASA No 3127, Houston, TX, 1991, pp. 282–293; also Lee, S., and Bejczy, A.K., "Redundant Arm Kinematic Control Based on Parametrization," *Proceedings of the IEEE International Conference on Robotics and Automation,* Sacramento, CA, April 1991, pp. 458–465.

[9]Bejczy, A. K., and Szakaly, Z. F., "Universal Computer Control System (UCCS) for Space Telerobots," *Proceedings of the IEEE Int'l Conference on Robotics and Automation,* Raleigh, NC, March 30–April 3, 1987; also Szakaly, Z., and Fleischer, G., "JPL Advanced Teleoperation Control System Critical Path Performance," Jet Propulsion Lab, JPL Memo 3470-90-332, Pasadena, CA, 1990.

[10]Bejczy, A. K., and Szakaly, Z., "A Harmonic Motion Generator for Telerobotic Applications," *Proceedings of IEEE Int'l Conference on Robotics and Automation* (Sacramento, CA), 1991, pp. 2032–2039.

[11]Bejczy, A. K., Kim, W. S., and Venema, S., "The Phantom Robot: Predictive Display for Teleoperation with Time Delay," *Proceedings of IEEE International Conference on Robotics and Automation,* Cincinnati, OH, May 1990, pp. 546–550.

[12]Bejczy, A. K., and Kim, W. S., "Predictive Displays and Shared Compliance Control for Time Delayed Telemanipulation," *Proceedings of IEEE International Workshop on Intelligent Robots and Systems* (IROS '90), Tsuchiura, Japan, July 1990, pp. 407–412.

[13]Kim, W. S., and Bejczy, A. K., "Graphics Displays for Operator aid in Telemanipulation," *Proceedings of IEEE International Conference on Systems, Man and Cybernetics,* Charlottesville, VA, Oct. 1991, pp. 1059–1067.

[14]Kim, W. S., "Graphical Operator Interface for Space Telerobotics," *Proceedings of IEEE International Conference on Robotics and Automation,* Atlanta, GA, May 1993, p. 95.

[15]Fiorini, P., Bejczy, A. K., and Schenker, P., "Integrated Interface for Advanced Teleoperation," *IEEE Control Systems Magazine,* Vol. 13, No. 5, Oct. 1993, pp. 15–20.

[16]Kim, W. S., and Bejczy, A. K., "Demonstration of a High-Fidelity Predictive/Preview Display Technique for Telerobotics Servicing in Space," *IEEE Transactions on Robotics and Automation, Oct. 1993, Special Issue on Space Telerobotics,* pp. 698–702; also Kim, W. S., Schenker, P. S., Bejczy, A. K., Leake, S., and Ollendorf, S., "An Advanced Operator Interface Design with Preview/Predictive Displays for Ground-Controlled Space Telerobotic Servicing," SPIE Conference No. 2057: Telemanipulator Technology and Space Telerobotics, Boston, MA, Sept. 1993.

[17]Kim, W. S., "Virtual Reality Calibration for Telerobotic Servicing," *Proceedings of the IEEE International Conference on Robotics and Automation,* San Diego, CA, May 1994, pp. 2769–2775.

[18]Lee, P., Hannaford, B., and Wood, L., "Telerobotic Configuration Editor," *Proceedings of the IEEE International Conference on Systems, Man and Cybernetics,* Los Angeles, CA, 1990, pp. 121–126.

[19]Hannaford, B., Wood, L., Guggisberg, B., McAffee, D., and Zak, H., "Performance Evaluation of a Six-Axis Generalized Force-Reflecting Teleoperator," Jet Propulsion Lab., JPL Pub. 89-18, Pasadena, CA, June 15, 1989.

[20]Hannaford, B., Wood, L., Guggisberg, B., McAffee, D., and Zak, H., "Performance Evaluation of a Six-Axis Force-Reflecting Teleoperation," *IEEE Transaction on Systems, Man and Cybernetics,* Vol. 21, No. 3, 1991.

[21]Das, H., Zak, H., Kim, W. S., Bejczy, A. K., and Schenker, P. S., "Performance Experiments with Alternative Advanced Teleoperator Control Modes for a Simulated Solar Max Satellite Repair," Proceedings of Space Operations, Automation and Robotics Symposium (SOAR '91) NASA No. 3127, Johnson Space Center, Houston, TX, July 9–11, 1991, pp. 294–301.

[22]Das, H., Zak, H., Kim, W. S., Bejczy, A. K., and Schenker, P. S., "Performance with Alternative Control Modes in Teleoperation," *PRESENCE: Teleoperators and Virtual Environments,* MIT Press Pub., Massachusetts Inst. of Technology, Cambridge, MA, Vol. 1, No. 2, 1993, pp. 219–228.

[23]Fiorini, P., Giancaspro, A., Losito, S., and Pasquariello, G., "Neural Networks for Segmentation of Teleoperation Tasks," *PRESENCE: Teleoperators and Virtual Environments,* MIT Press Pub., Massachusetts Inst. of Technology, Cambridge, MA, Vol. 2, No. 1, 1993, pp. 66–81.

[24]Hannaford, B., and Lee, P., "Hidden Markov Model Analysis of Force-Torque Information in Telemanipulation," *International Journal of Robotics Research,* Vol. 10, No. 5, 1991.

[25]Jau, B. M., "Man-Equivalent Teleopresence Through Four Fingered Human-Like Hand System," *Proceedings of the IEEE International Conference on Robotics and Automation* (Nice, France), IEEE Press, Los Alamitos, CA, 1992, pp. 843–848.

[26]Jau, B. M., Lewis, M. A., and Bejczy, A. K., "Anthropomorphic Telemanipulation System in Terminus Control Mode," *Proceedings of ROMANSY '94* (Gdansk, Poland), 1994, Springer-Verlag, Berlin, Germany (to appear).

[27]Corker, K., "Investigation of Neuromotor Control and Sensory Sampling in Bilateral Teleoperation," Ph.D. Thesis, Univ. of California, Los Angeles, CA, 1984.

[28]Bejczy, A. K., and Corker, K., "Manual Control Communication in Space Teleoperation," *Proceedings of Fifth CISM-IFTOMM Symposium on Theory and Practice of Robots and Manipulators,* R. Kogan Page Ltd., London, England, UK, and Hermes Pub., Paris, France, 1985, pp. 223–232.

[29]Brooks, T. L., and Bejczy, A. K., "Hand Controllers for Teleoperation: A State-of-the-Art Survey and Evaluation," Jet Propulsion Lab., JPL Pub. 85-11, Pasadena, CA, 1985.

[30]Mishkin, A. H., and Jau, B. M., "Space-Based Multifunctional End Effector Systems," Jet Propulsion Lab., JPL Pub. 88-16, Pasadena, CA, 1988.

[31]Diner, D.B., and von Sydow, M., "Stereo Depth Distortions in Teleoperation," Jet Propulsion Lab., JPL Pub. 87-1, Rev. 1, Pasadena, CA, 1988.

[32]Diner, D. B., and Fender, D. H., "Human Engineering in Stereoscopic Viewing Devices," Jet Propulsion Lab, JPL D-8186, Pasadena, CA, 1991.

[33]Diner, D. B., "A New Definition of Orthostereopsis for 3-D Television," *Proceedings of the IEEE International Conference on Systems, Man, and Cybernetics,* Charlottesville, VA, Oct. 1991, pp. 382–389.

[34]Fijany, A., and Bejczy, A. (eds.), *Parallel Computation Systems for Robotics; Algorithms and Architectures,* World Scientific Pub., 1992. Chap. 4.

Chapter 6

Supervised Autonomy for Space Telerobotics

Paul G. Backes*
*Jet Propulsion Laboratory, California Institute of Technology,
Pasadena, California 91109*

I. Introduction

TELEROBOTICS continues to provide a means for extending human presence in space as an integral part of the control architecture of unmanned spacecraft. Scientific data return is maximized within a fixed mission budget by properly separating the spacecraft and instrument control into ground- and spacecraft-based segments. Because the spacecraft must be reliable and fail-safe, its design is optimized to provide only the required onboard capabilities for communication and control of the spacecraft and instruments to achieve mission success. The Earth-based segment of the system generates command sequences which are telemetered to the remote spacecraft. Command generation on the ground, based on updated data from the spacecraft, provides the needed system flexibility to achieve mission success. Extensive human and computational resources are available on the ground compared to those on the spacecraft. The spacecraft is able to execute command sequences which have been telemetered from Earth as well as react to anomalous situations. Ground-based control of remote unmanned spacecraft is an application of supervised autonomous control. The same basic control architecture can be used for other types of space robots such as unmanned rovers[1] and the focus of this chapter, space manipulators. Further reference to telerobotics will imply control of a manipulator unless otherwise stated.

Manipulators in space can provide valuable extensions of human capability for task execution. A near-term application of space telerobotics is on Space Station Freedom. The required astronaut extravehicular activity (EVA) time required to perform all of the projected maintenance activities on the Space Station is projected to be above the expected available EVA time. Additionally, there will be an extended man-tended phase when the Space Station is manned only part of

Copyright © 1994 by the American Institute of Aeronautics and Astronautics, Inc. The U.S. Government has a royalty-free license to exercise all rights under the copyright claimed herein for Governmental purposes. All other rights are reserved by the copyright owner.

*Technical Group Leader, Robotic Systems and Advanced Computer Technology Section.

the time. Telerobotics can be used to perform some maintenance, assembly, and inspection tasks to relieve the astronauts of some duties. Also, through the utilization of ground control, many tasks can be performed using telerobotics when there are no astronauts on board the Space Station.

Telerobotics methods can be separated into three types: manual control, supervisory control, and fully automatic control.[2-4] The distinction between these methods is briefly described here. Sheridan's text[4] provides a good historical perspective and literary review on these approaches to telerobotics. The term teleoperation may be used generically to describe all telerobotics methods but is used here in its more common connotation of manual control. In manual control, all robot motion is specified by continuous input from a human, with no additional motion caused by a computer. In supervisory control, robot motion may be caused by either human inputs or computer generated inputs. In fully automatic control, all robot motion is caused by computer generated inputs.

There are two primary subsets of supervisory control: supervised autonomy and shared control. The distinction between them is the nature of the inputs from the operator. In shared control, operator commands are sent during execution of a motion and are merged with the closed-loop motion generated automatically.[5-7] Therefore, in shared control, all inputs from the operator are not known a priori to execution of a motion as inputs during execution are also used. In supervised autonomy, autonomous commands are generated through human interaction, but sent for autonomous execution.[8] A command can be sent immediately or iteratively saved, simulated, and modified before it is sent for execution on the real robot. Also, individual commands can be complete descriptions of the motion[9] or module commands specifying only modifications to the control or monitoring of a specific module of the remote system.[10]

The ability to iteratively save, simulate, and modify commands before sending them for execution is a critical feature of supervised autonomy which distinguishes it from other forms of supervisory control. For safety purposes it is valuable to be able to simulate task execution before sending command sequences to the manipulator for task execution. With supervised autonomy, a command or command sequence can be saved, simulated, and modified before it is sent for execution on the real robotic system. Safety is achieved by verifying the commands before sending them for execution on the real robots and through real-time monitoring. Commands can be modified and simulated until they are acceptable for execution by the robot. Individual commands can be concatenated into a command sequence which can then be iteratively simulated and modified and inserted into yet a larger sequence. Sequence generation for autonomous spacecraft is a formal process because dangerous or incorrect commands could result in serious damage, loss of unique scientific opportunities (e.g., during a planetary flyby), or loss of the entire spacecraft. In shared control, operator commands are sent immediately to be merged with the autonomous execution. Safety in shared control is achieved either by relying on the operator to input safe motions, or by having real-time autonomous monitoring and modification of the motion specified by the operator.

The term telerobotics implies a separation between the operator and the robot. This separation gives rise to a partitioning of a telerobotics system into two components: the local-site where the operator resides, and the remote site where the robots reside.[2] The separation between the operator and robot causes a communication time delay. Time delay is another factor which makes supervised autonomy an important approach for space telerobotics. For Space Station Freedom applications, the projected round-trip communication time delay between an Earth-based local site and Space Station-based remote site is expected to be approximately 7 s (mostly in data processing and relay)[11] while round-trip time delay to a planetary spacecraft or vehicle is measured in tens of minutes to hours.[1] Teleoperation and shared control become increasingly difficult with time delays due to the continuous real-time inputs.[2, 4] Supervised autonomy overcomes the time delay problem by providing closed-loop control at the remote site based on autonomous commands generated at the local site.

An important feature of supervised autonomy is bounded behavior execution.[8] Bounded behavior execution allows task execution to diverge from the nominally planned motion within a specified bound. For safe operation it is desirable to know a priori exactly what the manipulators will do during execution of a task. But, because the remote execution environment cannot be known a priori exactly, real-time execution will rely on both the a priori planned trajectory and perturbations due to remote sensed data. The safety of execution within a specified bound can be tested a priori at the local site. The remote system can then autonomously monitor execution in real time to ensure that the state of the motion is within the specified bounds, e.g., deviation from the a priori trajectory or contact force thresholds. If execution moves out of the specified bounds, then an automatic reflex action is invoked and further local-site commands are awaited.

The local and remote components of a supervised autonomy system can be divided into subcomponents. The local site includes sequence generation, sequence analysis, monitoring, and telemetry. The remote site includes telemetry, command parsing, sequence control, real-time control, monitoring, and reflex. Different implementation approaches may be desired for different application domains. Space applications impose important constraints on the telerobotic system architecture with the flight component on the system usually providing the most stringent constraints. Flight systems require robust flight qualified software running in limited computing environments (limited compared to the ground system). Modification of flight software during flight, although possible, requires an extensive and costly qualification process. This leads to the solution taken for unmanned robotic spacecraft control: command sequences are composed of command types and associated data which specify the desired spacecraft and instrument control behavior.[12, 13] The flight software is fixed but provides general command types which can be parameterized to generate a wide range of specific control behaviors. This supervised autonomy architecture for programming a remote spacecraft can be used for control of remote space manipulators.

Sequence generation is the process of generating a command sequence which can be telemetered to a remote autonomous robot control system. An operator interface is provided which the operator uses to specify the desired commands. Computer aids can help in the specification of tasks, commands, and parameterization. Computer aids include modeling, visualization, and task planning. Computer modeling provides a model of the manipulated object or task execution environment. Visualization provides a graphical representation of the scene. An accurate representation of the task execution scene is important to ensure that the a priori simulation is a valid representation of the execution that will occur on the real robot. One way to verify that the modeled environment matches the real environment is to overlay a graphical representation of the model on real images of the remote scene. The model can be modified to match the remote scene using data returned from the remote environment.[14, 15] Autonomous task planning aids can provide suggested task commands and parameterization to the operator. Automatic task planning associated with fully automatic control is not yet feasible. To aid the operator, an interface could provide suggested parameterization for specified individual command types[16] and let the operator specify the specific parameterization and sequence of commands. State transition graphs which provide the sequence of commands and automatic parameter selection for a selected task[17] could be provided, but on failure, reliance on interactive task description is again necessary. Automatic generation of low-level command primitives based on analysis of an operator's actions while interacting with a graphical interface has been suggested,[15] but the low-level commands do not include context information such as object mass properties and termination conditions which would be provided in a supervised autonomy system.

Sequence analysis determines the expected result of executing a generated sequence and the level of confidence in achieving that result. Graphical simulation is provided so that the powerful analysis capabilities of the human operator can be used. Automatic analysis by the computer may provide tests for dynamic loading, collisions, valid range of motion, and valid commanded velocities and accelerations. It is valuable to have as accurate a model of the remote system as possible to increase confidence in the sequence analysis results. Local-site monitoring analyzes the reports from the remote site to test for valid execution and system health. The local site will usually have much greater diagnostics capabilities than the remote site due to the greater human and computational resources available. Local-site telemetry provides the communication of command sequences to the remote site and receiving of status and data from the remote site.

Remote-site telemetry receives command sequences from the local site and sends status and data to the local site. Command sequences are parsed at the remote site into individual commands for execution. Sequence control provides the transitioning of commands. This includes transitioning to the next command in a command sequence on expected termination and transition to reflex action on a reflex monitor event. Real-time control provides the closed-loop servo control of the remote-site mechanisms. The control is based on commands generated at the local site. Remote-site monitoring is the analysis of remote-site execution to provide information on whether to transition the state of execution. Reflex is the ability to respond to monitored conditions. The most common reflex is to transition to the next command in a command sequence based on a monitor event which indicates that the previous command has been successfully completed. An equally important reflex is the ability to transition to a safety reflex action based on an unexpected monitor event.

The features and capabilities of a system providing supervised autonomy of a remote manipulator system are described in the following sections through the description of an operational laboratory system.

II.   Example Local-Remote System

The local-remote system architecture of the example supervised autonomy system is shown in Fig. 1. The same system also provides shared control and force reflecting teleoperation,[8, 16, 18] but those capabilities are not within the scope of this chapter. The primary operator interface workstation (Sun 3/60) provides interactive task description, sequence generation, and status display. A graphics workstation (Silicon Graphics IRIS) provides a stereo graphics overlay on stereo video as well as an interactive designation of objects or destinations. The remote-site simulator simulates remote-site execution with execution status displayed on the primary workstation and motion displayed on the graphics workstation. The remote site provides two control systems, one for independent, coordinated, or cooperative control of two task execution manipulators, and one for control of a third manipulator for positioning a suite of four cameras. The executive provides communication with the local site and initiates task commands as specified by the local site. Task primitives provide joint and task space control and monitoring of single or dual cooperating manipulators.

A.   Remote-Site System

The remote-site system design of a space telerobotic system has more constraints imposed on it than the local-site system. The resulting capabilities of the remote site will drive the design of the local-site. A primary remote-site constraint is flight qualification of the software. This creates the need for fixed

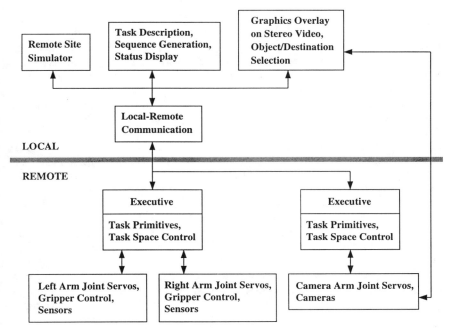

Fig. 1   Laboratory local-remote system block diagram.

flight software which has been validated before flight (or modified, validated, and uplinked infrequently). Fixed flight software precludes custom optimized programs for each mission task. Rather, the fixed flight software must provide sufficient functionality to complete both expected and unexpected mission tasks. The solution provided in this laboratory system is a family of parameterizable task primitives, each with general functionality for a class of manipulation tasks. Separate commands provide other needed capability such as databased update, status request, and execution interrupt. Task execution primitives are self-contained programs which provide manipulator control capability with behavior as specified by an input parameter set. The control capability is provided via the parameterization but the implementation details are hidden. A natural interface between the local- and remote-site systems is then the parameter lists for the various task primitives.

1. Executive

The executive provides functionality similar to that of a spacecraft command and data subsystem.[12, 19] It receives commands from the local site, parses the commands to determine command types, and initiates execution of the commands by executing task primitives or other commands with the parameterization given in the command data sets. The executive also returns system state information to the local site.

2. Commands

The interface commands that can be sent to the remote site by the local site include database commands, a status command, and execution commands. The database command has parameters specifying the arm and database datatype followed by the specific database parameters. Remote-site database parameters that can be modified by a database command include a transform specifying the position of the robot's base, force-torque, joint, and singularity safety limit thresholds, grasped object mass properties, and rate to report status to the local site. The database parameters are used by the task primitives along with the task primitive parameters when executing a task. Database parameters are separate from task primitive parameters because they are expected to change less frequently than task primitive parameters or provide system information. The status command requests that the remote site return the state of the arm specified in the command. This is useful when no task is being executed and no status information is otherwise being returned. There are six autonomous execution commands (with corresponding remote-site task execution primitives): Cartesian guarded motion, joint guarded motion, move to touch, single-arm generalized compliant motion, dual-arm generalized compliant motion, and grasp.

Guarded motion task primitives provide free space motion with monitoring for collisions. The Cartesian guarded motion primitive[16, 19] performs a single-arm Cartesian interpolated motion and stops on the destination position or sensed force or torque thresholds. Inputs to the primitive include which robot, time- or velocity-based motion, time or velocity to perform motion in, force and torque thresholds, coordinate frame to sense collision forces in, coordinate frame in which to perform Cartesian interpolation, and position destination via points to go through. The joint guarded motion primitive is the same as the Cartesian guarded motion primitive except that joint interpolation is used instead of Cartesian interpolation.

The move to touch primitive[16, 19] performs a single-arm move with Cartesian interpolated motion until the specified destination is reached or until a force or torque threshold is exceeded. If a force or torque threshold is reached, then the arm moves back toward its initial position until the force and torque magnitudes are below reverse thresholds, above safety thresholds, or the arm has returned to its initial position. Inputs include which arm, the Cartesian destination, the frame in which to perform Cartesian interpolation, the forward, reverse, and safety force and torque thresholds, and the forward and reverse velocities.

Generalized compliant motion provides general autonomous task execution capability for motion in contact with the environment (as well as shared control). There are both single- and dual-arm generalized compliant motion task primitives. The single-arm generalized compliant motion primitive[6, 9] performs general single-arm Cartesian space compliant motion tasks. Inputs to the primitive include which arm, destination coordinate frame, frame in which to perform Cartesian interpolated motion, frame in which to perform Cartesian force control, frame in which to generate Dither position commands, time- or velocity-based motion, time or velocity for positional motion, dither magnitude and period, position-force selection vector to select position and force DOF in the control frame, comply selection vector to select which position DOF also have compliance, force control gains, gains for stiffness control, force-torque and position-orientation thresholds, a parameter selecting which termination conditions to test for, and termination conditions including maximum errors in position, orientation, force, and torque and their rates of change. The motion in any DOF can have inputs from the position trajectory generator, dither, sensor based control (force, stiffness, etc.), or any simultaneous combination of sources. This approach is similar to impedance control[20] where resultant motion in any DOF is based on a combination of both the position setpoint and interaction forces, as contrasted with hybrid position-force control[21] where position and force DOF are separate. Nominal motion generates motion based on a Cartesian trajectory generator. Sensor-based motion (force, stiffness, and dither here) perturbs the nominal motion. Force control modifies the Cartesian setpoint to control contact forces. Stiffness control modifies the Cartesian setpoint to pull the motion back toward the nominal motion, thus counteracting the effects of force control. The dual-arm generalized compliant motion primitive provides all of the capabilities of the single-arm generalized compliant motion capability and provides cooperative dual-arm control.[7]

There are two segments of motion in the generalized compliant motion primitive, the nominal motion segment and the ending motion segment. When the primitive starts, it executes the nominal motion segment with the specified Cartesian interpolated motion and all other sensors. Motion stops if a monitor event is triggered or Cartesian interpolated motion is completed. If the nominal motion segment is completed (Cartesian interpolated motion is completed), then the end motion segment begins. Exactly the same control occurs except there is no Cartesian interpolated motion; only the sensor-based motion is active. But, whereas during the nominal motion segment the termination conditions were not being tested, they are tested during the ending motion and the motion can stop on a monitor event, time, or a termination condition (monitor events and time can also be considered termination conditions).

The grasp primitive opens or closes the gripper while performing generalized compliant motion. Inputs include the generalized compliant motion inputs as well as finger speed and finger separation. The gripper opens or closes with specified speed and to the specified finger separation and stops upon reaching either a finger threshold force or the specified separation. Generalized compliant motion provides compliance to relieve internal forces during the grasp, or apply specific forces during the grasp, e.g., comply in all DOFs except apply a force against the surface you are grasping.

Dual-arm equivalent guarded motion and move to touch primitives were not implemented because the chosen evolutionary path for task execution was to fold all capability into a common modular execution environment. This is discussed later in Sec. V.

B. Local-Site System

The remote-site system design specifies the interface that the local site can use to control the remote manipulators. The local-site system is then designed to provide the remote-site capability to the operator. There may be multiple local sites for one remote site such as for Space Station Freedom where local sites on Earth and the Space Station could control a common manipulation system.[8] The local sites could share responsibilities, e.g., the Earth-based local site could generate command sequences and telemeter them to the Space Station-based local site where an astronaut would initiate them. An Earth-based local site is valuable because it has much greater human and computational resources than a space-based local site. This section assumes a single local site is communicating with a single remote site.

Task description and sequence generation are provided by the user macro interface (UMI).[16] The UMI abstracts away the details of the local-remote interface and provides the operator with more natural menus for specifying tasks and parameterization. The resulting inputs from the operator are converted to equivalent commands and parameterization to be communicated to the remote site. An iconic graphical interface implementation of the UMI is presently under development, as discussed later in Sec. V. The operator has the option of running either the real remote-site robots or simulating the motion at the local site by sending commands to the remote-site simulator and observing the results on the graphics display. The graphical results are displayed both with wire frame graphics on the primary operator workstation or on the video-graphics workstation. Stereo graphics overlay on stereo video on the video-graphics workstation is shown in Fig. 2. Simulation mode is selected as a parameter in the UMI environment menu. The remote-site simulator runs identical control software as in the remote-site system and sends joint angle data to the UMI graphics displays.

The operator describes a task by utilizing the interactive UMI menu system. The UMI is a hierarchical menu system which guides the operator from general motion types at the top of the hierarchy to the specific at the bottom of the hierarchy. The UMI eventually specifies to the local-site executive task primitives and their parameterization to perform the specific tasks desired by the operator. The operator does not need to know the specific task primitives which will be used. Instead the operator specifies a generic motion type, e.g., guarded motion, move to

Fig. 2   Video-graphics workstation stereo graphics overlay on stereo video.

contact, compliant motion, or grasp, and then the interface provides a new macro menu with interaction germane to the specific motion type. For example, if the operator specifies compliant motion, the compliant motion menu will appear with hinge, slide, screw, insert, level, push, and general macro options. The operator will then select one of these and a new menu will appear with inputs pertaining only to that type of motion. The insert menu would allow the operator to specify the insertion direction, force, thresholds, etc. The operator then has only a small number of decisions at any point in the hierarchy but can specify a specific task.

The operator may save a specific parameterization of a task as a task command. For example, the operator can specify a door opening motion but internally it is actually the generalized compliant motion primitive with specific parameterization. The operator builds the door opening command and can save it by giving it a name, e.g., door_open. The operator can then use the door_open command later either with the simulator or on the real arms. The operator can also string several commands together creating a sequence and save the sequence with a name for later use. Relative and absolute motion commands are useful. A relative motion command has the same relative motion from its starting point as when taught even though the absolute starting point changes. This is useful for tasks being executed relative to their environment such as bolt turning and grasping. Absolute motion commands have an absolute position destination independent of where they start their motion. These are useful for moving to an absolute position before beginning a relative motion command.

The sequence editor allows the operator to edit a sequence which has been previously built. The operator selects the sequence menu and then the sequence editor menu and finally the sequence to edit. The parameterization for each command can then be modified by the operator.

The system environment is another branch of the tree where the operator specifies parameters which change rarely or are used globally by the different task macros, e.g., which arm (right, left, dual, camera), time delay to invoke, and safety force-torque thresholds.

The status of the remote-site system is updated on the local-site operator control station monitor whenever a system status (updated at a settable rate, usually approximately 1 Hz) or command result is returned from the remote site. This includes the graphics simulation which is automatically updated with the remote-site arm positions and the joint and force values which are updated on the graphical displays at the bottom of the interface, as well as the gripper positions.

The stereo graphics overlay on stereo video, shown in Fig. 2, is valuable for simulation and interactive destination selection.[8] Tasks are simulated by selecting simulation mode in the environment menu of the UMI. Then any command or command sequence that is sent is routed to the remote-site simulator for execution. The remote-site simulator sends commanded joint angles to the simulator for a graphics update. Interactive destination selection is achieved through the use of a spaceball input device which specifies motion of a graphical cursor in the graphics display. The operator selects with a UMI menu to return to the UMI the cursor position or the position of the object closest to the cursor. Selecting to return the cursor position allows the operator to interactively generate a path. The operator moves the cursor, presses a button on the spaceball, the motion is simulated, and then the operator can send the same motion for execution on the real robots. This provides a safe interactive means for generating a trajectory through a constrained workspace. The operator can also use the cursor to select an object in the environment. The operator moves the cursor near the object and presses a button on the spaceball. The coordinate frame, which is the internal positional representation of the object, closest to the cursor, is highlighted (turned red) to indicate the selected object and position. The operator can then graphically simulate a motion to an approach point above the coordinate frame. This command can then be sent to the real robot for execution. Any of these motions can also be saved for later execution or modification.

Graphics overlay on video is used to confirm that the geometrical model of the environment is valid. Operator coached machine vision (OCMV)[14] is used to match the geometric model with the video images returned from the remote environment. With OCMV, the operator provides a rough estimate of an object's position by manipulating a graphics environment which is overlayed on the returned video. The object's modeled position is then adjusted for more accuracy with machine vision. Once the model is validated, the video image can be turned off and the graphical scene can be used by itself. This graphical representation of the scene can now be used without the limitations of the fixed viewing characteristics of the real video. The eyepoint can be changed to a more useful point for a given task. This can be very important when the surfaces of interest are occluded from view. For example, during insertion of an orbital replacement unit (ORU) on the Space Station, there will likely be no way to see the mating points on the bottom of the ORU with the given camera locations. With a validated graphical representation of the scene, the eyepoint can be moved so that the motion behind the ORU can be seen graphically. This approach to manipulation will change the requirements for cameras in space from what a human would need to accomplish a task to what is necessary to generate a valid geometric model of the environment.

III. Sequence Control

Sequence control is control of the transition between commands in a sequence or transition to a reflex action. Two important parts of sequence control are run-time binding[8] before execution of a command and testing of termination conditions at the end of a command. Run-time binding binds parameters to a task command just before its execution is initiated. Parameters bound at run time may not be known at the time the command is built. Some examples of run-time binding include binding the current safety parameters, speed factor, and the reporting period to the parameter list. Each command in a task sequence completes due to satisfaction of a termination condition (including safety conditions). If the termination condition is one of the acceptable termination conditions specified in the command, then the next command of the sequence is issued. If not, then a safety reflex action is initiated, and a new command sequence must be sent. Sequences can have interspersed commands for all three remote-site robots and dual-arm cooperative control.

Transition between commands in a sequence can occur at either the local or remote site but transitioning to a reflex action should be done autonomously at the remote site. For sequence control at the local site, a delay at least as long as the round-trip time delay will occur between execution of each command in a sequence because the local site must then receive the remote-site status indicating that the command has been terminated successfully before sending the next command in the sequence. Earth-based sequence control may be feasible for Space Station or lunar applications where round-trip time delay will likely be less than 10 s, but for exploration of the rest of the solar system, the large time delays will make sequence control at the remote site more feasible. Remote-site sequence control has been used for unmanned spacecraft missions. In the laboratory system described here, sequence control occurs at the local site with reflex to safety commands done at the remote site.

IV. Example Task

Many tasks have been executed using the local-remote system described earlier.[7,22] A door opening task sequence is described here to illustrate supervised autonomy for a specific task. The task is to approach, grasp, open, and close a door. The manipulator opening the door is shown in Fig. 3. The task sequence, door_task, is made up of the commands shown in Table 1.

The manipulator is initially at a staging location. The gmove_knob_approach is an absolute motion command to move the manipulator gripper to an approach location above the door. The acceptable termination condition is stopped on position reached. All of the commands except the guarded motion (gmove_) commands are relative motion commands meaning they could be used for any location of the door. The guarded motion commands move the arm to the specific location of the door and then away after the task is completed. The mtouch_z100 mm command

Fig. 3   Door opening and closing task.

moves the gripper into contact with the doorknob using a Cartesian motion of 100 mm along the TOOL frame (attached to the gripper) Z axis until contact occurs. Actual contact and command termination occurs after approximately 50 mm. The acceptable termination condition is stopped on reverse force which indicates that the motion stopped when moving back toward the starting point and the contact force magnitude fell below the specified reverse force threshold. The grasp_close command closes the gripper while applying force control to null any internal forces due to misalignment of the gripper over the knob.

The door_open_30_deg and door_close_32_deg require the most sophisticated control and monitoring of all the tasks in the sequence so they will be discussed

Table 1   Commands comprising the task sequence door_task

Command	Associated task execution primitive
gmove_knob_approach	Cartesian guarded move
mtouch_z100 mm	Move to touch
grasp_close	Compliant grasp
door_open_30_deg	Generalized compliant motion
door_close_32_deg	Generalized compliant motion
grasp_open	Compliant grasp
gmove_knob_approach	Cartesian guarded move

in more detail. In the door_open_30_deg command, the generalized compliant motion primitive is specified along with specific input parameters. The nominal motion frame (NOM) for Cartesian interpolated motion is specified to have its Z axis along the hinge axis. The force control frame (FORCE) is specified to be on the knob. A Cartesian interpolated motion of 30 deg about the NOM Z axis is specified. Force control with zero setpoints is specified for all six DOF of the FORCE frame. Force control is necessary to correct for the difference between the physical motion constrained by the door and the a priori planned nominal motion based on the model. Stiffness control was specified in all DOF of the FORCE frame. Stiffness control is necessary to offset the force control based motion when forces are reduced. The acceptable termination condition was specified to be a low orientation error of 0.1 deg. This is triggered when the actual orientation of the nominal motion frame is within the specified bound during the ending motion. The results are shown in Fig. 4. The door was successfully opened 30 deg.

The value of the stiffness control is shown by executing the same task but without the use of stiffness control. The results are shown in Fig. 5. In this case the door opened a maximum of only 21.6 deg. The maximum rotation occurred when the trajectory generator finished. After that, the ending motion time segment began and the door slowly began closing due to its gravity weight. The ending condition of 0.1 deg from the 30-deg goal was never satisfied so the command stopped on the time timeout condition. The reason that the door did not open all of the way is that force control in the FORCE frame caused motion to resist the

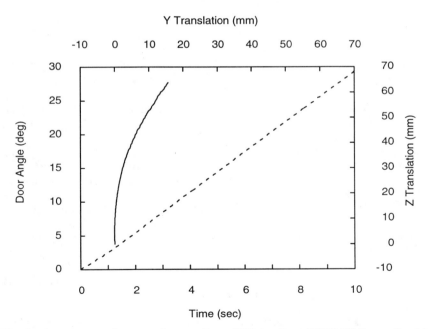

Fig. 4 Autonomous door opening results: solid is motion of FORCE frame (knob); dashed is rotation of NOM frame (hinge axis).

Y Translation (mm)

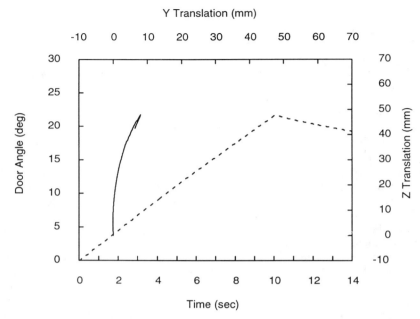

Fig. 5 **Autonomous door opening results (no stiffness control): solid is motion of FORCE frame (knob); dashed is rotation of NOM frame (hinge axis).**

nominal trajectory generator motion and there were no virtual springs to offset this motion.

The door closing command, door_close_32_deg, used the same parameters as were used for the door opening task, including stiffness control, except that the nominal motion was negative 32 deg and different termination conditions were used. A 32-deg motion was used to be sure to have the door close completely. If the a priori model was known to be accurate, then a 30-deg specified rotation might have been sufficient, but there will often be a difference between the modeled and physical environments and commands should be robust to this disparity. The specified acceptable termination condition was a combination of low orientation error (3 deg), low translational error velocity (1 mm/s), and low orientational error velocity (0.1 deg/s). The translational and orientational error velocities are the rates of change of the translation and orientation errors relative to the nominal motion trajectory. The results are shown in Fig. 6. The figure shows that the door was successfully closed 30 deg. The motion is nearly linear until the door makes contact and is closed at 30 deg. Then the rotation stops which triggers the termination condition.

The rest of the sequence then continues (assuming the termination conditions are always valid) with grasp_open to open the gripper, and gmove_knob_approach to move away from the knob.

The door opening and closing commands were shown in more detail to demonstrate the need for the various remote-site control features. The door closing task is actually the most difficult because it shows the potential difficulty of determining

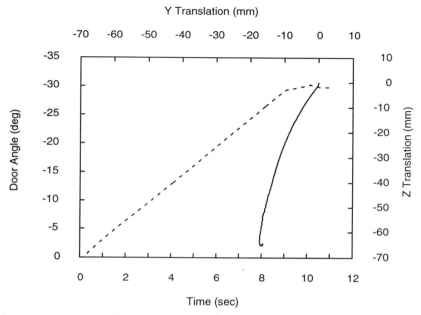

Fig. 6 **Autonomous door closing: solid is motion of FORCE frame (knob); dashed is rotation of NOM frame (hinge axis).**

when a command is completed successfully at a remote task execution site. We could not know exactly how many degrees to close the door. Closed was not a precise distance and orientation state, but rather a distance and orientation state in combination with a motion state constrained by the physical environment. Visual feedback will not be sufficient for many space tasks to determine successful completion. Processed sensory data, such as rate of change of position and orientation errors as used in this example, will be needed to determine successful task completion.

V.   Evolution of the Local-Remote System

The system just described provides the basic capabilities of a supervised autonomy system for space telerobotics. This type of system could be used for near-term applications such as telerobotics for maintenance of Space Station Freedom. As telerobotics evolves, more advanced technologies for implementing the capabilities will be provided. A new system is now under development for Space Station Freedom ground-remote telerobotics. Some of the technologies and implementation approaches for the new system are described here.

A.   Remote-Site Evolution

Three approaches have been considered for increasing the capability of the remote-site system: more general task primitives, an interpretive programming language, and data driven execution. The previous evolutionary path of the system described earlier was to continuously reduce the number of task primitives by increasing the capability of the generalized compliant motion primitive. It would be possible to enhance the generalized compliant motion primitive to include the capability of the two guarded motion primitives and the grasp primitive, thereby further reducing the number of task primitives. This could have two benefits: simplification of the system design and increased capability. The system would be simpler due to the reduced number of primitives, lines of code, and interface specifications. The system could be more capable because the permutations of the capabilities of the four primitives would be available in one primitive. Including the move to touch primitive capability could be more difficult because multiple trajectory segments are included.

An alternative approach to increasing the remote-site system capability would be to provide a programming language interface to the local site. This is a common approach for terrestrial robotics.[23, 24] This could potentially provide greater capability to the remote site, but flight qualification is likely to be a problem with this approach. The preceding task primitives approach has the advantage that a primitive's logic, control algorithms, and software can be flight qualified once and then used repeatedly with the only change being the specific parameterization (which of course will have to be verified for safety). Verification of a new program written in a robot control language, before sending it for execution at the remote site, could be difficult and costly. The flexibility gained may not be worth the risk in the approach, particularly for the first generation of space manipulators.

The approach selected to increase the remote-site system capability is data driven task execution with fixed software modules.[10] The system design is similar to the remote-site architecture for sequence control on the Galileo unmanned spacecraft.[12, 13] The resultant system capability is the permutations of capability of the various control modules (or subsystems in spacecraft control terminology). The implementation for space manipulator control utilizes impedance based generalized compliant motion with extensions for redundant arm control.[25, 26] Any number of control sources, e.g., trajectory generator, force error, and visual servoing, can be used simultaneously. The resultant behavior of the combination of control sources is the task execution. The remote site receives command sequences and stores them in a task command queue or reflex command queues. The task command queue holds the task command sequence to execute the task. The reflex command queues have task sequences which are executed on monitor events which are not specified as acceptable task termination conditions.

B. Local-Site Evolution

The local-site capability is being upgraded by providing automation aids for model calibration and task planning and an iconic interface for task planning and sequencing integrated with increased graphical interaction for sequence generation. As machine vision capabilities increase, the amount of operator interaction required in operator coached machine vision will decrease. Laser range scanner data will be merged with vision data to further enchance model building and update. Task simulation will be enhanced by adding dynamics and sensor-based control effects. For example, arm flexibility and environment contact dynamics will be modeled.

The sequence generation process with the UMI described earlier involved creating individual commands from macro templates and then concatenating these commands together into command sequences. An important part of creating individual commands is the generation of context-dependent parameterization. This step can be automated if the contextual data is provided to the interface before the command is created. This early step of inputting contextual task parameterization is called the knowledge insertion step and the UMI is being enhanced to allow insertion of these data. Thus, context-specific commands and parameterization are input into the operator interface during the knowledge insertion step and the operator specifies the sequence of context-specific commands for a task during sequence generation. The new system has a knowledgebase (a more general version of the previous database) which holds contextual task execution commands and parameterization. The task-specific data may originate from various places such as manufacturer specifications or empirical experimentation. Much of this contextual data would not be known by the operator generating the tasks, but it is automatically provided by the knowledgebase. During sequence generation, the task state must be known so that the contextual information can be automatically generated. Specialists in robotics and task specification can be used for the knowledge insertion phase while a person with different skills, e.g., an astronaut, could perform the sequence generation phase.

Integration of the iconic interface with the graphical environment will allow the operator to better utilize the graphical environment for sequence generation. The previous system provided graphical simulation, object selection, and destination selection. The new system will allow the operator to also select commands within the graphical environment from options automatically generated dependent on the task context. If the context is clear enough, commands could be selected automatically. The teleprogramming methodology was recently proposed where sequences of low-level primitive commands are automatically generated by interpreting the actions of an operator interacting with a graphical environment.[15] The approach may be valuable for unmodeled environments, but it lacks contextual information and the selection of specialized commands which are available with a supervised autonomy system. In the supervised autonomy approach, higher level commands are generated, either via the graphical environment or iconic menus. The operator specifies context by selecting objects, destinations, and task types. Commands specific to the task context are generated. These commands are automatically decomposed into specific remote-site task execution commands. For the new remote-site task execution system described in Sec. V.A, the sequence of module commands and parameterization is automatically generated. If multiple command options are possible, such as grasp compliantly or grasp while pushing against the object, the interface queries the operator for the specific command and any undefined parameterization. This interaction can occur within the graphical environment or on the iconic interface.

The utilization of a priori knowledge insertion and interactive sequence building is now described with a bolt turning example. After inserting an ORU into the Space Station truss, a bolt may be need to be tightened to secure the ORU. A specific torque associated with the ORU will be required. The specific required torque for that ORU bolt will have been inserted into the interface during the earlier knowledge insertion phase. During sequence generation, the operator selects the high-level command to turn the bolt and the operator interface automatically generates a command with detailed parameterization including the required torque read from the knowledgebase based on the known context of the specific ORU.

VI. Conclusions

Space applications provide both an important application domain for telerobotics and many important constraints on the implementation approach. Successful application of supervised autonomy methods to the remote control of unmanned spacecraft demonstrates the viability of the approach to that class of space robots. Supervised autonomy is also a viable near-term approach for the remote control of space manipulators. Safety is achieved by generating commands with parameters specific to the task and through a priori simulation. Effects of time delay are eliminated by providing closed-loop control, monitoring, and reflex at the remote site. Remote-site computation requirements are limited by providing only task execution and reflex at the remote site while all task planning is done at the local, e.g., Earth, site.

Acknowledgments

The work described in this paper was performed at the Jet Propulsion Laboratory, California Institute of Technology, under contract with NASA.

References

[1]Wilcox, B. H., "Robotic Vehicles for Planetary Exploration," *Journal of Applied Intelligence,* Vol. 2, 1992, pp. 181–193.

[2]Ferrell, W. R., and Sheridan, T. B., "Supervisory Control of Remote Manipulation," *IEEE Spectrum,* Oct. 1967, pp. 81–88.

[3]Brooks, T. L., III, and Sheridan, T. B., "Superman: A System for Supervisory Manipulation and the Study of Human/Computer Interactions," Massachusetts Inst. of Technology, TR MITSG 79-20, Cambridge, MA, July 1979.

[4]Sheridan, T., *Telerobotics, Automation, and Human Supervisory Control,* M.I.T. Press, Massachusetts Inst. of Technology, Cambridge, MA, 1992.

[5]Hayati, S., and Venkataraman, S. T., "Design and Implementation of a Robot Control System with Traded and Shared Control Capability," *Proceedings of the IEEE International Conference on Robotics and Automation,* 1989, pp. 1310–1315.

[6]Backes, P. G., "Generalized Compliant Motion with Sensor Fusion," *Proceedings of the 1991 ICAR: Fifth International Conference on Advanced Robotics, Robots in Unstructured Environments* (Pisa, Italy), June 19–22, 1991, pp. 1281–1286.

[7]Backes, P. G., "Dual-Arm Supervisory and Shared Control Space Servicing Task Experiments," *Proceedings of the AIAA Space Programs and Technologies Conference,* AIAA Paper 92-1677, Huntsville, AL, March 24–27 1992.

[8]Backes, P. G., "Ground-Remote Control for Space Station Telerobotics with Time Delay," *Proceedings of the AAS Guidance and Control Conference,* AAS Paper No. 92-052, Keystone, CO, Feb. 8–12 1992.

[9]Backes, P. G., "Generalized Compliant Motion Task Description and Execution Within a Complete Telerobotic System," *Proceedings IEEE International Conference on Systems Engineering,* Aug. 9–11 1990.

[10]Backes, P. G., Long, M. K., and Steele, R. D., "Designing Minimal Space Telerobotics Systems for Maximum Performance," *Proceedings of the AIAA Aerospace Design Conference,* Irvine, CA, Feb. 3–6 1992.

[11]Aster, R., de Pitahaya, J. M., and Deshpande, G., "Analysis of End-to-End Information System Latency for Space Station Freedom," Jet Propulsion Lab. Internal Doc. D-8650, Pasadena, CA, May 1991.

[12]Anon., Galileo Project, "Galileo Program Description Document—Command and Data Subsystem, Phase 9.1," Jet Propulsion Lab. TR 625-355-06000, D-535 Rev. G, Pasadena, CA, May 1989.

[13]Anon., Galileo Project, "Galileo Flight Operations Plan—Galileo Command Dictionary," Jet Propulsion Lab. TR PD 625-505, D-234, Pasadena, CA, Sept. 1989.

[14]Bon, B., Wilcox, B., Litwin, T., and Gennery, D., "Operator-Coached Machine Vision for Space Telerobotics," *SPIE Symposium on Advances in Intelligent Systems, Conference on Cooperative Intelligent Robots in Space,* Boston, MA, Nov. 1990; also, SPIE Doc. No. A92-51707 22-54, pp. 337–342.

[15]Funda, J., Lindsay, T. S., and Paul, R. P., "Teleprogramming: Toward Delay-Invariant Remote Manipulation," *Presence,* Vol. 1, No. 1, 1992, pp. 29–44.

[16]Backes, P. G., and Tso, K. S., "UMI: An Interactive Supervisory and Shared Control System for Telerobotics," *Proceedings of the IEEE International Conference on Robotics and Automation,* (Cincinnati, Ohio), 1990, pp. 1096–1101.

[17]Balaram, J., and Stone, H., "Intelligent Robotic Systems For Space Exploration," *Automated Assembly In The JPL Telerobot Testbed,* Kluwer Academic Publ., 1992.

[18]Hayati, S., Lee, T., Tso, K., and Backes, P. G., "A Testbed for a Unified Teleoperated-Autonomous Dual-Arm Robotic System," *Proceedings of the IEEE International Conference on Robotics and Automation,* (Cincinnatti, OH), pp. 1090–1095. 1990.

[19]Tso, K. S., Backes, P. G., Lee, T. S., and Hayati, S., "A Multi-Arm Tele/Autonomous Executive System," *Proceedings of the International Symposium on Robotics and Manufacturing,* Burnaby, B.C., Canada, July 18–20, 1990, pp. 845–852.

[20]Hogan, N., "Impedance Control: An Approach to Manipulation: Part I—Theory," *ASME Journal of Dynamic Systems, Measurement, and Control,* Vol. 107, March 1985, pp. 1–7.

[21]Raibert, M. H., and Craig, J. J., "Hybrid Position/Force Control of Manipulators," *ASME Journal of Dynamic Systems, Measurement, and Control,* Vol. 102, June 1981, pp. 126–133.

[22]Backes, P. G., "Supervised Autonomous Control, Shared Control, and Teleoperation for Space Servicing," *Proceedings Space Operations, Applications, and Research Symposium,* Houston, TX, Aug. 4–6 1992.

[23]Anon., "User's Guide to Val II: Programming Manual," UNIMATION, Inc. TR 398AGI, Danbury, CT.

[24]Nackman, L. R., Lavin, R. H., Taylor, R. H., Dietrich, W. C., and Grossman, D. D., "AML/X: A Programming Language for Design and Manufacturing," *Proceedings of the Joint Computer Conference,* Nov. 1986, pp. 145–159.

[25]Backes, P. G., and Long, M. K., "Merging Concurrent Behaviors on a Redundant Manipulator," *Proceedings of the IEEE Conference on Robotics and Automation,* (Atlanta, GA), May 1993, pp. 638–645.

[26]Long, M. K., and Backes, P. G., "Impedance Based Shared Control of a Redundant Robot," *IASTED International Conference on Control and Robotics,* (Vancouver, Canada) 1992, pp. 106–109.

Part 3. Planning and Perception

Chapter 7

Automatic Planning in Robotic Applications

Luiz S. Homem-de-Mello[*]

*Jet Propulsion Laboratory, California Institute of Technology,
Pasadena, California 91109*

I. Introduction

T HE research on automatic planning aims at creating computer systems that are able to construct networks of tasks that accomplish given goals. The input to a planner consists of descriptions of the current and the desired states of a given system. The output is one network of tasks, referred to as a plan, which, if properly executed, achieves the desired goal. For some problems, the network solution is a linear sequence of tasks to be executed one at a time.

One example application for a planning system is the generation of assembly sequences for large tetrahedral truss structures such as that shown in Fig. 1, which is a schematic representation of the NASA Langley Research Center robotic assembly facility. A robot arm is mounted on a base which is mounted on a carriage that can translate along one direction. The base where the robot is mounted can translate along a direction orthogonal to the carriage translations. These two motions allow the positioning of the robot arm in a Cartesian coordinate system. The truss structure is mounted on a base that can rotate. If necessary, before a strut is assembled, or disassembled, the structure is turned and both the base of the robot arm and the carriage are translated. The facility can be used to assemble, disassemble, and repair truss structures. The assembly process starts with all struts stored in pallets that are stacked on the same base where the robot arm is mounted. It ends when all struts are properly joined to form the whole structure. Ideally, after struts have been added, they are not removed until the end of the assembly process. (Therefore, in the assembly process the number of tasks is equal to the number of struts in the structure.) There must be a collision-free path for the robot to bring the strut from the pallet where it is stored to its position in the structure; and it must be possible for the robot to lock the joints that attach the strut to its

---

Copyright © 1994 by the American Institute of Aeronautics and Astronautics, Inc. The U.S. Government has a royalty-free license to excercise all rights under the copyright claimed herein for Governmental purposes. All other rights are reserved by the copyright owner.

[*]Member of Technical Staff, Robotic Systems and Advanced Computer Technology Section.

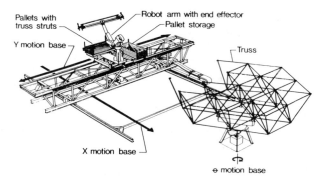

Pallets with
truss struts
Robot arm with end effector
Pallet storage
Y motion base
Truss
X motion base
θ motion base

Fig. 1   Robotic assembly facility at the NASA Langley Research Center.

nodes. In this application, the disassembly process can be viewed as the reverse of the assembly process. The repair process includes both additions and removals of struts. Typically when planning repairs, the length of the solution, that is, the total number of tasks, must be minimized.

The benefits of a computerized planning system for the truss-structure assembly application are clear. Because of the size and complexity of the structures, even trained humans may fail to detect dead-end sequences until a lot of work has been wasted. In the case of a repair in which a faulty strut is to be replaced, an ill-planned disassembly sequence may lead to an irreparable collapse of the whole truss structure. In addition to the difficulty humans have in guaranteeing correctness in the planning process, they often fail to notice which possibilities for the sequences are the most efficient. This difficulty is further aggravated by constant changes in the measure of the efficiency of the assembly sequence. For example, the efficiency may be measured by the total time it takes to complete the assembly in one case, and by the total energy in another case. Moreover, humans typically are slow in generating assembly sequences. There are many situations in which the planning must also be expeditious. Speed in plan generation is particularly important in the case of a repair in which a faulty strut is to be replaced. It is virtually impossible to preplan for every conceivable repair that may be needed.

Another example application for a planning system is the generation of a route for a mobile robot to go from one point to another. This problem is sometimes referred to as gross motion planning; it does not include the initial and final maneuvers during which the vehicle is in contact with obstacles. There are several versions of the mobile robot navigation problem depending on whether the environment is static or dynamic, and on whether or not information on the environment and on the state of the robot is available to the vehicle's control system. The state of a mobile robot includes its position and orientation and their derivatives. Given the initial and goal position of the robot, the planning system must generate a sequence of path segments to be followed. The path segments in this application correspond to the tasks in the assembly process. In the navigation of a mobile robot, however, the tasks can never be executed in parallel.

In addition to being useful for synthesizing engineering tools to be used off line, automatic planning techniques are also useful for developing control systems that will make robots more autonomous. Most architectures for robotic control encompass a hierarchy of functional levels. Typically, the lowest level deals with the servo control of the actuators, and the highest level deals with the planning decisions such as which part to mount next in the assembly application, or where to go next in the navigation application. The highest level also deals with the recovery from unexpected events that cause the execution of a plan to deviate from the intended course of action. In the assembly application, one part or subassembly may fall after it had been successfully mounted. In the navigation application, a planned turn may be missed. There are many situations in which adhering to the original plan, authough feasible, is not the most efficient approach to error recovery. (And there are situations in which adhering to the original plan is not feasible.) If a part P_i falls from an assembly, it may not be necessary to remove all the parts that were mounted after P_i, and then resume the original plan. It is often more efficient to replan the assembly with the initial state being the configuration of the parts just after the unexpected event. For some assemblies, the removal of some parts may not even be feasible. Likewise, if a mobile robot misses a planned turn, it may be more efficient to follow an alternative plan than to go back and resume the planned path; in some applications, backing may not be allowed.

II.   Domain-Independent Planning

The priority of the artificial intelligence research on planning has been to develop efficient general-purpose procedures that can find at least one plan in a wide variety of situations. The artificial intelligence approaches to solving the planning problem have concentrated around domain-independent planners. The central idea is to have one general-purpose inference engine that can be used for any application. In addition to the initial and goal states, the input of a domain-independent planner also includes the set of operators, each one being a description of a task that causes changes in the state of the system.

Most of the research on domain-independent planning evolved from Green's question answering system,[1] which used theorem proving by resolution. The operators were represented as implications, and a situation variable was introduced into the well-formed formulas that described a given state of a robot environment. A sequence of operations that transforms an initial state into a goal state was constructed by Green's question answering system to answer the question of the situation in which a given goal fact was true.

That approach has the advantage of generality, but it has serious disadvantages. One disadvantage, known as the combinatorial explosion, is the rapid growth of the search trees, which hinders the solution of nontrivial problems. Another disadvantage, known as the halting problem, is that the search is not guaranteed to terminate unless there actually exists a sequence of operators that converts the initial state into the goal. If no plan is generated after a long search, one does not know whether the task is indeed unfeasible, or the shortest solution is longer than those explored. These disadvantages are further aggravated by the frame problem, which is the need to express, in frame axioms, all of the relations that are not affected by the application of an operator.

Most of the artificial intelligence (AI) research on planning that followed Green's work focused on solving the combinatorial explosion, halting, and frame problems. (See Ref. 2.) The emphasis was on developing powerful control schemes to guide the search. The priority of that research was to develop general-purpose procedures to be used in many domains.

III.   Domain-Dependent Planning

In contrast, most of the robotics research on planning focused on developing domain-dependent approaches, which take advantage of the characteristics of a particular domain. Examples include Fahlman's BUILD[3] computer program, which generated plans for building specified structures out of simple objects, and Hayes' MACHINIST[4] program, which generates a sequence of operations to produce a part in a computer controlled machining center.

Enabling computers to solve problems like planning typically relies on the computational formalism known as production system.[5] To use that formalism, a representation of the problem in terms of a global database, a set of rules, and a control strategy must be specified. This process is often referred to as representing the problem.

When a driver uses a map to navigate a car in an urban environment, the problem can be represented as a graph search, and it is a straightforward process to convert a conventional map into a graph. The global database is a subgraph (i.e., a partial path); the rules specify how to extend a subgraph by adding links after the last node, and the control strategy guides the selection among all the rules that apply at a given point in the problem-solving process. Typically, in car trips, the navigator (sometimes, but not always, the driver is also the navigator) plans the route for the whole trip by searching the map before starting off. In addition, if a turn is missed, or a road is closed, error recovery can also be accomplished by searching the map.

In mobile robot applications like manufacturing, where the environment is structured, the map used by the robot is analogous to a conventional map used in driving a car. In applications like planetary exploration, where the environment is not structured, a map can be constructed, for example, by using Voronoi diagrams.[6]

The distinction between the construction of the map, and its use is important. Unlike driving a car in an urban environment, for most robotic applications the map is not available and, therefore, must be created. In off-line planning, it is often desirable to generate the map as it is searched to avoid generating portions of the map that are not necessary. In real-time control it is often desirable to use a map that was constructed beforehand to speed up the search computation.

As it will become clear in the next sections, other applications such as robotic assembly also involve activities that are analogous to the construction of a map, and to the search over that map.

IV.   Planning for Robotic Assembly

This section focuses on a basic approach to automatic planning that applies to a broad class of mechanical products. (A number of well thought-out implementations along the line of what is presented here have appeared in the literature. For details, see Ref. 7 and the proceedings of the recent robotics conferences.) The next section focuses on the truss structures, and shows how to take advantage of the characteristics of a special subclass of assemblies to gain the efficiency of the planning process.

The problem of generating the assembly plans for a product can be transformed into the problem of generating the disassembly plans for the same product. Since assembly operations are not necessarily reversible, the equivalence of the two problems requires that each disassembly task be defined as the reverse of a feasible assembly task, regardless of whether or not the reverse task itself is feasible. For the purpose of this discussion, the expression disassembly task is used to refer to the reverse of a feasible assembly task. In the disassembly problem, each task splits one subassembly into two or more smaller subassemblies, maintaining all contacts between the parts in any of the smaller subassemblies. A decomposition approach can be used to solve the disassembly problem. In this approach the problem of disassembling one assembly is decomposed into two or more distinct subproblems, each being to disassemble one subassembly. Every decomposition must correspond to a disassembly task. If solutions to the subproblems that result from the decompositions are found, a solution to the original problem can then be obtained by combining the solutions to the subproblems and the task corresponding to the decomposition. For subassemblies that contain one part only, a trivial solution containing no assembly task always exists.

This approach can be implemented as a decomposition production system[5] and lends itself to an AND/OR graph representation of assembly plans.[8] The nodes in the AND/OR graph correspond to connected stable subassemblies, and the and-arcs correspond to the assembly tasks. For the flashlight shown in Fig. 2, the AND/OR tree representation of one possible assembly plan is shown in Fig. 3, and the network representation of the same plan is shown in Fig. 4. Whereas the network representation is more intuitive and better known, the AND/OR graph representation[8] has the advantage that it can encompass all plans in one graph[8] because it can explicitly represent mutually exclusive plans.

Figure 5 describes the algorithm AOGRAPHSEARCH, which corresponds to the decomposition production system. The input to this algorithm is a representation of the whole assembly, and the output is an AND/OR graph representation of an assembly plan.

The representation of an assembly must include the geometric description of each part, and the configuration in which the parts fit together to form the final product. In addition, most assemblies use attachments that apply forces to constrain the relative motion among parts and stabilize the final assembly. Although attachments such as screws and clips are geometrically represented in the assembly description, it is sometimes impossible to infer from their geometric description

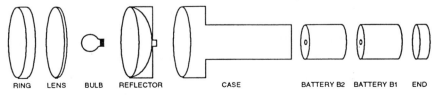

RING   LENS   BULB   REFLECTOR   CASE   BATTERY B2   BATTERY B1   END

Fig. 2   Simple mechanical assembly.

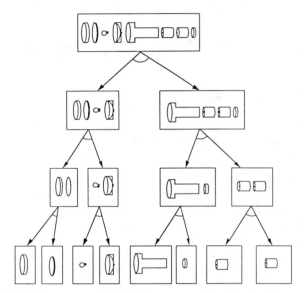

Fig. 3   AND/OR tree representation of one assembly plan for the assembly shown in Fig. 2.

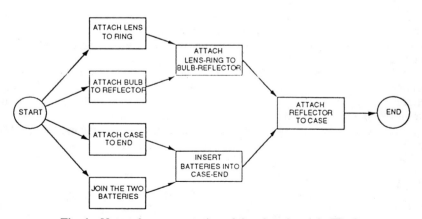

Fig. 4   Network representation of the plan shown in Fig. 3.

```
procedure AOGRAPHSEARCH(s)
begin
  create a partial solution consisting of start node s
  compute the heuristic function for this partial solution
  put this partial solution (with its value) on list OPEN
  while OPEN is not empty do
    begin
      remove the best partial solution from OPEN
      if all the terminal nodes in this partial solution correspond to
        one-part subassemblies, return it
      select a terminal node that does not correspond to a one-part
        subassembly and call it n
      expand n generating all its leaving and-arcs and all its successor
        nodes
      create one partial solution for each and-arc leaving n and compute
        their heuristic functions
      put these partial solutions (with their values) on OPEN
    end
  return FAIL
end
```

Fig. 5   Algorithm AOGRAPHSEARCH.

what the physical role of these parts might be as an attachment for the assembly. This difficulty can be overcome by representing the attachments explicitly.

One possibility for the representation of assemblies is a relational model that includes three types of entities: parts, contacts, and attachments. It also includes a set of relationships between entities. This relational model can be represented by an attributed graph. For example, Fig. 6 shows the relational model graph for the flashlight shown in Fig. 2. The relational model of a subassembly can be readily derived from the model of the whole assembly.

The data structures used to implement the relational model include complete geometric descriptions of the parts, which can be obtained from computer-aided design software. In addition, it is useful to be able to view the assembly at different levels of description: a very detailed level, which includes the geometry of all parts; an intermediate level, which includes the geometry of the contacts and their attachments; and a simplified description, called the graph of connections, which defines the connectivity of the parts. The graph of connections has one node for each part and one arc for each pair of parts that are in contact in the final assembly.

At each iteration of AOGRAPHSEARCH, one leaf node of a partial solution is expanded, that is, all its decompositions are generated. One way this can be implemented is by generating all cut-sets of the graph of connections of the assembly or subassembly associated to the node. Each decomposition is tested and those that do not correspond to feasible assembly tasks are discarded. The test of a decomposition includes the computation of two predicates: TASK-FEASIBILITY, and SUBASSEMBLY-STABILITY. The former returns TRUE when it is geometrically and mechanically feasible to join the subassemblies to produce the larger subassembly. The latter returns TRUE when the parts in the subassemblies that

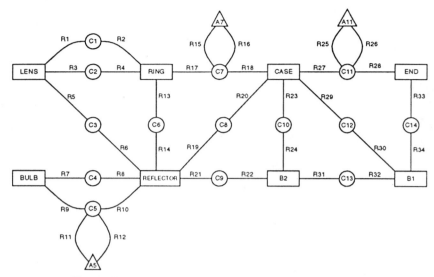

Fig. 6   Relational model for the flashlight shown in Fig. 2.

are joined do not fall due to gravity. The SUBASSEMBLY-STABILITY predicate must return TRUE for all the subassemblies that are joined.

The computation of the feasibility predicates involves complex geometric and physical reasoning. A number of techniques are currently being explored to deal with the computational complexity of assembly planning.

The high-level control of an assembly robot must have access to a precomputed set of alternative plans to avoid the need to perform time-consuming computation. This precomputed set is used by the control system much like a road map is used by the driver of a car. It enables the control system to figure out recovery plans when the execution of the assembly deviates from the intended course of action. Whenever unexpected events occur during the assembly process, the high-level control of the autonomous robot only searches the road map for a recovery plan.

V.   Planning the Assembly of Truss Structures

The approach presented in the previous section applies to a broad class of assemblies. But it is possible to develop more efficient planners for special subclasses of assemblies. For tetrahedral truss structures, such as the one in Fig. 1, a planner can avoid the complex geometric reasoning that is necessary to decide whether or not a candidate assembly task is feasible by using a representation that incorporates the geometry of the parts, and a set of production rules that apply to tetrahedral truss structures only (for more details, see Refs. 9 and 10). Moreover, the size of the search space can be significantly reduced, without loss of completeness, by using a multihierarchical representation for the truss structures.

Because struts will be added to the truss structure one at a time, there is no advantage in using a decomposition approach for planning the assembly of truss structures. But the production system formalism still applies. In this discussion, the generation of assembly plans is assumed to be carried in forward mode, which is easier to understand but not necessarily the most efficient computationally.

A. Representation of Tetrahedral Truss Structures

The representation to be introduced next is based on viewing tetrahedrons and octahedrons as the building blocks of a tetrahedral truss structure. Figure 7 shows a small truss structure, and Fig. 8 shows its building blocks. In addition, the octahedrons can be viewed as the composition of two pentahedral units. There are three distinct ways to subdivide an octahedron unit and these are shown in Fig. 9.

A tetrahedral truss structure can be represented as a graph in which the vertices correspond to volumetric units, and the edges correspond to face contacts between adjacent units. Figure 10 shows a graph representation for a 102-strut truss structure.

The geometry of this graph parallels that of the truss structure. Because of the regularity of the structures, their graph representation constitute a hexagonal grid. In addition, the hexagonal grid can be mapped into a rectangular grid as shown in Fig. 10, where the lines and columns are labeled with their indices.

There are three types of vertices represented, respectively, by hexagons, triangles, and half hexagons. Hexagon vertices correspond to octahedrons, half hexagons to pentahedrons, and triangles to tetrahedrons. Triangles pointing down in the figure correspond to tetrahedrons that have three nodes on the top plane of the truss structure and one node on the bottom plane; and triangles pointing up correspond to tetrahedrons that have three nodes on the bottom plane and one node on the top plane.

The mapping of the graph representation into a rectangular grid gives rise to a data structure for a computer implementation: a two-dimensional array in which each element may contain information about one building block of the truss structure. The indices of the array element indicate the position of the building block.

The edges in the graph shown in Fig. 10 are only implicitly encoded into the two-dimensional array data structure. In addition, the contacts between units that share only one strut, or only one node, are also implicitly encoded into the array. For example, a tetrahedron up at cell (i, j) (i.e., line i, column j) shares one strut with the tetrahedron down at cell $(i + 2, j)$, another strut with the tetrahedron

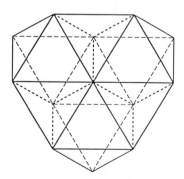

Fig. 7   Small tetrahedral truss structure.

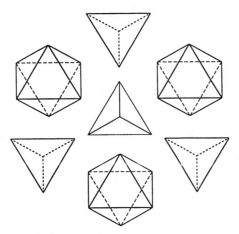

Fig. 8 **Building blocks of the structure shown in Fig. 7.**

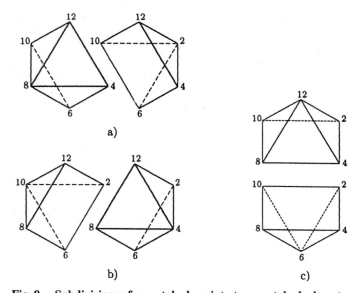

Fig. 9 **Subdivisions of an octahedron into two pentahedral parts.**

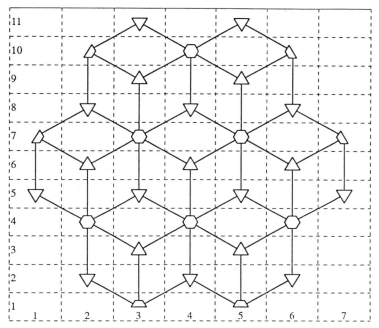

Fig. 10   Graph representation for a tetrahedral truss structure with 102 struts and its mapping into a rectangular grid.

down at cell $(i - 1, j - 1)$, and another strut with the tetrahedron down at cell $(i - 1, j + 1)$. As another example, an octahedron at cell (i, j) shares one node with the tetrahedron at cell $(i + 2, j - 2)$.

This graph and its corresponding data structure should be constrasted with the relational model graph introduced in the previous section. While extensive geometric and mechanical information is explicitly incorporated in a relational graph, such as the one shown in Fig. 6, there is no such information explicitly incorporated to the graph shown in Fig. 10.

B.   Production Rules

The production rules contain the conditions for the execution of an assembly task and the changes that occur in the state of the truss structure when that task is executed. A production rule has two parts: precondition and effect. The precondition specifies the situations in which the rule can be applied. The effect describes the changes that occur in the global database when the rule is applied.

The simplest way to introduce the production rules is by an example. Figure 11 shows one production rule. It corresponds to the assembly task that finishes up one octahedron, starting with one of its pentrahedral halves already assembled. This production rule is associated with the case in which only the base of the missing pentahedron is in place. Therefore, the assembly task will include the assembly of the four struts from the base to the apex node.

- Precondition:

    1. Cell (i, j) currently contains a pentahedron Pk.

    2. Goal is one octahedron in cell (i, j).

    3. All cells (x, y) for which $L(x, y, i, j, k) > 0$, where

    $$L(x, y, i, j, k) \quad = \quad -\cos \alpha \cdot x + \sin \alpha \cdot y + i \cdot \cos \alpha - j \cdot \sin \alpha,$$

    are empty

- Effect:

    1. Adjust the angle of the truss structure and the xy position of the robot arm according to the position of cell (i, j).

    2. Install pentahedron Pk' in cell (i, j), where $k' = \text{rem}(6 + k, 12)$.

Fig. 11   Production rule sample; the variables α and k reflect which of the six pentahedral halves is already in place; the variable k' reflects which pentahedron is missing.

The first two preconditions simply verify that the goal is an octahedron in a cell (i, j) that currently has a pentahedron.

The third precondition verifies that no collision will occur between the truss structure and the carriage where the base of the robot is mounted. It requires that all cells on the same side as the pentahedral to be added with respect to the line $L(x, y, i, j, k) = 0$ be empty. (The variable k is the index of the apex node of the pentahedron that is already in place. See Fig. 9.) This line has an orientation, like an axis, and goes through cell (i, j). It should be noted that this third precondition, which is a very simple to compute, corresponds to feasibility tests of the previous section, which are very complex to compute. The reduction in computation stems from the use of the graph representation for the truss structure, which incorporates the geometric information directly in the topology of the graph instead of as a pointer to a computer-aided design data structure.

Figure 12 shows a state that satisfies the precondition of the production rule shown in Fig. 11, for cell $(7, 5)$ and $k = 10$. The line $L(x, y, 7, 5, 10) = 0$ is also shown.

The effect of this production rule is the installation of pentahedron Pk' in cell (i, j). (The variable k' is the index of the apex node of the pentahedron that is missing. See Fig. 9.) Before this can be executed it may be necessary to turn the truss structure and to move the carriage and the base of the robot arm. This adjustment depends on the position of cell (i, j). The actual installation of struts follows a precompiled sequence of subtasks each of which is the addition of one strut. This precompiled sequence, which includes the motions of the robot arm, is independent of the position of the cell (i, j).

For cell $(7, 5)$ and $k = 10$ in Fig. 12, the effect of the production rule shown in Fig. 11 is the addition of a pentahedron that is part of an unfinished octahedron. It actually corresponds to four assembly tasks, each being the assembly of one strut.

The production rule shown in Fig. 11 is used when the generation of assembly sequences goes in forward fashion. It is straightforward to write a corresponding production rule for generating assembly sequences in backward fashion, or disassembly sequences.

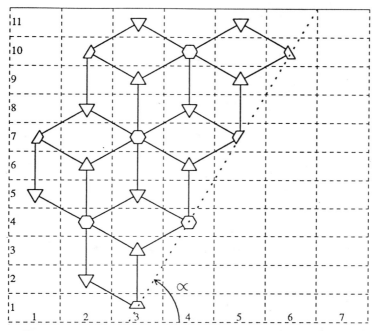

Fig. 12 **Graph representation for a tetrahedral truss structure with 102 struts and its mapping into a rectangular grid.**

For each possible geometric configuration that a cell can take, there is a production rule similar to the one in Fig. 11. Since there are only a few geometric configurations for a cell, the total number of production rules is small.

VI. Conclusion

The applications described in this chapter illustrate the main issues in planning for robotic applications. The computational formalism known as production system[5] can be used to enable computers to solve planning problems. One key to using this technique to developing an efficient planning system is a good representation of the problem. Typically, the actual generation of a solution is based on searching a graph much like a driver searches a map to find a route from one place to another. But for many applications, the map must be created. For off-line planning the generation of the map is usually carried out as it is searched, to generate only the needed portion of the map.

The real-time control system of an autonomos robot sends the appropriate signals to drive the actuators and cause the robot to follow the plan. But the control system must also deal with the recovery from unexpected events that cause the execution of a plan to deviate from the intended course of action. The recovery from errors can be accomplished by searching the graph, like in the generation of the initial plan. Because the computation involved in creating the map is usually very time consuming, the control system must have access to a precomputed graph.

References

[1]Green, C., "Theorem-Proving by Resolution as a Basis for Question-Answering Systems," *Machine Intelligence 4,* American Elsevier, 1969, pp. 183–205.

[2]Allen, J., Hendler, J., and Tate, A., *Readings in Planning,* Morgan Kaufmann, San Mateo, CA, 1990.

[3]Fahlman, S. E., "A Planning System for Robot Construction Tasks," *Artificial Intelligence,* Vol. 5, No. 1, 1971, pp. 189–208.

[4]Hayes, C. C., "Using Goal Interactions to Guide Planning: The Program Model," The Robotics Inst., CMU-RI-TR-87-10, Carnegie Mellon Univ., Pittsburgh, PA, 1987.

[5]Nilsson, N., *Principles of Artificial Intelligence,* Springer-Verlag, New York, 1980.

[6]Takahashi, O., and Schilling, R. J., "Motion Planning in a Plane Using Generalized Voronoi Diagrams," *IEEE Transactions on Robotics and Automation,* Vol. 5, No. 2, 1989, pp. 143–150.

[7]Homem-de-Mello, L. S., and Lee, S., *Computer-aided Mechanical Assembly Planning,* Kluwer, Norwell, MA, 1991.

[8]Homem-de-Mello, L. S., and Sanderson, A. C., "AND/OR Graph Representation of Assembly Plans," *IEEE Transactions on Robotics and Automation,* Vol. 6, No. 2, 1990, pp. 188–199.

[9]Homem-de-Mello, L. S., "Multihierarchical Representation of Tetrahedral Truss Structures for Assembly Sequence Planning," *Proceedings of the 1992 IEEE International Conference on Robotics Automation,* IEEE, Piscataway, NJ, 1992, pp. 2392–2403.

[10]Homem-de-Mello, L. S., "Artificial Intelligence Approach to Planning the Robotic Assembly of Large Tetrahedral Truss Structures," *Telematics and Informatics,* Vol. 9, No. 3/4, 1992, pp. 313–329.

Chapter 8

Techniques for Collision Prevention, Impact Stability, and Force Control by Space Manipulators

Richard Volpe*

*Jet Propulsion Laboratory, California Institute of Technology,
Pasadena, California 91109*

I. Introduction

T HE field of space robotics can be readily divided to planetary and zero-gravity operations. While the harsh environments of other planets will surely require robust robotic hardware, the algorithms controlling this hardware are not likely to be different in kind from Earth-based controllers. Therefore, it is usually the arena of zero-gravity operations where special control algorithms are developed for space robotics.[1-5] Among the pertinent issues that this research addresses are six-degree-of-freedom mobility, zero-friction motion, energy (thrust) minimization, large inertias, flexible structures, etc. Many of these have terrestrial analogs, especially in the field of underwater robotics (as is demonstrated by the utility of buoyancy tanks for astronaut mission training). Particularly, robot manipulation in space has a large overlap with its terrestrial counterpart.

Within this overlap of zero-gravity and terrestrial robotics, there are three main issues: unconstrained motion, stability during the contact transition, and force controlled manipulation of the environment. If a space robot is unattached to its environment, the first two of these research areas map closely to the problems of mobile ground robots and underwater vehicles. In this case, the main problems are path planning, obstacle avoidance, and rendezvous and docking. Force control is not pertinent, because any forces exerted between the robot and its environment will tend to repel each away from the other. Whereas this is especially true in space, it can also be a practical matter for mobile robots on land and in water. Therefore, if force control is to be applied, the robot should attach itself to the environment, making a continuous kinematic chain. (Constant force could also be applied by thrusters, wheels, or propellers, but it is inefficient and will cause a

Copyright © 1994 by the American Institute of Aeronautics and Astronautics, Inc. The U.S. Government has a royalty-free license to exercise all rights under the copyright claimed herein for Governmental purposes. All other rights are reserved by the copyright owner.

*Member of Technical Staff, Robotic Systems and Advanced Computer Technology Section.

net acceleration if the environment is not grounded.) The attachment of the robot to the environment is typically achieved through slow docking followed by base attachment, or grasping by one arm of a multiarm system.

Once the robot and its environment are coupled, the manipulation control issues are essentially the same for space, ground, or water robots. In this case all three problems of robot collision-free motion, contact transition, and force control are important. Collision-free motion is often more difficult with the constraint of the attached base.[6] The attachment does, however, allow the control of interaction forces, in the form of impact control and accurate force trajectory following on the contacted surfaces.

This chapter discusses several control strategies which have successfully addressed these problems of real-time collision-free motion through the environment, reduced velocity approach of surfaces to be contacted or docked with, impact control, and force control. Each of these techniques will be reviewed, and analysis, simulation, and experimentation will be presented. Finally, there will be a complete discussion of the implementational issues for all of these strategies.

Section II discusses the use of artificial forces provided by superquadric artificial potential functions. These functions can assume the shape of a wide variety of objects and provide complete repulsion from objects, or just enough repulsion to slow an approach for safe impact. To address the control of the robot after this impact, Sec. III reviews a method of impact control that provides stable, bounceless contact. A full understanding of this control strategy requires a presentation of the system model, which is provided in Sec. IV. Further, the impact controller analysis and experimentation indicates that two seeming disparate control schemes, second-order impedance control and proportion gain explicit force control, are essentially equivalent. This issue is discussed in detail in Sec. V and is followed by a discussion of other explicit force control schemes in Sec. VI. Finally, Sec. VII provides a detailed discussion of the implementational consideration needed to understand the behavior of all of these control schemes, and to make them work in practice.

II.   Artificial Forces

Moving amid the Space Station is essentially the same problem as moving through planetary terrain, or moving through a factory workcell. All require the avoidance of obstacles to reach the goal location. Research in obstacle avoidance can be broadly divided into two classes of methods: global and local. Global methods rely on the description of the obstacles in the configuration space of a manipulator.[7-9] Local methods rely on the description of the obstacles and the manipulator in the Cartesian workspace.[10-12]

Global methods require that two main problems be addressed. First, the obstacles must be mapped into the configuration space of the manipulator.[8] Second, a path through the configuration space must be found for the point representing the manipulator. Two techniques are used to generate these paths: geometric searches and artificial forces. The geometric search technique relies on an exhaustive search of the unoccupied configuration space for a continuous path from the start point to the goal point.[7, 13-16] If a path exists, it will be found. If multiple paths are found, the best may be chosen. The artificial force technique surrounds the configuration space obstacles with repulsive potential energy functions, and places the goal point at an global energy minimum.[9, 17–19] The point in configuration space representing the manipulator is acted on by a force equal to the negative gradient of this potential field, and driven away from obstacles and to the minimum.

Global methods have several disadvantages. The algorithms necessary for global methods are computationally intensive. Also, the computational costs increase quickly as a function of the manipulator's degrees of freedom: at least exponentially for geometric search techniques, and at least quadratically for artificial force techniques.[9] Thus, they are suited only for off-line path planning and cannot be used for real-time collision avoidance. An immediate consequence is that global algorithms are difficult to use for collision avoidance in dynamic environments, where the obstacles are moving in time. Also, using global algorithms it is very difficult to describe complicated motion planning tasks such as those arising when two manipulators cooperate.

A viable alternative to global methods is provided by local ones.[10–12, 20] Local methods also employ the use of artificial forces like those discussed previously. However, unlike configuration space forces, local forces are expressed in the Cartesian workspace of the manipulator. Collisions with objects are prevented by surrounding them with repulsive potential functions, and the goal point is surrounded by an attractive well. These potentials are added to form a composite potential which imparts forces on a model of the manipulator in Cartesian space. Torques equivalent to these forces cause the motion of the real manipulator.

The main advantage of local techniques is that they are less computationally demanding than global ones, permitting their use in real-time control. Further, they provide the necessary framework to deal with changing environments and real-time collision avoidance. When used with a teleoperated manipulator, local artificial forces also provide low-level collision avoidance, while high-level path planning of the manipulator is performed by the human operator.

A.   Model-Based Potential Energy Functions

The major interest in artificial force models has been in realizing obstacle avoidance schemes.[6, 12, 21–23] These schemes require the addition of attractive and repulsive potential energy functions. An attractive potential well is generally a bowl-shaped energy well which drives the manipulator to its center if the environment is unobstructed. However, in an obstructed environment, repulsive potential energy hills are added to the attractive potential well at the locations of the obstacles, as in Fig. 1. The addition of attractive and repulsive potentials provides the artificial forces which enable obstacle avoidance.

The assignment of potential energy values to the isopotential surfaces determines the repulsive nature of the function. Two possibilities exist: avoidance and approach functions. The avoidance function has a potential energy value at the surface of the object which is larger than the initial kinetic energy of the manipulator. Thus, an energy barrier is established which cannot be surmounted. The easiest way to ensure that the potential energy barrier is large enough is to set the potential function to infinity at the object surface.

We have previously proposed a second type of artificial potential energy function for approach.[6] Instead of a potential of infinity at the object surface, the function

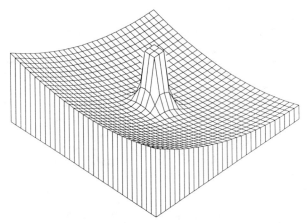

Fig. 1   Repulsive potential added to an attractive well.

goes smoothly to a finite value less than the kinetic energy of the manipulator. As
the manipulator moves toward the object, it gains potential energy, looses kinetic
energy, and slows down. Thus the approach potential provides deceleration forces
that ensure a safe contact velocity at the surface. Once stable contact has been
established, force control of the manipulator may begin.

Many proposed repulsive potentials have spherical symmetry.[6,11,21] These potentials are useful for surrounding objects with spherical symmetry, as well as
singularities in the workspace. Also, when added to a spherically symmetric
attractive well they will not create a local minimum.[6] But a spherically symmetric repulsive potential does not follow the contour of polyhedral objects.
For instance, an oblong object surrounded by a sphere effectively eliminates
much more volume from the workspace than is necessary or desirable. Potentials that follow the object shape were proposed to address the insufficiency of
radially symmetric. potentials.[6,10,12] A review of one of these schemes will be
provided next.

B.   Superquadric Isopotential Contours

We have proposed an artificial potential scheme based on the superquadric, a
mathematical function which is employed in computer vision and object modeling
techniques.[24,25] This scheme provides obstacle avoidance capability for manipulators in an environment of stationary or moving objects, preventing end effector and
link collisions with these objects. This local avoidance scheme provides obstacle
avoidance capability without creating local minima.

The superquadric is a deformable parametric surface and is used in this scheme
as the isopotential surface for the potential function. Since it is deformable, isopotential surfaces near the object may closely model the object, whereas surfaces
further away can be spherical. These spherical surfaces prevent the formation of
local minima when this function is added to a larger spherical attractive potential well.

To obtain isopotential contours that follow the object shape near the surface an object is surrounded with a superquadric[24,25]:

$$K = \left[\left(\frac{x}{a} \right)^{2n} + \left(\frac{b}{a} \right)^2 \left(\frac{y}{b} \right)^{2n} \right]^{\frac{1}{2n}} - 1 \tag{1}$$

Figure 2 shows a plot of K at regular intervals with n varying from a very large value to a value near unity.

Since the parameter n must vary from infinity to one while K varies from zero to infinity, n is defined as:

$$n = \frac{1}{1 - e^{-\alpha K}} \tag{2}$$

where α is an adjustable parameter. Other definitions of n are possible, but this form is useful because it is related to the magnitude of the potential (as will be shown).

C. Repulsive Artificial Force Functions

The artificial force experienced is dependent on the form of the potential energy function assigned to the isopotential contours defined previously. We have utilized two types of repulsive energy functions: the avoidance function, and the approach function.

1. Avoidance Function

The avoidance function surrounds an object and prevents a manipulator from touching the object. This is true, independent of the manipulator's kinetic energy,

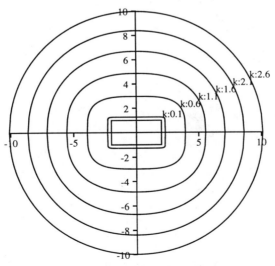

Fig. 2   Isopotential contours for $K = 0.1$ to $K = 2.6$, and $\alpha = 1.5$.

and is ensured by setting the magnitude of the potential at the surface to infinity. Away from the surface, the energy values behave like natural potentials (e.g. electrostatic, gravitational, etc.) in their inverse dependence on distance. This is done with the Yukawa potential[26] which has K^{-1} dependence for short-distance repulsion, but exponential decay at larger distances:

$$U(K) = A\frac{e^{-\alpha K}}{K} \qquad (3)$$

Figure 3 shows this function with $\alpha = 1$ and $A = 1$ for a rectangle. The parameter α determines how rapidly the potential rises near the object and falls off away from the object. This rate is tied to the rate at which the n-ness of the ellipse changes as expressed in Eq. (2). The parameter A acts as an overall scale factor for the potential. Large values of A will make the object have a spherical field of repulsive forces at large distances. Small values of A will allow the object to be approached much more closely. At this closer range, the isopotential contours will have large values of n and will approximate the shape of the object. Typically A is unity.

2.   Approach Function

The approach function surrounds an object and decreases the approach speed of the manipulator as it moves toward the object. This is achieved by setting the value of the potential energy at the surface of the object to be slightly less than the initial kinetic energy of the manipulator. As the manipulator moves toward the object its kinetic energy is transformed to potential energy, and its velocity decreases. Setting the magnitude of the potential function at the surface less than the initial kinetic energy ensures that the manipulator will always reach the surface.

An appropriate approach potential has all of the attributes of the avoidance potential, but goes to a finite maximum value at the surface of the object. Therefore, far from the object, the form of the avoidance potential is used. However, closer to the surface the potential is Gaussian in shape, the slope smoothly changing to

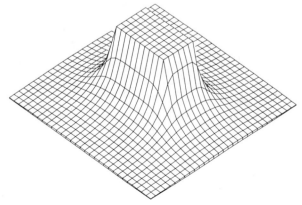

Fig. 3   **Avoidance potential for a rectangle with $\alpha = 1$ and $A = 1$; large values have been truncated.**

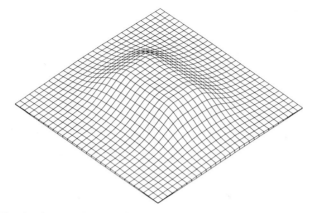

Fig. 4   Approach potential function for a rectangle with α = 1.

zero at the surface so that no artificial force is experienced when real contact with the environment is established. A general function form which remains valid for all values of α is

$$U(K) = \begin{cases} \dfrac{A}{K} e^{-\alpha K}, & K \geq 1 \\[2mm] A \exp\left(-\alpha K^{1+\frac{1}{\alpha}}\right), & 1 > K \geq 0 \end{cases} \tag{4}$$

Figure 4 shows this function with $\alpha = 1$ for a rectangle.

D.   Simulation and Experimentation

The collision prevention and approach capabilities provided by these functions are demonstrated by simulations of two-and-three link manipulators. Figure 5 shows simulation of successful avoidance of several objects by a three-link manipulator. Figure 6 show simulation of smooth approach of an obstacle by a two-link manipulator. Figure 7 shows experimental data of successful avoidance of several objects by the end effector of a SCARA-type manipulator.

These results rely on knowledge of the spatial relationship between the robot and the obstacle. For the test completed, this was calculated from a priori knowledge of the object position. However, the formulation directly supports measurement of the relationship by cameras or proximity sensors. This type of sensor-based collision avoidance has been explored by several researchers.[27–32] It remains an area of active research since the distance measurements tend to be sparse, noisy, inaccurate, and difficult to physically, electrically, and computationally integrate into the system. However, sensor-based knowledge of the environment promises to be the best way to avoid obstacles or prepare for intended impacts with it.

III.   Impact Control

Even if the manipulator has been slowed by a repulsive approach force, switching from free-space motion to constrained force control has the significant problem

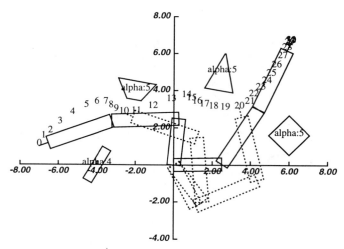

Fig. 5   Successful navigation around four obstacles using superquadratic avoidance potentials and a modified conical attractive well; the dotted manipulators are intermediate configurations.

of impact forces.[33] These forces can be very large, and can drive an otherwise stable controller into instability. Although detrimental for terrestrial operations, in space large impact forces can have added severity. Large forces can repel the environment, or excite its modes of oscillation. Further, lightweight and delicate space hardware is more susceptible to damage, difficult to repair, and more costly to replace. Given these problems, robust impact control is extremely valuable.

Typically, it is the force control strategy that must deal with this transient phenomenon, since the large force does not occur until after contact has occurred. However, the natural elasticity of the impact, or the response of the force controller

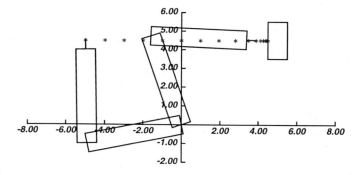

Fig. 6   Successful approach and contact with a rectangle surrounded by the proposed approach potential. For this simulation there was no attractive point, but the end effector was position controlled in the y direction. The initial velocity was 1 unit/s in the x direction. The contact velocity was 0.06 unit/s.

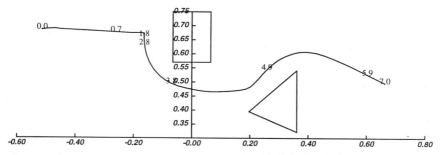

Fig. 7 **Experimental data showing a path taken to successfully navigate between two objects; for the rectangle, the potential parameters are: $\alpha = 6, A = 0.1$; for the triangle, the potential parameters are: $\alpha = 3, A = 0.01$. The numbers along the paths indicate the time in seconds.**

to the transient, can cause the manipulator to rebound from the environment. Thus, the manipulator is once again unconstrained. This phenomenon can establish oscillatory behavior or worse, drive the manipulator unstable. Obviously it is the goal of any controller to pass through this transitory period successfully, and to have the manipulator stably exerting forces on the environment. The controller must, therefore, pass through the impact phase by attempting to maintain contact with the environment until all of the energy of impact has been absorbed. To maintain stability and contact during this phase, strategies for impact control will be reviewed in this section.[34]

Previous research in force control has treated the impact phase as a transient that is dealt with by the same controller used to follow commanded force. The form of the force controller is typically an explicit force or impedance controller.[35, 36] In this section it will be shown that the best implementation of these strategies for force following is insufficient for impact control. But the impact controller presented here still fits into the same framework. To understand this, the previous schemes will be briefly discussed and their weaknesses revealed. Later, our previously proposed impact control strategy will be presented in the context of explicit force control and impedance control. An analysis will explain how the strategy provides stability, and experimental results will demonstrate its effectiveness.

A. Maximal Active Damping

One proposed method of dealing with the impact problem is to employ maximal damping during the impact phase.[37] Any force controller may be used; proportional control was used in this reference. The goal of this strategy is to damp out the oscillations caused by the transition. Although this may be successful for soft environments, stiff environments have oscillations with small amplitudes and high frequencies. This makes damping difficult for three reasons. First, changes in position of the environmental surface may be smaller than the resolution the manipulator's position measurement devices. In this case, no velocity will be sensed. Second, for fast oscillations the calculated velocity signal will lag its ideal value, and the damping force may cause instability by being applied out of phase with the true velocity of the surface.[36] Third, flexion in the links because of impact can slightly change the arm structure, thereby making the kinematics and velocity signal computation erroneous. These problems are compounded by the fact that a stiff environment which causes them will also cause a larger impact force and need stablizing compensation all the more. Thus, this scheme may fail when most needed.

B.   Passive Compliance and Damping

Another method for absorbing the shock of impact is to use passive compliance, either on the end effector or in the environment. Some researchers have proposed the use of soft force sensors or compliant skin covering for the force sensor.[38-41] These methods provide stable impact in two ways. First, the material used naturally provides passive damping that helps absorb some of the energy of impact, without the resolution or time lag problems of active damping. Second, the compliance of the material effectively lessens the stiffness of the system composed of the material and the environment. Following from the argument of the previous paragraph, this lessening of the stiffness helps active damping work. Because the end effector remains in contact with the environment over a larger range of displacement for the same experienced force, the displacement will not be below the resolution of the arm's position (and therefore velocity) measurement devices. Also, the frequency of oscillation will be lower, reducing the phase lag of the computed velocity which is needed for active damping.

But there are problems with passive compliance. First, it may not be modified without physical replacement of the material. Second, it limits the effective stiffness of the manipulator during position control. Third, it eliminates precise knowledge of the position of the environment. And fourth, it limits the forces that may be applied—beyond acertain range of operation the compliant material is not linear and is prone to physical failure.

C.   Integral Explicit Force Control

Integral force control acts as a low pass filter.[42-44] Thus, for impact transients, the high frequency components are filtered effectively. For impacts with low energy or with an inelastic environment this may be sufficient.[42] Otherwise, bouncing and possible instability will occur.[34] This is because of the nonlinear loss of contact with the surface and subsequent integrator windup which cause severe hopping on the surface.

D.   Impedance Control and Proportional Explicit Force Control

It has been analytically and experimentally demonstrated that impedance control against a stiff environment is equivalent to proportional gain explicit force control with feedforward.[34, 36, 45] Although both schemes have been tried by many researchers, the gains in these implementations are typically not tuned for the best impact response.[37, 40, 46-48] For explicit force controllers the gain is tuned for optimal command following once contact has been established. Equivalently, the mass ratio of impedance control is chosen to obtain the desired inertia for free-space motion or force exertion, but not impact. The result is an oscillatory system in which bouncing occurs after impact. This is consistent with simulation and experimental results.[34,40,49] A solution to this problem is to use a different proportional gain for the impact phase. To understand the proper choice for the gain values it is necessary to analyze both explicit force control and impedance control schemes with a proper system model. This model is reviewed next, followed by a review of the force control strategies.

IV.   Arm/Sensor/Environment Model

The physical system employed in the study of the robot impacting the environment is depicted in Fig. 8. Note that the arm and the environment are part of the same kinematic chain, since they are both attached to mechanical ground. For many of our experimental tests, the environment is a cardboard box with an aluminum plate resting on it. The measured stiffness of this environment is $\sim 10^4$ N/m. The box is resting on a table that is considerably stiffer than the box, and is therefore considered the ground for these tests. The force sensor is mounted on link six of the manipulator, the Carnegie Mellon University Direct Drive (CMU DD) Arm II. It has a measured stiffness of 5×10^6 N/m. Attached to the force sensor is a steel probe with a brass weight on its end. The brass weight serves as an end effector substitute and provides a flat stiff surface for applying forces on the environment. Previous analysis has indicated that a fourth-order model of this arm/sensor/environment is necessary and sufficient for force control. This section presents a review of the development of this model. Full details may be found in Refs. 36, 50, and 51.

The dynamics of an n-DOF, serial link manipulator are described by a set of nonlinear, coupled differential equations.[52] Included in this description are Coriolis and centripetal forces as well as viscous damping and gravitational loading. However, these elements of the description may not always be significant. For instance, Coriolis and centripetal forces are not present when the manipulator is statically exerting force on the environment; viscous damping is not present in direct drive motors; and gravitational loading is not present for a space-based robot. Further, active compensation can remove the torques caused by these physical effects. For instance, calculation of the inverse dynamics of the arm removes the effects of gravity loading and Coriolis and centripetal forces,[53] and negative damping gains can remove the effects of viscous friction.[54]

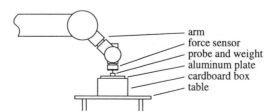

arm
force sensor
probe and weight
aluminum plate
cardboard box
table

Fig. 8   Experimental setup for force control experiments.

Therefore, for the purpose of this discussion the most important component of the dynamic description of the manipulator is its inertia. All other nonlinear components of the description will be ignored, and assumed to be insignificant or compensated for. Further, by the appropriate transformation the description of the manipulator dynamics may be represented in Cartesian space.[37] If the task frame in Cartesian space is aligned with the principle axes of the inertia tensor, the dynamic description becomes fully decoupled. In this case, a single degree of freedom may be considered independently.

The most basic one-DOF model of a manipulator is a second-order model that has a single mass, damping, and stiffness for the manipulator. The mass is configuration dependent and represents the effective manipulator inertia in that degree of freedom. The damping, if it exists, is a combination of the projection of the viscous joint damping into Cartesian space, and the active damping which may be performed directly in Cartesian space. The stiffness is caused by the combination of mechanical and actively applied stiffnesses. The mechanical stiffness can come from either the links or the actuators. For now, we will ignore the link stiffness and consider the links to be pure transmitters of force. Actuator stiffness typically comes from gearing which is nonbackdriveable. Many manipulators are backdriveable and do not exhibit mechanical stiffness. The CMU DD Arm II has no joint friction or gearing and, therefore, damping and stiffness will only be present if provided actively.[53]

Having introduced a model of the manipulator, it is necessary to discuss an environmental model. Some researchers have made no assumptions about the structure of the environment, and have assumed instead that interaction with it will produce measurable forces.[55-59] Other researchers, usually those working with a compliant system or sensor, have modeled the environment as a mechanical ground.[39, 60] Still others have recognized that the environment has some compliance, and therefore have modeled it as a simple stiffness.[35, 37, 46, 61-68] Finally, some researchers have modeled the environment as a complete second-order system with components of mass and damping, as well as stiffness.[42, 49, 50, 58] This last form of the environmental model recognizes that the environment has oscillatory modes of its own, but simplifies the overall analysis by only considering the first mode. Thus, the second-order model is more restrictive than just a general environment that exerts measurable force on the arm. But the specific representation of the model's dynamic components will permit a better understanding of the interaction between the arm and the environment.

Between the arm and the environment exists the force sensor. While a very stiff force sensor may not always exhibit its dynamics, under certain circumstances they may become important. The use of a stiff robot position controller, contact with a stiff environment, or impact with the environment may excite the sensor dynamics. Therefore, it is sometimes necessary to include the sensor in our model. A second-order model of the sensor dynamics can be added to the preceding models of the arm and environment by placing a spring and damper between the masses of these two second-order systems.

Finally, it is necessary to return to the subject of link stiffness and higher order arm dynamics. It has been recognized by some researchers that the arm has higher order dynamics that may need to be modeled.[38, 42, 49, 50] This is particularly

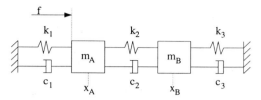

Fig. 9 General fourth-order model of the arm, sensor, and environment system.

true if the environment and sensor are stiff. Inclusion of a second-order approximation for the link stiffness makes the composite arm model fourth order, and the arm/sensor/environment model sixth order. However, if the link and sensor dynamic characteristics are similar, they may be lumped together. For instance, a typical force sensor is composed of strain gauges mounted on aluminum. If such a sensor is mounted on an aluminum robot arm, there is no clear distinction between the end of the last link and the beginning of the sensor. Modeling just the first mode of vibration of this entire assembly requires only a second-order model for both the arm links and the force sensor. (This concatenation of the stiffness and damping of both components reduces the total stiffness and damping of the link-sensor by a geometric proportionality factor that depends on the arm and sensor designs.[36]) Thus the entire model for the arm-actuator/arm-linkage-and-sensor/environment system can be reduced from sixth to fourth order. This model is shown in Fig. 9. The transfer function of this system is

$$\frac{F_m}{F} = \frac{(m_B s^2 + c_3 s + k_3)k_2}{(m_B s^2 + (c_2 + c_3)s + (k_2 + k_3))(m_A s^2 + c_1 s + k_1) + (m_B s^2 + c_3 s + k_3)(c_2 s + k_2)} \quad (5)$$

where $F_m = k_2(X_B - X_A)$ is the measured force; x_A is the measured position of the arm; x_B is the position of the environment; and m, k, and c are the mass,

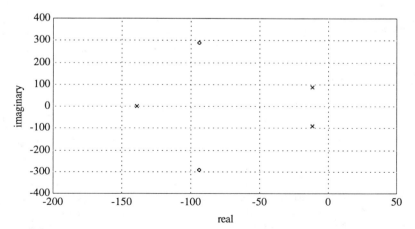

Fig. 10 Pole and zero locations for the fourth-order model, using the experimentally extracted parameters; not shown is the leftmost pole which is at $-28,000$ on the real axis.

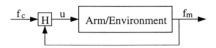

Fig. 11   Explicit force control block diagram.

stiffness, and damping parameters of the fourth-order model. This is similar to the model presented in Ref. 49.

We have experimentally extracted parameter values for the components of this model for the described experimental configuration. Theoretical and experimental details can be found in Ref. 51. The pole/zero locations indicated by the extracted values differ greatly from those assumed by other researchers.[49,50] Figure 10 shows all but the leftmost pole, which is at $-28,000$ on the real axis. The complex pole pair is caused mainly by the environment. The real pole pair (the real pole shown plus the other not shown) is caused mainly by the sensor dynamics. These pole pairs will be called the environment and sensor poles, respectively. It can be seen that the sensor poles are fairly far removed from the environmental ones, and are located farther into the left half plane. The leftmost sensor pole will be ignored since it is negative real and far removed from the others.

V.   Explicit Force Control and Impedance Control

The system modeled in the previous section is the plant of the controller used for environmental interaction. Two main conceptual choices have emerged for the choice of this controller structure: explicit force control and impedance control. It has been shown both theoretically and experimentally that second-order impedance control against a stiff environment is essentially equivalent to proportional gain explicit force control with feedforward.[36,45] The argument supporting this conclusion will only be reviewed here.

First, it is necessary to present the block diagrams of the explicit force and impedance controllers, as in Figs. 11 and 12. Next, it is important to recognize that the linear impedance controller may be separated into a position component and a force component, as in Fig. 13. Further, Fig. 14 shows that because there is no external reference force signal, the force loop may be considered an internal explicit force controller. The type of this internal explicit force controller can be extracted by looking at the impedance control law [23,36]:

$$\tau = J^T \Lambda M^{-1} [C(\dot{x}_c - \dot{x}_m) + K(x_c - x_m) - f_m]$$
$$-J^T \Lambda \dot{J}\dot{\theta} + h + g + J^T f_m \qquad (6)$$

where Λ is the manipulator inertia matrix in Cartesian space; M, C, and K are the second-order impedance matrices; h is the vector of Coriolis and centripetal forces; g is the gravitation force vector; J is the manipulator Jacobian; f, τ, x, and $\dot{x}$ are vectors of force, torque, position, and velocity; and subscripts c and m indicate commanded and measured quantities. The terms that compensate for velocity-dependent forces and gravity can be considered feedforward terms, and ignored for the remainder of this discussion. What is left is an equation for torque of the form:

$$\tau = J^T \left[H'(f_c - f_m) + f_m - K_v \dot{x}_m \right] \qquad (7)$$

Fig. 12 Impedance control block diagram.

$$f_c = K(x_c - x_m) + C\dot{x}_c \tag{8}$$

$$H' = \Lambda M^{-1} \tag{9}$$

$$K_v = H'C \tag{10}$$

The active damping provided by K_v may be added to the passive damping in the plant [c_1 in Eq. (5)] and removed from further consideration in the control equations.

Thus, the internal explicit force controller in impedance control can be represented by the block diagram in Fig. 15, where G is the plant given by Eq. (5). In this figure, the positive feedback loop acts as a reaction force compensation. If the sensor dynamics are ignored, the physical reaction force loop may be directly extracted from the plant.[36] As seen in Fig. 16, this creates a new plant, G', and a negative feedback loop of the physical reaction force. Further, this figure shows an equivalent expression of the proportional gain as $H' = H + 1$. The transfer function for this controller is

$$\frac{F_m}{F_c} = \frac{H'G'}{1 + H'G'} \tag{11}$$

$$= \frac{(H+1)G'}{1 + (H+1)G'} \tag{12}$$

It can be seen directly that an equivalent block diagram of this system may be constructed as in Fig. 17. This is a proportional gain explicit force controller with feedforward force and serves as the inner force loop in the impedance controller.

Therefore, the impedance controller has the form of a proportional-derivative (PD) position controller surrounding a proportional force controller. But when in contact with a stiff environment, the position of the environment can be set as the origin ($x_m = 0$). Also, the commanded velocity is usually zero ($\dot{x}_c = 0$). Thus, the external position loop of the impedance controller provides a command force that is simply: $f_c = Kx_c$. The external position loop becomes, in effect, functionless. We conclude that for the case of a stiff environment, impedance control is equivalent to proportional gain explicit force control with feedforward. Experimentation has validated this conclusion.[45]

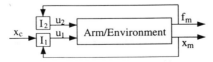

Fig. 13 Impedance control block diagram with the controller divided into its position part, I_1, and its force part, I_2.

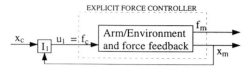

Fig. 14   Impedance control block diagram redrawn to show the inner explicit force controller.

It is interesting to look at what this equivalence implies for gain value selection. (In this discussion only the one-dimensional or diagonal matrix case will be considered.) First, the stability of the impedance controller is guaranteed for $H' \geq 0$. This is equivalent to the condition $\Lambda M^{-1} \geq 0$. Assuming a constant manipulator inertia Λ, gain H' varies as the inverse of the target impedance mass, M. Zero gain means infinite mass, and large gain means small mass. For the proportional force controller, stable gain values are $H \geq -1$ since $H = H' - 1 = \Lambda M^{-1} - 1$. Negative proportional force control gains down to -1 are stable. Further, it will be seen in the next section that they are desirable for impact control.

A.   Impact Control Without Sensor Dynamics

The model of the arm/environment plant that neglects sensor dynamics results in pole and zero locations similar to those shown in Fig. 10, except that the sensor poles are not present. For a proportional gain explicit force controller with this plant, the root locus is shown in Fig. 18 ($H \geq -1$). (The poles shown in the middle of the locus are for $H = 0$ and correspond to the environmental pole locations in Fig. 10.) Note that one pole will go into the right half plane for $H < -1$ as predicted. Observing this root locus it is immediately apparent that the most stable gain is the one that places the two poles at the point where the roots leave the real axis. Ignoring the sensor dynamics, an approximate value of this gain may easily be determined.[36] The double root of the characteristic equation occurs for a value of the proportional gain close to negative one. There are three equivalent ways to view or interpret this result.

1.   Proportional Force Control with Reaction Force Compensation

This is the controller in Fig. 16. In this case, the controller does not utilize the force error signal since $H' \approx 0$. However, the reaction force of the impact is directly negated by a feedback signal. Viewed this way, the impact controller does not bounce because the oscillations in the commanded force and those in the experienced force are equal and opposite. Thus the surface is at a node of two interfering pressure waves. No net force means no net acceleration. Any initial oscillation is damped out by natural and active damping.

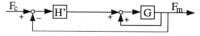

Fig. 15   Block diagram of a force-based explicit force controller with proportional gain and positive feedback for reaction force compensation.

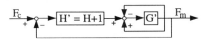

Fig. 16   Block diagram of a force-based explicit force controller with proportional gain and extra feedback for reaction force compensation; the plant G has be expanded into its components, and the sensor dynamics have been ignored.

2.   Proportional Force Control with Negative Gain and a Feedforward Signal

This is the controller in Fig. 17. While this controller looks different than before, it has been shown previously that it is equivalent. In this case the controller multiplies the force error by $H = H' - 1 \approx -1$. There is also a feedforward signal of the commanded force.

3.   Impedance Controller with a Large Target Mass

As discussed previously, an impedance controller is equivalent to an explicit force controller when in contact with a stiff environment. Impedance controllers employ a proportional gain, ΛM^{-1}, where Λ is the arm inertia and M is the desired inertia. Viewed in this way, the impact controller matches the apparent mass of the arm to the stiffness and damping of the environment such that the resultant system is critically damped. More imprecisely, it can be said that the arm is made to appear so massive that it cannot bounce.

B.   Impact Control with Sensor Dynamics

Including the sensor dynamics changes the preceding analysis somewhat by introducing an additional set of poles. Obviously, if the sensor poles are far from the environmental poles they will have little effect, and the preceding results will remain the same. However, the fourth-order model that was previously developed has one pole relatively close to the environmental poles and zeros. Figure 19 shows the root locus for this system for proportional gain values of $-1 \leq H < \infty$ or $0 \leq H' < \infty$. The points of closest approach of the locus to the real axis correspond to gain values of $H \approx -0.8$ or $H' \approx 0.2$. These are the best values for impact control.

It is important to point out that this locus also indicates that positive gain proportional force control, as well as impedance control, will become unstable for this system. The instability of these schemes has been confirmed experimentally. [36, 43, 45, 69] The points on the locus in the right half plane correspond to values of $H > 1$ or $H' > 2$, and have been shown to be unstable. While the root locus suggests that very high gains would again be stable, experimentation has indicated that the system model breaks down for these large parameter values.

Fig. 17   Block diagram of a force-based explicit force controller with proportional gain and unity feedforward; the G plant has be expanded into its components, and the sensor dynamics have been ignored.

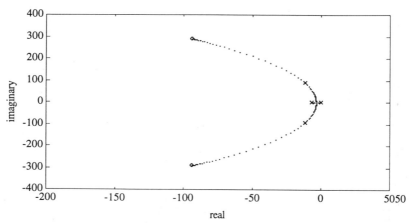

Fig. 18   Root locus for the second order model for $H' = H + 1$; the double root occurs for $H' \approx 1.5 \times 10^{-3}$; the poles shown in the middle of the locus are for $H = 0$ and correspond to the environmental pole locations in Fig. 10.

C.   Impact Experimental Results

Figures 20 show the results of impact tests on the modeled environment using impact control. The solid line is the measured force; the dashed line is the reference force, which is nonzero only after impact; and the dotted line is the scaled measured velocity. As can be seen, the impacts occur at a speed of 0.75 m/s. Figure 20a shows the stability provided by negative proportional gains. This is in contrast to the open-loop response shown in Fig. 20b. For positive gains the bouncing becomes more severe, until instability results.[34]

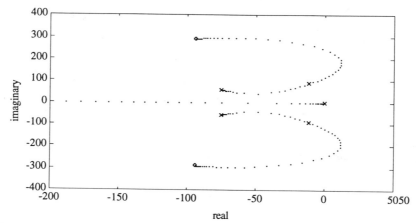

Fig. 19   Root locus for the fourth-order model for $-1 \leq H < \infty$ or $0 \leq H < \infty$; the poles shown in the middle of the locus are $H = 0$ and correspond to the environmental pole locations in Fig. 10.

a)

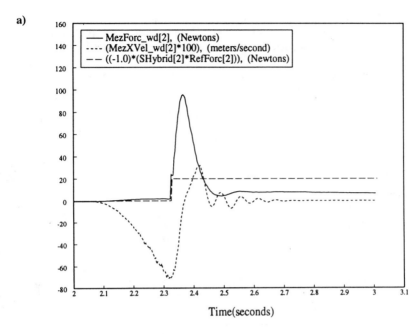

Time(seconds)

b)

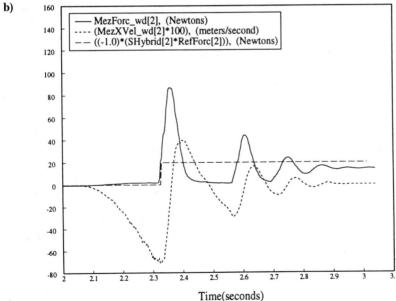

Time(seconds)

Fig. 20 Experimental data comparing the response of impacts on the test environment using proportional force control with feedforward force: a) result of using a negative proportional gain of –0.75; b) open-loop response to the impact.

Figures 21 show the equivalent response of impedance control to the impact transient, as predicted. In these figures, the solid line is measured force; the dashed line before the impact is commanded position; the dashed line after impact is the commanded position minus the environment position, multiplied by the impedance stiffness parameter; and the dotted line is the measured velocity. As predicted, Figs. 20 and 21 show that impedance control provides a response equivalent to that of proportional gain control with feedforward force.

To further test the impact controller, collisions with a very stiff steel environment were performed.[34] The results of these tests are shown in Fig. 22. Again, the impact controller eliminates hopping and provides stability.

VI.   Explicit Force Control

After impact control has successfully provided stable contact transition for the manipulator, it is desirable to provide accurate force control for the system. This section will review commonly proposed techniques and experimental data evaluating them. Also, since its equivalence to proportional gain force control has already been shown, impedance control will not be reviewed further here.

Explicit force control strategies utilize direct evaluation of desired and measured forces to determine the control signal. Many basic forms of this type of controller have been proposed, and this section will review several of them.[43] Two types of output are possible from force controllers: forces or positions. In the former, forces are commanded directly, and then translated into manipulator joint torques. In the latter, position setpoints are given to an inner-loop position controller. We have previously shown that position-based methods may be recast as force-based methods.[70] Therefore, only the force-based methods are reviewed here.

Force-based explicit force control describes a method that compares the reference and measured force signals, processes them, and provides an actuation signal directly to the plant. The reference force may also be fedforward and added to the signal going to the plant, described in Sec. IV. To control this plant some subset of proportional-integral-derivative (PID) control (i.e., P, I, PD, etc.) is usually chosen. From a computational perspective, all are approximately equal in complexity. The strategies presented here are generalizations of schemes previously proposed, as indicated hereafter. In all cases, the joint torques commanded by these schemes are obtained through the transpose of the Jacobian, and gravity compensation is employed for terrestrial operation. The parameters f and $\dot{x}$ are Cartesian force and velocity. K is a gain for either proportional (subscript fp), integral (fi), or derivative (fd) force control. K_v is the velocity gain and provides active damping that is incorporated directly into the plant.[70] The subscripts c and m denote commanded and measured quantities. The variables s and a are Laplace domain complex numbers.

1) Proportional control[28, 37, 40, 46, 49, 50, 62]:

$$f = f_c + K_{fp}(f_c - f_m) - K_v\dot{x}_m \tag{13}$$

2) Integral control[42, 44, 71, 72]:

$$f = K_{fi} \int (f_c - f_m)\,\mathrm{d}t - K_v\dot{x}_m \tag{14}$$

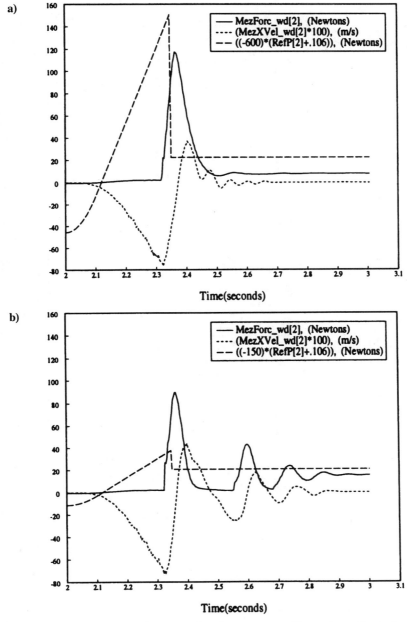

Fig. 21 Experimental data comparing the response of impacts on the test environment using impedance control, with mass ratios of 0.25 and 1.0. The response is the
same as that shown in Fig. 20, indicating the equivalence of impedance control with
proportional gain force control.

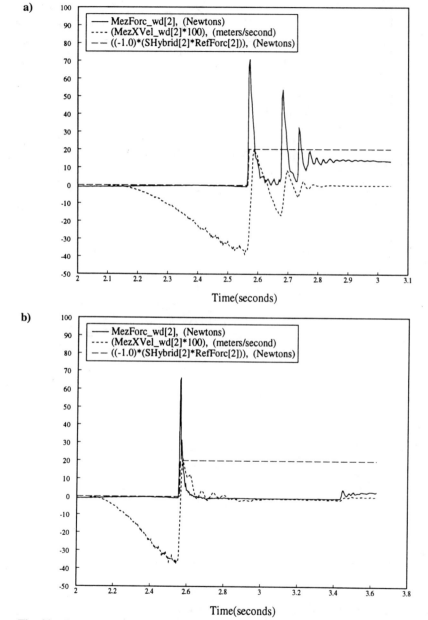

Fig. 22   Impact control on the steel pedestal in the z direction: comparison of $H = 0$ (open loop) and $H = -0.85$

3) Proportional-integral control[49, 64, 73]:

$$f = K_{fp}(f_c - f_m) + K_{fi} \int (f_c - f_m)\, dt - K_v \dot{x}_m \qquad (15)$$

4) Proportional-derivative control[39, 44, 49]:

(Unfiltered)

$$f = f_c + K_{fp}(f_c - f_m) + K_{fd} \frac{d}{dt}(f_c - f_m) - K_v \dot{x}_m \qquad (16)$$

(Filtered)

$$F(s) = F_c(s) + \left[K_{fp} + K_{fd}s\right]$$
$$\times \left[F_c(s) - \left(\frac{a}{s+a}\right) F_m(s)\right] - K_v s X_m(s) \qquad (17)$$

5) Second-order low pass filter control[36, 70]:

$$F(s) = \frac{K_{fp}}{s(s+a)} [F_c(s) - F_m(s)] + K_v s X_m(s) \qquad (18)$$

A.   Experimental Evaluation of Force Control

Of the proposed force control strategies, integral force control has proven superior to the others.[43] This is mainly because of the noisy nature of force signals, which the low-pass integral controller effectively filters. The proportional controller is not strongly effected by the noise, but the PD controller is driven unstable by it. (As will be seen later, even filtering could not improve the response of the PD controller.) Also, the best response of the second-order low-pass filter occurs when one pole dominates and the behavior is like the integral controller.

Therefore, the two main force controllers of interested have proven to be the proportional and integral gain controllers. Figures 23 show the best responses obtained with each controller in contact with the previously discussed environment. The solid and dashed lines are the measured and reference forces, respectively. Note that the proportional controller has a steady-state error, which cannot be decreased without increasing overshoot and instability.[43] Alternatively, the integral controller has lag, but follows the form of the reference and has no steady-state error. Finally, the response of the PI controller is intermediate to these two controllers and remains inferior to the integral controller.

B.   Blending Impact and Force Control

The transition from position control to impact control is abrupt, and triggered by the impact force spike. But the transition method from impact control to integral force control is less obvious. One method of making this transition is to have a transition period in which the proportional gain and force feedforward of the

a)

b)

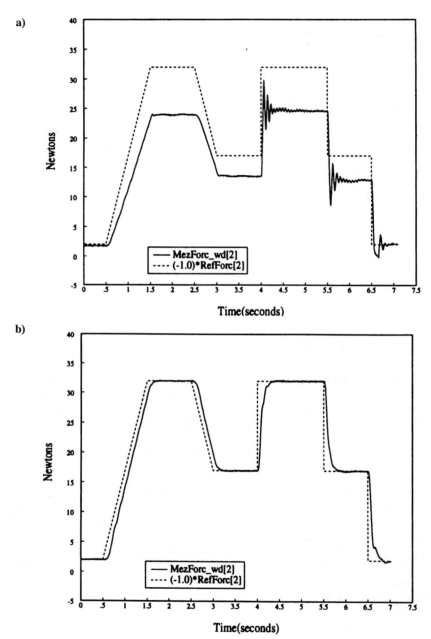

Fig. 23   Experimental data of explicit force control schemes; these are the best results for a) proportional control with force feedforward and b) integral gain control.

impact controller are brought to zero, while the integral gain is increased to its best value. We propose a linear switch of all gain values after the impact force pulse diminishes. Determination of the times to begin and end the transition is currently obtained empirically. The beginning and ending gain values are determined from tests of the impact and integral gain force controllers interacting with the same environment.[69]

Figure 24 shows the results of this strategy. The impact control phase lasts for 0.15 s (about the width of the impact spike) after the beginning of the impact. This is followed by a period of transition from impact control to integral gain force control which lasts for 0.15 s. During the transition phase H is varied linearly from -0.75 to 0; K_{ff}, the feedforward gain, is varied linearly from 1 to 0; and K_{fi} is varied linearly from 0 to 15. After this transition period, integral force control is continued with the gain at 15. As can be seen, this simple strategy provides stability through the impact period and excellent position and force control before and after the impact.

VII. Algorithm Implementation Considerations

In addition to the preceding control issues, there are many problems and issues associated with the implementation of the previously reviewed algorithms. Some are only minor annoyances, whereas others can effect the stability or range of

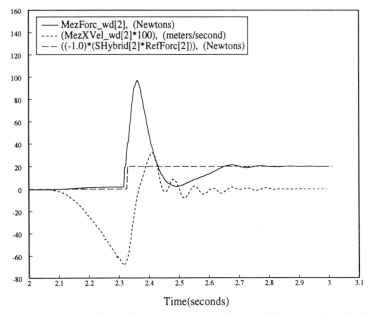

Time(seconds)

Fig. 24 Experimental data of impact control with transition to integral gain force control. The impact control phase lasts for 0.15 s after the beginning of the impact; this is followed by a period of transition from impact control to integral gain force control which lasts 0.15 s; beyond 0.3 s after impact, integral gain force control is used.

operation of a particular controller. Among the issues discussed in this section are position measurement resolution, velocity signal calculation, force signal noise and filtering, the force signal derivative, hybrid control switching, and impact transient handling.

Much of the following discussion utilizes graphed data to illustrate and validate the topic. In the legends of the graphs, the reference value of the applied force is called "RefForc"; the measured value of the experienced force is called "Mez-Forc_wd"; the filtered value of this measured force is called "Filter_Forc"; the derivative of this filtered force is called "Fdot"; the measured value of the Cartesian velocity is called "MezXVel_wd"; the measured value of the end effector position is call "MezP"; and the hybrid control selection parameter is called "SHybrid".[63] All of these variables are vectors and the indices follow the conventions of the C computer language.

A.   Position Signal

The position signal is a 16-bit absolute position value obtained from pancake resolvers located at each joint. Therefore the resolution of each joint is $\approx 10^{-4}$ rad. Since the CMU DD Arm II has a reach of approximately 1 m, the position error incurred from all six joints is about 1 mm. Therefore, the workspace of the manipulator can be thought of as quantized into small volumes, 1 mm in diameter. Obviously, the mapping from joint space error to task space error does not generate consistently shaped volumes, but the more important conceptual fact of quantized position measurements remains valid.

This intrinsic position error is important for both obstacle avoidance and force control, for the same reason. Both the physical world and the computed values of the potential functions, are continuous at the millimeter scale. But the measured positions obtained from the arm are effectively discontinuous at this scale. Therefore, if forces from the potential function or physical environment change drastically from one millimeter volume to the next, instability is likely. Typically this will appear as chattering at the boundary between zero force and a large force. This phenomenon can occur for controllers in contact with stiff environments, or steep potential functions. In both cases, this extra constraint is essentially a step function to the arm controller. If the environment is made less stiff, or the potential function less steep, then the spatial quantization effects will be reduced and the chattering will disappear. When implementing the superquadric potential functions, this criterion places a qualitative upper bound on the value of α, which effects the potential function steepness.

B.   Velocity Signal

The angular velocity signal for the joints of the CMU DD Arm II is obtained by differencing and averaging the angular position signal. During every control cycle, the position is obtained and placed in a stack. A velocity signal averaged over the past n control cycles can be obtained by simple differencing of the current position with the one n cycle before it:

$$
\begin{aligned}
v_{\text{avg}} &= \frac{1}{n} [v(t) + v(t - T) + \cdots + v(t - nT)] \\
&= \frac{1}{n} \left\{ \frac{p(t) - p(t - nT)}{T} + \frac{p(t - T) - p(t - 2T)}{T} \right. \\
&\qquad \left. + \cdots + \frac{p[t - (n - 1)T] - p(t - nT)}{T} \right\} \\
&= \frac{1}{nT} [p(t) - p(t - nT)]
\end{aligned}
\tag{19}
$$

Good results are obtained for the CMU DD Arm II with $3 \le n \le 10$. The lower number provides a velocity signal with less lag and more noise, and the higher number does just the opposite. For free-space motion with the CMU DD Arm II, the natural frequency of the system is determined by the stiffness provided by the position gain. This frequency is usually low enough that the velocity signal lag is not significant. However, when the arm is in contact with the environment, the natural frequency of the system is largely determined by the environmental stiffness. This frequency is much higher than in the free-space motion case. Therefore, the velocity signal lag can become a major portion of the oscillation cycle. As the delay approaches 90 deg the velocity signal will be in phase with the position signal. In this case, the velocity gain will not damp, but rather will add to the already large stiffness of the system, driving it toward instability.

For the tests conducted in contact with the environment, a velocity averaging factor of $n = 3$ was used for joints 4, 5, and 6. A factor of $n = 5$ was used for joints 1, 2, and 3. For tests involving free-space motion, the natural frequency is smaller and a factor of $n = 10$ for joints 1, 2, and 3 is usually used. The value $n = 3$ for the last three joints tends to be sufficient at all times.

Since the Cartesian velocity signal contains components from all of the joint signals, the delay will be between three and five cycles. For the control rate of 300 Hz, the delay is between 0.01 and 0.016 s. Figure 25 shows the velocity and position signals during proportional gain explicit force control ($K_{fp} = 0.75$), after a step input. The delay of the velocity signal is about 0.01 s, or a 45-deg phase lag for the 12-Hz oscillation. This also explains why an averaging factor of $n = 10$ is unacceptable. This delay would put the velocity signal in phase with the position signal.

Note that active damping must be used with caution when the manipulator is in contact with the environment. The time delay from the velocity calculation/filtering is always present. If this delay is a significant part of the natural frequency of the system, then the velocity signal will act as a position signal and add to instability. Further, stiffer environments have higher oscillation frequencies, making the velocity signal least reliable when it would be most useful. Therefore, the damping intrinsic to impedance control, and sometimes used in explicit force control, is always suspect.

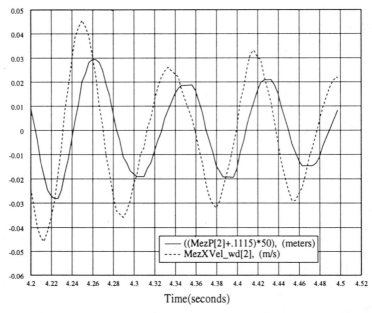

Time(seconds)

Fig. 25 Velocity phase lag caused by averaging; it can be seen that the velocity signal lags its ideal value by about 0.16 s, or 45 deg.

Finally, it is worth mentioning that these problems with delay only apply to active damping. Passive damping, as supplied by some soft sensors or end effector covers, will provide damping without time delay.[39,40] These devices also lower the natural frequency of the system, making active damping possible.

C. Force Signal

A Lord 15–50 force sensor was used in all of the experiments. In its factory configuration, it supplies eight strain gauge values at 416 Hz. However, the controllers used often ran at only 300 Hz, as previously discussed. Since the Lord sensor controller has its own internal clock, there is no way to easily change the update rate. Alternatively, individual requests for data can be made, but only with a maximum rate of 250 Hz because of clock skew. Therefore, we chose to receive the data at the faster rate of 416 Hz and ignore one of every four sets. This has the added effect that each data set could be as old as one 416-Hz cycle, or 2.4 ms. This asynchronous sampling has no appreciable effect on the stability of the controllers since force oscillations were an order of magnitude slower than the control rates and sampling time.

What does drastically effect control stability is the noise in the resultant force signal. As shown in Fig. 26, this can be substantial. Filtering of the force signal is partially effective, but introduces lag as can be seen in Fig. 27, where the measured force signal is a solid line and the filtered force signal is a short dash line. This effect is very detrimental for PD force control,[69] as will be reviewed in Sec. VII.D.

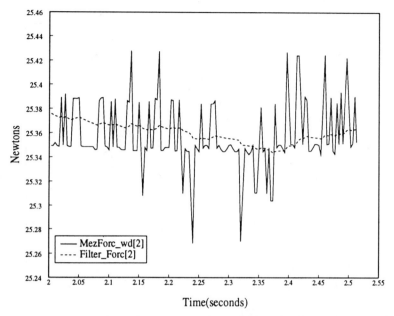

Fig. 26   Filtered and unfiltered force signals.

The force signal noise has several contributors, discussed below: intrinsic noise, kinematic fluctuations, kinematic inaccuracies, and inertial effects. All of these factors contribute to a noise amplitude of ~ 0.1 N. This is an order of magnitude above the sensor resolution.

1. Intrinsic Noise

This is present in the analog and digital electronics of the sensor system as well as the joint position resolvers. The joint position measurement noise contributes because of the need for transformation of the measured force to the control frame. Considering the fact that the CMU DD Arm II has 16-bit absolute positioning resolvers, fluctuation in the last bit typically causes angular errors on the order of $2(2\pi/2^{16}) = 96$ μrad. The additive effect of the six joints will then cause a worst case error of 0.57 rad in orientation. Multiplying this by the 1.25-m reach of the arm, gives a position error of 0.71 mm.

2. Kinematic Fluctuations

The kinematics of the arm are based on the assumption that the links are completely rigid. However, bending or oscillations in the arm structure lead to erroneous calculations of the sensor frame position and orientation, and therefore the measured force. This is especially true when there are forces exerted on the arm, such as during impact and force control.

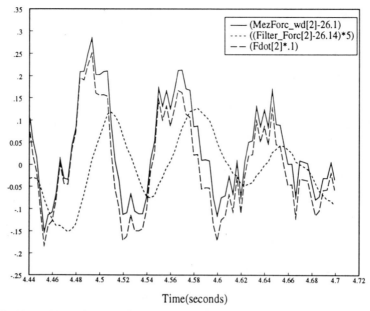

Fig. 27   Lag of the filtered force causes the force derivative to be in phase with the measured force.

3.   Inertial Effects

Inertia causes the measured force not to equal the applied force if the sensor is accelerating. Since most environments are stationary, zero acceleration usually implies that the arm has zero velocity as well.

Many manipulators (including the CMU DD Arm II) are capable of rapid acceleration, and the inertial forces can be considerable. We have observed the inertia of the end effector causing problems with both impedance and explicit force control schemes. For impedance control, the manipulator drifts, since it is attempting to apply an impedance to an external force when one is not actually present. For explicit force control, the hybrid controller will switch from position to force control in the direction of the inertial force.

D.   Derivative of the Measured Force

The previous section described the noise that is present in the force signal. This noise makes it essentially impossible to use PD force control. Even with low-pass filtering, the system was unstable for appreciable derivative gain values. Figure 28 shows the response of the system (solid), as well as the reference force (short dash), and filtered force (long dash), for $K_{fp} = 0.5$, $K_{fd} = 0.01$, $K_v = 10$, and $a = 10$ in Eq. (17). The results are not much better than for proportional gain alone.[69] As will be described; improvements in the performance of this controller cannot be made by varying the gains given here.

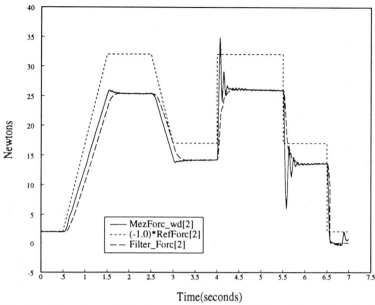

Fig. 28   Experimental data from PD control with $K_{fd} = 0.5$ and $K_{fd} = 0.01$.

First, increasing the derivative gain does not improve the response of the system because the amplified low-frequency noise can still drive the system unstable. While Fig. 28 seems to show a fairly smooth filtered force signal, a close-up view of the same data has already been shown in Fig. 26. Much of the noise has been removed, but with a large enough gain the noise will dominate. Making the cutoff frequency of the filter lower ($a < 10$) will eliminate this noise, but it introduces a more serious problem of lag.

Figure 29 shows that the calculated derivative (solid curve) appears accurate. (The dotted curve is the measured force.) However, it is apparent from this figure and Fig. 28 that there is lag introduced by the filtering process. This lag becomes extremely important when it is a significant portion of the period of oscillation of the system. Figure 27 shows the original force signal (solid), the filtered force signal (short dash), and the derivative of the filtered signal (long dash). For this oscillation frequency, the filtering process causes the filtered force to lag the measured force by one quarter cycle. This makes the force signal 180 deg out of phase with the ideal derivative signal. Thus, the proportional gain acts as a destabilizing negative derivative gain. Further, the derivative of the filtered signal leads it by one-quarter cycle. Thus, the derivative is in phase with the originally measured force and the derivative gain acts as a proportional gain. Increasing the derivative gain causes greater oscillations exactly when the effective damping is being reduced by the proportional gain. This obviously will cause the system to go unstable.

It can be concluded from this discussion that the filter pole should be significantly larger than the natural frequency of the system, However, it also must be small enough to effectively filter the noise of the force sensor. These two

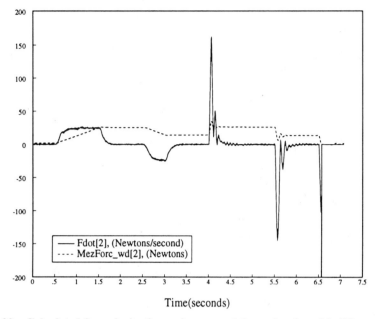

Fig. 29 Calculated force derivative and measured force signal used in PD control.

criteria could not be met with our system. To be fair, most systems will never meet this criteria. Force-controlled systems are most challenged by stiff environments that have high natural frequencies. It is unlikely that a sensor can be built that has noise only at frequencies much greater than the natural frequencies of these environments.

One solution, however, is to use a soft force sensor or compliant covering on the sensor. The compliance acts as a low-pass filter with no time delay. In this way, the derivative of the force signal may be used under the condition that the time necessary to calculate it is not significant. In this case, without a noisy force signal, simple differencing of the current and most recent force samples will usually suffice. Thus, all that is required is that the force sampling frequency not be of the same order of magnitude as the natural frequency of the system. Successful PD force control with a soft force sensor has been reported elsewhere.[39] Soft sensors, however, have other drawbacks such as smaller operation range, mechanical fatigue, unactuated degrees of freedom, nonlinearities, etc.

E. Hybrid Control Switching

For an explicit force controller it is necessary to switch from position to force control when using a hybrid control framework. One way to achieve this is to switch to force control when a measured force threshold is exceeded. To prevent force signal noise from causing the switch, a value of 2 N was used for the threshold value for switching to force control. Also, since the noise still exists while in force control mode, the measured force may drop below 2 N inadvertently. Thus, a lower threshold value of 1 N was chosen for switching from force control. The switching strategy was implemented in a joystick process running at 150 Hz. Because the switching was done by the joystick controller, the joystick values could be interpreted as commanded velocity (free-space motion), or commanded force (constrained motion).

Another aspect of the switching strategy is that it could be made unidirectional— permitting only switching to force control. When unidirectional, force control will remain in effect even if the measured force is reduced below the threshold, as when the manipulator leaves the surface. The behavior of the controller for this case of contact loss can be quite interesting and illustrative.[69,74] Experience showed that some of the controllers tested were sure to become unstable when surface contact was lost. To prevent damage to the system, bidirectional switching was usually used. (If the end effector lost contact with the environment, the controller reverted to position mode, as can be seen at the tail the data in Fig. 28.) To prevent the manipulator from losing contact with the environment, impact control proved extremely effective.[74]

F. Impact Transient

The transition from free-space motion to contact with the environment provides the greatest test to the stability of the chosen control strategy. This is because of the almost instantaneous exertion of reaction forces upon the arm. Some researchers have addressed this problem by utilizing soft force sensors, or a soft skin over the force sensor surface.[39,40] The introduction of extra compliance extends the period of impact and absorbs some of the energy. As mentioned in Sec. VII.D we have chosen not to use passive compliance because of its intrinsic problems such as mechanical fatigue and nonlinearities.

Considering the case of hard-surface-to-hard-surface impact, the transient time is very short. For a manipulator end effector moving at 1 m/s impacting a surface with stiffness of 10^4 N/m, the force will initially increase at a rate of 10^4 N/s. If a resolution of 1 N is considered adequate for control, the sampling rate must be 10^4 Hz. This is 25 times faster than our sampling rate. Further, a required sampling rate of over 10^6 Hz would be necessary for robustness to some of the surfaces/speeds that were tried in our experimentation.

This impact transient is further complicated by the available torque of the actuators. Even if the sampling rate were fast enough to adequately detect the rise in external forces caused by the impact, the joint torque necessary to substantially soften the impact is not available. In other words, the arm cannot stop itself instantaneously to prevent the impact. Assuming a maximum allowed impact force of 10 N, the arm would have to stop in 1 mm. To stop this suddenly, an acceleration of 500 m/s^2 is required (over 50 g). For a manipulator with an effective Cartesian space mass of 1 kg, 500 N is required. At least 98% of this force must be provided by the arm itself. Obviously this is not feasible for conventional actuators and manipulators.

A third problem is the kinematic fluctuations of the arm during impact. This is mainly because of the compression and flexion caused by the impact forces. For instance, vibrations in the links will cause changes in the end effector position, although no change in joint position is measured. This can cause problems for schemes that rely on accurate measurement of the surface position or velocity, such as stiff impedance controllers[55] or impact damping strategies.[37]

VIII. Conclusions

This chapter has presented solutions that address the three main issues of both space and terrestrial manipulation: collision-free motion through the environment, stable transition through the contact phase, and accurate force control on the surface. To move through the environment successfully, an artificial force function based on superquadrics has been reviewed. This function can provide collision avoidance, or surface approach at safe velocities. After impact has occurred, stability may be maintained with the impact controller proposed. Subsequent force-controlled manipulation is best achieved with integral gain force control. Also, through analysis of the impact and force control strategies, the essential equivalence of second-order impedance control and proportional gain force control has been demonstrated. Finally, an in-depth discussion of the implementational consideration for these schemes has been provided.

Acknowledgments

This research was performed at Carnegie Mellon University and supported by an Air Force Graduate Laboratory Fellowship (for Richard Volpe), DARPA under contract DAAA-21-89C-0001, the Department of Electrical and Computer Engineering, and The Robotics Institute. The writing and publication of this chapter was supported by the preceding organizations and the Jet Propulsion Laboratory, California Institute of Technology, under a contract with NASA. The views and conclusion contained in this document are those of the author and should not be interpreted as representing the official policies, either expressed or implied, of the U.S. Air Force, DARPA, or the U.S. Government. Reference herein to any specific commercial product, process, or service by trade name, trademark, manufacturer, or otherwise, does not constitute or imply its endorsement by the United States Government or the Jet Propulsion Laboratory, California Institute of Technology.

References

[1]Vafa, Z., and Dubowsky, S., "On the Dynamics of Space Manipulators Using the Virtual Manipulator, with Applications to Path Planning," *The Journal of the Astonautical Sciences,* Vol. 38, No. 4, 1990, pp. 441–472.

[2]Papadopoulos, E., and Dubowsky, S., "On the Nature of Control Algorithms for Free-Floating Space Manipulators," *IEEE Transactions on Robitics and Automation,* Vol. 7, No. 6, 1991, pp. 750–758.

[3]Yamada, K., "Arm Path Planning for a Space Robot," *Proceedings of the IEEE/RSJ Conference on Intelligent Robots and Systems,* 1993, pp. 2049–2055.

[4]Yoshida, K., and Sashida, N., "Modeling Impact Dynamics and Impulse Minimization for Space Robots," *Proceedings of the IEEE/RSJ Conference on Intelligent Robots and Systems,* 1993, pp. 2064–2069.

[5]Xu, Y., and Kande, T. (eds.), *Space Robotics: Dynamics and Control,* Kluwer Academic Pub., 1993.

[6]Volpe, R., and Khosla, P., "Manipulator Control with Superquadric Artificial Potential Functions: Theory and Experiments," *IEEE Transactions on Systems, Man, and Cybernetics; Special Issue on Unmanned Vehicles and Intelligent Systems,* Vol. 20, No. 6, Nov./Dec. 1990.

[7]Udupa, S., "Collision Detection and Avoidance in Computer Controlled Manipulators," Proceedings of the 5th Joint International Conference on AI, 1977.

[8]Lozano-Perez, T., "Spatial Planning: A Configuration Space Approach," *IEEE Transactions on Computers,* Vol. C-32, No. 2, 1983, pp. 102–120.

[9]Rimon, E., and Koditschek, E., "The Construction of Analytic Diffeomorphisms for Exact Robot Navigation on Star Worlds," *Proceedings of the IEEE Conference on Robotics and Automation,* 1989, pp. 21–26.

[10]Khatib, O., "Real-Time Obstacle Avoidance for Manipulators and Mobile Robots," *The International Journal of Robotics Research,* Vol. 5, No. 1, 1986, pp. 90–98.

[11]Andrews J. R., and Hogan, N., "Impedance Control as a Framework for Implementing Obstacle Avoidance in a Manipulator," *Control of Manufacturing Processes and Robotic Systems,* edited by D. E. Hardt and J. Wayne, American Society of Mechanical Engineers, 1983, pp. 243–251.

[12]Krogh, B., "A Generalized Potential Field Approach to Obstacle Avoidance Control," *Proceedings of the SME Confernce in Robotics Research: The Next Five Years and Beyond,* Paper MS 84–484, Bethehem, PA, Aug. 1984.

[13]Lozano-Perez, T., and Wesley, M., "An Algorithm for Planning Collision Free Paths Among Polyhedral Objects," *Comm. of ACM,* Vol. 22, No. 10, 1979, pp. 560–570.

[14]Lozano-Perez, T., "A Simple Motion Planning Algorithm for General Manipulators," AI Lab. Memo, AI Labs, Massachusetts Institute of Technology, Cambridge, MA, 1986.

[15]Schwartz, J., and Sharir, M., "On the Piano Movers Problem, Part II," TR 41, Courant Inst. of Mathematical Sciences, New York, 1982.

[16]Faverjon, B., "Obstacle Avoidance Using an Octree in the Configuration Space of a Manipulator," *Proceedings of the IEEE Conference on Robotics and Automation* (Atlanta, GA), 1984, pp. 504–512.

[17]Okutomi, M., and Mori, M., "Decision of Robot Movement by Means of a Potential Field," *Advanced Robotics,* Vol. 1, No. 2, 1986, pp. 131–141.

[18]Newman, W., "Automatic Obstacle Avoidance at High Speeds via Reflex Control," *Proceedings of the IEEE Conference on Robotics and Automation,* 1989, pp. 1104–1109.

[19]Warren, C., "Global Path Planning Using Artificial Potential Fields," *Proceedings of the IEEE Conference on Robotics and Automation,* 1989, pp. 316–321.

[20]Volpe, R., and Khosla, P., "A Strategy For Obstacle Avoidance and Approach Using Superquadric Potential Functions," *Proceedings of the 5th International Symposium of Robotics Research* (Tokyo, Japan), MIT Press, Massachusetts Inst. of Technology, Cambridge, MA, 1989, pp. 93–100.

[21]Koditschek, D. E., "Exact Robot Navigation by Means of Potential Functions: Some Topological Considerations," *IEEE International Conference on Robotics and Automation,* Raleigh, NC, March 31–April 3, 1987.

[22]Newman, W. S., and Hogan, N., "High Speed Control and Obstacle Avoidance Using Dynamic Potential Functions," *IEEE International Conference on Robotics and Automation,* Raleigh, North Carolina, March 31–April 3, 1987.

[23]Hogan, N., "Impedance Control: An Approach to Manipulation Parts I, II, and III," *Journal of Dynamic Systems, Measurement, and Control,* Vol. 107, March 1985.

[24]Bajcsy, R., and Solina, F., "Three Dimensional Object Representation Revisited," *First International Conference on Computer Vision* (London, England, U.K.), 1987, pp. 231–240.

[25]Barr, A. H., "Superquadrics and Angle-Preserving Transformations," *IEEE Computer Graphics and Applications,* Vol. 1, Jan. 1981, pp. 11–23.

[26]Cohen-Tannoudji, B., Diu, C., and Laloe, F., *Quantum Mechanics,* Vol. 2, John Wiley and Sons, New York, 1977, p. 957.

[27]Cheung, E., and Lumelsky, V., "Motion Planning for Robot Arm Manipulators with Proximity Sensing," *Proceedings of the IEEE International Conference on Robotics and Automation* (Philadelphia, PA), 1988, pp. 740–745.

[28]Wedel, D., and Saridis, G., "An Experiment in Hybrid Position/Force Control of a Six DOF Revolute Manipulator," *Proceedings of the IEEE Conference on Robotics and Automation,* 1988, pp. 1638–1642.

[29]Novak, J., and Feddema, J., "A Capacitance-Based Proximity Sensor for Whole Arm Obstacle Avoidance," *Proceedings of the IEEE Conference on Robotics and Automation,* 1992, pp. 1307–1314.

[30]Nguyen, C., Antrazi, S., and Campbell, C., "Virtual Force Control of a Six Degree of Freedom Parallel Manipulator," *Robotics and Manufacturing: Recent Trends in Research, Education, and Applications,* ASME Press, 1992, pp. 357–363.

[31]Boddy, C., and Taylor J., "Whole-Aim Reactive Collision Avoidance Control of Kinematically Redundant Manipulators," *IEEE International Conference on Robotics and Automation* (Atlanta, GA), 1992, pp. 382–387.

[32]Allotta, B., et al., "Controlling Contact by Integrating Proximity and Force Sensing," Second International Conference on Experimental Robotics, Toulouse, France, June 25–27, 1991.

[33]Paul, R., "Problems and Research Issues Associated with the Hybrid Control of Force and Displacement," *Proceedings of the IEEE Conference on Robotics and Automation,* 1987, pp. 1966–1971.

[34]Volpe, R., and Khosla, P., "A Theoretical and Experimental Investigation of Impact Control for Manipulators," *International Journal of Robotics Research,* Vol. 12, No. 4, 1993, pp. 351–365.

[35]Whitney, D., "Historical Perspective and State of the Art in Robot Force Control," *Proceedings of the IEEE Conference on Robotics and Automation,* 1985, pp. 262–268.

[36]Volpe R., "Real and Artificial Forces in the Control of Manipulators: Theory and Experiments," Ph.D. Thesis, Carnegie Mellon Univ., Pittsburgh, PA, 1990.

[37]Khatib O., and Burdick, J., "Motion and Force Control of Robot Manipulators," *Proceedings of the IEEE Conference on Robotics and Automation,* 1986, pp. 1381–1386.

[38]Roberts, R. K., "The Compliance of End Effector Force Sensors for Robot Manipulator Control," Ph. D. Thesis, Purdue Univ. 1984.

[39]Xu, Y., and Paul, R., "On Position Compensation and Force Control Stability of a Robot with a Compliant Wrist," *Proceedings of the IEEE Conference on Robotics and Automation,* 1988, pp. 1173–1178.

[40]An, C., and Hollerbach, J., "Dynamic Stability Issues in Force Control of Manipulators," *Proceedings of the IEEE Conference on Robotics and Automation,* 1987, pp. 890–896.

[41]Akella, P., Siegwart, R., and Cutkosky, M., "Manipulation with Soft Fingers: Contact Force Control," *Proceedings of the IEEE Conference on Robotics and Automation,* 1991, pp. 652–657.

[42]Youcef-Toumi, K., and Gutz, D., "Impact and Force Control," *Proceedings of the IEEE Conference on Robotics and Automation,* 1989, pp. 410–416.

[43]Volpe, R., and Khosla, P., "A Theoretical and Experimental Investigation of Explicit Force Control Strategies for Manipulators," *IEEE Transactions on Automatic Control,* Vol. 38, No. 11, 1993, pp. 1634–1650.

[44]Vischer, D., and Khatib, O., *Design and Development of Torque-Controlled Joints,* Springer-Verlag, Berlin/Heidelberg, Germany, 1990, pp. 274–286.

[45]Volpe, R., and Khosla, P., "The Equivalence of Second Order Impedence Control and Proportional Gain Explicit Force Control: Theory and Experiments," Proceedings of the Second Annual International Symposium on Experimental Robotics, Toulouse, France, June 1991.

[46]Youcef-Toumi, K., "Force Control of Direct-Drive Manipulators for Surface Following," *Proceedings of the IEEE Conference on Robotics and Automation,* 1987, pp. 2055–2060.

[47]Kazerooni, H., "Robust, Non-Linear Impedance Control for Robot Manipulators," *Proceedings of the IEEE Conference on Robotics and Automation,* 1987, pp. 741–750.

[48]Hogan, N., "Stable Execution of Contact Tasks Using Impedance Control," *Proceedings of the IEEE Conference on Robotics and Automation,* 1987, pp. 1047–1054.

[49]Eppinger, S., and Seering, W., "Understanding Bandwidth Limitations on Robot Force Control," *Proceedings of the IEEE Conference on Robotics and Automation* (Raleigh, NC), 1987, pp. 904–909.

[50]Eppinger, S., and Seering, W., "On Dynamic Models of Robot Force Control," *Proceedings of the IEEE Conference on Robotics and Automation,* 1986, 29–34.

[51]Volpe, R., and Khosla, P., "Theoretical Analysis and Experimental Verification of a Manipulator/Sensor/Environment Model for Force Control," Proceedings of the IEEE International Conference on Systems, Man, and Cybernetics, Los Angeles, CA, Nov. 1990.

[52]Fu, K., Gonzalez, R., and Lee, C., *Robotics: Control, Sensing, Vision, and Intelligence,* McGraw-Hill, New York, 1987.

[53]Khosla, P., "Real-Time Control and Identification of Direct Drive Manipulators," Ph.D. Thesis, Carnegie Mellon Univ., Pittsburgh, PA, 1986.

[54]Colgate, J., and Hogan, N., "Robust Control of Dynamically Interacting Systems," *International Journal of Control,* Vol. 48, No. 1, 1988, pp. 65–88.

[55]Salisbury, J. K., "Active Stiffness Control of a Manipulator in Cartesian Coordinates," *IEEE Conference on Decision and Control,* 1980, pp. 95–100.

[56]Hogan, N., and Cotter, S. L., *Cartesian Impedance Control of a Nonlinear Manipulator,* American Society of Mechanical Engineers, New York, 1982, pp. 121–128.

[57]Maples, J., and Becker, J., "Experiments in Force Control of Robotic Manipulators," *Proceedings of the IEEE Conference on Robotics and Automation,* 1986, pp. 695–702.

[58]Kazerooni, H., Sheridan, T., and Houpt, P., "Robust Compliant Motion for Manipulators, Parts I and II," *IEEE Journal of Robotics and Automation,* Vol. RA-2, No. 2, 1986, pp. 83–105.

[59]Goldenberg, A., "Implementation of Force and Impedance Control in Robot Manipulators," *Proceedings of the IEEE Conference on Robotics and Automation,* 1988, pp. 1626–632.

[60]Sharon, A., Hogan, N., and Hardt, D., "Controller Design in the Physical Domain (Application to Robot Impedance Control)," *Proceedings of the IEEE Conference on Robotics and Automation,* 1989, pp. 552–559.

[61]Whitney, D., "Force Feedback Control of Manipulator Fine Motions," *Journal of Dynamic Systems, Measurement, and Control,* June 1977, pp. 91–97.

[62]Paul, R., and Wu, C., "Manipulator Compliance Based on Joint Torque," *IEEE Conference on Decision and Control,* 1980, pp. 88–94.

[63]Raibert, M., and Craig, J., "Hybrid Position/Force Control of Manipulators," *Journal of Dynamic Systems, Measurement, and Control,* Vol. 103, No. 2, 1981, pp. 126–133.

[64]De Schutter, J., "A Study of Active Compliant Motion Control Methods For Rigid Manipulators Based on a Generic Scheme," *Proceedings of the IEEE Conference on Robotics and Automation,* 1987, pp. 1060–1065.

[65]De Schutter, J., "Improved Force Control Laws For Advanced Tracking Applications," *Proceedings of the IEEE Conference on Robotics and Automation,* 1988, pp. 1497–1502.

[66]Kahng, J., and Amirouche, F., "Impact Force Analysis in mechanical Hand Design— Part I," *International Journal of Robotics and Automation,* Vol. 3, No. 3, 1988, pp. 158–164.

[67]Lawrence, D., "Impedance Control Stability Properties in Common Implementations," *Proceedings of the IEEE Conference on Robotics and Automation,* 1988, pp. 1185–1190.

[68]Ishikawa, H., Sawada, C., Kawase, K., and Takata, M., "Stable Compliance Control and Its Implementation for a 6 D. O. F. Manipulator," *Proceedings of the IEEE Conference on Robotics and Automation,* 1989, pp. 98–103.

[69]Volpe, R., and Khosla, P., "An Experimental Evaluation and Comparison of Explicit Force Control Strategies for Robotic Manipulators," Proceedings of the IEEE International Conference on Robotics and Automation, Nice, France, May 10–15, 1992.

[70]Volpe, R., and Khosla, P., "An Analysis of Manipulator Force Control Strategies Applied to an Experimentally Derived Model," Proceedings of the IEEE/RSJ International Conference on Intelligent Robots and Systems, Raleigh, NC, July 7–10, 1992.

[71]Townsend, W., and Salisbury, J., "The Effect of Coulomb Friction and Stiction on Force Control," *Proceedings of the IEEE Conference on Robotics and Automation,* 1987, pp. 883–889.

[72]Colgate, E., and Hogan, N., "An Analysis of Contact Instability In Terms of Passive Physical Equivalents," *Proceedings of the IEEE Conference on Robotics and Automation,* 1989, pp. 404–409.

[73]Miyazaki, F., and Arimoto, S., "Sensory Feedback for Robot Manipulators," *Journal of Robotic Systems,* Vol. 2, No. 1, 1985, pp. 53–71.

[74]Volpe, R., and Khosla, P., "Experimental Verification of a Strategy for Impact Control," Proceedings of the IEEE International on Robotics and Automation, Sacramento, CA, April 1991.

Chapter 9

Versatile and Precise Vision-Based Manipulation

Steven B. Skaar* and Emilio Gonzalez-Galvan†

University of Notre Dame, Notre Dame, Indiana 46556

Introduction

I MPLEMENTATION of robotics in space entails a wide variety of challenges. Of particular interest here is the matter of bringing about precise, autonomous positioning of one rigid body (say a tool or a structural member which is to be added to an assembly) relative to a second object. Although the required degree of precision for such tasks will depend on the geometries involved, as well as the availability of force control and passive compliance, some degree of positioning precision will inevitably be required for the maneuver to succeed.

It is important when considering this kind of operation to emphasize robustness as well as precision. A number of disturbances having to do with heat, motion, and the requirement for light weight are likely in practice to alter the nominal geometry of the system. The means used to produce the required positioning precision must therefore continue to perform despite a variety of disturbances. One of the advantages of directing the manipulator's movement via human-guided teleoperation lies precisely in this area; with teleoperation, the precision of positioning which may be obtained, and hence terminal maneuver success itself, does not depend on the retention of carefully calibrated geometric relationships. The human controller is able to use his own perception of the adjustment required to close the gap, in his own subjective frame of reference, and hence to culminate satisfactorily the maneuver.

By contrast, the most common approach to the use of vision for the purpose of autonomous guidance relies on a carefully calibrated and pre-established mapping of three-dimensional physical space into the two-dimensional image plane (or camera space) of each participating camera. The degree of difficulty entailed by implementation of this strategy clearly depends on the requirements for precision of the particular task. Nevertheless, generally speaking, this approach is very demanding in terms of its intolerance of those heat- and motion-induced geometrical variations that are likely to occur in space operations. The source of difficulty

Copyright © 1994 by the American Institute of Aeronautics and Astronautics, Inc. All rights reserved.

*Associate Professor, Department of Aerospace and Mechanical Engineering.
†Graduate Assistant and Ph.D. Candidate, Department of Engineering Science and Mechanics.

is twofold. First, the calibrated mapping of three-dimensional physical space into two-dimensional camera space must not be disturbed. Thus, the relative position of the cameras in relation to the platform of each manipulator, once established, must not shift. Even small disturbances produce large errors in assessed workpiece positions. Likewise, the optical characteristics of each camera must remain unchanged by temperature variations. The second issue relates to the kinematics of the manipulator itself. Again, global characterization of these kinematics can be difficult to establish with much precision. And once they are established, the kinematics must not be altered by heat- or motion-induced deformation.

Even in settings such as a factory, where the temperature and other conditions might be held nearly constant, this calibration approach is seldom used in the three-dimensional case. If calibration were to be used in a factory, however, disabling disturbances might readily be accommodated and adjusted by a technician, in contrast with remote operations in space where such access is difficult. Moreover, the variety of contingencies which may arise in space increases the desirability of a method whereby terminal precision may be adjusted and improved to nearly any required degree by making appropriate use of that visual information which is acquired during the approach of the manipulated body toward its intended objective.

It is therefore desirable to avoid system brittleness. It is also desirable to have on-line measures available that permit the full use of all video data and joint-rotation data which are acquired during a given maneuver to maximize terminal precision. Nominal or as-built geometries should be exploited to the degree that they are known, but maneuver accuracy must not be limited by the ongoing fidelity of these quantities.

Camera-space manipulation (CSM)[1-4] is largely free from the brittleness that attends calibration-dependent methods. Its robustness stems from the use of actual approach data for the purpose of planning the remainder of the approach through to termination. By adjusting the speed of arm movement, and hence the density of video and joint-rotation samples, the expectation of terminal precision can be adjusted as required.

One of the ideas on which camera-space manipulation is based is that of identifying the conditions for maneuver success directly in terms of the reference frame of each camera sensor. This makes the terminal result significantly less sensitive to those errors which are caused by calibration relative to a fixed, common, physical frame of reference. Another approach to visually guided manipulation which likewise works directly with the image plane is visual servoing.[5,6] Here, the increments in joint rotation which are required to produce desired increments in camera-space end-effector position are first identified (often using nominal manipulator kinematics and some kind of camera calibration.) Then, small action is used to close the gap in the image plane between desired and actual position. This approach relies upon the differential Jacobian, and produces only first-order corrections. By contrast, CSM uses a nonlinear algebraic description of the local relationship between joint coordinates and the camera-space locations of points on the manipulated body.

The algebraic description of the relevant manipulator input to camera-space response offers several advantages. Because each of a minimum of two participating cameras has at any given time its own current best estimate of this algebraic relationship, finite joint rotations which, always according to the best, most current information, are projected to terminate the maneuver successfully, may be computed at any juncture during the approach. Estimates of the required relationship between joint rotations and image-plane response are based on the recent sequence of video and joint-rotation samples leading up to the current position. They are thus local in terms of joint space, camera space, physical space, and time.

The density of samples on the basis of which the commanded, joint-rotation sequence is calculated can be increased by slowing down the rate of approach of the arm. This is one reason camera-space manipulation affords the advantage of adjustable precision. Another source of precision enhancement, as well as robustness, is related to the absence of a need to retain a fixed (calibrated) physical position and orientation of participating cameras relative to the manipulator base. This flexibility permits the repositioning of cameras (e.g., pan/tilt) as well as focal-length adjustment. In fact, as discussed later in this chapter, one very useful aspect of the method lies in its ability to apply video information acquired prior to a camera move to the relevant parameter estimates following the move. This feature adds greatly to the convenience and speed with which postmove estimates may be reobserved. It has been shown that the method is able to achieve reliable, orchestrated, simultaneous movement of cameras and manipulator which results in a very wide operating range in addition to high reliability, robustness, and precision.

Following this introduction, the method itself is outlined. Reasons for its robustness are discussed, and versatility and the potential for high precision are shown in this context. Next is a discussion of specific test results. These results are explained partially in terms of an analysis involving the location, and the numbers, of cameras. A precision-enhancing measure is put forth in the next section, and this is followed by a section which details a useful extension of the approach to servoable cameras. Finally, a summary and conclusion section is given.

Three-Dimensional Rigid-Body Positioning Using Camera-Space Manipulation

The mapping from the three-dimensional physical space to the two-dimensional image plane is approximately a perspective projection of the three-dimensional space onto the image plane. The perspective projection is described by the pinhole camera model, which projects points referred to a camera-fixed physical reference frame XYZ onto the image plane with two-dimensional coordinates (x_c, y_c) according to

$$x_c = f\frac{X}{Z} \tag{1a}$$

$$y_c = f\frac{Y}{Z} \tag{1b}$$

where f is the effective focal length of a camera, and X and Y are parallel to x_c and y_c, respectively. The Z axis is in the direction of the camera focal axis. The points specified in the camera-fixed reference frame can be referred to a second physical reference frame xyz using the relation

$$\left\{\begin{matrix} X \\ Y \\ Z \end{matrix}\right\} = T \left\{\begin{matrix} x \\ y \\ z \end{matrix}\right\} + \left\{\begin{matrix} X_0 \\ Y_0 \\ Z_0 \end{matrix}\right\} \tag{2a}$$

where

$$T = \begin{bmatrix} e_0^2 + e_1^2 - e_2^2 - e_3^2 & 2(e_1 e_2 + e_0 e_3) & 2(e_1 e_3 - e_0 e_2) \\ 2(e_1 e_2 - e_0 e_3) & e_0^2 - e_1^2 + e_2^2 - e_3^2 & 2(e_2 e_3 + e_0 e_1) \\ 2(e_1 e_3 + e_0 e_2) & 2(e_2 e_3 - e_0 e_1) & e_0^2 - e_1^2 - e_2^2 + e_3^2 \end{bmatrix} \tag{2b}$$

The parameters e_0, e_1, e_2, and e_3 are the four Euler parameters satisfying the constraint $e_0^2 + e_1^2 + e_2^2 + e_3^2 = 1$, and (X_0, Y_0, Z_0) locates the origin of the coordinate system xyz relative to the camera-based coordinate system XYZ. Thus the pin-hole camera model can be seen to contain seven independent parameters: X_0, Y_0, Z_0, f, and three of the four Euler parameters.

If the variation of Z in the region of the physical space of interest is very small compared with the average distance from the camera to the region, then we may approximate the quantity f/Z by a single constant. Consequently, f/Z may be combined with e_0, e_1, e_2, e_3, X_0, and Y_0 in the expressions for X and Y in Eqs. (2) to form a new parameter vector $\mathbf{C} = [C_1, C_2, C_3, C_4, C_5, C_6]^T$. The pin-hole camera model may thus be approximated by

$$\begin{aligned} x_c &= (C_1^2 + C_2^2 - C_3^2 - C_4^2)x + 2(C_2 C_3 + C_1 C_4)y + 2(C_2 C_4 - C_1 C_3)z + C_5 \\ &\equiv f_x(x, y, z; \mathbf{C}) \end{aligned} \tag{3a}$$

$$\begin{aligned} y_c &= 2(C_2 C_3 - C_1 C_4)x + (C_1^2 - C_2^2 + C_3^2 - C_4^2)y + 2(C_3 C_4 - C_1 C_2)z + C_6 \\ &\equiv f_y(x, y, z; \mathbf{C}) \end{aligned} \tag{3b}$$

Equations (3) represent an asymptotic limit of Eqs. (1) as Z becomes very large. They define the orthographic camera model. If we assume a perfect perspective projection, then the image-plane samples would fit the pin-hole camera model exactly, while the orthographic model would be in error. If the samples are confined to a small region relative to Z, then they will be well accommodated by the orthographic model.

In camera-space manipulation, Eqs. (3) are combined with the nominal manipulator kinematics to form the local nonlinear relationship between the manipulator joint rotations and camera-space positions of manipulable cues. The nominal manipulator kinematics relate the joint rotations to the positions of cues on the grasped object referred to the manipulator Cartesian reference frame, which are assumed to project into camera-space according to Eqs. (3).

The model of Eqs. (3) is used as the basis for establishing, and refining locally, the nonlinear relationship between the vector of internal joint rotations $\Theta \equiv [\theta_1, \ldots, \theta_6]^T$ and the appearance of the ith of n cues, or distinctly recognizable points, on the manipulated body. It may be noted that, with a similar, modified formulation, lines or other features could be used in lieu of these points. Denoting the physical coordinates of one cue by (x_i, y_i, z_i), there is generally available the nominal kinematic relationship

$$x_i = r_x^i(\Theta)$$
$$y_i = r_y^i(\Theta) \tag{4}$$
$$z_i = r_z^i(\Theta)$$

Referring to Eqs. (3), the six-element vector $\mathbf{C}^j$, which is specific to the jth of a minimum of two cameras, relates the appearance of the ith cue in the image plane (x_c^i, y_c^i) to Θ according to

$$x_c^i = f_x[r_x^i(\Theta), r_y^i(\Theta), r_z^i(\Theta); \mathbf{C}^j] \tag{5}$$
$$y_c^i = f_y[r_x^i(\Theta), r_y^i(\Theta), r_z^i(\Theta); \mathbf{C}^j] \tag{6}$$

The procedure for determining $\mathbf{C}^j$ is summarized briefly here. An initial guess for $\mathbf{C}^j$ of camera j is first assumed. Once a sufficiently large number and spread of manipulable-cue measurements and simultaneous joint-coordinate measurements are acquired at instants of time denoted here by $t_1, t_2, \ldots t_k$, a parameter correction vector $\Delta\mathbf{C}^j$ is computed according to Ref. 7.

$$\Delta\mathbf{C}^j = (A^T W A)^{-1} A W R \tag{7}$$

where the $2k$ elements of the vector of residuals R are given by

$$R_{2i-1} = x_c(t_i) - f_x(\Theta(t_i); \mathbf{C}^j) \tag{8}$$
$$R_{2i} = y_c(t_i) - f_y(\Theta(t_i); \mathbf{C}^j) \qquad (i = 1, 2, \ldots, k) \tag{9}$$

where $[x_c(t_i), y_c(t_i)]$ are the camera-space cue measurements at $t = t_i$ and where $\Theta(t_i)$ are the corresponding sampled joint coordinates. [If more than one manipulable cue is detected at a given instant, then consecutive $\Theta(t_i)$ vectors would be unchanged for those measurements. Note that it is assumed that the proper cues are identified.] The $2k \times 6$ matrix $\mathbf{A}$ has elements given by

$$A_{2j-1,i} = \frac{\partial f_x}{\partial C_i^l}[\Theta(t_j); C^l] \tag{10}$$

$$A_{2j,i} = \frac{\partial f_y}{\partial C_i^l}[\Theta(t_j); C^l] \tag{11}$$

where C^l represent the current estimates for the lth camera. The $2k \times 2k$ matrix W is evaluated in a way that slightly favors the most-recent samples in accordance with a suboptimal scheme outlined in Ref. 2.

In practice, the challenge of utilizing new measurements to locally improve $\mathbf{C}^l$ estimates lies primarily in the areas of rapid and reliable cue-centroid detection and cue identification, that is, distinguishing among the various cues and identifying them once they are located in the image plane. Although it is the case that in our recent tests the cue detection procedure is identical for both the manipulable and nonmanipulable cues, the subsequent identification procedure differs significantly for the two types of cues. With target positions of each manipulable cue, and with current $\mathbf{C}^l$ estimates, both established for all participating cameras, joint-coordinate selection for the relative position of interest is performed as follows: Over all joint rotations Θ, the function $J(\Theta)$ is minimized, where J is given by

$$J(\Theta) = \sum_{j=1}^{NC} \sum_{i=1}^{NQ} \left\{ X_i^j - f_x[r^i(\Theta); \mathbf{C}^j] \right\}^2 + \left\{ Y_i^j - f_y[r^i(\Theta); \mathbf{C}^j] \right\}^2 \qquad (12)$$

where NQ is the number of positioned points, and where NC (= 2 or 3 in our tests) is the number of participating cameras. X_i^j, Y_i^j are the target coordinates of the ith manipulable cue in the image plane of the jth camera. Determination of these target positions is discussed later.

A standard Newton-Raphson algorithm is used to solve the six necessary conditions

$$\frac{\partial J}{\partial \theta_i} = 0, \qquad (i = 1, 2, \ldots, 6) \qquad (13)$$

for the six elements θ_i that would then be the commanded position. The problem of convergence to the wrong roots (or failure to converge at all) is avoided by solving for closely-spaced consecutive targets, and using as the initial Θ the converged results from the previous intermediate target position.

Camera-Space Maneuver Objectives

The objectives of the maneuver must be specified in terms of desired relative positions of manipulable and nonmanipulable cues in camera space.

Consider for example the desired intermediate position of the two pieces as illustrated in Fig. 1. The insertion piece is poised to enter the hole. It is properly aligned and rotated such that selected faces of the two cubes are parallel to one another. The distance d between the foremost face of the entering piece and the approached surface containing the hole is specified. Insertion can be viewed as a process of reducing d to zero and entering with gradually more negative d values, all the while retaining proper alignment.

It is useful to distinguish among the various cues by numbering the five manipulable cues m_1 through m_5 and the 14 nonmanipulable cues n_1 through n_{14}. A coordinate system is attached to each of the two bodies as indicated in Fig. 1. For any given value of d, points specified relative to the receptacle's three-dimensional reference frame can be related to the tool's three-dimensional reference frame according to

$$\begin{bmatrix} x^W \\ y^W \\ z^W \\ 1 \end{bmatrix} = \overline{T} \begin{bmatrix} x^T \\ y^T \\ z^T \\ 1 \end{bmatrix} \qquad (14)$$

where x^W, y^W, z^W are Cartesian coordinates of some point referred to the receptacle's frame of reference, x^T, y^T, z^T are Cartesian coordinates of the same

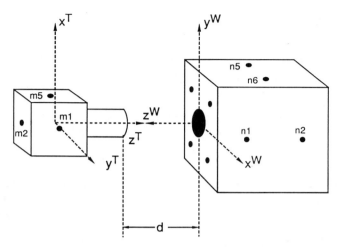

Fig. 1   Cue positions on the entering and receiving pieces.

point referred to the tool coordinate systems, and the homogeneous transformation matrix $\overline{\mathbf{T}}$ is given by

$$\overline{\mathbf{T}} = \begin{bmatrix} 0 & 1 & 0 & 0 \\ 1 & 0 & 0 & 0 \\ 0 & 0 & -1 & d \\ 0 & 0 & 0 & 1 \end{bmatrix} \tag{15}$$

Note that, because proper rotation and alignment of the two pieces is already built into $\overline{\mathbf{T}}$, it is a function of d only.

The matrix $\overline{\mathbf{T}}$ as previously described contains all of the information regarding the desired position and orientation of the tool in relation to the workpiece at a given juncture during the maneuver. It serves as the basis for the establishment of the camera-space objective for each of the manipulable cues at each intermediate stage of the maneuver. Target camera-space positions for the manipulable cues are determined from the known, desired $\overline{\mathbf{T}}$ and appearances of nonmanipulable cues in the camera of interest as follows:

First, all nonmanipulable cues that are adequately exposed to a particular camera are detected and identified. It has been observed in previous work that there should be a minimum of four such nonmanipulable cues identified, and that the identified cues should not all be contained on a single physical plane. Assuming that n nonmanipulable cues which satisfy these requirements are found, we denote the camera-space coordinates of their centroids by (X_i, Y_i), $i = 1, 2, \ldots, n$.

Six parameters, $\overline{C}_1, \ldots, \overline{C}_6$, can be identified which relate, within a local region, points in physical space to their positions in camera space. In particular, if x_i^W, y_i^W, z_i^W denote the known (permanent) physical coordinates of the

ith nonmanipulable cue centroids relative to the workpiece frame of reference, then

$$X_i = (\overline{C}_1^2 + \overline{C}_2^2 - \overline{C}_3^2 - \overline{C}_4^2)x_i^W + 2(\overline{C}_2\overline{C}_3 + \overline{C}_1\overline{C}_4)y_i^W$$
$$+ 2(\overline{C}_2\overline{C}_4 - \overline{C}_1\overline{C}_3)z_i^W + \overline{C}_5$$
$$\equiv f_x(x_i^W, y_i^W, z_i^W; \overline{C}) \tag{16}$$
$$Y_i = 2(\overline{C}_2\overline{C}_3 - \overline{C}_1\overline{C}_4)x_i^W + 2(\overline{C}_1^2 - \overline{C}_2^2 + \overline{C}_3^2 - \overline{C}_4^2)y_i^W$$
$$+ 2(\overline{C}_3\overline{C}_4 - \overline{C}_1\overline{C}_2)z_i^W + \overline{C}_6$$
$$\equiv f_y(x_i^W, y_i^W, z_i^W; \overline{C}) \qquad (i = 1, 2, \ldots, n) \tag{17}$$

Equations (3) and (4) will not hold precisely for all measured (X_i, Y_i) camera-space pairs, even in a contained region of space, due to measurement error. Thus, best estimates for $\overline{C}$ are found based on the minimization over all $\overline{C}$ of J given by

$$J(\overline{C}) = \sum_{i=1}^{n} \left[X_i - f_x(x_i^W, y_i^W, z_i^W; \overline{C})\right]^2$$
$$+ \left[Y_i - f_y(x_i^W, y_i^W, z_i^W; \overline{C})\right]^2 \tag{18}$$

with the functions f_x and f_y defined in Eqs. (3) and (4).

With the $\overline{C}$ estimates in hand, target camera-space positions of the manipulable cues are easily determined. Consider, for example, the jth manipulable cue with known permanent coordinates x_j^T, y_j^T, z_j^T in its own (tool) body-fixed reference frame. These are now referred to the workpiece frame at the desired tool/workpiece configuration using $\overline{T}$ in accordance with Eqs. (14) and (15). Finally, target camera-space coordinates (X_j^T, Y_j^T) in a given camera for the jth manipulable cue are found from

$$X_j^T = f_x(x_j^W, y_j^W, z_j^W; \overline{C}) \tag{19}$$
$$Y_j^T = f_y(x_j^W, y_j^W, z_j^W; \overline{C}) \tag{20}$$

To summarize, the objectives of the maneuver at each intermediate stage are described in terms of target camera-space objectives for all or some manipulable points. These targets must be established for each of a minimum of two participating cameras, and at least three separate manipulable cues should be involved. While it may be true that not all positioned points actually appear in all cameras, this does not prevent establishing and using target camera-space positions for them.

The sequence of operations is outlined in Fig. 2. Here, the need to establish some initial estimate of C is met by means of a preplanned trajectory. This sequence is a pre-established initialization process which is independent of the receiving-piece location, and which will not be required if, because of previous operation, significant data are already available for the current camera position. (As described in a later section, even if camera data are acquired in an earlier, now-altered camera position, these data can still be used to eliminate the preplanned trajectory provided knowledge exists of the camera axes about which pan/tilt rotations occurred.)

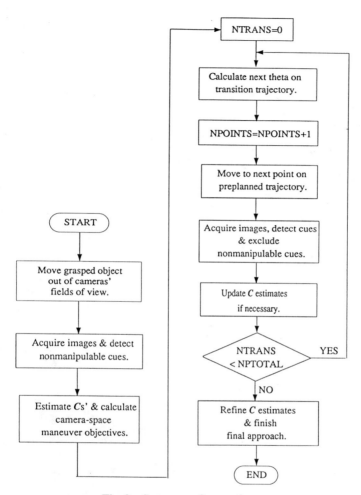

Fig. 2   Sequence of operations.

Following initialization of **C**, the transition trajectory ensues. As the name implies, this trajectory transitions the workpiece from its current location to the first point (in our experiments corresponding to $d = 100$ mm) in the straight-line approach. Points along the transition trajectory may be determined using any one of a wide range of possible interpolation formulae. One requirement of this sequence of intermediate points is that it must connect the initial, actual joint-coordinate vector with the algorithm's estimate of the vector which initializes the terminal, straight-line sequence. The second requirement is that is should allow for significant video sampling, and simultaneous joint-rotation sampling, at a variety of junctures along the way to permit updating of **C** estimates in accordance with Eqs. (4–11). This updating is critical for maneuver success because it introduces to the estimates data which are specific to the joint-coordinate region, camera-space region, and physical-space region in which the maneuver will culminate.

Experimental Results

Figures 1 and 3 show the two members which are to be joined in this task. The tolerance of the fit may be varied in that the outer sleeve of the round insertion piece may be removed and replaced by a piece with a different outside diameter. The radius of curvature of each of the two approaching edges is essentially zero, so the role of passive compliance in the initial entry is very small. Because the rotation along the entering axis is specified and controlled, the task requires six degrees of freedom.

To enable rapid computer analysis of visual samples, readily detectable ring shapes, either dark on a light surface or light on a dark surface, are painted in precise locations onto the 76 (mm) receiving cube (nonmanipulable cues). The experiment is set up such that, between the two participating cameras, at least four of the cube's surfaces are in view. The inside diameter of each ring is 6.5 mm, and the outside diameter is 19.5 mm. A large number of nonmanipulable cues was chosen to take advantage of the positioning resolution that stems from this redundancy. The mix of black and white rings was chosen as a possible assist in the cue identification procedure.

The member that is inserted (tool) consists of a small 38 (mm) aluminum cube with a 38-mm-long cylinder of variable radius protruding from one side. Manipulable cues of the same size as the others are painted on the center of four of the small cube's surfaces as well as at the end of the cylinder. Only one of the

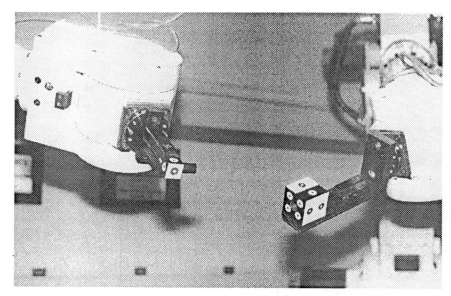

Fig. 3   Two members to be joined in the task.

rings on this manipulable body is black, while the rest are white. (See Ref. 2 for a discussion of the cue detection and cue identification procedure.)

The manipulator used is the 3500 lb, six-axis GMF S-400, with ac servomotor actuation. Two standard black-and-white video cameras—Panasonic model WV-1410—are connected to a Data Translation model DT-2851 frame-grabber board and Data Translation DT-2859 multiplexer board which are installed in a Zenith Z-248 computer. The Z-248 performs all image analysis, estimation, and trajectory calculations. It supplies, at a rate of about one every 3 s, points along a joint-coordinate sequence to the controller of the GMF S-400, which in turn executes these commands with Cartesian interpolation, PID control, and ac servomotor actuation.

Most of our experiments were performed using a sleeve size for the cylindrical tool 0.7 mm smaller (in radius) than that of the cylindrical hole of the receiving piece.

The most straightforward way of characterizing the results is in terms of the frequency of maneuver success. It was clear throughout our several hundred trials that the contribution of passive compliance to maneuver success was negligibly small. This claim is based first on the very small (≤ 0.1-mm) radius of curvature of approaching edges, and second on the observed absence of movement of the receiving piece or relative shifting within the grasper of the entering piece during a successful insertion procedure. Although difficult to quantify, it was observed that the method typically produced a very accurate relative orientation of the two pieces before and during insertion, and that owing to this accuracy, no instances of jamming[8] of the entering piece into the receiving piece were detected in these tests.

Unsuccessful approaches were detected in advance of an actual collision. With the tool held in its pre-entry position ($d = 10$ mm), the sleeve can be slid by hand along the smaller shaft toward the hole, and an assessment of the alignment can be made. In the rare event of an actual collision, the tool slips out of its grasp, and no damage is done.

The results are presented as follows: first, the experimental conditions under which the 0.7-mm-tolerance case almost uniformly succeeded are given in some detail. They are given along with an indication of the range of conditions within which consistent success could be relied on. Subsequent to this, various departures from the ideal conditions are examined and an attempt is made to understand the relationship between these departures and the degree and type of failure of the maneuver. Specifically, the factors adversely affecting maneuver success that are examined are: position of the cameras, lighting, the weighting of manipulable-cue measurements in determining $\mathbf{C}$, and the number of cameras. On this last item—the number of cameras—it was generally the case that the minimum of two was used. Where failure occurred, however, because of one of the other factors, the ability of an added third camera to regain maneuver success was sometimes tested. Finally, the results of modifying the camera model from a simple orthographic projection to a pinhole camera model are given.

Camera Positions

Whereas the precise locations of the cameras were never critical, adequate separation of the cameras was important. Focal axis lines separated by 85–95 deg worked well consistently (see "Analysis of Sensitivity"). Equally important was a reasonably large distance between each camera and the mating parts. Using a 75-mm focal length lens on a standard black-and-white video camera, a distance of between 20 and 25 ft worked well. Additionally, the cameras were each positioned such that at least two faces, and preferably three, of the stationary cube lay in adequate view for cues on those faces to be detected.

Lighting

As might be expected, good lighting was essential. In particular, if the lighting was inadequate to the extent of causing a failure to detect either manipulable or nonmanipulable cues, resolution of positioning would usually suffer. In addition to overhead lighting, we used three 100-W sources of side lighting, positioned 10–15 ft from the mating parts. No hard rule can be established regarding the minimum number of cues that needed to be detected for maneuver success, but it was found that with a full complement of detected cues, and with other conditions satisfied, the maneuver almost invariable succeeded.

Trajectory

The parameters $\mathbf{C}^j$ which are used in Eqs. (5) and (6) to identify the local relationship between joint positions and the camera-space locations of manipulable cues were, in our tests, recomputed at two junctures during the maneuver. The first juncture was the end of the preplanned trajectory, and hence the corresponding initial $\mathbf{C}^j$ estimates were unaffected by the actual cube position. With these early estimates in hand—one set for each camera—the transition trajectory to the beginning of the straight-line final approach was determined.

The issues associated with the nature of the approach motion that produces reliable placement resolution are important for general implementation and optimal efficacy. These matters are incompletely understood. We do understand, however, that the motion should involve large ($\geq$ 20-deg) rotation of at least one of the wrist joints, that it should not consist primarily of straight-line motion, and that a fairly large portion of it (perhaps the final 10–15 samples) should occur within reasonable translational proximity (<2 ft) and rotational proximity (<5 deg–10 deg) of the actual entering position.

Weighting of Measurements

The computation of the vector $\mathbf{C}$ for each participating camera is based on a systematic use of weighted cue measurements made during the preplanned trajectory and, for an update, the transition trajectory. Specifically, the weighting matrix W of Eq. (7) is adjusted to deweight early measurements in accordance with the total amount of joint rotation (i.e. sum of absolute values of each element of $\Delta\Theta$) separating the juncture at which a given factored-in measurement was acquired from the juncture at which the last available measurement (presumably the one closest to the target) was acquired. In these tests, it was found that a nearly equal weighting of all measurements, with only a slight skewing in favor of latter samples, worked best. The criterion for evaluating a given weighting formula is the resulting ability to forecast the relationship between future joint rotation and the camera-space position of manipulable cues, particularly as the tool approaches the cube.

Figure 4 indicates the measurement made from one particular camera of several manipulable cues over a particular preplanned trajectory sequence. Superimposed onto these measurements are the model's best fit with equal weighting in the first plot. The * denote subsequent target positions for the same cues. The ability to reach these camera-space target positions using information from the preplanned trajectory alone (measurements from the transition trajectory are not used here) is inferred from the proximity to these * of measured camera-space positions of cues after actual positioning occurred. The effect of a small weighting of the same measurements from the preplanned trajectory can be seen in Fig. 4.

Number of Participating Cameras

While two cameras were used for all previously reported experiments, a third camera was introduced in the few cases where physical positioning error from two cameras was large (between 3 and 6 mm). The objective was to assess the influence of this redundant third camera on error reduction in cases where the information from two cameras alone was clearly inadequate. In each instance where a third camera was used, its partial influence was assessed ceteris parabus since a new approach trajectory using information added by the third camera only was compared against the previous approach with a subset of that information (i.e., information from the two original cameras).

The focal axis of the third camera was separated from each of those of the other two by about 90 deg. Because of space limitations, however, its distance from the mating parts was only about 12 ft.

When we considered the effect of a third camera, the major source of a large error from two cameras only was regarded as an inadequate number of detected cues. Where the two-camera error was 3–6 mm (the largest error we observed in any of our tests), a third camera nearly always reduced this error by at least a factor of two. This error reduction was not always adequate to result in maneuver success, however. Nevertheless, we conclude that added cameras, and the redundancy they bring to the operation are potential safeguards against imprecise positioning from camera-space manipulation.

Analysis of Senstivity to Camera Position/Number

This section presents analytical work aimed at understanding some of the experimental findings previously described. While several operating factors have been considered in the previous section, the objective here is to study how relative camera orientation and the number of participating cameras govern positioning precision, given the assumption that the **C** parameter estimates available at the preterminal location can be characterized as possessing normally distributed errors. It is assumed further that all cameras are equal in all respects and that the corresponding parameters of each camera contain the same amount of error. The point of interest here is to determine how the cameras may be oriented relative to each other and how the use of more than two cameras may help to restrict the propagation of these errors into the terminal physical space position of a point on the end effector.

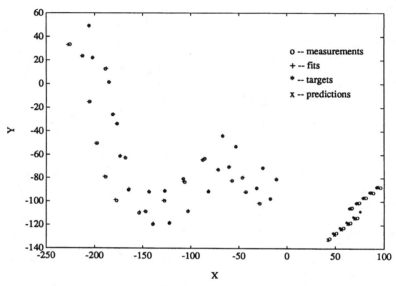

Fig. 4a   Planned trajectory, best fit, and target realization with equal weighting.

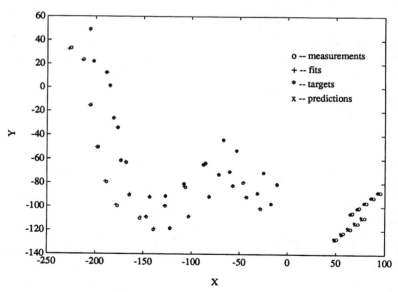

Fig. 4b   Planned trajectory, best fit, and target realization with skewed weighting.

In this work, a simplified camera-space manipulation task requiring three-dimensional positioning of the tip of the end effector of the three-axis manipulator shown in Fig. 5 is studied. The estimation equations are here expressed in terms of Cartesian coordinates rather than joint coordinates, and to simplify calculations, the errors in estimates of C_j^l ($j = 1, \ldots, 6$; for the lth camera) are assumed to be uncorrelated in addition to being normally distributed about the correct values with small variances that are assumed a priori. Because the errors are assumed to be normally distributed, the uncertainty of the parameter estimates is characterized by the assumed variances. If X_l^T and Y_l^T denote the target camera space location of the end effector tip in the lth camera space, and $f_x(x, y, z; \mathbf{C}^l)$ and $f_y(x, y, z; \mathbf{C}^l)$ [having the same form as the f_x and f_y defined in Eqs. (3) and (4)] represent the local mapping from the XYZ space into the lth camera space, and $\mathbf{C}^l$ denotes the latest estimates of the six parameters C_j^l, then if L cameras are being used, the desired terminal position of the end-effector tip is given by

$$\min_{(x,y,z)} J = \sum_{l=1}^{L} [(X_l^T - f_x(x, y, z; \mathbf{C}^l)]^2$$
$$+ [(Y_l^T - f_y(x, y, z; \mathbf{C}^l)]^2$$

The objective is to find the uncertainty in the terminal x, y, z evaluated from this equation, given an uncertainty in each of the six parameter estimates C_j^l. The terminal $r = [x, y, z]^T$ is evaluated by solving the necessary conditions corresponding to the preceding minimization,

$$\frac{\partial J}{\partial x} = 0, \qquad \frac{\partial J}{\partial y} = 0, \qquad \frac{\partial J}{\partial z} = 0$$

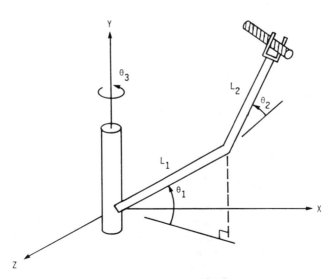

Fig. 5   Three-axis manipulator.

These yield the following nonlinear relations,

$$g_1(x, y, z, C_1^1, C_2^1, \ldots, C_6^L) = 0$$
$$g_2(x, y, z, C_1^1, C_2^1, \ldots, C_6^L) = 0$$
$$g_3(x, y, z, C_1^1, C_2^1, \ldots, C_6^L) = 0$$

Each camera is associated with six parameters, so that the functions g_i ($i = 1$–3) represent hypersurfaces in $6L + 3$ dimensions, so that,

$$\Delta g_i = \frac{\partial g_i}{\partial x} \Delta x + \frac{\partial g_i}{\partial y} \Delta y + \frac{\partial g_i}{\partial z} \Delta z$$
$$+ \frac{\partial g_i}{\partial C_1^1} \Delta C_1^1 + \cdots + \frac{\partial g_i}{\partial C_6^L} \Delta C_6^L = 0 \tag{21}$$

Written in matrix form this yields

$$A \Delta r = -B \Delta \mathbf{C}$$

where

$$A = \begin{bmatrix} \dfrac{\partial g_1}{\partial x} & \cdots & \dfrac{\partial g_1}{\partial z} \\ \dfrac{\partial g_1}{\partial x} & \cdots & \dfrac{\partial g_3}{\partial z} \end{bmatrix} \qquad B = \begin{bmatrix} \dfrac{\partial g_1}{\partial C_1^1} & \cdots & \dfrac{\partial g_1}{\partial C_L^6} \\ \dfrac{\partial g_1}{\partial C_1^1} & \cdots & \dfrac{\partial g_3}{\partial C_L^6} \end{bmatrix} \tag{22}$$

and

$$\Delta r = [\Delta x, \Delta y, \Delta z]^T \quad \text{and} \quad \Delta \mathbf{C} = [\Delta C_1^1, \ldots, \Delta C_L^6]^T$$

Whenever A is nonsingular,

$$\Delta r = -A^{-1} B \Delta \mathbf{C}$$

Multiplying both sides by Δr^T,

$$\Delta r \Delta r^T = A^{-1} B \Delta \mathbf{C} \Delta \mathbf{C}^T B^T A^{-T} \tag{23}$$

Now assuming that both sides of Eq. (23) are evaluated for a number of $\Delta \mathbf{C}$, one can compute an expected $\Delta r \Delta r^T$ such that,

$$E(\Delta r \Delta r^T) = A^{-1} B E(\Delta \mathbf{C} \Delta \mathbf{C}^T) B^T A^{-T}$$

or,

$$\Lambda_{rr} = A^{-1} B \Lambda_{CC} B^T A^{-T} \tag{24}$$

where Λ_{rr} and Λ_{CC} denote the covariance matrices corresponding to r and $\mathbf{C}$. Equation (24) relates the uncertainty of the estimates of parameters for camera 1, for instance, C_j^1, as it propagates to the terminal position r. As already pointed out, the functions f_x and f_y for each camera are of the same form as Eqs. (3). The parameters $C_1^1, \ldots, C_4^1$, are related to the Euler parameters corresponding to camera orientation, by

$$e_0 = C_1^1/\|\mathbf{C}^1\|^{1/2}, \quad e_1 = C_2^1/\|\mathbf{C}^1\|^{1/2}, \quad e_2 = C_3^1/\|\mathbf{C}^1\|^{1/2}, \quad e_3 = C_4^1/\|\mathbf{C}^1\|^{1/2}$$

where $\|\mathbf{C}^1\| = C_1^{1^2} + C_2^{1^2} + C_3^{1^2} + C_4^{1^2}$ depends on camera focal length, digitization scaling, etc., and the distance of separation between the camera and the end-effector tip, measured along the camera optical axis. The rotation matrix corresponding to a given camera orientation is given by

$$[c^l] = \begin{bmatrix} e_0^2 + e_1^2 - e_2^2 - e_3^2 & 2(e_1e_2 + e_0e_3) & 2(e_1e_3 - e_0e_2) \\ 2(e_1e_2 - e_0e_3) & e_0^2 - e_1^2 + e_2^2 - e_3^2 & 2(e_2e_3 + e_0e_1) \\ 2(e_1e_3 + e_0e_2) & 2(e_2e_3 - e_0e_1) & e_0^2 - e_1^2 - e_2^2 + e_3^2 \end{bmatrix}$$

where, in terms of the 3-1-3 Euler angles, ϕ, θ, and ψ,

$$e_0 = \cos\left(\frac{\phi + \psi}{2}\right)\cos\left(\frac{\theta}{2}\right)$$

$$e_1 = \cos\left(\frac{\phi - \psi}{2}\right)\sin\left(\frac{\theta}{2}\right)$$

$$e_2 = \sin\left(\frac{\phi - \psi}{2}\right)\sin\left(\frac{\theta}{2}\right)$$

$$e_3 = \sin\left(\frac{\phi + \psi}{2}\right)\cos\left(\frac{\theta}{2}\right)$$

Results

In the calculations studying the effect of relative orientation of two cameras, the Euler angles of camera 1 were assumed to be $\phi = -90$ deg, $\theta = 34$ deg, and $\psi = 90$ deg. The rotation matrix of camera 2 was obtained by multiplying the $[c^l]$ by

$$\mathbf{l}_{r\alpha} = \begin{bmatrix} \cos\alpha & 0 & \sin\alpha \\ 0 & 1 & 0 \\ -\sin\alpha & 0 & \cos\alpha \end{bmatrix}$$

where α, the in-plane separation of the two focal axes, was allowed to vary from 0 deg to 180 deg. When a third camera was added, camera 1 and camera 2 were assumed to be at $\alpha = 90$ deg relative to each other. The rotation matrix of camera 3 was obtained, first by multiplying $[c^l]$ by the matrix $\mathbf{l}_{r\alpha}$ for $\alpha = 0$ deg–90 deg keeping $\beta = 0$ deg, and next by holding its α relative to camera 1 fixed at 60 deg, letting its out-of-plane angle, β, vary from 0 deg to 90 deg and multiplying $[c^l]$ by

$$\mathbf{l}_{r\beta} = \begin{bmatrix} 1 & 0 & 0 \\ 0 & \cos\beta & -\sin\beta \\ 0 & \sin\beta & \cos\beta \end{bmatrix}$$

It is noted that all cameras are considered to have been placed so that the end-effector tip appears near the center of the image plane, so that a change in the physical position is implicit in the changes in orientation mentioned earlier. The $\|\mathbf{C}^l\|$ is taken to be the same for all cameras. A value of $\|\mathbf{C}^l\| = 0.24$ was used in the calculations, and C_5^l and C_6^l were chosen in each case to make the image-plane location of the end-effector tip be near the image-plane center. The $6L \times 6L$ covariance matrix Λ_{CC} was assumed to be $\Lambda_{CC} = \mathrm{diag}[0.000045, \ldots, 0.000045]$.

The 3×3 covariance matrix λ_{rr} evaluated according to Eq. (24) represents a "constant-likelihood ellipsoid" centered on the correct physical position of the end-effector tip. Thus, the smaller the extent of this ellipsoid, the smaller is the uncertainty in the position estimate. The eigenvalues (denoted λ_1, λ_2, and λ_3) of Λ_{rr} equal the lengths of the principal axes of the ellipsoid, while the eigenvectors correspond to its principal directions. The region of uncertainty in the estimated position can be characterized by the volume of the ellipsoid (given by $1/6\pi\lambda_1\lambda_2\lambda_3$), or by the trace of Λ_{rr}, $(\lambda_1 + \lambda_2 + \lambda_3)$, which remains invariant to rotational transformation.

Figure 6 plots the ellipsoid volume against the angle α separating the two cameras. It is seen that the position-estimate uncertainty is very large when the angle α between the two cameras is small. Furthermore, the uncertainty is seen to be approaching a minimum as α approaches 90 deg. The same inference can be drawn by plotting the trace of Λ_{rr} vs α. The uncertainty is not significantly greater than the minimum over the range of α values from 60 to 120 deg. It may be pointed out that the experimentally observed precision in the rigid-body mating task was found to show a similar dependence on relative camera orientation when two cameras were used. For a more complete discussion of the effects of camera position and number, see Ref. 9.

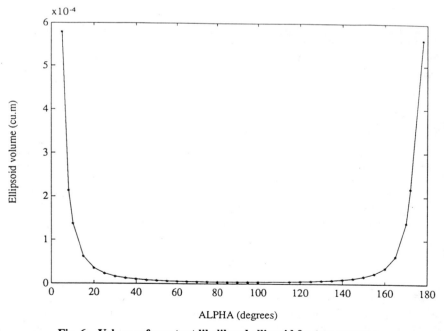

Fig. 6   Volume of constant likelihood ellipsoid for two cameras.

Precision-Enhancing Measure

In our experiments, we have found that the dominant source of deterministic error can be substantially corrected if there is some knowledge of the distance which separates each camera from the target region of interest. This correction, which we have called flattening, has been shown in our experiments to eliminate, on average, somewhat more than half of the remaining terminal error which is observed absent use of the measure, when knowledge of the aforementioned distance is within about 10% of the true value.

The idea behind flattening is to modify the camera-space samples such that they become consistent with our assumption of an orthographic projection. This is achieved by introducing to the raw camera-space samples of the manipulable cues a correction based on the pin-hole camera model.

To outline the concept of flattening, we let (x_c, y_c) on the left-hand sides of Eqs. (3) be the camera-space coordinates of the location of a sampled manipulable cue, and (X, Y, Z) be the physical-space coordinates of the location of the cue referred to the camera-fixed reference system. Multiplying both sides of the equation by Z/Z_r where Z is prescribed, yields

$$x_c \frac{Z}{Z_r} = f \frac{X}{Z} \frac{Z}{Z_r} = \frac{f}{Z_r} X \tag{25a}$$

$$y_c \frac{Z}{Z_r} = f \frac{X}{Z} \frac{Z}{Z_r} = \frac{f}{Z_r} Y \tag{25b}$$

The left-hand sides of Eqs. (25) are the flattened camera-space coordinates of a sampled cue. These can be seen to define a location proportional to the corresponding orthographic projection of the cue onto the XY plane (which is parallel to the image plane).

A straightforward application of flattening would involve the following steps: Once the $\mathbf{C}$ estimates are determined by minimizing the sum of the squared differences between the sampled image-plane coordinates of the manipulable cues and their coordinates as would be estimated by the model (see Ref. 1), approximate Euler parameters are found by normalizing C_1, through C_4 such that these four elements sum square to unity. Thus,

$$e_{i-1} \approx \frac{C_i}{(C_1^2 + C_2^2 + C_3^2 + C_4^2)^{1/2}} \quad i = 1, \ldots, 4 \tag{26}$$

The Euler parameters, in turn, are used to find the direction cosine matrix [Eqs. (2)] which quantifies the orientation of the camera-fixed reference frame (XYZ) relative to the manipulator Cartesian reference frame (xyz). The origin of the reference frame XYZ is taken to be the focal point of the camera lens, with X and Y parallel to the image-plane axes x_c and y_c, and Z directed along the focal axis. It is convenient, but not necessary, to formulate the manipulator nominal kinematics relative to the Cartesian reference frame with origin o which is located within the workspace such that the Z coordinate of point o, Z_o, is approximately equal to the distance between the origins of the two coordinate systems xyz and XYZ, and $X_o \approx Y_o \approx 0$. Using the nominal manipulator kinematics, the Euler parameters [from Eqs. (2)], and an approximate measurement of Z_o, then, the Z coordinates of each sampled manipulable cue (recall that "Z" denotes the Z coordinate referred to the camera-fixed reference frame) are found according to the third of Eqs. (2a). The image-plane coordinates of each cue are next multiplied by the quantity Z/Z_r, where Z_r is the anticipated value of Z for one of the manipulable cues at the maneuver termination, and is provided by the target camera-space position of that cue. The samples thus flattened are next used to re-estimate the six view parameters via Eqs. (3). This procedure is repeated in an iterative manner until successive parameter estimates converge, and the maneuver is completed using these converged estimates.

The reason this correction is applied to samples of the manipulable cues only is that the breadth of physical space spanned by these samples in the directions of the camera axes—and hence the shortcoming of the orthographic model—is typically much larger for the manipulable cues than for the stationary nonmanipulable cues. The scaling factor Z_r was chosen to be the anticipated value of Z of one of the cues at maneuver termination to ensure local compatibility between the camera-space target positions and the estimated relationship between the joint rotations and the camera-space response of manipulable cues. This allows that the correction factor at the point of maneuver termination is unity and that it is not necessary to make any modification to the target coordinates. Thus the correction factor is defined such that it gradually approaches unity as the manipulable cues approach the maneuver termination. In this work, Z_r can be re-estimated at each step of the procedure using the current parameter estimates and the joint coordinates at maneuver termination (as determined by the parameter estimates based on the raw samples of the manipulable cues). Thus, the correction factor is unity, strictly speaking, for that single target camera-space cue position for which the Z_r is computed. In this work, use was made of the fact that the resulting errors are small because these target positions are clustered in a small region. Recently, the flattening procedure has been extended to incorporate the Z_r variation among the targets by flattening them as well.

The method has been applied recently to the task of mounting an automobile wheel onto a brake plate, arbitrarily positioned and oriented. The required tolerance of about 1 mm, with minimal compliance was met easily and consistently, for a very wide range of positions and orientations of the receiving piece. As mentioned, the observed improvement in positioning has been reliably a factor of two. It is also worth noting that the use of this procedure reduces substantially the need to position cameras at distances from the workpiece which are large relative to the span of end-effector movement.

Camera Pan/Tilt

As discussed, identification of the six elements of the vector **C** requires that the arm be exercised through an extensive preplanned trajectory, a trajectory which generally has no purpose apart from parameter initialization. In principle, therefore, participating cameras can be panned and/or tilted to include any required subdomain of the robot's workspace while retaining the narrow field of view which may be essential for the required terminal precision. Reidentification of the parameters, however, would come at the cost of otherwise-extraneous arm movement unless the camera-specific information which is acquired prior to pan/tilt can be applied to the new camera position.

We have found that such an application is possible, and that the quality of parameter estimates can be fully recovered with one or two video samples acquired after the camera movement. The requirements for realizing this recovery of estimates are 1) knowledge of the camera axis about which rotation occurs, 2) knowledge of the angle of camera rotation about this axis to within approximately a tenth of a degree, and 3) use of flattening as earlier described.

The implications of the ability to direct long-focal-length cameras as required without undergoing additional otherwise-unnecessary arm movement for parameter reidentification are great, especially in the context of space robotics. The combination of versatility, robustness, and very high precision is, to our knowledge, otherwise unavailable.

Summary and Conclusions

The problem of achieving reliable and precise closure in the case of three-dimensional, rigid-body positioning has been historically considered problematic in the field of robotics. Extreme measures have been considered including the use of teleoperation, carefully calibrated optics and kinematics, eye-on-hand camera configurations, contact-based estimation, special hardware designs to exploit passive compliance, and so on.

Part of the difficulty seems to be with the use of kinematic and optical models. There has been a tendency, on the one extreme, to build very accurate equipment and to calibrate until these models are globally very exact and then to rely on them completely. At the opposite extreme, there has been a tendency to use the models only minimally, and to rely primarily on sensed information acquired very near the maneuver termination, information which can be used to close small gaps between current and desired positions. Neither of these extremes exploit the rich potential which has been recognized in estimation theory of using nominal mathematical models along with a sequence of measurements acquired in the broad vicinity of the terminal point of interest. The resulting precision and robustness can be very great.

Camera space manipulation builds on this philosophy of managing information. The innovation by which it accomplishes this is similar in some ways to the means by which humans realize remarkably versatile hand-eye coordination. Rather than referring maneuver objectives to an independent, fixed, physical coordinate system, the criteria for maneuver success are understood entirely in terms of the appearance of the manipulated object in the reference frame of the visual sensor. In the case of camera-space manipulation, two or more independent, two-dimensional image planes are used.

The aforementioned model-based estimation consists of the application of video samples and joint-rotation samples acquired at several instants during the maneuver approach. From these samples and from available (though globally imperfect) kinematics and camera models, estimates of the nonlinear relationship between internal joint rotations and the appearance in camera space of the manipulated body can be determined with extremely high accuracy in the region of interest.

Finally, using these current estimates along with camera-specific image-plane maneuver objectives, the sequence of commanded joint rotations is calculated by means of a least-squares resolution of all image-plane objectives. Importantly, samples acquired very near maneuver culmination are generally not necessary for even very high terminal precision. Similarly, ongoing movement of the arm is not dependent on acquisition of any particular sample. Instead, the commanded joint rotations at any juncture during the motion are based on estimates made using whatever subset of samples may be available at that juncture of interest. Therefore, smooth, uninterrupted, precise, and robust motion is possible.

References

[1]Skaar, S. B., Brockman, W. H., and Hanson, R., "Camera-Space Manipulation," *International Journal of Robotics Research,* Vol. 6, No. 4, 1987, pp. 20–32.

[2]Skaar, S. B., Brockman, W. H., and Jang, W. S., "Three-Dimensional Camera-Space Manipulation," *International Journal of Robotics Research,* Vol. 9, No. 4, 1990, pp. 22–39.

[3]Skaar, S. B., Chen, W. Z., and Miller, R. K., "High Resolution Camera-Space Manipulation," *Proceedings of the ASME Design Automation Conference,* Miami, FL, Sept. 1991.

[4]Skaar, S. B., Yaldo-Mooshabad, I., and Brockman, W. H., "Nonholonomic Camera-Space Manipulation," *IEEE Transactions on Robotics and Automation,* Vol. 8, No. 4, 1992, pp. 464–479.

[5]Tani, K., Abe, M., Tanie, K., and Ohno, T., "High-Precision Manipulator with Visual Sense," *Proceedings 7th International Symposium on Industrial Robots: Japan Industrial Robot Association* (Tokyo, Japan), 1977, pp. 561–568.

[6]Weiss, L. E.,, Sandersen, A. C., and Neuman, C. P., "Dynamic Sensor-Based Control of Robots with Visual Feedback," *IEEE Journal of Robotics and Automation,* Vol. RA-3, No. 5, 1987, pp. 404–417.

[7]Junkins, J. L., *An Introduction to Optimal Estimation of Dynamical Systems,* Sijthoff and Noordhoff, Alphen Aan Den Rijn, The Netherlands, 1978, pp. 29–33.

[8]Nevins, J. L., and Whitney, D. E., "Robot Assembly Research and Its Future Applications," *Computer Vision and Sensors-Based Robots,* edited by G.G. Dodd and L. Rossol, Plenum, New York, 1979, pp. 275–321.

[9]Chen, W. Z., Korde, U. A., and Skaar, S. B., "Position Control Experiments Using Vision," *International Journal of Robotics Research,* Vol. 13, June 1994, pp. 199–208.

Part 4.   Dynamics and Control

Chapter 10

Tutorial Overview of the Dynamics and Control of Satellite-Mounted Robots

Richard W. Longman*
Columbia University, New York, New York 10027

Introduction

D URING on-orbit operations it will become increasingly common to have robots on satellites manipulating loads with masses that are not negligible compared to that of the satellite. Such operations present various challenges. The first difficulty relates to robot path planning, a topic that receives considerable attention for ground-based robots, and one that is significantly more complicated in space. The complication arises from the fact that the robot is not mounted on an inertially fixed base, but rather on the satellite that can move in response to the robot motion. This affects what one must command the robot to do to get the end effector to the desired position in space. A related second class of difficulty is that the robot motions produce attitude and acceleration disturbances to the satellite. The attitude disturbances can be a serious problem for any experiments on board that require fine pointing of instruments. The translational accelerations can be a serious concern when the satellite houses experiments designed to take advantage of the microgravity environment offered by spaceflight. Translational acceleration of the spacecraft is of primary concern to this environment, but attitude angular velocity and angular acceleration are important as well. Yet a third class of difficulties for robotics in space is the robot structural flexibility that is introduced by the common simultaneous robot design requirements of low weight and large workspace.

To illustrate the first class of problems consider the remote manipulator system (RMS), mounted on a Shuttle of mass 67,000 kg, and manipulating a load whose mass is 30,000 kg—the maximum payload rating for the RMS (see Fig. 1). For simplicity, let the Shuttle and load be point masses. Then commanding the robot to move the load 6 m using the usual robot kinematic equations will result in 6 m of motion relative to the Shuttle. Conservation of the system center of mass location indicates that the Shuttle will move in the opposite direction. This results in an actual motion of the load of only 4.2 m, and in the robot missing its target by 1.8 m.

Copyright © 1994 by the American Institute of Aeronautics and Astronautics, Inc. All rights reserved.
*Professor of Mechanical Engineering.

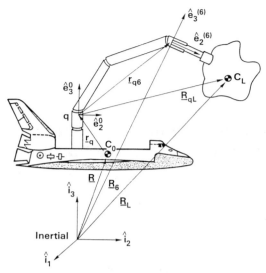

Fig. 1   Satellite-mounted robot with a massive load.

The fact that the Shuttle moves 1.8 m in the preceding illustration indicates that it has undergone acceleration. The size of this acceleration is determined by the path chosen for the robot maneuver, and the time history along this path. The illustration uses point mass models, and is therefore too simple to illustrate attitude disturbances to the Shuttle; but such disturbances are obvious when the point mass for the Shuttle is replaced by a rigid body.

To illustrate the third class of problems we simply comment that approximately one-third of the time spent by astronauts in operating the RMS is spent waiting for vibrations to decay to a required 2-in. level before grasping an object. Stated in more *graphic* terms, for every 6 h of operation of the RMS during a space flight, the astronauts spend 2 h waiting for the vibrations to decay.

This chapter gives an overview of these new problem areas in satellite operations and gives certain methods of treating them. The material presented summarizes results reported in Refs. 1–5.

Before going into detail, let us preview the material discussed, and at the same time indicate how the material relates to the references. Three basic topics in the field of robotics are the forward kinematics, the inverse kinematics, and the robot workspace. In Ref. 1, the author and co-workers introduced a new space-based kinematics for robots mounted on satellites that have an attitude control system in operation. Reference 2, and an expanded journal version appearing in Ref. 3, give this new space-based forward and inverse kinematics for the Shuttle RMS. The next section of this paper is devoted to an overview of these results.

The kinematics problems change character when the satellite attitude control system is turned off, as it often is during RMS operation on the Shuttle, and this is the topic of the following section. The satellite is then free not only to translate, but to rotate, as a result of robot motion. In this case, the problem of positioning the load is no longer a kinematics problem. Reference 4 introduces two new terms for this situation, the *forward kinetics problem* and the *inverse kinetics problem,* in place of the terms forward kinematics and inverse kinematics problems. The reference presents the first complete analytical solution to the difficult new inverse kinetics problem. The paper also develops the space-based robot workspace which is often larger than the corresponding ground-based workspace. In Ref. 6, Vafa and Dubowsky discuss similiar problems via their virtual manipulator approach, and present a numerical approach to the inverse kinetics problem.

Another aspect of the load positioning problem is that the robot operation applies forces and torques to the spacecraft that disturb its attitude and position. In the next section of this chapter, results from Ref. 1 are summarized that permit direct computation of the disturbing torques and forces produced by robot motion. Such computations could be used in a feedforward control to help maintain accurate satellite attitude. Reference 5 extends these results to the computation of disturbing torques and forces for flexible robots. Reference 6 develops ways to compute inverse kinetics solutions (for satellites without an operating attitude control system) that limit the robot disturbances to the satellite. The optimization of such path planning is discussed in Ref. 7.

The final section of this chapter discusses by example, new attitude control problems introduced by the flexibility of typical large, lightweight, space-based robots such as the RMS.[5] It demonstrates that maneuvers of a flexible robot generally produce robot vibrations at the end of a maneuver that can tumble the spacecraft.

The reader is referred to Refs. 8 and 9 for more information about these topics. Reference 8 is a special issue on robotics in space of *The Journal of The Astronautical Sciences* edited by the author and Lindberg. It includes Refs. 3–6 as well as other papers, such as Ref. 10, which present ground experiments in space robotics. Reference 9 is an edited book reprinting 10 important papers in the field. It includes Refs. 3, 4, and 6 as the first three papers of the book. An overview similar to the present work appears in Ref. 11.

Manipulation by Robots Mounted on Satellites with the Attitude Control System in Operation

Perhaps the three most basic problems in the field of robotics are 1) the forward kinematics problem, i.e., given all of the robot joint variables, find the inertial position and orientation of the load; 2) the inverse kinematics problem, i.e., given the desired inertial position and orientation of the load, find a set of robot joint variables that will accomplish this goal; and 3) the determination of the robot workspace. Various definitions of workspace are used; here we choose to find the set of points in inertial space that the robot can reach.

In this section each of these problems is generalized to satellite-mounted robots for the case when the satellite has its attitude control system (ACS) turned on. It is reasonable to assume that the satellite would have an active attitude control system in operation under most conditions. Generally one needs to keep a fixed satellite attitude for purposes of maintaining ground communications with directional antennae, or for maintaining solar panels pointing toward the sun, or maintaining the pointing of various sensors or experiments that observe the Earth or stars, etc. With the attitude control system on, the satellite is prevented from rotating in response to robot motion, but is free to translate in response to robot motion—the robot base is unconstrained in translation, but constrained in rotation. The problem of planning a robot maneuver to reach a desired location in inertial space must account for the fact that in moving the joints of the robot, the base of the robot which is fixed on the satellite will move in a correlated way. One must plan the robot joint motions to account for the motion they produce in the robot base to have the end effector reach the desired location in inertial space.

New Forward and Inverse Robot Kinematics Problems

In this section we introduce the basic concepts involved in these problems by considering the simplified case in which the satellite and the robot load are modeled as rigid bodies with mass, and the robot links themselves are considered to have negligible mass. This simplified situation is relatively easy to understand, and it contains all of the fundamental concepts. For the general situation where the mass of each robot link is included in the analysis, the reader is referred to Ref. 1 for a detailed derivation of the forward and inverse kinematics for a polar coordinate robot, and to Ref. 2 or its journal version, Ref. 3, for an analogous derivation for an elbow manipulator. Reference 3 is reprinted in the book on *Space Robotics: Dynamics and Control*,[9] as the introductory overview chapter. The reader is also referred to Ref. 6 for the virtual manipulator approach to this same problem.

Forward Kinematics Problem

For ground-based robotics the forward kinematics problem is a very straightforward kinematics problem. One fixes coordinate systems in the base of the robot, in the robot load, and in each of the robot links. One then uses the robot joint variables to determine the orientation, and the location of the origin, of each coordinate system relative to the coordinates in the neighboring link. This is easily done using the 4×4 matrices of the homogeneous transformation. Cascading these results for successive links, from one end of the robot to the other, gives the direction cosine matrix describing the load's orientation in base coordinates, and the vector locating the robot load relative to the robot base.

Now consider the satellite-mounted robot case, with the attitude control system turned on and maintaining a spacecraft orientation that is fixed relative to the fixed stars. The ground-based forward kinematics just described can be used to find the position and orientation of the load relative to the robot base coordinate system, i.e., relative to the spacecraft. One must then determine the location and orientation of the robot base in inertial space to know the location and orientation of the load.

Since the attitude of the spacecraft is being maintained relative to inertial space, the direction cosine matrix obtained earlier, giving the attitude of the load relative to satellite fixed coordinates at the robot base, also tells one the attitude of the load in an inertially fixed reference frame whose axes can be chosen parallel to the spacecraft axes.

The task of finding the inertial position of the load remains. It is assumed that the satellite-robot system is in free flight, and not subject to any external forces. In actuality, if the satellite is in a circular orbit around the Earth, the gravitational force from the Earth is counterbalanced by centrifugal effects, giving the appearance of zero applied force. With this assumption, summing Newton's law for each body in the system establishes that the center of mass of the system of bodies is fixed in inertial space. This model neglects small torques due to the gravity gradient (from the Earth and the moon), torques due to differential atmospheric drag from any residual atmosphere, and torques due to solar radiation pressure. Rotating reference frame effects are neglected in cases in which the system is pointing toward the Earth.

Suppose that at time $t = 0$ the robot is in a home position relative to the satellite. Then one can determine the location of the system center of mass relative to the satellite coordinates at the robot base. For the simplified situation considered here, there are only two masses, that of the satellite and that of the load, and the locations of these centers of mass are both known in satellite coordinates for this robot home position. The system's center of mass then lies on the line joining the two centers of mass, and is located along this line at a distance equal to the total distance times the ratio of the load mass to the total mass.

After the joint variables are changed to new values, the center of mass of the system remains at this same inertial location. This inertially fixed point is on the new line that joins the satellite's and load's mass centers, at the same percentage of the total distance from the spacecraft mass center. Using the ground-based forward kinematics to express the load position relative to the satellite tells one the length of this new line and its orientation relative to the spacecraft. The orientation of this line is then also known relative to inertial space because the satellite coordinate directions remain unaltered during the robot motion with the ACS on. Putting these facts together allows one to uniquely determine the inertial position of the load's center of mass, and this completes the forward kinematics problem for a robot mounted on a satellite with the ACS on.

Note that this is still a kinematics problem in the sense that knowledge of the present robot joint variables (and the load and satellite masses) allows one to know the present load position and orientation in inertial space. It is not a pure kinematics problem in the sense that the result depends on the inertial properties of the system, i.e., on the mass of the load and the mass of the satellite. Nonetheless, this inertia-dependent problem will be referred to with the word kinematics in this chapter.

References 1–3 show that when the mass of each of the robot links is not neglected one obtains the same kind of result. Kinematic equations very similar to the ground-based forward kinematic equations are obtained, but this time they contain various mass ratios in their coefficients.

Inverse Kinematics Problem

The ground-based inverse kinematics problem is much more difficult than its forward kinematics counterpart. The forward kinematics equations, developed as described earlier, give the load position and orientation relative to the robot base coordinates as a function of the robot joint variables. In the ground-based inverse kinematics problem these equations must be inverted. One is given the load position and orientation, and these complicated nonlinear equations must be solved for the corresponding robot joint variables. There is usually more than one solution.

Consider a robot mounted on a satellite with its ACS on, and again assume the masses of the robot links are negligible compared to the satellite and load rigid-body masses. In this special case, it is again possible to create a simple solution to the satellite-mounted inverse kinematics problem, provided the solution to the ground-based inverse kinematics problem is known.

At the start of a robot maneuver, one knows the location of the load relative to the satellite, and hence the location of the system's center of mass relative to the satellite. For the sake of being definite, we can fix an inertial reference frame at this system mass center, with axes parallel the satellite coordinates (i.e., robot base coordinates). One prescribes where one wants the load in inertial space at the end of the maneuver (i.e., relative to the inertial reference frame), and how it is to be oriented. Hence, one knows the final position of the load's center of mass relative to the inertial reference frame. One can then use the definition of the center of mass for this two-mass system to find the position, at the end of the maneuver, of the satellite's center of mass relative to the inertial frame.

The satellite attitude remains fixed throughout the maneuver, so that knowing the inertial position of the satellite's center of mass at the end of the maneuver tells one where the robot base is located at the end of the maneuver relative to inertially fixed coordinates. This, together with the prescribed desired final position and orientation of the load relative to inertial coordinates, allows one to determine the needed position and orientation of the load relative to the satellite's fixed coordinates, located at the base of the robot. We can now use the ground-based inverse kinematics solution to solve for the needed joint variables at the end of the maneuver.

Thus, we have reduced the problem to the mathematical problem that is solved by the ground-based inverse kinematics solution. The reduction process found the location of the satellite, and hence of the robot base, that is required by conservation of the system's center of mass position for the load to be at the desired position in inertial space.

These results can be generalized to the case were the masses of the robot links are not neglected in the solution of the space-based inverse kinematics problem. See Refs. 1–4 for the much more complex analysis for this case. Of course, the inverse kinematics equations involve mass ratios now, but are otherwise similar to the ground-based case.

Choice of Kinematics to Use in Space and the Matching Conditions

In the previous section we described how to develop the forward and inverse kinematics for robots mounted on satellites. However, this is only part of the story. First, the kinematic solutions developed involve mass ratios that include the mass of the load, and during robot operation, various different loads may be grasped and released. This means that the on-orbit forward and inverse kinematics keeps changing. Second, the kinematics problems discussed earlier are for positioning and orienting an object in inertial space. Some tasks will require placing the load in a position given relative to the satellite, rather than specified relative to inertial space. Again, the kinematics used changes from task to task.

The set of computations that must be performed on-orbit can be summarized as follows. At time $t = 0$ with the robot in some home position, one computes the location of the satellite-robot system's center of mass relative to a chosen inertial reference frame, as well as the satellite's center of mass position relative to this inertial reference frame. For this initialization, one can choose the inertial reference frame origin to coincide with the system's center of mass location, but this luxury disappears as soon as a load is released or grasped. The system's center of mass equations give the components on the inertial reference frame of the system's center of mass as a function of the satellite's center of mass position in these coordinates and the robot joint variables and the link and load masses. Throughout on-orbit robot operations, these equations are alternately solved for the satellite's center of mass position relative to inertial space, given the system's center of mass location in inertial space, and vice-versa.

Whenever the robot arm is being commanded to go to a position prescribed in inertial space, one must use the space-based kinematics described in the previous section. At the end of such a maneuver, one knows the system's center of mass position in inertial space and the robot joint variables. Thus, one can compute the location of the satellite's center of mass from the previously described system center of mass equation. Assume for the moment that a load is being grasped at the end of the maneuver, and that it has zero velocity in inertial space, as does the satellite. Once the load is grasped, the total mass of the satellite-robot system changes because of the nonzero value of the load mass, and this new mass within the system alters the location of the system's center of mass. Using the inertial location of the satellite's center of mass just computed, one uses the system center of mass equations to solve for the location of this new mass center in inertial space. This represents a new initialization of the kinematic equations. The equations for the two-step computation just described are termed the matching conditions in Ref. 1. The space-based kinematics described in the previous section are then used for any maneuvering of this grasped load, but with the new set of mass values included in the kinematic equations. If instead of grasping a load, a load is being released at the end of the robot maneuver, we assume that it is being released with zero velocity, and then all of the same comments again apply.

Certain robot operations in space require that the load be positioned relative to the satellite instead of positioned relative to inertial space. Suppose for example, that the Shuttle RMS grasps a massive satellite positioned at a fixed location in inertial space, and moves it to a position inside the Shuttle bay. Since the desired end position for the load is given relative to the satellite, i.e., relative to robot base coordinates, the usual ground-based kinematics apply instead of the space-based kinematics. However, in preparation for any future maneuvers to locations specified in inertial space, one must keep track of the satellite and the system's center of mass positions in inertial space. The matching conditions are slightly different at the end of such a maneuver—this time the total mass of the system does not change, but the mass of the load changes from that of the massive satellite grasped by the manipulator, to zero when it is place in the Shuttle bay and the robot end effector releases it. At the same time, the mass of the Shuttle is increased by the same amount, and the location of the Shuttle's center of mass changes relative to Shuttle's fixed coordinates at the robot base.

Note that the preceding statements assumed that any load grasped has zero velocity relative to the satellite, and that any load that is released from the robot arm is released with zero relative velocity. Otherwise, the new system's center of mass location after the grasping or releasing will have a constant velocity relative to the original inertial coordinate frame. If the load being grasped has rotation, it is the task of the satellite attitude control system to correct the resulting impulsive attitude disturbance. A more complicated situation arises when the load grasped has spinning internal parts such as momentum wheels or reaction wheels.

In the above discussion we assumed the ACS maintained an inertially fixed attitude, but in many applications an Earth-pointing attitude is maintained for purposes of ground communication or for sensing or monitoring of the Earth. Maintaining an Earth-pointing attitude can also be useful for eliminating attitude disturbances from gravity gradient torques. For low Earth orbit an Earth-pointing attitude corresponds to approximately one rotation of the satellite every 90 min. For maneuvers that take an amount of time that is significant relative to this 90 min, some adjustment of the preceding results would be necessary. However, these situations produce dynamics problems in place of the much simpler kinematics problems treated here.

New Robot Workspace

In the previous sections we described how one obtains solutions for the satellite-mounted robot inverse kinematics problem. As in the ground-based inverse kinematics problem, there will be no solution to this problem if the desired load position or orientation is outside the robot workspace. In this section, we study the way in which mounting the robot on a satellite with an active ACS affects the robot workspace. There are various definitions of workspace, such as the reachable workspace and the dextrous workspace. Here we limit ourselves to the reachable workspace in inertial space, meaning we wish to determine the set of points in inertial space that can be reached by the end effector of the robot. Again, we will limit ourselves to the situation where the masses of the robot links are negligible compared to the satellite's mass and the robot load.

The reachable workspace as seen in satellite coordinates is the same as that for a ground-based robot. More specifically, the radial distance from the satellite's center of mass to the boundary of the workspace in any chosen direction is the same as the corresponding distance for a robot mounted on an inertially fixed base. What is new is that the center of mass of the satellite moves as one extends the arm in any chosen direction, which implies that the robot base moves.

The satellite's center of mass, the load's center of mass, and the system's center of mass all lie on a line in this two-mass system. When the load's center of mass is extended to the limit in any chosen radial direction from the satellite's center of mass, the distance reached relative to this fixed point in the satellite base coordinates is the same distance that would be reached by a ground-based robot relative to this same point in the base coordinates. In the ground-based case, this point is an inertially fixed point. In the satellite-mounted case, this satellite's center of mass point moves relative to inertial space, i.e., relative to the system's center of mass, as the load is extended, and they move in opposite directions. Hence, the maximum robot reach in all directions from the satellite's center of mass point is decreased for all directions. The workspace has shrunk by a factor depending on the satellite's mass and the load's mass. The maximum reach in all directions from the inertially fixed system's center of mass point is decreased by a multiplicative factor given by the ratio of the satellite's mass to the total system mass (see Ref. 4 for the analytical derivation).

Note that the workspace is now a function of the mass of the load being manipulated. Furthermore, matching conditions are required. After grasping or releasing a load, the preceding multiplicative factor changes for the maximum reach in each direction from the system's center of mass location, and furthermore, the system's center of mass location moves to a new location in inertial space. The new workspace after grasping or releasing is given relative to this new inertial location.

Predicting and Correcting the Attitude Disturbances Induced by Robot Motion

It is the objective of the ACS to maintain the attitude in spite of any disturbances, such as the disturbances caused by robot motion. Formulas are derived in Ref. 1 which compute these torques directly without the need for a full dynamic analysis of the multibody system. This information can be used 1) to predict the disturbances for any chosen robot path, 2) to help size the attitude control actuators, and 3) as feedforward commands that improve the attitude control system performance.

Feedforward control aims to cancel the robot disturbance torques before they have a chance to affect the satellite attitude. Without feedforward control, the disturbances have to produce an attitude error before any corrective torque is applied, while feedforward control predicts the disturbing torque and applies a torque to cancel it at the same time the disturbance occurs. In the ideal case, this allows the ACS to perfectly maintain the desired attitude.

Cancellation of the disturbance torques requires a continuously variable control actuator such as reaction wheels or control moment gyros, as opposed to reaction control jets. Note that as long as the ACS can recover to the desired satellite attitude at the end of the maneuver, the desired end-effector position in inertial space is obtained when using the space-based robot kinematics discussed earlier. Maintaining zero attitude error during the maneuver is not important to this positioning problem. Of course, it can be important in path planning to avoid possible collisions during a maneuver.

In the special case when the bearings in the robot base joint happen to be mounted at the center of mass of the satellite it is only the torques transmitted from the robot to the satellite through these bearings that disturb the satellite attitude. However, when these bearings are mounted elsewhere, the forces transmitted from the robot to the satellite also produce torques about the satellite center of mass that disturb the satellite attitude. Hence, to predict attitude disturbances, it is necessary to compute both the torques and the forces applied by the robot to the satellite.

The reaction moment feedforward command M_r needed to cancel the attitude disturbance is given by $M_r = M_s^b + r_b \times F_s$, where M_s^b and F_s are the moments and forces exerted by the satellite on the robot through the base link when the satellite is free to translate but not to rotate. The moments are about point b at the center of the bearings of the base joint, and r_b is a vector from the satellite center of mass to b. In Ref. 1, it is shown that the moments and forces needed in this expression can be obtained as follows from the moments and forces, M_f^b and F_s, that would be applied by the robot to the base if the base were inertially fixed

$$M_s^b = M_f^b + [m_{07}(R_{cs} - R) - m_{17}r_b] \times \frac{{}^0\mathrm{d}^2R}{\mathrm{d}t^2} \tag{1}$$
$$F_s = (m_0/m_{07})F_f$$

Here R is the vector from the origin of the inertial coordinates to the center of mass of the satellite, R_{cs} is the vector from this same origin to the system center of mass, m_0 is the mass of the satellite, m_{07} is the mass of the system, and m_{17} is the mass of the robot arm with load. The presuperscript on the derivative indicates an inertial derivative. The forces and moments that apply to the inertially fixed base case can be computed directly from the time histories of the robot joint variables using the following formulas

$$M_f^b = \sum_i \left\{ m_i \rho_{ci} \times \frac{{}^0\mathrm{d}^2\rho_{ci}}{\mathrm{d}t^2} + \frac{{}^i\mathrm{d}^2}{\mathrm{d}t^2}(\mathbf{I}^{ci} \cdot \omega^{i0}) + \omega^{i0} \times (\mathbf{I}^{ci} \cdot \omega^{i0}) \right\}$$
$$F_f = \frac{{}^0\mathrm{d}^2}{\mathrm{d}t^2}\left(\sum_i m_i \rho_{ci} \right) \tag{2}$$

Here ρ_{ci} is the vector from point b in the center of the bearings of the robot base to the center of mass of the ith link, ω^{i0} is the angular velocity of body i with respect to inertial space (i.e., relative to the robot base, so that it is a pure function of robot joint angles and angle rates), and $\mathbf{I}^{ci}$ is the inertia dyadic of body i about its center of mass.

These results are generalized in Ref. 5 to include the extra attitude disturbances produced by structural vibrations in the robot arm. Such vibrations are significant in many space-based robot applications because of the need for a large work space and for light weight. It is seen that the usual effect of such vibrations is to produce disturbance torques that do not average to zero, and hence can tumble the spacecraft. This phenomenon is discussed later.

Limiting the Disturbances to the Microgravity Environment During Robot Operation

The multiple users of a satellite can have conflicting requirements. One of the important characteristics of space is the microgravity environment that it offers. On the Space Station, there can be users who need to maintain the microgravity environment for long periods of time, and robot operations produce serious disturbances. Robot trajectories can be planned that limit this disturbance in an optimal way, as described in Ref. 7. When one assumes that the ACS does a perfect job of eliminating the attitude disturbances, one is left with the translational motion of the satellite. For a microgravity environment, it is not the satellite displacement that matters, but the acceleration of the satellite. This acceleration can be kept below any desired threshold by making the robot motions sufficiently slow. On the other hand, given a desired robot load final position and orientation, and a fixed amount of time to get the load to this position in inertial space, one can find an optimal trajectory to minimize the accelerations of the spacecraft. Thus, one minimizes some cost functional of the disturbance force F_s given earlier.

As before, when we consider the robot links to have a negligible mass, we can easily obtain some understanding of the problem. Suppose that we wish to manipulate a load from a given initial position in inertial space to a given final inertial position, and do so in such a way that we minimize the maximum magnitude of the acceleration of the spacecraft—i.e., minimize the maximum disturbance to the microgravity environment during the maneuver. We should connect the initial and final load position by a straight line, since acceleration in any other direction will be of no assistance, and will produce extra disturbance. We view the straight line in satellite coordinates. The summation in Eq. (2) is reduced to one term, and the right-hand side becomes the moment of acceleration of the load relative to the satellite, along this line in satellite coordinates. Then to minimize the magnitude of the force in Eq. (1), which is proportional to the acceleration disturbance on the satellite, one should accelerate the load at a constant rate along this line for the first half of the distance to the desired end point, and then decelerate it at this same rate for the second half of the straight line trajectory. The acceleration and deceleration rates are chosen to be that value that allows the load to just reach the target point in the time allotted.

The problem becomes significantly more complicated in the case of a robot mounted on a satellite with the ACS off, see Ref. 7.

Manipulation with a Robot Mounted on a Satellite with the Attitude Control System Turned Off

The ACS is often turned off on the Shuttle when the RMS is in use. Some reasons to turn off the ACS include the following:

1) If the ACS uses gas thrusters it is necessarily an on-off system. When the robot is about to grasp a satellite or other load, and the ACS happens to decide that the attitude error has now reached the threshold where correction is necessary, then turning on the attitude control thrusters could easily cause a collision of the robot with the intended load.

2) Even when actuators are used that do not necessitate an on-off mode of operation, such as reaction wheels or control moment gyros, inaccuracies and transients (e.g., overshoot) in the ACS performance might still suggest robot operation with the ACS off.

3) In the case of the Shuttle, the ACS is often turned off to avoid ACS exhaust gas impingement on the manipulated load.

4) If one can be smart enough in planning the robot maneuver, the robot joint history can be chosen to correct its own attitude disturbances, thus saving ACS fuel.

When a robot is mounted on an inertially fixed base, specifying the current values of all of the joint variables uniquely determines the inertial position of the end effector at the tip of the robot arm, and also uniquely determines its inertial attitude or orientation. As shown earlier, when the robot is mounted on a satellite with the ACS in operation, specifying the current robot joint variables, and the masses of the satellite, each robot link, and the load uniquely determines the inertial position and orientation of the load. However, when the robot is mounted on a satellite that is free to both translate and rotate in response to the robot motion, then the current position and orientation of the end effector as seen in an inertial frame is no longer determined by knowledge of the current joint variables and system masses. It becomes a function of the whole history of all of the robot joint variables. Thus, in deciding how to get the robot load into the desired position one needs to determine a whole history of joint angles to use. This is in stark contrast to the ground-based robot case, where one simply aims at the needed final values of the joint variables, and can use any path leading to these values, such as interpolation in joint space or interpolation in Cartesian space. These two robot trajectories would generally produce different final inertial positions of the load when applied to a robot mounted on a satellite with the ACS off.

Illustrative Example

Here we demonstrate the dependence of the end-effector position on the history of robot joint angles using an example which is simple enough that one's intuition can predict the behavior. The associated mathematics can be found in Ref. 3. Consider a simple two-link robot mounted on a satellite as shown in Fig. 2. For simplicity, we consider that the satellite is a one-dimensional distribution of mass, that the second link of the robot has inertial properties identical to those of the satellite, and that the first link of the robot is massless. The joint angles are shown in Fig. 2 with joint variables ϕ_1 and ϕ_2 which both assume the value zero when the robot links are positioned as they are in Fig. 2. Note that ϕ_1 is the angle of rotation of robot link one relative to the satellite.

For a robot with an inertially fixed base, the statement that $\phi_1 = \phi_2 = 180$ deg would identify a unique inertial position of the end effector. Here we consider two maneuvers that go from $\phi_1 = \phi_2 = 0$ deg to these same final robot joint angles of $\phi_1 = \phi_2 = 180$ deg, and show that the final position of the end effector is very different for the two cases. Maneuver no. 1, illustrated on the left in Fig. 3 shows a sequence of end-effector positions when ϕ_1 is rotated first from 0 deg to 180

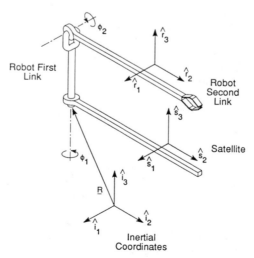

Fig. 2 Three-stick satellite-robot system.

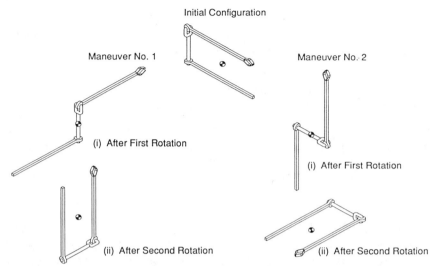

Fig. 3   Two maneuvers going from the initial robot joint angles to the same final robot joint angles, but producing a different end-effector position in inertial space.

deg, and then ϕ_2 is rotated from 0 deg to 180 deg. Maneuver no. 2 on the right of Fig. 2 reverses the sequence, rotating ϕ_2 first and then ϕ_1. The final satellite and end-effector positions shown at the bottom of the figure are very different. One can of course consider other maneuver histories for these same final joint angles, e.g., rotating the joints simultaneously, and these would give still other end-effector positions.

The following observations are sufficient to predict the motion shown in the figure. For simplicity, consider that link one of the robot has zero length. At time $t = 0$, the robot-satellite system is assumed to be at rest in inertial space. Then, rotating any joint of the robot produces a rotation of each of the bodies involved about a principal axis of that body. Therefore each separate rotation of an axis of the robot will produce a motion that is confined purely to an inertially fixed plane. Furthermore, since the satellite and robot body 2 have the same mass and inertia properties, when the joint angle rotates through an angle ϕ, each body will move through an angle $\phi/2$ in inertial space but in opposite directions. Also, since we assumed there are no external forces on the satellite-robot combination, the center of mass of the system must stay fixed in inertial space. This center of mass position is indicated in the figure.

This example shows that in spite of the fact that the satellite-robot system has zero angular momentum, it is still possible to accomplish a net rotation of the entire system as if it were a rigid body. One can see this by considering using maneuver no. 1 and following it by maneuver no. 2 done in reverse, to get back to the initial joint angles. Although the robot is back to its original position relative to the satellite, the satellite-robot system has rotated relative to inertial space, and this rotation has been accomplished with zero system angular momentum. This is always possible as demonstrated in Ref. 4.

A common example of this phenomenon is the ability of a cat, when dropped upside down, to turn over before landing. Astronauts are also trained in procedures of this kind. When they are in free flight either inside an orbiting space vehicle, or on a space walk outside the vehicle, such maneuvers allow them to turn around in spite of the fact that they are not in contact with some external body on which they could push or apply moments to turn themselves.

The same phenomenon will appear later in this chapter, and appears in Ref. 5, in a somewhat disturbing example. There it is demonstrated that structural vibrations in a typical large, lightweight robot such as the Shuttle RMS will usually produce torques on the satellite that can tumble the satellite after the robot maneuver is over, because of stored angular momentum in the robot vibrations which must be cancelled by angular momentum of the spacecraft when the ACS is off.

Second Example

Various researchers have done experiments in the gravitational environment on the Earth's surface to demonstrate the behavior and the difficulties of robot operations on a satellite with the ACS turned off. The closest thing that one can do to simulate the free-flying satellite in space is to support the satellite-robot on a film of air using an air table. This restricts the satellite movement to the plane of the table, but allows free translation and rotation in this plane. In space, satellite rotational motion in response to robot movements is restricted to a plane only under very specialized situations involving rotation about principal axes of both the satellite and robot, and requiring appropriate restrictions on the center of mass locations. Each individual rotation in the previous example falls into this category, by choice, so that the situation is sufficiently simple to allow one's intuition to predict the answer. Planar motion experiments are not rich enough to illustrate the full robot joint path dependency of the final satellite attitude, but they are rich enough to illustrate the path dependency through presenting a time-varying robot inertia to the robot base motor. One experimental paper of this kind is Ref. 10.

The phenomenon can be understood in terms of the following simple example. Suppose that the robot arm is mounted at the center of mass of the satellite, and that there is a point mass load at the end of the arm. With the robot arm fully extended in the plane of the table, consider rotating the arm in this plane through an angle of 180 deg relative to the satellite. Relative to inertial space the arm will rotate through some lesser angle, call it α, and the satellite will move in the opposite direction by the complementary amount 180 deg $-\alpha$. The ratio of the robot arm rotation with respect to inertial space to the satellite rotation relative to inertial space is determined by the reciprocal of the ratios of the inertias of the satellite and the robot with load. At the end of the rotation, the arm is pulled inward so that the end effector is, say, half the original distance from the satellite. Then the arm presents a much smaller inertia for rotation about its base. If we then rotate the arm back through angle -180 deg, and then extend it out to the original fully extended length, the robot arm is back to its initial configuration relative to the satellite. But during the return rotation of the arm the satellite will make a smaller magnitude rotation relative to inertial space than in the first arm rotation, because the ratios of the inertias of robot and satellite have changed. Thus, there is a net change of the end-effector position in inertial space during this maneuver, although the robot joint angles are back to their original values.

Because of the path dependency described in the preceding two examples, the forward kinematics problem is no longer a kinematics problem—it is a dynamics problem of predicting the motion of the multibody dynamic system when the time histories of the robot joint variables are given. For purposes of creating a parallel appearance, Ref. 4 uses the term kinetics in place of dynamics, and coins the term *forward kinetics problem* for this new type of robot problem. Similarly, the term *inverse kinetics problem* is coined to describe the new version of the inverse kinematics problem that arises for robots mounted on satellites that are free to rotate and translate in response to the robot motion. Each of these problems is treated in detail in Ref. 4, and each is summarized hereafter.

Satellite-Mounted Robot Forward Kinetics Problem

The forward kinetics problem is actually a straightforward problem in rigid-body dynamics. It is equivalent to predicting the motion of a body in free space as the result of motion of some internal moving parts. Given the initial position and orientation of the satellite, and given the initial robot joint variables which determine the initial positions and orientations of all robot links and the robot load, one can predict where the load and the satellite will go as the joint variables are changed as a prescribed function of time. One simply sets up the differential equations of motion of the multibody dynamic system and solves them. One can use any dynamics formalism, such as Newton-Euler, or Lagrange's equations, or Kane's equations.

The most common situation involves the satellite-robot system at rest in inertial space at time $t = 0$. Furthermore, it is reasonable to assume, as we have before, that there are no external forces or torques being applied to the satellite or robot. In this case, one can simplify this problem by making use of a first integral of the motion. A basic law of angular motion is that the inertial time rate of change of the angular momentum of a system is equal to the sum of all external moments on the system provided the angular momentum and the moments are defined about an inertially fixed point or about the system's center of mass. When done about some other point, for example, a point in the center of the bearings of the base joint of the robot mounted on the satellite, then one needs to add some extra terms. Since there are no external moments, one can integrate this expression once to obtain a first integral of the motion; and this can be solved for the inertial angular velocity of the spacecraft as a function of the angular velocity of each robot link and the robot load, relative to the spacecraft, and the robot base location in inertial space which is expressible in terms of the joint variables using the center of mass conservation. In other words, the angular velocity of the spacecraft is given algebraically as a function of the robot joint variables, and their derivatives. Such an expression is given as Eqs. (15) and (17) in Ref. 4.

Given the satellite angular velocity as a function of time in terms of the robot joint variable time histories, one obtains the satellite attitude history by integrating a set of first-order equations. These equations are kinematic equations relating the angular velocity of a reference frame to any chosen set of variables specifying reference frame orientation, such as direction cosines, or Euler angles, or quaternions. Given the satellite attitude, one uses the ground-based forward kinematics to determine the location and orientation of the load relative to the satellite. Then the orientation of the spacecraft together with the center of mass conservation, specifies the inertial position and orientation of all masses in the system as a function of time. In particular, it gives the inertial position and orientation of the load. See Ref. 4 for details.

Satellite-Mounted Robot Inverse Kinetics Problem

There is a fundamental difference between the ground based inverse kinematics problem and the inverse kinetics problem for a robot mounted on an unconstrained satellite base.

Reference 4 shows that by proper choice of the robot path going from one chosen joint angle configuration to another, one can cause the satellite attitude to assume any desired attitude once the robot arrives at its final joint angles. One implication of this result is that it is possible to reorient any load and have the robot base attitude be unchanged at the end of the maneuver. However, this is true only at the end of the maneuver (or if desired, it can hold at several points during the maneuver), and it is impossible to avoid attitude disturbances during the maneuver (or between these points). It is the subject of current research to find ways to easily plan robot joint angle histories that will minimize the satellite disturbances.[7]

The standard robot inverse kinematics problem is modified to the following in the case of a robot mounted on a satellite with the ACS off, One specifies: 1) not only the desired position and orientation of the load in inertial space, but also 2) the desired final satellite attitude.

Note that, in contrast to the satellite attitude being specifiable, conservation of the system's center of mass prevents one from arbitrarily specifying a desired satellite position at the end of the maneuver.

As an initial step in attacking this inverse problem, one can compute the final joint angles of the robot. Based on these specified quantities, one can solve the attitude-fixed satellite-mounted robot inverse kinematics Problem discussed earlier in the subsection by that name, to find the joint angles needed at the end of the maneuver. This kinematics problem acknowledges that the center of mass of the system must be fixed, so that the choice of the satellite attitude and of the desired load inertial position will dictate the position of the robot base in inertial space. To proceed with the solution of the inverse problem, one must determine a joint angle history, going from the given initial joint angles to these final joint angles, which is picked so that robot motion history produces satellite attitude motion that results in the chosen final attitude. This is generally a very difficult problem.

Note that the usual inverse kinematics problem in robotics has more than one solution, but only a small number of solutions. For example, in elbow robots one usually has both elbow up and elbow down solutions, although one of these may not be physically usable. On the other hand, the inverse kinetics problem discussed here usually has an infinite number of solutions. In spite of having so many solutions, the process of finding even one solution is unfortunately quite complicated, requiring solution of a very complicated boundary value problem in ordinary differential equations.

One feasible solution is given in Ref. 4, and this one is nearly an analytical solution to the very complex governing equations. The solution is described in Fig. 4 in terms of a seven-stage sequence of operations. Stage 1 starts with the current robot position and follows any chosen path to align the robot arm along a system principal axis. During this stage one solves the forward kinetics problem to determine the resulting configuration. The robot arm then executes a coning motion about this principal axis. This coning motion is analogous to spinning a wheel inside the spacecraft, and hence will rotate the spacecraft about this principal axis by any desired amount. The rotation resulting from this coning motion is a nonlinear effect, and is related to kinematic drift of an inertial platform. A second order analytic solution is given in Ref. 4. Stage 3 realigns the robot arm along another principal axis, and stage 4 makes a coning motion about this axis. Stages 5 and 6 do the same for the third principal axis. Then stage 7 moves the joints of the robot to the final desired angles by making any specific choice of path. For this stage one again uses the forward kinetics solution, but this time one solves the equations backwards to find the starting satellite orientation needed in order to end on the desired configuration. Then the initial satellite attitude for stage 2 is known, and the final satellite attitude for stage 6 is known. It is then necessary to determine how much coning to do in stages 2, 4, and 6 so that when these stages are combined with transfers from one principal axis to the next in stages 3 and 5, the end result is the needed satellite rotation for stages 2–6. Reference 4 gives an analytical solution to this problem governing stages 2–6, which happens to be an inverse kinematics problem. When combined with the second-order solution for the amount of coning needed along each principal axis, one obtains an analytical solution to the inverse kinetics problem.

This is the only analytical solution to the inverse kinetics problem available so far. Reference 6 gives a numerical approach to generating solutions, and it is the subject of current research to find reasonable methods for generating optimized solutions.[7]

Satellite-Mounted Robot Workspace in Inertial Space

Consider again the case of a point mass load, and consider a simple three-degree-of freedom elbow manipulator. By fully extending the arm, the distance between the load and the robot base joint is made as large as possible. If, in addition, we pick the direction of this extended arm so that it is along the direction of the vector from the center of mass of the satellite to the robot base joint, then the load and the satellite's center of mass are as far apart as possible. Under the assumption that the masses of the robot links are negligible compared to the load and the satellite, this means that the load is as far as possible from the system's center of mass. This distance is then the maximum distance from the system's mass center to the boundary of the robot workspace.

However, in the case where the satellite is free to not only translate but to rotate in response to the robot motion, we have seen earlier that the robot's motion history can be chosen to produce any desired final satellite attitude. Therefore, by

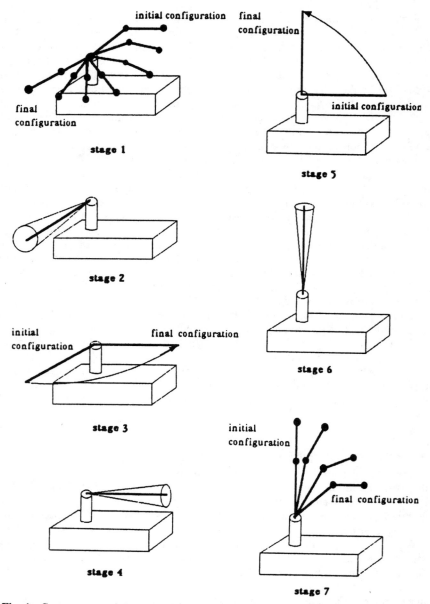

Fig. 4   Seven stages of the feasible solution to the inverse kinetics problem, viewed in satellite coordinates.

using all possible satellite attitudes, together with the robot joint angles determined earlier for maximum reach, it is possible to make the maximum reach apply in all directions in space. Therefore, the robot workspace is a sphere in inertial space, and the radius of the sphere is a monotonically decreasing function of the load mass.

By having turned off the ACS, the forward and inverse kinematics problems have become much more complicated forward and inverse kinetics problems, but there is a reward for this added complexity —a workspace that is a perfect sphere, and one which in general is much larger than the workspace of the attitude-fixed case, and usually much larger (depending on the mass of the load) than the workspace of the same robot mounted on an inertially fixed base.

Attitude Tumbling Caused by Flexibility in Satellite-Mounted Robots

Because of the rather large work space requirements for many satellite-mounted robots, and because of the expense of getting mass into orbit, future space manipulators will exhibit significant structural flexibility. Such oscillations will certainly affect the robot disturbances to the satellite attitude, and while the oscillations last, these disturbances will persist after the end of the robot maneuver. One might presume that robot arm structural vibrations would simply induce satellite attitude oscillations. In this case, use of the feedforward reaction moment compensation torques obtained earlier using a rigid-body model would compensate for robot motion disturbances in the sense that once the vibrations subside the satellite attitude would be correct.

In Ref. 5 it was demonstrated that such a presumption is absolutely false. The most common situation is that the structural vibrations of the robot arm can tumble the spacecraft.

This tumbling effect can be easily understood with the following example, which is treated in detail in Ref. 5. For simplicity consider a two-link robot such as that shown in Fig. 5, with a circular cross section for the flexible member producing identical vibration mode shapes and frequencies for vibrations in any two orthogonal planes through its center line. For the moment, let the robot be mounted on an inertially fixed base. Consider the vibrations induced by a maneuver that rotates robot angle α_2 about $\hat{s}_1$, starting from, say, 45 deg above the horizontal down to the horizontal plane, and then rotates robot angle α_1 about $\hat{s}_3$ through, say, 90 deg and comes to a stop. Immediately after the acceleration and deceleration of angle α_2, the robot beam will be oscillating in the α_2 plane. The amplitude and phase angle of these oscillations can be adjusted to any desired value by adjusting the rate of acceleration and deceleration, and by adjusting the total specified angle of travel. While the beam is oscillating in this manner, the rotation of angle α_1 takes place, and this superimposes an additional oscillation, this time in the horizontal plane, and again this oscillation can have any chosen amplitude and phase angle. Since structural vibrations for a long aluminum beam such as the Shuttle RMS arm are very lightly damped, these oscillations will persist after the end of the relatively short robot maneuver.

If the phase angle just happens to be exactly the same for both oscillations, then the combined oscillation is in a plane determined by the relative magnitudes of each. For all other phase angles the combined motion of the two oscillations will cause the tip of the robot arm to execute some form of ellipse. If the oscillations

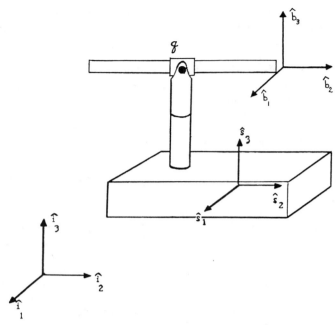

Fig. 5   Satellite-mounted two-link robot.

happen to be exactly out of phase and of equal amplitude, then the tip of the arm executes a circular motion. Except for the first very special case, the final circular or elliptical motion of the arm tip indicates that the arm motion contains angular momentum. This angular momentum decays relatively slowly in the lightly damped structure.

Now let the robot be mounted on a satellite. To make the situation as simple as possible, consider the α_2 joint for rotation in the vertical plane to be mounted at the center of mass of a satellite which has a spherically symmetric inertia ellipsoid. Furthermore, consider the beam of the satellite second link to be in such a position that the α_2 joint is at the beam's center of mass. Under these conditions, it is easy to show that the same types of robot oscillations occur in the satellite-mounted case as in the ground-fixed case, although there is a change in the frequency of the oscillations. For example, when the α_2 angle rotates, the satellite will rotate in the opposite direction. When mounted on a satellite with no external forces, the total angular momentum of the system must be zero. Therefore the angular momentum in the robot arm must be cancelled by the angular moment of the satellite—i.e., the satellite must tumble. One could think of the robot vibration averaged over revolutions to be the same as if there is a wheel rotating at the end of the robot arm. Rotating a wheel inside a spacecraft will cause the spacecraft to rotate—reaction wheels are a standard actuator to use in attitude control. We conclude, that if there is no damping in the structural vibrations of the robot links, then at the end of a typical maneuver, robot link vibrations will induce rotation of the satellite in inertial space, and this rotation will continue forever. When there is damping, the vibration and the rotation will both eventually stop, but at a satellite attitude that would be very difficult to predict without a very accurate damping model.

Here, the presentation has relied on one's physical intuition. A complete mathematical treatment of the problem can be found in Ref. 5.

Conclusions

As a somewhat amusing analogy, consider that usually the dog wags its tail. In the case of a satellite-mounted robot with rigid links manipulating a particularly large load, one might consider that the tail wags the dog (the tail being the load, and the dog being the satellite). But when the robot links are flexible, the tail tumbles the dog.

References

[1]Longman, R. W., Lindberg, R. E., and Zedd, M. F., "Satellite-Mounted Robot Manipulators—New Kinematics and Reaction Moment Compensation," *International Journal of Robotics Research,* Vol. 6, No. 3, 1987, pp. 87–103; also, *The Proceedings of the 1985 AIAA Guidance, Navigation, and Control Conference* (Snowmass, CO), AIAA, Washington, DC, 1985, pp. 278–290.

[2]Lindberg, R. E., Longman, R. W., and Zedd, M. F., "Kinematics and Reaction Moment Compensation for a Spaceborne Elbow Manipulator," AIAA Paper 86-0250, Jan. 1986.

[3]Lindberg, R. E., Longman, R. W., and Zedd, M. F., "Kinematic and Dynamic Properties of an Elbow Manipulator Mounted on a Satellite," *The Journal of the Astronautical Sciences,* Special Issue on Robotics in Space, Vol. 38, No. 4, 1990, pp. 397–421.

[4]Longman, R. W., "The Kinetics and Workspace of a Satellite-Mounted Robot," *The Journal of the Astronautical Sciences,* Special Issue on Robotics in Space, Vol. 38, No. 4, 1990, pp. 423–440; also *The Proceedings of the 1988 AIAA Guidance, Navigation and Control Conference* (Minneapolis, MN), AIAA, Washington, DC, 1988, pp. 374–381.

[5]Longman, R. W., "Attitude Tumbling due to Flexibility in Satellite-Mounted Robots," *The Journal of the Astronautical Sciences,* Special Issue on Robotics in Space, Vol. 38, No. 4, 1990, pp. 487–509; also, *The Proceedings of the 1988 AIAA Guidance, Navigation, and Control Conference* (Minneapolis, MN), AIAA, Washington, DC, 1988, pp. 365–373.

[6]Vafa, Z., and Dubowsky, S., "On the Dynamics of Space Manipulators Using the Virtual Manipulator, with Applications to Path Planning," *The Journal of the Astronautical Sciences,* Special Issue on Robotics in Space, Vol. 38, No. 4, 1990, pp. 441–472.

[7]Schulz, V., Longman, R. W., and Bock, H. G., "Optimal Path Planning of Satellite-Mounted Robot Manipulators," *Advances in the Astronautical Sciences,* Vol. 82, Pt. 1, 1993, pp. 311–329.

[8]Longman, R. W., and Lingberg, R. E. (eds.), *The Journal of the Astronautical Sciences,* Special Issue on Robotics in Space, Vol. 38, No. 4, 1990.

[9]Xu, Y., and Kanade, T. (eds.), *Space Robotics: Dynamics and Control,* Kluwer Academic Pubs., Boston, MA, 1992.

[10]Alexander, H. L., and Cannon, R. H., Jr., "An Extended Operational-Space Control Algorithm for Satellite Manipulators," *The Journal of the Astronautical Sciences,* Special Issue on Robotics in Space, Vol. 38, No. 4, 1990, pp. 473–486.

[11]Longman, R. W., "New Problems in Space Flight Operations Introduced by Satellite-Mounted Robots," *Proceedings of the 3rd International Symposium on Spacecraft Flight Dynamics,* European Space Agency, ESA SP-326, Paris, France, Dec. 1991, pp. 357–362.

Chapter 11

Reorientation of Free-Flying Multibody Structure Using Appendage Movement

Ranjan Mukherjee[*]

Naval Postgraduate School, Monterey, California 93943

and

Yoshihiko Nakamura[†]

University of Tokyo, Tokyo, Japan

Nomenclature

$^I A_B \in R^{3\times3}$ = rotation matrix from the inertia frame to the manipulator base frame

$^I A_k \in R^{3\times3}$ = rotation matrix from the inertia frame to the kth body frame (the vehicle frame for $k = 0$, the kth link frame of the manipulator for $k = 1, \ldots, n$)

$E_i \in R^{i\times i}$ = $i \times i$ identity matrix

frame B = manipulator base frame

frame E = manipulator end-effector frame

frame K = kth body frame; the kth link frame of the manipulator for $k = 1, \ldots, n$; the nth link frame is identical to the end-effector frame; we denote the vehicle frame with $k = 0$

frame I = inertia frame

frame V = vehicle frame

$^I I_k \in R^{3\times3}$ = inertia matrix of the kth body about its center of mass in the inertia frame (kgm^2)

$^k I_k \in R^{3\times3}$ = inertia matrix of the kth body about its center of mass in the kth body frame, a constant matrix (kgm^2)

$J_2^k \in R^{6\times n}$ = Jacobian matrix of the position and orientation of the center of mass of kth body ($k = 1, \ldots, n$) in the inertial frame (m)

m_k = mass of the kth body (kg); the 0th body is the vehicle; the kth body ($k \geq 1$) is the kth link of the manipulator

Copyright © 1994 by the authors. Published by the American Institute of Aeronautics and Astronautics, Inc., with permission. Released to AIAA to publish in all forms.

*Assistant Professor, Mechanical Engineering Department.
†Professor, Department of Mechano-Informatics.

n $=$ degrees of freedom of the manipulator

${}^B r_k \in R^3$ $=$ position vector from the origin of the manipulator base frame to the center of mass of the kth body represented in the manipulator base frame (m)

${}^I r_k \in R^3$ $=$ position vector from the origin of the inertia frame to the center of mass of the kth body represented in the inertia frame (m)

${}^I z_k \in R^3$ $=$ unit vector in the direction of the z axis of the kth link coordinates represented in the inertial frame (m)

α, β, γ $=$ z-y-x Euler angles (rad)

$\dot{\theta}_1 \in R^6$ $=$ linear velocity of the center of mass and angular velocity of the vehicle in the inertia frame (m/s, rad/s)

$\theta_2 \in R^n$ $=$ joint variables $(q_1, \ldots, q_n)$ of the manipulator (rad)

${}^I \omega_k \in R^3$ $=$ angular velocity of the kth body in the inertia frame (rad/s)

I. Introduction

W ITH advances in space applications, robotics has been recognized as a key to making space missions safe and cost effective. A free-flying robot consisting of two or more manipulators mounted on a space vehicle, has been conceived by NASA for performing various tasks in space. The kinematics and dynamics of such robotic systems have intrinsic features because of microgravity and momentum conservation. Additionally, the preciousness of jet fuel necessitates the control of the system without the use of reaction jets.

Alexander and Cannon[1] discussed and experimentally demonstrated the computation of joint torques for manipulator endpoint control while assuming that the thrust forces of the vehicle were known. Vafa[2] and Vafa and Dubowsky[3] proposed a novel concept to simplify the kinematics and dynamics of space robot systems. A virtual manipulator is an imaginary manipulator that has a kinematic and dynamic structure which is similar to the real vehicle/manipulator system, but which is fixed at the center of mass of the whole system. By deriving the motion of the virtual manipulator for a given end effector motion, the motion of the actual system can be obtained in a straightforward manner. Umetani and Yoshida[4] proposed a method which would continuously control the end effector without actively controlling the vehicle thrust forces. The momentum conservations for linear and angular motion were explicitly represented and used as constraint equations to eliminate the dependent variables and obtain the generalized Jacobian matrix that relates the end-effector motion to the joint motion. Longman et al.[5] also discussed the coupling of manipulator motion and vehicle motion. Miyazaki et al.[6] discussed a sensor feedback scheme using the transposed generalized Jacobian matrix. Mukherjee and Nakamura[7] and Yamada[8] developed efficient algorithms for the computation of the dynamics of robot manipulators in space.

The linear and angular momentum conservation equations have been used by various researchers to eliminate dependent variables.[4,6] Although both of the equations are represented by equations of velocities, the linear momentum conservation is exhibited by the motion of the center of mass of the whole system, and can therefore be represented by equations of positions instead of velocities. This implies that the linear momentum conservation equations are integrable and hence are holonomic constraints. On the other hand, the angular momentum conservation equations cannot be integrated, which means that they are nonholonomic constraints. Vafa and Dubowsky[3] proposed cyclic motion of the manipulator joints to change the vehicle orientation. This illustrated the possibility of utilizing the non-holonomic mechanical structure of space vehicle/manipulator systems. However, this scheme has to assume small cyclic motion to neglect nonlinearity of an order greater than two, and therefore requires many cycles to make even a small change in vehicle orientation.

For an n-DOF manipulator on a vehicle, the motion of the end effector is described by $n + 6$ variables, n of the manipulator and six of the vehicle. By eliminating the holonomic constraints of linear momentum conservation, the total system is formulated as a nonholonomic system of $n + 3$ variables including three dependent variables. Although only n variables out of the $n + 3$ can be independently controlled, with an appropriate path planning scheme it would possible to converge all of the $n + 3$ variables to their desired values because of the nonholonomic mechanical structure. A similar situation is experienced in our daily life. Although an automobile has two independent variables to control, i.e., wheel rotation and steering, it can be parked at an arbitrary place with an arbitrary orientation in a three-dimensional space. This is possible because it is a nonholonomic system.

In this chapter, we propose a new path planning scheme to control both the vehicle orientation and the manipulator joints by actuating the manipulator joints only. First, the formulation of the nonholonomic mechanical structure of space robot systems[9] is briefly overviewed in Sec. II. Second, a rigorous mathematical proof of the nonholonomic mechanical structure is provided by using the Frobenius's Theorem in Sec. III. A path planning scheme for nonholonomic systems is then proposed by use of a Lyapunov function in Sec. IV. The scheme is called the bidirectional approach. This approach deals with the total nonlinearity of space vehicle/manipulator systems without neglecting the nonlinearity of higher order, with no algorithmic limitation on the allowable change. The results obtained while using the bidirectional approach clearly imply that with different choices of the spatial joint trajectories, the end effector can reach a given position and orientation with many different values of the nine states even if there are only six joints. Although this indicates the presence of redundancy, it is different from ordinary kinematic redundancy in the sense that it cannot be manifested locally through self-motions or null-space motions. On the contrary, it only exhibits itself only after a global motion. In Sec. III.D, we term this redundancy as nonholonomic redundancy, and we establish a method to utilize it while planning a trajectory for the end effector in Sec. V. While the end effector converges to its desired configuration, obstacles and joint limits are avoided via the utilization of nonholonomic redundancy.

II.   Differential Kinematics and Momentum Conservation of Space Robots

The basic kinematic equations of space robot systems are developed in this section.[9] Figure 1 shows the model of a space robot system. Five kinds of frames, the inertia frame, the vehicle frame, the manipulator base frame, the kth link frames, and the manipulator end-effector frame, are represented by I, V, B, K,

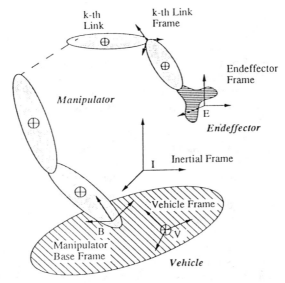

Fig. 1   Five coordinate frames for the space vehicle/manipulator system.

and E, respectively. The link frames of the manipulator are defined by Denavit-Hartenberg convention.[12] The vehicle frame is assumed to be fixed at the center of mass of the vehicle.

Assuming zero linear and angular momentum of the system at the initial time, the linear and angular momentum conservation equations are represented by

$$\sum_{k=0}^{n} m_k\,{}^I\dot{r}_k = 0 \tag{1}$$

$$\sum_{k=0}^{n} ({}^I I_k\,{}^I \omega_k + m_k\,{}^I r_k \times {}^I \dot{r}_k) = 0 \tag{2}$$

The vehicle and manipulator motions are described by $\dot{\theta}_1$ and θ_2, defined as

$$\dot{\theta}_1 = \begin{pmatrix} {}^I\dot{r}_0 \\ {}^I\omega_0 \end{pmatrix} \tag{3}$$

$$\theta_2 = \begin{pmatrix} q_1 \\ \vdots \\ q_n \end{pmatrix} \tag{4}$$

${}^I\dot{r}_k$ is computed from the relation

$$
\begin{aligned}
{}^I\dot{r}_k &= {}^I\dot{r}_0 + {}^I\omega_0 \times ({}^I r_k - {}^I r_0) + (E_3 \quad 0)J_2^k\dot{\theta}_2 \\
&= (E_3 \quad -{}^I R_{0k})\dot{\theta}_1 + (E_3 \quad 0)J_2^k\dot{\theta}_2
\end{aligned} \tag{5}
$$

where ${}^I\boldsymbol{R}_{0k}$ is defined by

$$
{}^I\boldsymbol{R}_{0k} \overset{\triangle}{=} \begin{pmatrix} 0 & -{}^Ir_{0kz} & {}^Ir_{0ky} \\ {}^Ir_{0kz} & 0 & -{}^Ir_{0kx} \\ -{}^Ir_{0ky} & {}^Ir_{0kx} & 0 \end{pmatrix} \tag{6}
$$

$$
{}^I\boldsymbol{r}_{0k} \overset{\triangle}{=} {}^I\boldsymbol{r}_k - {}^I\boldsymbol{r}_0 \overset{\triangle}{=} \begin{pmatrix} {}^Ir_{0kx} \\ {}^Ir_{0ky} \\ {}^Ir_{0kz} \end{pmatrix} \tag{7}
$$

We also have the following relations for ${}^I\boldsymbol{I}_k$ and ${}^I\boldsymbol{\omega}_k$

$$
{}^I\boldsymbol{I}_k = {}^I\boldsymbol{A}_k \, {}^k\boldsymbol{I}_k \, {}^I\boldsymbol{A}_k^T \tag{8}
$$

$$
{}^I\boldsymbol{\omega}_k = \begin{cases} (0 \;\; \boldsymbol{E}_3)\,\dot{\boldsymbol{\theta}}_1 & \text{for } k = 0 \\ {}^I\boldsymbol{\omega}_0 + \sum_{j=1}^k {}^I\boldsymbol{z}_{j-1}\,\dot{q}_j & \text{for } k = 1,\dots,n \end{cases} \tag{9}
$$

By substituting Eqs. (5) and (8) into Eqs. (1) and (2) and summarizing them in a matrix form, the linear and angular momentum conservations are represented by the following equation

$$
\boldsymbol{H}_1\dot{\boldsymbol{\theta}}_1 + \boldsymbol{H}_2\dot{\boldsymbol{\theta}}_2 = 0 \tag{10}
$$

where

$$
\boldsymbol{H}_1 \overset{\triangle}{=} \begin{pmatrix} \sum_{k=0}^n m_k \boldsymbol{E}_3 & -\sum_{k=0}^n m_k\,{}^I\boldsymbol{R}_{0k} \\ \sum_{k=0}^n m_k\,{}^I\boldsymbol{R}_k & \sum_{k=0}^n {}^I\boldsymbol{A}_k\,{}^k\boldsymbol{I}_k\,{}^I\boldsymbol{A}_k^T - \sum_{k=0}^n m_k\,{}^I\boldsymbol{R}_k\,{}^I\boldsymbol{R}_{0k} \end{pmatrix} \tag{11}
$$

$$
\boldsymbol{H}_2 \overset{\triangle}{=} \begin{pmatrix} \sum_{k=1}^n m_k (\boldsymbol{E}_3 \;\; 0)\boldsymbol{J}_2^k \\ \sum_{k=1}^n m_k (\boldsymbol{E}_3 \;\; 0)\boldsymbol{J}_2^k + \boldsymbol{P} \end{pmatrix} \tag{12}
$$

and

$$
{}^I\boldsymbol{R}_k \overset{\triangle}{=} \begin{pmatrix} 0 & -{}^Ir_{kz} & {}^Ir_{ky} \\ {}^Ir_{kz} & 0 & -{}^Ir_{kx} \\ -{}^Ir_{ky} & {}^Ir_{kx} & 0 \end{pmatrix}, \qquad {}^I\boldsymbol{r}_k \overset{\triangle}{=} \begin{pmatrix} {}^Ir_{kx} \\ {}^Ir_{ky} \\ {}^Ir_{kz} \end{pmatrix} \tag{13}
$$

$$
\boldsymbol{P} \overset{\triangle}{=} (\boldsymbol{P}_1 \;\; \boldsymbol{P}_2 \;\; \cdots \;\; \boldsymbol{P}_n) \tag{14}
$$

$$
\boldsymbol{P}_i \overset{\triangle}{=} \left(\sum_{k=i}^n {}^I\boldsymbol{A}_k\,{}^k\boldsymbol{I}_k\,{}^I\boldsymbol{A}_k^T \right) {}^I\boldsymbol{z}_{i-1}
$$

The pure geometrical relationship between the end-effector motion, and $\dot{\boldsymbol{\theta}}_1$ and $\dot{\boldsymbol{\theta}}_2$ is described in the following form:

$$
\dot{\boldsymbol{x}}_E \overset{\triangle}{=} \begin{pmatrix} {}^I\dot{\boldsymbol{r}}_E \\ {}^I\boldsymbol{\omega}_E \end{pmatrix} \overset{\triangle}{=} \begin{pmatrix} {}^I\dot{\boldsymbol{r}}_n \\ {}^I\boldsymbol{\omega}_n \end{pmatrix} = \boldsymbol{J}_1\dot{\boldsymbol{\theta}}_1 + \boldsymbol{J}_2\dot{\boldsymbol{\theta}}_2 \tag{15}
$$

where J_1 and J_2 are the Jacobian matrices that do not take into account the momentum conservations. In Eq. (10), $H_1 \in R^{6 \times 6}$ is always nonsingular. This can be explained as follows: for $\dot{\theta}_2 = 0$, the momentum of the system would be given as $H_1 \dot{\theta}_1$. Now, for nonzero $\dot{\theta}_1$, it is physically impossible for the momentum to be zero. This physically signifies that the matrix H_1 has no null space and is hence invertible. Therefore, Eq. (10) is identical to

$$\dot{\theta}_1 = -H_1^{-1} H_2 \dot{\theta}_2 \tag{16}$$

Substituting Eq. (16) into Eq. (15) offers

$$\dot{x}_E = \left(-J_1 H_1^{-1} H_2 + J_2 \right) \dot{\theta}_2 = \tilde{J} \dot{\theta}_2 \tag{17}$$

Umetani and Yoshida[4] named the coefficient matrix of the preceding equation the generalized Jacobian matrix. In this derivation, the momentum conservations of Eq. (10) were used as constraint equations to eliminate the dependent variables.

If $h \in R^6$ represents the position and the z-y-x Euler angles (α_E, β_E, γ_E) of the end effector, then the end-effector motion can be conveniently described in the following way:

$$\dot{h} = \hat{J} \dot{\theta}_2 \tag{18}$$

$$\hat{J} \triangleq \begin{pmatrix} E_3 & 0 \\ 0 & N_E^{-1} \end{pmatrix} \tilde{J}, \qquad N_E \triangleq \begin{pmatrix} 0 & -\sin \alpha_E & \cos \alpha_E \cos \beta_E \\ 0 & \cos \alpha_E & \sin \alpha_E \cos \beta_E \\ 1 & 0 & -\sin \beta_E \end{pmatrix}$$

It is important to note here that Eq. (18) is not a state equation because it does not have a view of the three dependent nonintegrable variables, namely, the z-y-x Euler angles of vehicle orientation.

III.   Nonholonomic Mechanical Structure

A.   Physical Meaning

The linear momentum conservation represented by Eq. (1) can be analytically integrated as follows:

$$\int_0^t \sum_{k=0}^{n} m_k {}^I \dot{r}_k \, dt = \sum_{k=0}^{n} m_k {}^I r_k(t) - \sum_{k=0}^{n} m_k {}^I r_k(0) = 0 \tag{19}$$

The above equation physically means that the center of mass of the whole system does not move. Here, ${}^I r_k$ is computed by

$$^I r_k = {}^I A_B {}^B r_k + {}^I r_0 \tag{20}$$

where ${}^I A_B$ is a function of the vehicle orientation and ${}^B r_k$ is a function of the joint variables of the manipulator. Knowing the vehicle orientation, the joint variables, and the initial position of the total center of mass, the vehicle position ${}^I r_0$ can be obtained by substituting Eq. (20) into Eq. (19). Therefore, the linear momentum conservation is considered a holonomic constraint because it is integrable.

Although Eqs. (1) and (2) are both represented by velocities, Eq. (2) can not be analytically integrated, and therefore it is a nonholonomic constraint. The rigorous mathematical definitions and proof of the nonholonomic constraints are given in Sec. III.C. The physical characteristic of the nonholonomic constraints are exhibited by the fact that even if the manipulator joints return to their initial configuration after a sequence of motion, the vehicle orientation may not be the same as its initial value. In Sec. IV, we propose to control both the independent and dependent variables by directly controlling the independent variables only.

B. Nonholonomic Equation

The basic system equation is obtained by taking the vehicle orientation and θ_2 as the state variable and the $\dot{\theta}_2$ as the input variable. First, the coefficient matrix of Eq. (16) is divided into a top $3 \times n$ matrix and a bottom $3 \times n$ matrix as follows:

$$H = \begin{pmatrix} H_r \\ H_\omega \end{pmatrix} = -H_1^{-1} H_2 \tag{21}$$

The state variable x and the input variable u are defined by

$$x = \begin{pmatrix} \alpha \\ \beta \\ \gamma \\ \theta_2 \end{pmatrix} \in R^{n+3} \tag{22}$$

$$u = \dot{\theta}_2 \in R^n \tag{23}$$

α, β, and γ are the z-y-x Euler angles of the vehicle with respect to the inertia frame. The relationship between the Euler angles and $^I\omega_0$ is given by

$$^I\omega_0 = N \begin{pmatrix} \dot{\alpha} \\ \dot{\beta} \\ \dot{\gamma} \end{pmatrix} \tag{24}$$

where

$$N = \begin{pmatrix} 0 & -\sin\alpha & \cos\alpha\cos\beta \\ 0 & \cos\alpha & \sin\alpha\cos\beta \\ 1 & 0 & -\sin\beta \end{pmatrix}$$

The system equation becomes

$$\dot{x} = Ku \tag{25}$$

where

$$K = \begin{pmatrix} N^{-1}H_\omega \\ E_n \end{pmatrix} \in R^{(n+3)\times n} \tag{26}$$

C. Mathematical Proof

A mathematical proof for the system of Eq. (25) being nonholonomic is provided in this subsection. As preparation, we refer to the following three definitions and two theorems from nonlinear control theory (see, for example, Ref. 11) and classical mechanics.

Definition 1 (Lie bracket).   If $X_1(x)$ and $X_2(x)$ are two vector fields on R^n, then the Lie bracket of X_1 and X_2, denoted by $[X_1, X_2]$, is a third vector field defined by

$$[X_1, X_2] = \frac{\partial X_2}{\partial x} X_1 - \frac{\partial X_1}{\partial x} X_2 \tag{27}$$

A linear space Δ spanned by a set of vector fields is said to be a distribution. Involutivity for a distribution is defined next.

Definition 2 (involutivity).   A distribution Δ is said to be involutive if and only if the Lie bracket of any pair of vector fields in Δ is also in Δ.

Because of the characteristic nature of Lie brackets, involutivity of a distribution defined by $\{X_1, \ldots, X_m\}$ can be concluded by verifying the following equation for any choice of i, and j

$$[X_i, X_j] = \sum_{k=1}^{m} \alpha_{ijk} X_k \tag{28}$$

where α_{ijk} is a scalar function.

Theorem 1 (Frobenius's theorem).   A distribution is said to be completely integrable if and only if it is involutive.

In classical mechanics, holonomy is defined[12] as follows in Definition 3.

Definition 3 (holonomy).   A system is said to be holonomic if and only if its motion is constrained by a set of algebraic equations involving only the generalized coordinates and time.

When the system behavior is represented in terms of a set of first-order linear differential equations, the feasible velocity of the generalized coordinates span a linear space. Integrability (namely, holonomy) of a system is answered by finding vector fields that define the linear space and by applying Theorem 1. The preceding discussion is summarized next in the form of a theorem.

Theorem 2 (holonomy and nonholonomy).   Let $\{X_1, \ldots, X_m\}$ be the set of vector fields defining the linear space of the feasible velocity of a system. The system is holonomic if and only if the set of vector fields is completely integrable everywhere in the generalized coordinates. Otherwise, the system is nonholonomic.

For the system defined by Eq. (25), the column vectors of matrix K define the linear space of the feasible velocity of the system. Accordingly, we can choose them as a set of vector fields. From Eq. (26) K has the following form

$$K \triangleq \begin{pmatrix} \alpha_1 & \cdots & \alpha_n \\ \beta_1 & \cdots & \beta_n \\ \gamma_1 & \cdots & \gamma_n \\ & E_n & \end{pmatrix} \in R^{(n+3)\times n} \tag{29}$$

The set of vector fields become $\{X_1, \ldots, X_n\}$ where

$$X_i = (\alpha_i \quad \beta_i \quad \gamma_i \quad e_i^T)^T \tag{30}$$

and $e_i \in R^n$ is a unit vector with one as the ith entry. It is clear from Eq. (28) that X_i, $(i = 1, 2, \ldots, n)$ are linearly independent. Consequently, the Lie brackets of the set of vector fields take the following form:

$$[X_i, X_j] = (p_1 \quad p_2 \quad p_3 \quad 0 \quad \cdots \quad 0)^T \in R^{n+3} \tag{31}$$

Since the Lie bracket has zeros after the third component, it is obvious from Definition 2 that the system is involutive if and only if p_i are all zero. Therefore, to prove the nonholonomic nature, it is sufficient from Definition 3 to show that at least one p_i has a nonzero value at least at one configuration. We computed the values of p_i for a numerical model of space robot with nontrivial mass distribution. Figure 2 shows the structure and configuration of the space robot whose kinematic and dynamic parameters are given in Tables 1–3. The manipulator has a PUMA-type structure with six DOF. The values of p_i are given in Table 4, where i and j represent the combination of Lie brackets. From the table we can conclude that the space robot model has a nonholonomic mechanical structure. We checked out the values of p_i at several different configurations and they always had many nonzeros.

A natural question one may ask is whether or not a system can attain any value of the generalized coordinates starting from any initial value of it. This is a question of controllability. For a symmetric and affine nonlinear system like Eq. (25), controllability is confirmed if the set of vector fields that include $\{X_1, X_2, \ldots, X_m\}$, and all the possible Lie brackets of arbitrary order obtained

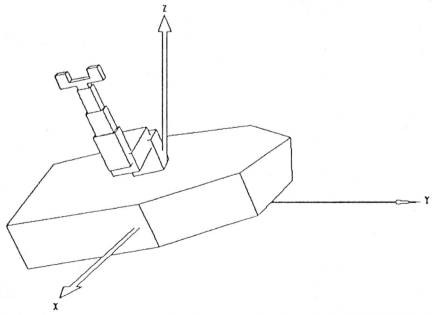

Fig. 2   Structure and configuration of the space robot system (a six-DOF PUMA-type manipulator mounted on a space vechicle) chosen to prove its nonholonomic nature.

Table 1 Kinematic (Denavit-Hartenberg) parameters[a]

k =	0	1	2	3	4	5	6
$\bar{\alpha}_k$ (deg)	0.00	$-\pi/2$	0.00	$\pi/2$	$-\pi/2$	$\pi/2$	0.00
a_k (m)	0.00	0.00	0.50	0.00	0.00	0.00	0.00
r_k (m)	0.50	0.00	0.00	0.00	0.50	0.00	0.35

[a]Here, $\bar{\alpha}_k$ is the twist of the kth link (deg), a_k is the length of the kth link (m), and r_k is the distance between the $(i-1)$th and ith links (m).

from $\{X_1, X_2, \ldots, X_m\}$, span a linear space of the same dimension as that of the generalized coordinates, everywhere in the generalized coordinates. This could have been verified directly for the system defined by Eq. (25) if the system had not been so complex. In this paper, we will not proceed further with our discussion on controllability. In Sec. IV, we discuss trajectory planning while assuming the existence of a feasible trajectory connecting the initial and final generalized coordinates.

D. Nonholonomic Redundancy

An important property of the space robot system can be readily understood from Eqs. (18) and (25). By integrating Eq. (25) with respect to time, we have

$$x(t_2) = x(t_1) + \int_{t_1}^{t_2} K(\theta_2)\dot{\theta}_2 \, dt = x(t_1) + \sum_{i=1}^{6} \int_{\theta_2(t_1)}^{\theta_2(t_2)} K_i(\theta_2) \, d\theta_{2i} \qquad (32)$$

where K_i is the ith column of K, θ_{2i} is the ith element of θ_2, and t_1 and t_2 are the initial and final times, respectively. The functional dependence of the matrix K is on θ_2 only and not on θ_1 (Ref. 13). Equation (32) implies that the trajectory of x is dependent on the trajectory of θ_2, but is independent of time. Therefore, once the trajectory of θ_2 is determined, changing the velocity along the trajectory does not alter the trajectory of x. This stems from the fact that the system maintains zero momentum.

In Sec. V, we will plan trajectories for the end-effector variables h using Eq. (18). Since different end-effector paths imply different paths of θ_2 and x, as evident

Table 2 Dynamic parameters (inertia tensor)[a]

	(i, j)	$k = 0$	$k = 1$	$k = 2$	$k = 3$	$k = 4$	$k = 5$	$k = 6$
	(1,1)	25.0000	0.3170	0.0496	0.1910	0.0487	0.02050	0.0145
	(2,1)	0.0000	0.0000	0.0000	0.0000	0.0000	0.00000	0.0000
^kI_k	(3,1)	0.0000	0.0000	0.0000	0.0000	0.0000	0.00000	0.0000
(kg-m²)	(2,2)	25.0000	0.1840	0.4410	0.1910	0.0308	0.02050	0.0145
	(3,2)	0.0000	0.0000	0.0000	0.0000	0.0000	0.00000	0.0000
	(3,3)	25.0000	0.3170	0.4410	0.0821	0.0487	0.00771	0.0103

[a]^kI_k is the inertia matrix of the kth body about its center of mass in the kth body frame (kg-m²).

Table 3 Dynamic parameters (mass and position of center of mass)[a]

		$k=0$	$k=1$	$k=2$	$k=3$	$k=4$	$k=5$	$k=6$
	(x)	0.0000	0.000	−0.250	0.000	0.000	0.000	0.000
S_k(m)	(y)	0.0000	0.150	0.000	0.000	0.100	0.000	0.000
	(z)	−0.4000	0.000	0.000	0.150	0.000	0.100	−0.075
m_k(kg)		500.00	30.00	20.00	20.00	10.00	5.00	5.00

[a]Here, S_k is the position vector of the center of mass of the kth link from the origin of the kth link coordinates (m), and m_k is the mass of the kth body (kg).

from Eqs. (18) and (25), the inverse kinematic solutions to the final end-effector configuration are dependent on the history of the end-effector motion. Because of the large number of paths that can be chosen for the end effector, the inverse kinematics at the final time admits a large number of solutions. Clearly, we have redundancy in choosing the nine state variables of x for a particular end-effector configuration at the final time.

This redundancy, which we term nonholonomic redundancy, is different from ordinary kinematic redundancy and cannot be exhibited through self motion. This is clear from the fact that $\dot{h} = 0$ implies $\dot{\theta}_2 = 0$ for a nonsingular $\hat{J}$, from Eq. (18). The inherent difference between nonholonomic redundancy and ordinary kinematic redundancy can be further explained as follows. As defined before, $u, x,$

Table 4 Values of p_i for the space robot system in Fig. 2

Orientation (deg):	7.500	10.000	15.000
Joint angle (deg):	0.000	0.000	0.000
	0.000	0.000	0.000

i	j	p_1	p_2	p_3
1	2	−0.00178931	0.01554863	0.00466567
1	3	0.00081895	−0.00683658	−0.00206614
1	4	−0.00003016	−0.00000399	0.00000536
1	5	0.00009580	−0.00077568	−0.00023897
1	6	−0.00000636	−0.00000084	0.00000113
2	3	−0.01212043	−0.00349256	0.00940485
2	4	0.00000235	−0.00003863	−0.00000914
2	5	−0.00183629	−0.00043520	0.00106273
2	6	0.00000050	−0.00000815	−0.00000193
3	4	0.00000533	−0.00005760	−0.00001574
3	5	−0.00012598	0.00001182	−0.00008245
3	6	0.00000113	−0.00001215	−0.00000332
4	5	−0.00000150	0.00002255	0.00000665
4	6	0.00000000	0.00000000	0.00000000
5	6	0.00000111	−0.00001302	−0.00000346

and h denote the input space, the generalized coordinates, and the end-effector variables, respectively. Both kinematically redundant and nonholonomically redundant manipulators exhibit redundancy because the generalized coordinates are of higher dimension than the end-effector variables. The difference between the two types of redundancy is, however, attributed to the fact that the dimension of the input space is equal to that of the generalized coordinates for ordinary kinematic redundancy whereas, for nonholonomic redundancy, the input space is smaller than the dimension of the generalized coordinates.

IV.   Bidirectional Motion Planning

A.   Lyapunov Function

Vafa and Dubowsky[3] proposed using the cyclic motion of manipulator joints to change the vehicle orientation without changing the manipulator joint configuration. This scheme assumed small cyclic motion to neglect higher order nonlinearity, and therefore required many cycles to make even a small change in vehicle orientation.

In this section, the input variable u is synthesized based on the Lyapunov's direct method[14, 15] so that the vehicle orientation and the joint variables converge to their desired values. This approach deals with the total nonlinearity of space robot systems without neglecting nonlinearity of any order, and without any algorithmic limitation on the allowable change.

The following function is chosen as a candidate for the Lyapunov function.

$$v \triangleq (1/2)\Delta x^T A \Delta x \tag{33}$$

$$\Delta x \triangleq x_d - x \tag{34}$$

where A is a positive definite constant matrix. Clearly, $v = 0$ is attained only when $x_d = x$. Here, x_d is the goal of the state variable. The time derivative of v becomes

$$\dot{v} = -\Delta x^T A \dot{x} = -\Delta x^T A K u \tag{35}$$

where Eq. (25) was substituted. To guarantee that $\dot{V}$ is negative (semidefinite), we proposed the following determination of the input[9]

$$u = (AK)^T \Delta x \tag{36}$$

Consequently, the rate of change of the Lyapunov function becomes

$$\dot{v} = -u^T u \leq 0 \tag{37}$$

If Eq. (37) is negative definite with respect to Δx, then Lyapunov's theorem[14] can conclude its global stability. However, this is not the case in our problem since $(AK)^T$ in Eq. (36) has a three-dimensional null space and any Δx lying within this space results in $\dot{v} = 0$. It may be useful to recollect LaSalle's theorem[15] at this juncture. According to the theorem, we can conclude that the system is stable if the maximum invariant set consists of only the trivial solution. Clearly, the maximum invariant set is comprised of the entire null space of $(AK)^T$. Therefore, LaSalle's theorem cannot be used to confirm stability of the system.

We have tried many different approaches to avoid the null space within the framework mentioned earlier. However, the numerical simulation was always drawn to the null space and was unable to move from there except for some trivial cases. Finally, we establish a new method which we call the bidirectional approach.

B.   Bidirectional Approach

We virtually assume two space robot systems that have the same mechanical structure. The first one is at the initial configuration of our original problem while the second one is at the desired configuration. The system dynamics for these two robot systems are therefore given as

$$\dot{x}_1 = K_1 u_1 \quad \text{with } x_1(0) = x(0)$$
$$\dot{x}_2 = K_2 u_2 \quad \text{with } x_2(0) = x_d \tag{38}$$

We use the same Lyapunov function as Eq. (33) and redefine Δx in the following way

$$v \overset{\Delta}{=} (1/2)\Delta x^T A \Delta x \tag{39}$$

$$\Delta x \overset{\Delta}{=} x_1 - x_2 \tag{40}$$

Note that v becomes zero only when $x_1 = x_2$. The rate of change of the Lyapunov function is

$$\dot{v} = \Delta x^T A \hat{K} \begin{pmatrix} u_1 \\ u_2 \end{pmatrix} \tag{41}$$

where

$$\hat{K} = (K_1 \quad -K_2) \in R^{(n+3)\times 2n} \tag{42}$$

We choose the inputs as follows:

$$\begin{pmatrix} u_1 \\ u_2 \end{pmatrix} = -(A\hat{K})^{\#}\Delta x \tag{43}$$

where $(A\hat{K})^{\#}$ is the pseudoinverse of $A\hat{K}$. From the properties of the pseudoinverse, we can show that the choice of inputs in Eq. (43) results in

$$\dot{v} = -\Delta x^T A \hat{K}(A\hat{K})^{\#}\Delta x$$
$$= -\Delta x^T \{A\hat{K}(A\hat{K})^{\#}\}^T A\hat{K}(A\hat{K})^{\#}\Delta x \leq 0 \tag{44}$$

Accordingly, if we can show that the equality of the preceding equation holds only when $\Delta x = 0$, it is guaranteed that the two robot systems necessarily meet from Lyapunov's theorem.

Since the null space of $(A\hat{K})^{\#}$ is the same as that of $(A\hat{K})^T$, we can claim the negative definiteness of $\dot{v}$ if $(A\hat{K})^T \in R^{2n\times(n+3)}$ has no dependent column vectors. Since A is positive definite, the preceding condition is equivalent to requiring that $\hat{K}^T$ has no dependent column vectors.

From Eqs. (26) and (42), $\hat{K}$ is represented by

$$\hat{K} = \begin{pmatrix} K_{01} & -K_{02} \\ E_n & -E_n \end{pmatrix} \tag{45}$$

where $K_{01}, K_{02} \in R^{3 \times n}$ are the submatrices consisting of the first three row vectors of K_1 and K_2, respectively. From the structure of Eq. (45), it is understood that rank$(\hat{K})$ is $n + \text{rank}(K_{01} - K_{02})$, which implies that the dimension of the null space of $(A\hat{K})^T$ is less than three if K_{01} and K_{02} are not equivalent. The null space becomes trivial even if rank$(K_{01} - K_{02})$ is three. It is expected that K_{01} and K_{02} shoud have different null spaces if two state variables x_1 and x_2 are not the same. Therefore, we claim that Eq. (43) has less of a chance of getting stuck at the null space than Eq. (36). We once again refer to LaSalle's theorem here. In this context we note that the maximum invariant set with the particular choice of input as in Eq. (43), comprises of the null space of $(A\hat{K})^\#$, which has less than three dimensions.

If the Lyapunov function of Eq. (39) converges to zero at $t = t_f$, using the obtained trajectories of the two systems, namely, $x_1(t)$ and $x_2(t)$ $(0 \le t \le t_f)$, we can synthesize the trajectory of the original single space robot system that connects the initial state and the goal state, as follows:

$$x(t) = \begin{cases} x_1(t) & \text{for } 0 \le t \le t_f \\ x_2(2t_f - t) & \text{for } t_f \le t \le 2t_f \end{cases} \tag{46}$$

The corresponding input is

$$u(t) = \begin{cases} u_1(t) & \text{for } 0 \le t \le t_f \\ -u_2(2t_f - t) & \text{for } t_f \le t \le 2t_f \end{cases} \tag{47}$$

C. Numerical Simulation

To investigate the effectiveness of the bidirectional approach we have performed a numerical simulation using the same model as that in Fig. 2. We examined a number of different cases, and here we present one interesting example. For this example, the initial and the desired values of the state variables were

$$\begin{aligned} x(0) &= (0 \quad 0 \quad 0 \quad -90 \quad 0 \quad 30 \quad 15 \quad 0 \quad 45)^T \\ x_d &= (5 \quad 5 \quad 15 \quad -90 \quad 0 \quad 30 \quad 15 \quad 0 \quad 45)^T \end{aligned} \tag{48}$$

where the unit is degrees.

In the preceding example, the initial and the desired values of vehicle orientation are different, and those of the joint angles are the same. The joints of the robot are therefore expected to move along some closed trajectory that will produce the desired change in the vehicle orientation. The matrix A was selected as $A = \text{diag.}(10, 10, 10, 1, 1, 1, 1, 1, 1)$. Because the pseudoinverse sometimes generates an unrealistically large motion, we introduced a saturation function which proportionally reduces the magnitude of u_1 and u_2 if they become larger than a set value.

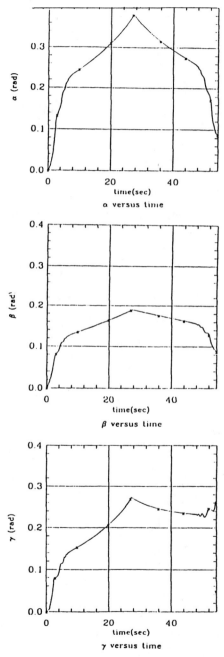

Fig. 3a   Synthesized trajectory of the vehicle orientation for the bidirectional approach.

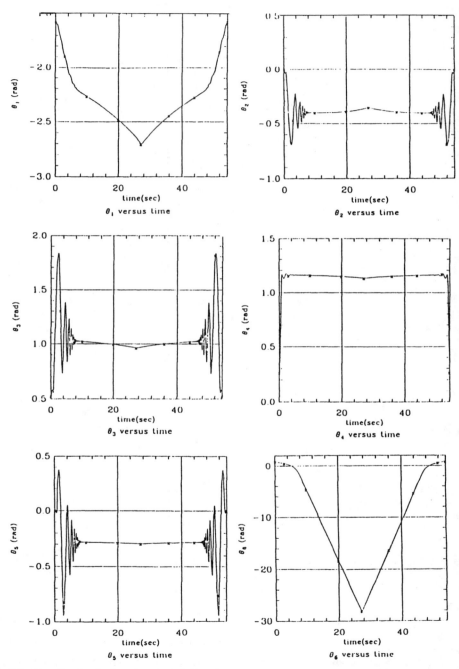

Fig. 3b   Synthesized trajectory of the joint variables for the bidirectional approach.

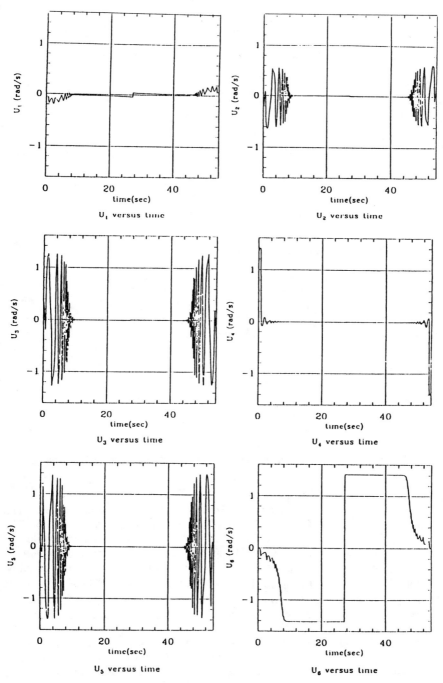

Fig. 4   Input variables for the bidirectional approach.

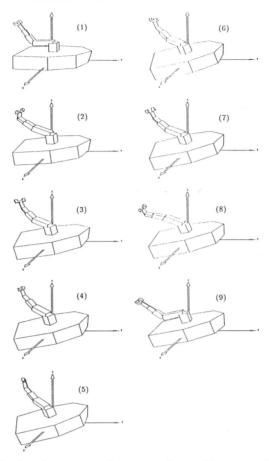

Fig. 5   Behavior of the space robot system for the bidirectional approach.

Figures 3a and 3b show the synthesized trajectory of nine state variables. The input is shown in Fig. 4. The behavior of the system is observed in Fig. 5. The nine figures correspond to the nine marks on the graphs of Fig. 3, parts a and b. Figure 6 represents the change of the Lyapunov function. It took approximately 26 s to converge, and it was observed that the Lyapunov function converged smoothly. Figure 7 is the curve of the inverse of the condition number of $(A\hat{K})^T$. The figure indicates that rank $(A\hat{K})$ was always nine although the matrix was ill-conditioned. This fact implies that at every point of the trajectory the bidirectional approach did not have a nontrivial null space.

We know from our discussion in Sec. III.D that a proportional change of the input variables results in the change of the velocity along the planned trajectory and causes no inconsistency to the nonholonomic nature. Therefore, it is possible to change the velocity along the trajectory so that the input becomes smoother.

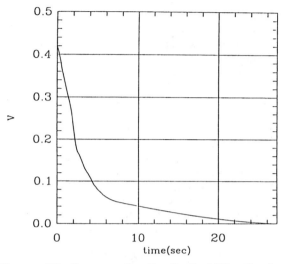

Fig. 6 Change of the Lyapunov function for the bidirectional approach.

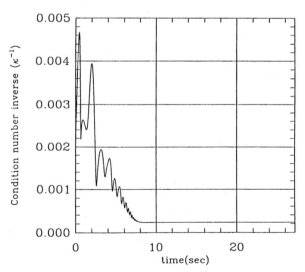

Fig. 7 Inverse of the condition number of $(A\widehat{K})^T$ for the bidirectional approach.

V.   Utilization of Nonholonomic Redundancy

A.   Lyapunov-Like Functions in Hierarchy

Our free-flying space robot has only six DOF and no local redundancy. Hence, if its end-effector trajectory is specified, there are no remaining DOF that can be utilized. We assumed that our space robot has six DOF only to emphasize the presence of nonholonomic redundancy in the absence of ordinary kinematic redundancy. The approach we are about to discuss can, however, be consistently applied to space robots with more than six DOF.

We modify the task-priority-based control[6, 17] within its framework as follows. We place the first priority on the convergence of the end-effector position and orientation to a desired value. The second priority is given to the other requirements such as remaining within joint limits, avoiding obstacles, and so on. A Lyapunov-like function[14, 15] is used for each subtask with different priorities. We then synthesize the joint velocity considering the different functions and their priorities. The choice of the primary Lyapunov-like function is not unique. It is defined as a function that has the global minimum at the destination of the end effector and no local minimum. The general form that this function takes is as follows:

$$v_1 = v_1(\boldsymbol{\theta}_1, \boldsymbol{\theta}_2) \tag{49}$$

The time derivative of v_1 is given as

$$\dot{v}_1 = -\boldsymbol{p}_1^T \boldsymbol{u}, \qquad \boldsymbol{p}_1 \overset{\triangle}{=} -\left(\frac{\partial v_1}{\partial \boldsymbol{\theta}_1}\boldsymbol{H} + \frac{\partial v_1}{\partial \boldsymbol{\theta}_2}\right)^T \tag{50}$$

and where Eqs. (16) and (21) and $\boldsymbol{u} = \dot{\boldsymbol{\theta}}_2$ were substituted. The secondary Lyapunov-like function is defined as a function whose gradient causes repulsive motion when the system reaches the obstacles or the constraints. Its general form is given by

$$v_2 = v_2(\boldsymbol{\theta}_1, \boldsymbol{\theta}_2) \tag{51}$$

Similar to Eq. (50) we have

$$\dot{v}_2 = -\boldsymbol{p}_2^T \boldsymbol{u}, \qquad \boldsymbol{p}_2 \overset{\triangle}{=} -\left(\frac{\partial v_2}{\partial \boldsymbol{\theta}_1}\boldsymbol{H} + \frac{\partial v_2}{\partial \boldsymbol{\theta}_2}\right)^T \tag{52}$$

We specify the convergence rates of v_1 and v_2 as $\dot{v}_1 = -\boldsymbol{p}_1^T \boldsymbol{K}_1 \boldsymbol{p}_1$ and $\dot{v}_2 = -\boldsymbol{p}_2^T \boldsymbol{K}_2 \boldsymbol{p}_2$, where $\boldsymbol{K}_1$ and $\boldsymbol{K}_2$ are symmetric positive definite matrices. Then we have a pair of simultaneous equations from Eqs. (50) and (52) which are summarized in the following form:

$$\boldsymbol{C}\boldsymbol{u} = \boldsymbol{d}$$
$$\boldsymbol{C} \overset{\triangle}{=} \begin{pmatrix} \boldsymbol{p}_1^T \\ \boldsymbol{p}_2^T \end{pmatrix}, \qquad \boldsymbol{d} \overset{\triangle}{=} \begin{pmatrix} \boldsymbol{p}_1^T \boldsymbol{K}_1 \boldsymbol{p}_1 \\ \boldsymbol{p}_2^T \boldsymbol{K}_2 \boldsymbol{p}_2 \end{pmatrix} \tag{53}$$

We can find an exact solution to Eq. (53) iff rank$\boldsymbol{C}$ = rank$[\boldsymbol{C}\ \boldsymbol{d}]$. This condition is necessarily satisfied when $\boldsymbol{p}_1$ and $\boldsymbol{p}_2$ are independent. When $\boldsymbol{p}_1$ and $\boldsymbol{p}_2$ are parallel, the matrix C does not remain full rank, and Eq. (53) may not have an exact solution. When C is on the verge of loosing rank, the solution to Eq. (53) tends to have a large magnitude. This is a singularity problem. This problem is commonly encountered because $v_1 = 0$ and $v_2 = 0$ can seldom be satisfied simultaneously.

To avoid the aforementioned singularity problem, we introduce the concept of hierarchy and propose to choose the input vector u as follows:

$$u = K_1 p_1 + (E_6 - p_1 p_1^\#) w_1 \tag{54}$$

where, $p_1^\#$ is the pseudoinverse of p_1, $E_6 \in R^{6 \times 6}$ is the identity matrix, w_1 is an arbitrary vector, and K_1 is a symmetric, positive definite, constant matrix. Substituting Eq. (54) into Eq. (50), and using the symmetry of the matrix $(E_6 - p_1 p_1^\#)$, and the relation $p_1 = p_1 p_1^\# p_1$ from the property of pseudoinverses, we see that

$$\dot{v}_1 = -p_1^T K_1 p_1 \tag{55}$$

We define v_1 such that p_1 becomes zero only at the destination. Then $\dot{v}_1$ is guaranteed to be negative definite under the choice of the input given by Eq. (54), for any choice of the arbitrary vector w_1. The input therefore monotonously drives the system to its primary goal.

Geometrically, Eq. (54) can be explained as follows. The first term gives the direction of u to reduce the first Lyapunov-like function. The second term represents the direction of u that does not change v_1. We call this second term an equipotential motion. The exact nature of the equipotential motion depends on the choice of w_1.

Note that because p_1 is a vector, the rank of the coefficient matrix of w_1 in Eq. (54) is five, and we therefore have five DOF in choosing the equipotential motion. This was made possible by considering the convergence of the end effector to its desired configuration, and not the trajectory of it, as the first priority task.

The arbitrary vector w_1 in Eq. (54) is now chosen as

$$w_1 = k_2 p_2 \tag{56}$$

where, k_2 is a positive scalar constant. Substituting Eqs. (54) and (56) into Eq. (52), the secondary Lyapunov-like function is found to behave as follows:

$$\dot{v}_2 = -p_2^T K_1 p_1 - k_2 p_2^T (E_6 - p_1 p_1^\#) p_2 \tag{57}$$

When the system reaches the destination, p_1 becomes zero and $\dot{v}_2$ becomes negative definite. In such a situation, the secondary Lyapunov-like function decreases monotonously. If the system is not at its destination, the first term of Eq. (57) may be positive, negative, or even zero. The second term is, however, negative semidefinite since $(E_6 - p_1 p_1^\#)$ is a positive semidefinite matrix and it attempts to reduce the secondary Lyapunov-like function. The first term is a disturbance to the convergence of v_2 and it arises because of the motion of the system that guarantees the monotonous convergence of v_1. The presence of the first term does not guarantee the negative definiteness of $\dot{v}_2$. Generally, when the constraints are close, the second term takes a large value and the effect of the first term is reduced. However, the presence of the first term may force the system to violate a constraint in critical cases. This is a fundamental limitation of this approach. In the search of a feasible solution, it is our opinion that our approach is better than solving the set of simultaneous equations, given by Eq. (53), which will have frequent singularities.

B. Primary Lyapunov-Like-Function

The following function is chosen as a candidate of the primary Lyapunov-like function:

$$v_1 \triangleq \frac{1}{2} \Delta h^T A \Delta h, \qquad \Delta h \triangleq h_d - h \qquad (58)$$

where A is a symmetric positive definite constant matrix, and h is defined by Eq. (18). $v_1 = 0$ is attained only when $h_d = h$, where h_d is the constant goal of the end effector. The time derivative of v_1 becomes

$$\dot{v}_1 = -\Delta h^T A \dot{h} = -p_{1e}^T u, \qquad p_{1e} \triangleq (A \widehat{J})^T \Delta h \qquad (59)$$

where Eq. (18) and $u = \dot{\theta}_2$ were substituted. If we compute the input by Eq. (54), then from Eq. (55) the rate of change of the Lyapunov-like function becomes

$$\dot{v}_1 = -p_{1e}^T K_1 p_{1e} \leq 0 \qquad (60)$$

Equation (60) shows that $\dot{v}_1$ is negative semidefinite and p_{1e} becomes zero when $\widehat{J}$ becomes singular and Δh lies in the null space. This is a singularity problem. However, since we use the transposed Jacobian as in Eq. (59), the singularity does not cause an abrupt change of u near the singularity. The difficulty occurs when p_{1e} becomes zero because of the singularity and p_2 happens to be zero, simultaneously. In this case the system will be stuck somewhere other than the goal. Except for this case, the convergence of the first Lyapunov function is guaranteed. We will not address this type of singularity problem in this paper any further. A discussion on the singularities of $\tilde{J}$ can be found in Ref. 18.

C. Secondary Lyapunov-Like Function

While planning the end-effector trajectory, we have to guarantee its physical feasibility by avoiding joint limits and collision between the manipulator links and obstacles. We consider the physical feasibility of this using the secondary Lyapunov-like function. The primary Lyapunov-like function is consistently used with the secondary Lyapunov-like function by defining a hierarchy between them. In this section, examples of the secondary Lyapunov-like function are given for joint limit avoidance and for obstacle avoidance.

1. Joint Limit Avoidance

We first assume that each joint of θ_2 has the joint limit $|\theta_{2i}| \leq \theta_{2i\ \max}(i = 1, 2, \ldots, 6)$, where $\theta_{2i\ \max}$ is a positive constant. We choose the following function as the Lyapunov candidate:

$$v_{2J} = \sum_{i=1}^{6} \left(\frac{\theta_{2i}}{\theta_{2i\ \max}} \right)^2 \qquad (61)$$

v_{2J} is always positive, and is zero only when $\theta_2 = 0$. The time derivative of Eq. (61) becomes

$$\dot{v}_{2J} = -p_{2J}^T u, \qquad p_{2J} \triangleq -2 \operatorname{col.} \left(\frac{\theta_{2i}}{\theta_{2i \, max}^2} \right) \tag{62}$$

where $u = \dot{\theta}_2$, and col.$(*_i)$ implies a column vector with $*_i$ as the ith component. Equation (62) corresponds to Eq. (52). Reducing v_{2J} implies bringing the manipulator back to its home position.

For the sake of simplicity, we assumed the joint limits to be symmetric. Equation (61) can be easily extended to accommodate asymmetric joint limits.

2. Obstacle Avoidance

We now consider a simple case to show the possibility of planning a collision-free nonholonomic path. The obstacle is a sphere of radius r, whose center is located at $(x_0 \quad y_0 \quad z_0)^T$ in the inertia frame. We assume that we know the critical point on the manipulator where the obstacle and the manipulator may collide with each other. The surface of the spherical obstacle is given by

$$g(x, y, z) = \left(\frac{x - x_0}{r} \right)^2 + \left(\frac{y - y_0}{r} \right)^2 + \left(\frac{z - z_0}{r} \right)^2 - 1 = 0 \tag{63}$$

When the critical point $x_C = (x \quad y \quad z)^T$, is outside the sphere, $g(x_C)$ gives the squared distance between the critical point and the surface of the sphere. Therefore, we choose the following Lyapunov candidate

$$v_{2O} = 1/g(x, y, z) \tag{64}$$

Note that v_{2O} is positive definite as long as the critical point stays outside the sphere, and becomes zero when x_C goes to infinity. Because the robot can move within a closed limited region, v_{2O} never becomes zero.

The time derivative of Eq. (64) becomes

$$\dot{v}_{2O} = -\frac{2}{g^2} [(x - x_0)/r^2 \quad (y - y_0)/r^2 \quad (z - z_0)/r^2] \dot{x}_C \tag{65}$$

where, $\dot{x}_C$ and $u = \dot{\theta}_2$ are related by the following equation similar to Eq. (18).

$$\dot{x}_C = \hat{J}_C u \tag{66}$$

Substituting Eq. (66) into Eq. (65), we get the form of Eq. (52)

$$\dot{v}_{2O} = -p_{2O}^T u, \qquad p_{2O} \triangleq \frac{2}{g^2} \hat{J}_C^T \begin{pmatrix} (x - x_0)/r^2 \\ (y - y_0)/r^2 \\ (z - z_0)/r^2 \end{pmatrix} \tag{67}$$

$\hat{J}_C$ in Eq. (66) possibly has the same type of singularity that we discussed in the earlier subsection. For a fixed-base terrestrial robot, if we choose a critical point on the first link, for instance, the joint motions of the second through sixth joints do not affect the motion of the critical point. Therefore, the rank of the coefficient matrix of Eq. (66) becomes one. For a space robot, even if the critical point is on the first link of the manipulator or on the vehicle, the motion of the second or higher joints do affect the motion of the critical point through dynamic coupling between the manipulator and the vehicle. Therefore it is likely that $\hat{J}_C \in R^{3\times 6}$ in Eq. (66) has rank three in most cases regardless of the choice of the critical point.

The Lyapunov-like function of Eq. (64) is a simple form of the artificial potential function used by Khatib[19] for obstacle avoidance of a manipulator.

D. Numerical Simulation

1. End-Effector Trajectory Planning

We present here the simulation results of one particular case. The initial state and the final end-effector configuration were set at

$$x = (150.0 \quad 0.0 \quad 0.0 \quad -165.0 \quad 10.0 \quad 15.0 \quad 0.0 \quad -15.0 \quad 0.0)^T \quad (68)$$

$$x_E = (0.217 \quad 0.460 \quad 1.121 \quad -12.45 \quad -39.40 \quad -11.22)^T \quad (69)$$

where the units in Eq. (68) are in degrees and those in Eq. (69) are in meters and degrees. The critical point was chosen as the elbow of the manipulator, i.e., the location of the third joint. We first planned for the end effector to reach its desired configuration without imposing any joint limits and in the absence of obstacles. We considered only the primary Lyapunov-like function v_1 given by Eq. (58) for path planning, and therefore used only the first term of the input given by Eq. (54). We chose p_1 in Eq. (54) as p_{1e} defined by Eq. (59). The matrix A was chosen as the identity matrix.

The vector p_{1e} can become very small in magnitude when the matrix $\hat{J}$ is close to becoming singular and the vector Δh is in the degenerate direction of $\hat{J}$. In such situations, the input to the system becomes very small. We chose $K_1 = k_1 E_6$, $k_1 > 0$, and used k_1 to tackle this problem. We used a value of 500.0 for k_1. This k_1, however, results in a very large input to the system when the matrix $\hat{J}$ is well conditioned. To avoid this, we place an upper limit on the norm of the input. If $u^T u > 1.0$, we proportionally reduce the input, by changing k_1, such that $u^T u = 1.0$. Changing k_1 does not cause any inconsistency with the theory we developed in this section.

Figure 8 shows the trajectory of the elbow coordinates of the manipulator for the simulation in the absence of joint limits and in the absence of obstacles. Figure 9 shows the trajectory of the joint variables. The behavior of the system for this case is observed in Fig. 10 at eight different intermediate stages. The convergence time for the simulation was approximately 4.65 s. The actual time taken for the simulation was approximately 7.5 min on a SUN 4/260 computer.

2. End-Effector Trajectory Planning with Joint Limits

As the next step, we imposed joint limits and using the Lyapunov-like functions v_1 of Eq. (58) and v_{2J} of Eq. (61) in hierarchy, we carried out a simulation for the same initial and desired configuration given by Eqs. (68) and (69). The joint limit for all the joints was set at 180.0 deg i.e., $\theta_{2i \max} = \pi$ rad for $(i = 1, 2, \ldots, 6)$.

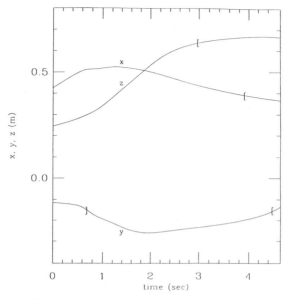

Fig. 8   Trajectory of the elbow of the manipulator for the simulation in the absence of joint limits and obstacles; the region on the curves to the right of the "[" mark and to the left of the "]" mark is where the particular coordinate would violate the presence of the obstacle.

From Fig. 9 it is clear that the first joint would violate its limit if the trajectory is planned with the primary Lyapunov-like function alone.

We chose the input according to Eqs. (54) and (56), and the value of p_{2J} defined by Eq. (62) was chosen as p_2 for Eq. (56). The vectors p_1 and p_2 in Eqs. (54) and (56) may be almost parallel in some situations. In such cases the second term of the input of Eq. (54) becomes small. In the case of path planning with joint limits, we alleviate this problem by using a value of 49.3 for the constant k_2. However, when the vectors p_1 and p_2 are almost perpendicular, this particular value of k_2 magnifies the second term of the input to a large extent. We therefore place an upper bound on the norm of the second term. When the norm is greater than 3.0, the second term is proportionally reduced to have a norm of exactly 3.0. The gain and the saturation for the primary Lyapunov-like function discussed for end-effector trajectory planning in the absence of obstacles and joint limits was used concurrently.

Figure 11 shows the trajectory of the joint variables for the simulation in the presence of joint limits only. It can be seen from Fig. 11, which is very different from Fig. 9, that none of the joint limits has been violated. The system behavior is shown in Fig. 12 at eight different intermediate stages. The convergence criterion was kept the same as in the first case, and the time for convergence was noted to be approximately 1.5 s. The actual time taken for the simulation was approximately 2.5 min on a SUN 4/260 computer.

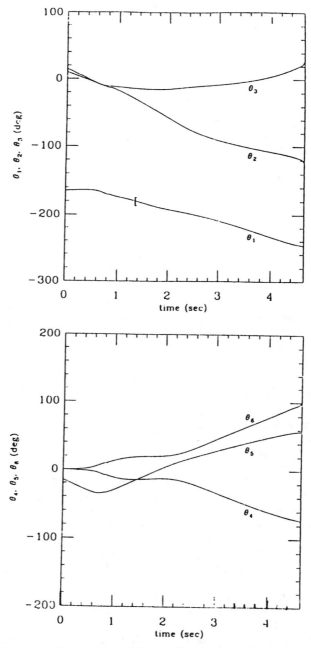

Fig. 9   Trajectory of the joint variables for the simulation in the absence of joint limits and obstacles; the region on the curves to the right of the "[" mark and to the left of the "]" mark is where the particular joint would violate the joint limit.

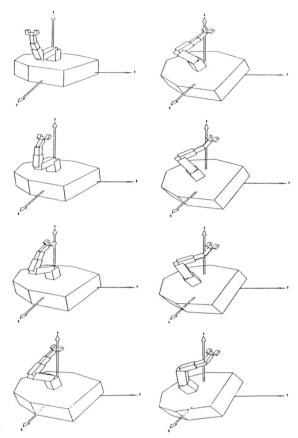

Fig. 10   System behavior for the simulation in the absence of joint limits and obstacles.

3.   End-Effector Trajectory Planning in Presence of Obstacles

Finally we simulated the case in the absence of joint limits but in the presence of an obstacle. The same initial and desired configurations given by Eqs. (68) and (69) were used. The spherical obstacle was assumed to have a radius of $r = 0.025$ (m), and located at the coordinates $x_0 = 0.363$ (m), $y_0 = -0.127$ (m), and $z_0 = 0.663$ (m). It is evident from Fig. 8 that the elbow shall collide with this obstacle if the path is planned with the primary Lyapunov-like function alone. We therefore used the Lyapunov-like functions v_1 and v_{2O} given by Eqs. (58) and (64) in hierarchy. We chose p_2 in Eq. (56) as p_{2O} given by Eq. (67).

We used a value of $k_2 = 0.75$. When the elbow comes close to the obstacle surface, the second term of the input in Eq. (54) tends to become very large. This is clear from the definition of the Lyapunov-like function v_{2O}. If the norm is greater than 3.0, the second term is proportionally reduced to have a norm of exactly 3.0.

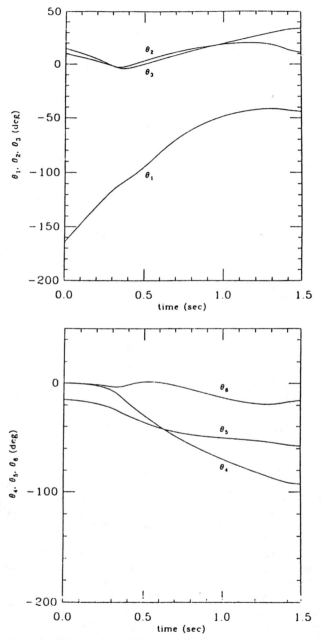

Fig. 11 **Trajectory of the joint variables for the simulation in the presence of joint limits; none of the joints violate their limits.**

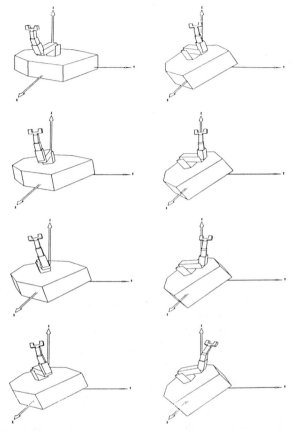

Fig. 12   System behavior for the simulation in the presence of joint limits.

The elbow trajectory for the obstacle avoidance case is shown in Fig. 13. It is clear from Fig. 13 that the elbow avoids the obstacle. The system behavior is shown in Fig. 14 at eight intermediate stages. The convergence criterion was the same as in the first case and the convergence time was noted to be approximately 3.16 s. The actual time taken for the simulation was approximately 5 min on a SUN 4/260 computer.

The successful utilization of nonholonomic redundancy is clear from Figs. 10, 12, and 14. The final configurations in each of these figures achieve the same end-effector position and orientation but the vehicle orientation and the joint variables are quite different from one another.

VI.   Conclusion

This chapter discussed the nonholonomic mechanical structure of space robot systems. A rigorous mathematical proof of the nonholonomic nature of free-flying space robot systems was provided using Frobenius' theorem. A method for the

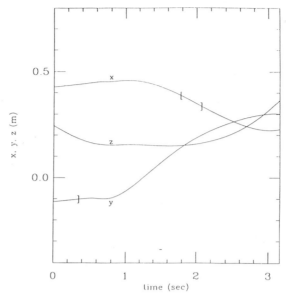

Fig. 13 Trajectory of the elbow of the manipulator for the simulation in the presence of obstacles; the region on the curves to the right of the "[" mark and to the left of the "]" mark is where the particular coordinate would violate the presence of the obstacle—clearly, the elbow never collides with the obstacle.

nonholonomic motion planning for space robot systems was then established. A six-DOF manipulator mounted on a space vehicle was described by nine state variables: the six joint angles of the manipulator and the three Euler angles of the vehicle orientation. Using the bidirectional approach for the motion planning, it was possible to converge all nine states to their desired values using only the six inputs. This approach also significantly reduced the chance of the computation getting stuck at the null space.

We also discussed the presence of nonholonomic redundancy in a free-flying space robot with six DOF. Nonholonomic redundancy is very different from ordinary kinematic redundancy. Unlike ordinary kinematic redundancy that can be exhibited locally through self-motions, nonholonomic redundancy manifests itself only after a global motion. The utilization of nonholonomic redundancy will play an important role in broadening the capability of space robot systems. We developed a trajectory planning scheme for the end effector of a space robot for the utilization of nonholonomic redundancy. This scheme was established by introducing the concept of hierarchical Lyapunov-like functions. We showed through simulations that nonholonomic redundancy can be effectively used to avoid joint limits and obstacles.

References

[1]Alexander, H. L., and Cannon, R. H., "Experiments on the Control of a Satellite Manipulator," Proceedings of the 1987 American Control Conference, 1987.

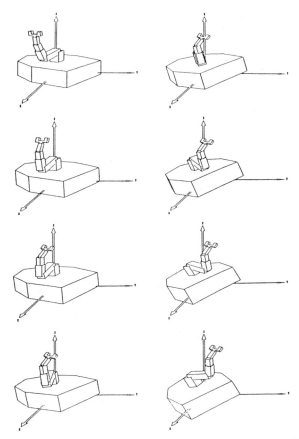

Fig. 14   System behavior for the simulation in the presence of obstacles.

[2]Vafa, Z., "The Kinematics, Dynamics and Control of Space Manipulators: The Virtual Manipulator Concept," Ph.D. Dissertation, Massachusetts Inst. of Technology, Cambridge, MA, 1987.

[3]Vafa, Z., and Dubowsky, S., "On the Dynamics of Manipulators in Space Using the Virtual Manipulator Approach," IEEE International Conference on Robotics and Automation, Raleigh, NC, 1987, pp. 579–585.

[4]Umetani, Y., and Yoshida, K., "Continuous Path Control of Space Manipulators Mounted on OMV," *Acta Astronautica,* Vol. 15, No. 12, 1987, pp. 981–986.

[5]Longman, R. W., Lindberg, R. E., and Zedd, M. F., "Satellite-Mounted Robot Manipulators: New Kinematics and Reaction Moment Compensation," *International Journal of Robotics Research,* Vol. 6, No. 3, 1983, pp. 87–103.

[6]Miyazaki, F., Masutani, Y., and Arimoto, S., "Sensor Feedback Using Approximate Jacobian," *Proceedings of the USA-Japan Symposium on Flexible Automation,* 1988, pp. 139–145.

[7]Mukherjee, R., and Nakamura, Y., "Formulation and Efficient Computation of Inverse Dynamics of Space Robots," *IEEE Transactions on Robotics and Automation,* Vol. 8, No. 3, 1992, pp. 400–406.

[8]Yamada, K., "Formulation of Space Multi-Body Systems and its Application to Control," Ph.D. Dissertation, Univ. of Tokyo, Tokyo, Japan, 1989 (in Japanese).

[9]Nakamura, Y., and Mukherjee, R., "Nonholonomic Path Planning of Space Robots via a Bi-Directional Approach," *IEEE Transactions on Robotics and Automation,* Vol. 7, No. 4, 1991, pp. 500–514.

[10]Denavit, J., and Hartenberg, R. S., "A Kinematic Notation for Lower Pair Mechanisms Based on Matrices," *Journal of Applied Mechanics,* Vol. 22, 1955.

[11]Isidori, A., "Nonlinear Control Systems: An introduction," *Lecture Notes in Control and Information Sciences,* Springer Verlag, New York, 1985.

[12]Goldstein, H., *Classical Mechanics,* 2nd ed. Addison Wesley, Reading, MA 1980.

[13]Marsden, J. E., Montgomery, R., and Ratiu, T., "Reduction, Symmetry, and Phases in Mechanics," *Memoirs of the AMS,* Providence, RI, No. 436, 1990.

[14]Lyapunov, A. M., "On the General Problem of Stability of Motion," Kharkov Mathematical Society, Russia, 1892 (in Russian).

[15]LaSalle, J., and Lefschetz, S., *Stability by Lyapunov's Direct Method with Applications,* Academic Press, New York, 1961.

[16]Nakamura, Y., and Hanafusa, H., "Optimal Redundancy Control of Robot Manipulators," *International Journal of Robotics Research,* Vol. 6, No. 1, 1987, pp. 32–42.

[17]Nakamura, Y., Hanafusa, H., and Yoshikawa, T., "Task-Priority Based Redundancy Control of Robot Manipulators," *International Journal of Robotics Research,* Vol. 6, No. 2, 1987, pp. 3–15.

[18]Papadopoulos, E., and Dubowsky, S., "On the Dynamics Singularities in the Control of Free-Floating Space Manipulators," *ASME Winter Annual Meeting: Dynamics and Control of Multibody/Robotic Systems with Space Applications,* DSC-Vol. 15, 1989, pp. 45–51.

[19]Khatib, O., "Real-Time Obstacle Avoidance and Mobile Robots," *Proceedings of the 1985 IEEE International Conference on Robotics and Automation,* 1985, pp. 500–505.

Chapter 12

Transfer Functions of Flexible Beams and Implications of Flexibility on Controller Performance

Sabri Cetinkunt[*]

University of Illinois at Chicago, Chicago, Illinois 60680

and

Wayne J. Book[†]

Georgia Institute of Technology, Atlanta, Georgia 30332

I. Introduction

E VERY mechanical system has some degree of structural flexibility. In mechanical motion control systems, if the application requirements are such
that the inaccuracies induced as a result of flexibility of the components are insignificant, then the system may be considered as a perfectly rigid system for
controller design and operation. However, if the perfomance requirements are so
high that the inaccuracies caused by the flexibility must be taken into account,
then the flexible behavior must be included in models used for controller design.
Generally, high speed, high accuracy, large payload, and long, slender structural
requirements necessitate the consideration of the flexibility.

The structure of space robotic manipulators is significantly different than that
of industrial robotic manipulators. Industrial manipulators are designed to be very
bulky to provide high rigidity. In space manipulators, however, the same rigidity
cannot be provided since the manipulators must have orders of magnitude larger
workspace. Therefore, the structure of space robots will be slender and flexible.
Furthermore they will have to manipulate very large payloads (Fig. 1).

The structural flexibility is a much more significant factor in space robotics than
it is in industrial robotics. To be specific, we must have a measure to quantify the
significance of flexibility. If the manipulator performance (which may be quantified
by the bandwidth and the accuracy) is satisfactory under a controller which ignores

Copyright © 1994 by the American Institute of Aeronautics and Astronautics, Inc. All
rights reserved.

[*]Associate Professor, Department of Mechanical Engineering.
[†]Professor, George Woodruff School of Mechanical Engineering.

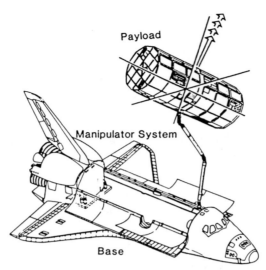

Fig. 1 Example of robotic manipulator application in space; long, slender manipulator structure and large payloads are typical characteristics.

flexibility, then the flexibility is insignificant for all practical purposes. However, if the manipulator performance is not satisfactory, i.e., sufficiently fast settling does not occur to within desired final accuracy because of vibrations, then the controller must take the structural flexibility into account. First, the manipulator strength must be sufficient to guarantee that the total deformation caused by static and dynamic loads stays within the elastic range. Then it can be moved with a controller which is designed for a slow enough closed-loop system (CLS) bandwidth relative to the manipulator stiffness such that the flexibility is not significant and can be ignored. Therefore, the practical engineering question for space robotics is the following: given a manipulator with a very long, slender structure, and a desired manipulation bandwidth, can the manipulator flexibility be ignored, or does it have to be taken into account in the controller design to achieve the desired manipulation bandwidth?

For a particular manipulator, there is a critical bandwidth that characterizes the flexibility. If the desired CLS bandwidth is larger than that critical value, the flexibility must be taken into account. The critical bandwidth values are determined by the lowest natural frequency with locked actuators at a particular configuration. Studies in the control of manipulators taking flexibility into account have primarily concentrated on the single link flexible arm example. The dynamics of the beam is modeled with a Bernoulli-Euler model. The influence of actuator and sensor locations on the stability of flexible systems under output feedback was first explained by Gevarter.[1] If an input control and an output measurement influence coefficient of a mode have the same sign, the CLS will be stable, otherwise it will be unstable as the loop gain increases (Fig. 2). One way of guaranteeing that the influence coefficients have the same sign (phase relationship) is to place the actuators and sensors at the same location. Hence, a flexible system under

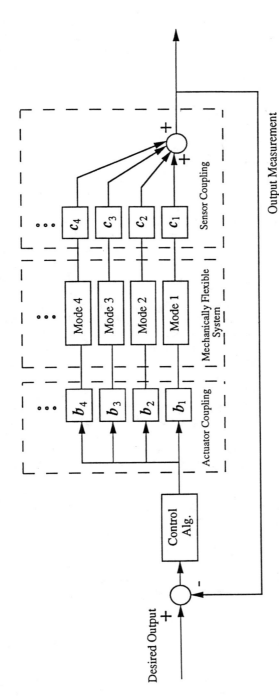

Fig. 2   Model dynamics of a mechanically flexible system illustrating actuation (input) and sensing (output) coupling between the dynamic modes and control system actuators/sensors.

the control of collocated actuators and sensors stays stable as loop gain varies. If the sensor to actuator information flow path involves information flow from sensor(s) to actuator(s) which are not collocated (noncollocated) then the CLS would become unstable as loop gain increases. If the feedback control algorithm is based on state feedback (i.e., modal control) as opposed to output feedback control, it must be finite dimensional. The dynamic behavior is described by an infinite number of normal modes. A practical modal controller can control only a finite number of modes because actuator and sensor bandwidth limitations and limited sampling time available in real-time digital control. The interaction between control action and uncontrolled (residual) modes (the so-called control spillover), and the contribution of residual modes to the measured output (the so-called observation spillover) can lead to instability in the closed-loop. The control spillover alone would not cause closed-loop instability.[2] The spillover problem is a result of finite-dimensional state feedback control. It will exist for collocated as well as noncollocated actuator-sensor configurations. Whether the spillover causes a CLS to be unstable or not depends on the particular case. This problem can also be posed as a robustness problem as follows. Given a particular finite-dimensional state feedback controller, a robustness measure of the CLS stability against residual dynamics can be determined for the particular controller. If the residual dynamics uncertainty stays within the robustness measure, the CLS will stay stable despite the spillover effects. Otherwise, the CLS will be unstable because of spillover.

Noncollocation results in a time delay in the dynamic relationship between the control action at the actuation point and its observed effects at the sensor point. This time delay is related to the time period needed for the deformation waves to propagate from the actuator point to the sensor point through the elastic medium. Mathematically, this effect shows up as zeros in the right-hand and left-hand plane symmetrically about the imaginary axis.[3-5] Collocated actuator-sensor arrangement in the control system is attractive because of its robust closed-loop stability property. However, the highest CLS bandwidth that can be achieved by the collocated control approach is limited by the lowest cantilever natural frequency of the beam.[6] If a higher CLS bandwidth is desired, a noncollocated control approach must be taken. Well-designed noncollocated controllers can achieve a bandwidth that is two to three times faster than the lowest cantilever frequency. The ultimate upper limit of bandwidth is, of course, set by the time delay associated with the propagation of bending waves from actuation point to sensing point.

In the remainder of this chapter, we discuss the untruncated and truncated transfer functions and their pole-zero structure as a function of actuator and sensor locations. Then we give a physical explanation of zeros of the transfer function, particularly the nonminimum phase zeros, and their implication on the control system performance limitations. We discuss the accuracy issues of finite-dimensional models developed using different mode shapes from a controller point of view.

II.   Nontruncated Dynamic Models of a Flexible Beam

Consider the dynamics of the flexible beam shown in (Fig. 3). Neglecting rotary inertia and shear deformation effects, the dynamics of the position of any point along the beam may be described by the Bernoulli-Euler beam equation

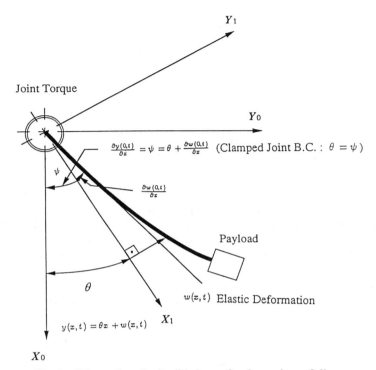

Joint Torque

$\frac{\partial y(0,t)}{\partial x} = \psi = \theta + \frac{\partial w(0,t)}{\partial x}$ (Clamped Joint B.C. : $\theta = \psi$)

ψ

$\frac{\partial w(0,t)}{\partial x}$

Payload

θ

$w(x,t)$ Elastic Deformation

$y(x,t) = \theta x + w(x,t)$ X_1

X_0

Fig. 3   Schematics of a flexible beam for dynamic modeling.

$$EI\frac{\partial^4 y(x,t)}{\partial x^4} + \rho a \frac{\partial^2 y(x,t)}{\partial t^2} = 0 \tag{1}$$

where $0 < x < l$, and the associated boundary conditions are, at $x = 0$

$$y(0,t) = 0 \tag{2a}$$

$$EI\frac{\partial^2 y}{\partial x^2} + u = J_h \frac{d^2}{dt^2}\left(\frac{\partial y}{\partial x}\right) \tag{2b}$$

at $x = l$

$$EI\frac{\partial^2 y}{\partial x^2} = -J_p \frac{d^2}{dt^2}\left[\frac{\partial y(l,x)}{\partial x}\right] \tag{2c}$$

$$EI\frac{\partial^3 y}{\partial x^3} = m_p \frac{d^2}{dt^2}[y(l,t)] \tag{2d}$$

where J_h is joint moment of inertia, ρa is mass per unit length, EI is rigidity of the beam, and m_p, J_p represents payload mass and moment of inertia about its center of mass, respectively. Control torque u is included as part of the boundary conditions in the formulation Eq. (2b). As the control algorithm changes (i.e., gain adaptation), the boundary condition (2b) changes. The general solution for the displacement of any point along the beam can be found using the Laplace transform

$$y(\xi, s) = A \sin \beta \xi + B \sinh \beta \xi + C \cos \beta \xi + D \cosh \beta \xi \tag{3}$$

where

$$\xi = \frac{x}{l}$$

$$\beta^4 = -s^2 \left(\frac{\rho a l^4}{EI} \right)$$

$$y(x, s) = l y(\xi, s)$$

$$y^{(i)} = l^{1-i} y^{(i)}(\xi, s)$$

where $^{(i)}$ means partial derivative order with respect to the first argument. A, B, C, D are functions of β and s in general, and are determined from the boundary conditions Eqs. (2a–2d). The Laplace transformed and non-dimensionalized boundary conditions with respect to the spatial variable are, at $\xi = 0$,

$$y(0, s) = 0 \tag{4a}$$

$$y''(0, s) + \left(\frac{l}{EI} \right) u(s) = -\beta^4 \left(\frac{1}{\rho a l^3} \right) J_h y'(0, s) \tag{4b}$$

at $\xi = 1$

$$y''(1, s) = \beta^4 \left(\frac{1}{\rho a l^3} \right) J_p y'(1, s) \tag{4c}$$

$$y'''(1, s) = -\beta^4 \left(\frac{1}{\rho a l} \right) m_p y(1, s) \tag{4d}$$

From boundary condition (4a), $C + D = 0 \rightarrow C = -D$; then the general solution Eq. (3) becomes

$$y(\xi, s) = A \sin \beta \xi + B \sinh \beta \xi + C(\cos \beta \xi - \cosh \beta \xi) \tag{5}$$

A. Open-Loop Transfer Functions

Application of the remaining three boundary conditions (4b–4d) on the general solution (5) results in the solution for the A, B, C coefficients. We need to consider two cases to obtain the eigenvalue problem and open-loop transfer functions.

1) If unforced motion of the beam is considered, then the input torque is zero, $u = 0$. This leads to the solution of the open-loop eigenvalue problem. The application of boundary conditions (4b–4d) yields the following equation:

$$\begin{pmatrix} n_{11} & n_{12} & n_{13} \\ n_{21} & n_{22} & n_{23} \\ n_{31} & n_{32} & n_{33} \end{pmatrix} \begin{pmatrix} A \\ B \\ C \end{pmatrix} = \begin{pmatrix} 0 \\ 0 \\ 0 \end{pmatrix} \tag{6}$$

For the solution of Eq. (6) to be nontrivial, the determinant of the 3×3 matrix must be zero $(\det|N_1| = 0)$. This is the characteristic equation of transcendental form giving $\{\beta_i, i = 1, \ldots, \infty\}$ characteristic roots, and associated characteristic vectors $[A_i, B_i, C_i, -C_i]$. This is the same solution as those of Barbieri and Özgüner,[7] which is a special case of this analysis, and of Book,[8] which is more general than this analysis. Notice that the unforced motion consideration here represents a pinned boundary condition at the actuation end. If the payload does not exist ($m_p = J_p = 0$), then the solution of β_is and $[A_i, B_i, C_i, -C_i]$ will be identical to the solution of the eigenvalue problem for the pinned-free beam. No assumptions are made regarding the type of values of β_i's and $[A_i, B_i, C_i, -C_i]$. They may have complex values as a result of appropriate boundary conditions. A feedback controlled flexible beam in general may have complex conjugate poles (modes with damping) and complex mode shapes. Therefore, the mode shapes may have nonstationary nodes.

2) If the input torque is not zero, this will result in a nontruncated transfer function from the joint torque to the chosen output variable. This relationship can be obtained by the application of boundary conditions (4b–4d):

$$
\begin{pmatrix} n_{11} & n_{12} & n_{13} \\ n_{21} & n_{22} & n_{23} \\ n_{31} & n_{32} & n_{33} \end{pmatrix} \begin{pmatrix} A \\ B \\ C \end{pmatrix} = \begin{pmatrix} 0 \\ 0 \\ u \end{pmatrix}
\tag{7}
$$

Thus, the coefficient vector can be obtained as

$$
(A \quad B \quad C)^T = [N_1]^{-1} (0 \quad 0 \quad 1)^T u
\tag{8}
$$

Hence a transfer function between input torque, u, and an output is easily found using Eqs. (5) and (8) as follows:

i) Hub angle as the output, $\partial y(0, s)/\partial x = y'(0, s) = \theta + \partial w(0, s)/\partial x$, (Fig. 3),

$$
y'(0, s) = \beta (1 \quad 1 \quad 0) [N_1]^{-1} \begin{pmatrix} 0 \\ 0 \\ 1 \end{pmatrix} u(s)
\tag{9}
$$

ii) Displacement at a position along the beam as the output, $0 < \xi_0 < 1$,

$$
y(\xi_0, s) = [\sin \beta\xi_0, \sinh \beta\xi_0, (\cos \beta\xi_0 - \cosh \beta\xi_0)][N_1]^{-1} \begin{pmatrix} 0 \\ 0 \\ 1 \end{pmatrix} u(s)
\tag{10}
$$

iii) Tip displacement as the output,

$$
y(1, s) = [\sin \beta, \sinh \beta, \cos \beta - \cosh \beta][N_1]^{-1} \begin{pmatrix} 0 \\ 0 \\ 1 \end{pmatrix} u(s)
\tag{11}
$$

Notice that the transfer functions between control torque at the base and various outputs have the same pole location, but different zero locations, as displayed by denominator and numerator polynomials. More explicity, it can be shown that the open-loop transfer function for a hub angle is

$$
\frac{y'(0, s)}{u(s)} = \frac{N_\theta(\beta)}{D_\theta(\beta)}
\tag{12a}
$$

where

$$N_\theta(\beta) = \frac{-2l}{EI\beta}[1 + \cos(\beta)\cosh(\beta) + \eta\beta(\cos\beta\sinh\beta - \sin\beta\cosh\beta)] \quad (12b)$$

$$D_\theta(\beta) = 2\{(\sin\beta\cosh\beta - \cosh\beta\sinh\beta + 2\eta\beta\sin\beta\sinh\beta)$$
$$+ \epsilon\beta^3[1 + \cos\beta\cosh\beta + \eta\beta(\cos\beta\sinh\beta - \sin\beta\cosh\beta)]\} \quad (12c)$$

for

$$\epsilon = \frac{J_h}{\rho a l^3}, \qquad \eta = \frac{m_p}{\rho a l}, \qquad J_p = 0$$

Similarly, the open-loop transfer function for the tip position can be written as

$$\frac{y(1, s)}{u(s)} = \frac{N_{y_{tip}}(\beta)}{D_{y_{tip}}(\beta)} \quad (13a)$$

where

$$N_{y_{tip}}(\beta) = \frac{-2l}{EI\beta^2}(\sin\beta + \sinh\beta) \quad (13b)$$

$$D_{y_{tip}}(\beta) = 2\{(\sin\beta\cosh\beta - \cos\beta\sinh\beta + 2\eta\beta\sin\beta\sinh\beta)$$
$$+ \epsilon\beta^3[1 + \cos\beta\cosh\beta + \eta\beta(\cos\beta\sinh\beta - \sin\beta\cosh\beta)]\} \quad (13c)$$

The open-loop transfer functions between the control torque at the base and various outputs (hub angle, displacement of any point along the beam, tip displacement) are exact in the sense that there is no modal truncation. The roots of the denominator (a transcendental function with infinitely many roots representing infinitely many modes) can be found by numerical root finding algorithms. They are along the imaginary axis (Figs. 4a). The zeros of the collocated open-loop transfer function are also along the imaginary axis and alternate with the poles. The location of the zeros correspond to the location of the poles for the clamped-joint boundary condition (Fig. 4a). The zeros of the transfer function between the joint torque and the tip position [Eq. (11)] are all on the real axis and symmetric about the origin (Fig. 4c). The zeros of the transfer function between the joint torque and the displacement of any point along the beam has some real zeros and some along the imaginary axis (Fig. 4b). As the output sensor moves from collocated position towards the tip of the beam, zeros start to migrate up along the imaginary axis and start to pop up on the real axis (Figs. 4a–4c). When the output sensor is at the other boundary of the beam, all of the zeros are on the real axis and symmetric about origin (Fig. 4c).

The migration of zeros from the shape shown in Fig. 4a to the shape in Fig. 4b, and eventually to the shape in Fig. 4c, as output sensor location is moved from the base toward the tip, can be easily verified by interpreting the behavior of modes in frequency response as a function of the sign of the modes at a particular sensor location.[9-11]

B. Closed-Loop Transfer Functions

So far we have considered the open-loop transfer functions. The joint torque will in fact be the function of a feedback control algorithm. This means the boundary

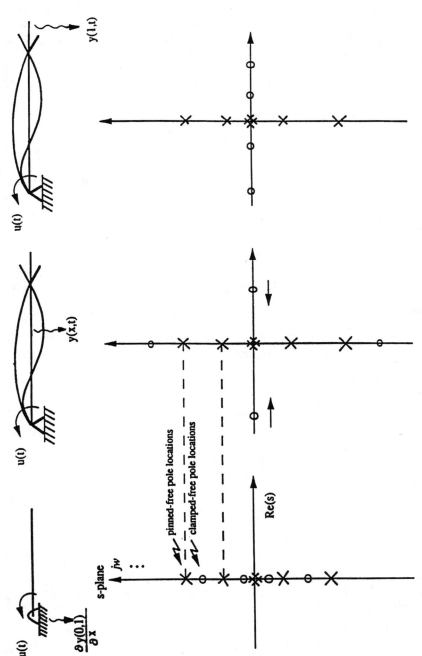

Fig. 4 Pole zero locations of open-loop transfer functions of a flexible beam from joint torque to various sensor locations.

condition (4b) will change as a function of the feedback gains; therefore, the $[N_1]$ matrix will be a function of the type and parameters of the feedback control algorithm. Let us arrange Eq. (6) such that the first two rows of $[N_1]$ are formed by the application of boundary conditions (4c) and (4d), which are not affected by feedback. Hence the first two row elements of $[N_1]$ are not affected by the control. The last row of $[N_1]$ is formed by the application of boundary condition (4b), which involves the feedback control $u(s)$, hence the elements of the last row of $[N_1]$ are functions of the feedback gains. From boundary condition (4b),

$$-\left(\frac{EI}{l}\right)\left[\beta^5\left(\frac{1}{\rho a l^3}\right)J_h\right]A - \left(\frac{EI}{l}\right)\left[\beta^5\left(\frac{1}{\rho a l^3}\right)J_h\right]B + \left(\frac{EI}{l}\right)(2\beta^2)C = u(s)$$

(14)

Let us look at three cases to illustrate how feedback changes the closed loop dynamics.

1) Tip position proportional plus derivative (PD) controller

$$u(s) = -(k_{vy}^* s + k_{py}^*)[y(l, s) - y_d(l, s)]$$

(15)

$$u(s) = -(k_{vy} s + k_{py})[y(1, s) - y_d(1, s)]$$

(16)

where $k_{vy} = lk_{vy}^*$, $k_{py} = lk_{py}^*$, $y_d(1, s)$ is the desired tip position.

$$u(s) = -(k_{vy} s + k_{py})(A \sin \beta + B \sinh \beta + C(\cos \beta - \cosh \beta))$$
$$+ (k_{vy} s + k_{py})y_d(1, s)$$

(17)

Substituting Eq. (17) into Eq. (14),

$$\left\{-\left(\frac{EI}{l}\right)\left[\beta^5\left(\frac{1}{\rho a l^3}\right)J_h\right] + (k_{vy} s + k_{py}) \sin \beta\right\}A$$

$$+\left\{-\left(\frac{EI}{l}\right)\left[\beta^5\left(\frac{1}{\rho a l^3}\right)J_h\right] + (k_{vy} s + k_{py}) \sinh \beta\right\}B$$

$$+\left\{\left(\frac{EI}{l}\right)(2\beta^2) + (k_{vy} s + k_{py})(\cos \beta - \cosh \beta)\right\}C$$

$$= (k_{vy} s + k_{py})y_d(1, s)$$

(18)

where n_{31}, n_{32}, n_{33}, coefficients of A, B, C in Eq. (18), are functions of feedback gains k_{vy}, k_{py}. Rewriting Eq. (7) for this feedback controller will result in the following equation, which can be solved for (A, B, C).

$$[N_2]\begin{pmatrix} A \\ B \\ C \end{pmatrix} = \begin{bmatrix} 0 \\ 0 \\ (k_{py} + sk_{vy})y_d(s) \end{bmatrix}$$

(19)

Accordingly, the transfer function from the desired tip position command to the actual tip position is given by

$$\frac{y(1, s)}{y_d(1, s)} = (\sin \beta, \quad \sinh \beta, \quad \cos \beta - \cosh \beta) [N_2]^{-1} \begin{bmatrix} 0 \\ 0 \\ (k_{py} + sk_{vy}) \end{bmatrix} \quad (20)$$

Since the last row of $[N_2]$ is function of feedback gains, the poles of the closed-loop transfer function (TF) between $y_d(1, s)$ and $y(1, s)$ will be different than the poles of the open-loop TF. This displays the way feedback changes the CLS poles. It also displays a well-known fact regarding feedback control, namely, that feedback does not change the location of open-loop zeros, rather the controller zeros are merely added to the existing zeros. The zeros of Eq. (20) are given by

$$(\sin \beta, \quad \sinh \beta, \quad \cos \beta - \cosh \beta) \begin{pmatrix} h_{11} & h_{12} & h_{13} \\ h_{21} & h_{22} & h_{23} \\ h_{31} & h_{32} & h_{33} \end{pmatrix} \begin{pmatrix} 0 \\ 0 \\ 1 \end{pmatrix} (k_{vy}s + k_{py}) = 0 \quad (21)$$

where $[h]$ is adj(N_2). It is a simple algebraic operation to show that h_{13}, h_{23}, h_{33} are all independent of the feedback gains.

2) The hub angle PD controller:

$$u(s) = -(k_{v\psi}s + k_{p\psi})[y'(0, s) - y'_d(0, s)] \quad (22)$$

Following the same steps it can be shown that

$$\frac{y'(0, s)}{y'_d(0, s)} = \beta (1 \quad 1 \quad 0) [N_3]^{-1} \begin{pmatrix} 0 \\ 0 \\ 1 \end{pmatrix} (k_{v\psi}s + k_{p\psi}) \quad (23)$$

where the last row of the $[N_3]$ matrix row is a function of $k_{v\psi}$ and $k_{p\psi}$, hence illustrating the effect of feedback on closed-loop poles. It can similarly be shown that open-loop zeros are not affected; the zero of the PD controller is simply added to the existing open-loop TF zeros.

Furthermore, besides the change of eigenvalues as a function of feedback, one can observe the change of mode shapes as a function of feedback by setting the reference input to zero and solving for the closed-loop eigenvectors of (A_i, B_i, C_i) of Eqs. (20) and (23).

3) The tip position PD plus hub angle PD controller. The input torque, $u(s)$, is given by

$$u = (k_{py} + sk_{vy})[y_d - y(1, s)] - (k_{p\psi} + sk_{v\psi})y'(0, s) \quad (24)$$

Note that setting $k_{p\psi} = 0$ results in hub rate proportional feedback control. Similarly, Eq. (17) becomes

$$\frac{y(1, s)}{y_d(1, s)} = (\sin \beta, \quad \sinh \beta, \quad \cos \beta \quad - \cosh \beta) [N_4]^{-1} \begin{bmatrix} 0 \\ 0 \\ (k_{py} + sk_{vy}) \end{bmatrix} \quad (25)$$

where

$$[N_4]\begin{pmatrix} A \\ B \\ C \end{pmatrix} = \begin{bmatrix} 0 \\ 0 \\ (k_{py} + sk_{vy})y_d \end{bmatrix} \tag{26}$$

The elements of $[N_4]$ are given as follows:

$$n_{11} = -\cos\beta + \left(\frac{m_p}{\rho al}\right)\beta\sin\beta$$

$$n_{12} = \cosh\beta + \left(\frac{m_p}{\rho al}\right)\beta\sinh\beta$$

$$n_{13} = (\sin\beta - \sinh\beta) + \left(\frac{m_p}{\rho al}\right)\beta(\cos\beta - \cosh\beta)$$

$$n_{21} = -\sin\beta - \left(\frac{J_p}{\rho al^3}\right)\beta^3\cos\beta$$

$$n_{22} = \sinh\beta - \left(\frac{J_p}{\rho al^3}\right)\beta^3\cosh\beta$$

$$n_{23} = (-\cos\beta - \cosh\beta) + \left(\frac{J_p}{\rho al^3}\right)\beta^3(\sin\beta + \sinh\beta)$$

$$n_{31} = \left(\frac{EI}{l}\right)\left(-\frac{J_h}{\rho al^3}\right)\beta^5$$
$$+ (k_{p\psi} + sk_{v\psi})\beta + (k_{py} + sk_{vy})\sin\beta$$

$$n_{32} = \left(\frac{EI}{l}\right)\left(-\frac{J_h}{\rho al^3}\right)\beta^5 + (k_{p\psi} + sk_{v\psi})\beta$$
$$+ (k_{py} + sk_{vy})\sinh\beta$$

$$n_{33} = 2\beta^2\left(\frac{EI}{l}\right) + (k_{py} + sk_{vy})(\cos\beta - \cosh\beta)$$

It is easy to verify that $[N_1]$, $[N_2]$, and $[N_3]$ are special cases of $[N_4]$. Notice that $\det(N_1)$ describes the open-loop eigenvalue problem, and $\det(N_2)$, $\det(N_3)$, $\det(N_4)$, describe the closed-loop eigenvalue problem for the transfer functions described by Eqs. (20), (23), and (25).

III. Truncated Dynamic Models of a Flexible Beam

We now outline the truncated finite-dimensional dynamics model derivation using different mode shapes. The structural differences of dynamic models obtained using different mode shapes are illustrated. The standard assumptions associated with Bernoulli-Euler beam theory are made in what follows (Fig. 3).

Let any position along the beam (Fig. 3) be defined by

$$y = \theta x + \sum_{i=1}^{\infty} \phi_i(x) q_i(t) \tag{27}$$

where ϕ_i are assumed to be mode shapes forming a complete set, and q_is are generalized modal coordinates describing the flexible deformations. Here, y is physically the distance travelled shown in Fig. 3. If θ is measured from a coordinate frame fixed to joint, $\phi_i(x)$s are used as pinned-free mode shapes. If θ is measured from a coordinate frame that is always tangent to the beam at the base, then $\phi_i(x)$ must be clamped-free mode shapes.

The standard procedure in developing the mathematical model using the Lagrangian approach is to write the kinetic and potential energies of the total system after the kinematic description is decided on.

For the flexible arm of Fig. 3, the kinetic energy is of the form

$$T = \frac{1}{2} \sum_{i=0}^{n} \sum_{j=0}^{n} \dot{q}_i \dot{q}_j \left[\int_0^l \rho a(x) \phi_i(x) \phi_j(x) \mathrm{d}x + J_h \phi_i'(0) \phi_j'(0) \right.$$

$$\left. + J_p \phi_i'(l) \phi_j'(l) + m_p \phi_i(l) \phi_j(l) \right] \tag{28}$$

where $q_0 = \theta$, $\phi_0(x) = x$ represents the rigid-body mode.

Similarly, the elastic potential energy can be expressed as

$$P = \frac{1}{2} \sum_{i=0}^{n} \sum_{j=0}^{n} q_i q_j \left[\int_0^l EI(x) \phi_i''(x) \phi_j''(x) \mathrm{d}x \right] \tag{29}$$

The generalized force is derived from the virtual work as a result of the joint torque

$$\delta W = \sum_{i=0}^{n} Q_i \delta q_i = u \delta \left(\frac{\partial y(0, t)}{\partial x} \right) = u \sum_{i=0}^{n} \phi_i'(0) \delta q_i(t) \tag{30}$$

where $Q_0 = u$, $Q_i = \phi_i'(0)u$, for $i = 1, 2, \ldots, n$, and u is the input torque at the base.

If we let M_{ij} equal the terms in the bracket of Eq. (28) and let K_{ij} equal the terms in the bracket of Eq. (29), the Lagrangian formulation results in the model of form

$$[M_{ij}]\ddot{\underline{q}}_i + [K_{ij}]\underline{q}_i = \underline{b}u \tag{31}$$

where $\underline{b}^T = [1\phi_1'(0) \ldots, \phi_n'(0)]$. Note that $\phi_0(x) = x$, $\phi_0''(x) = 0$, $k_{00} = 0$, which corresponds to the rigid-body mode.

Equation (31) can be put in the state space form

$$\dot{\underline{x}} = F\underline{x} + Gu$$
$$\underline{y} = H\underline{x} \tag{32}$$

where

$$F = \begin{pmatrix} \underline{0} & I \\ -M^{-1}K & \underline{0} \end{pmatrix}$$

$$G = \begin{pmatrix} \underline{0} \\ M^{-1}\underline{b} \end{pmatrix}$$

$$H = \begin{bmatrix} \underline{0} & 1 & \phi_1'(0) & \cdots & \phi_n'(0) \\ \underline{1} & \phi_1(l) & \cdots & \phi_n(l) & \underline{0} \end{bmatrix}$$

where $\underline{0}$ is an $1 \times n$ zero vector, and measured outputs $y^T = [y'(0, t), y(l, t)]$ are hub angle and tip position as indicated by the H matrix.

It is clear from Eqs. (28), (29), and (31) that the coupling between rigid and flexible modes depends on the choice of mode shapes. If one chooses the mode shapes such that expressions in brackets of Eqs. (28) and (29) equal zero for $i \neq j$ (orthogonal) for the system of Fig. 3, the resultant equations will be decoupled.

If clamped-free mode shapes are used, the flexible modes are excited through the coupling with rigid-body mode in the first row and column of the mass matrix ($[M_{ij}]$) since clamped-free mode shapes $\{\phi_i(x); i = 1, \ldots, \infty\}$ alone form a complete basis and

$$x = \sum_{i=1}^{\infty} v_i \phi_i(x) \tag{33a}$$

$$v_i = \int_0^l x\phi_i(x)\mathrm{d}x \tag{33b}$$

where v_i (Fourier coefficients) are different from zero.

Therefore $[M_{ij}]$ will have the form

$$[M_{ij}] = \begin{pmatrix} m_{11} & m_{12} & m_{13} & m_{14} & \cdots & m_{1n} \\ m_{21} & m_{22} & & & & \\ m_{31} & & m_{33} & & \bigcirc & \\ m_{41} & & & m_{44} & & \\ \vdots & & \bigcirc & & \ddots & \\ m_{n1} & & & & & m_{nn} \end{pmatrix} \tag{34a}$$

Since clamped-free mode shapes are such that $\{\phi_i'(0) = 0, i = 1, \ldots, \infty\}$, the input matrix will be of the form

$$\underline{b}^T = [1 \quad 0 \quad \cdots \quad 0] \tag{34b}$$

On the other hand, if pinned-free mode shapes are used, the rigid-body mode is part of the basis of modal functions, therefore,

$$\int_0^l x\phi_i(x) \, \mathrm{d}x = 0; \quad i = 1, \ldots, \infty \tag{35}$$

and $[M_{ij}]$ is a diagonal matrix.

$$[M_{ij}] = \mathrm{Diag}\{M_{ii}\} \tag{36a}$$

There is no coupling between rigid and flexible modes through mass and stiffness matrices. The flexible modes are directly excited by the input since the input matrix is

$$\underline{b}^T = \begin{bmatrix} 1 & \dfrac{\mathrm{d}\phi_1(0)}{\mathrm{d}x} & \cdots & \dfrac{\mathrm{d}\phi_n(0)}{\mathrm{d}x} \end{bmatrix} \tag{36b}$$

where $\mathrm{d}\phi_i(0)/\mathrm{d}x \neq 0$, $i = 1, \ldots, n$, for pinned-free modes.

Notice that if a mode shape is chosen for a specific value of payload, that when the actual payload is different than that the modes will no longer be decoupled in this model. The payload will result in dynamic coupling through $[M_{ij}]$ in Eq. (31). $[K_{ij}]$ is diagonal for both clamped-free and pinned-free mode shapes. Equation (31) can be used to model a flexible arm with pinned- or clamped-joint BC by simply using the appropriate mode shapes in the definition of M_{ij}, K_{ij}, and b_i terms. Let us note at this point that almost all control studies to date retain two or three flexible modes in their work.

IV.   Results and Discussion

A.   Pole-Zero Structure

The pole-zero structure of the collocated transfer functions of untruncated and truncated models have essentially the same structure, with the exception of some minor accuracy differences. However, the zeros of a noncollocated transfer function of truncated and untruncated models are significantly different. The zeros of the untruncated noncollocated transfer function are all real and symmetric about the origin (Fig. 5a). The zeros of the truncated noncollocated model are complex conjugate and symmetric about the imaginary axis (Fig. 5b). Figures 5a and 5b show the open-loop pole-zero locations and the locus of roots as a function of a tip position PD feedback controller gain. Although the zero locations of truncated and untruncated noncollocated transfer functions seem very different mathematically, their physical effect is not very different. They represent the upper limit of a CLS bandwidth because of the time period needed for the bending waves to propagate from joint to tip.

B.   Limitation of Control Bandwidth Because of Flexibility

Mathematical analysis shows that the zeros of open-loop transfer functions represent the upper limit of a CLS bandwidth which can be achieved for an output feedback by increasing the loop gain to infinitely large values. Even if perfect actuators (no time delay, no saturation limits) were available, the zero locations are the values of the highest possible CLS bandwidth.

The physical explanation of this upper limit of a CLS bandwidth imposed by the distributed flexible nature of the mechanical system may be given by interpreting the zeros by their equivalent time delay approximations. The physical source of such a time delay is related to the time period needed for the bending waves to propagate from the actuator location to the controlled output (sensor) location. A wave is a mechanism that propagates in a medium (solid, fluid, or gas) and transfers energy from one region to another. The speed of wave propagation depends on the material properties and the nature of the wave, i.e., longitudinal, torsional, or bending waves. The dynamic behavior of the following problems

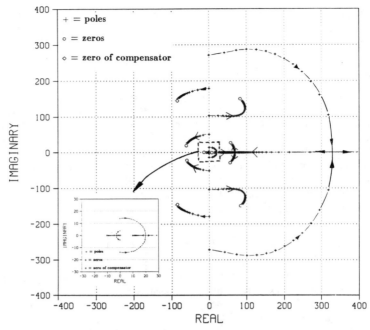

Fig. 5a **Root locus of finite-dimensional model with pinned-free modes under a noncollocated PD controller (tip position and velocity feedback), without hub rate feedback.**

are mathematically equivalent: longitudinal and torsional vibration of beams, and transverse vibration of a string under tension,

$$\frac{\partial^2 w}{\partial x^2} = \frac{1}{c^2}\frac{\partial^2 w}{\partial t^2} \tag{37}$$

where w is the appropriate displacement variable for the problem, x and t are spatial and time variables, and c is the speed of wave propagation in the medium.

$$c^2 = \begin{cases} E/\rho & ; & \text{for longitudinal vibration} \\ G/\rho & ; & \text{for torsional vibration} \\ S/m & ; & \text{for string under tension } S \end{cases} \tag{38}$$

where E is the Young's modulus, G the shear modules, S the tension on the string, ρ the mass density, and m the mass per unit length. In general a wave function of the form

$$u(x, t) = u(t - x/c) \tag{39}$$

satisfies the wave equation [Eq. (37)]. This equation states that the particle at position x will experience the same motion as the particle at position $x = 0$, x/c time period later. However, Eq. (39) would not satisfy bending vibration equations of a beam [Eq. (1)]. This indicates that the nature of wave propagation is

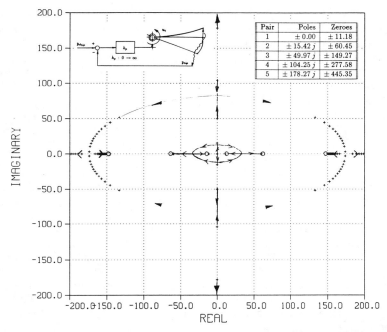

Fig. 5b   Root locus of infinite-dimensional model under a noncollocated PD controller (tip position feedback only), without hub rate feedback.

different in bending than it is in the longitudinal and torsional vibration of beam and the transverse vibration of a string under tension. An elementary theory of wave propagation predicts that dispersion will occur in bending waves, whereas longitudinal waves propagate without changing form. That is, the speed of wave propagation in the longitudinal and torsional case is constant. This theory is valid if the wavelength is long compared with the lateral dimensions of the beam. In bending vibrations the speed of wave propagation becomes a function of the wavelength. This is called dispersion. For a very long wavelength, the bending wave propagation speed can be approximated by

$$c = 2\pi c_0 K / \Lambda \qquad (40)$$

where Λ is the wavelength, $c_0 = \sqrt{E/\rho}$, and K is the radius of gyration of the cross section about the neutral axis of the beam. If the wavelength becomes smaller and comparable to the lateral dimensions of the beam, dispersion occurs at all of the vibration types previously considered. The wave propagation speeds in bending approach that of Rayleigh surface waves.[12] In fact, in this case, elementary theory predictions fail, and rotary inertia and shear deformation effects must be taken into consideration for accurate predictions of wave propagation phenomenon.

Although the physics of wave propagation phenomenon are quite complicated, the approximate descriptions for special cases (i.e., long wavelengths compared

to lateral dimensions of the beam) do provide useful engineering insights. Specifically, they help our understanding of the physical source of the zeros and limitations
they impose on the CLS bandwidth. For a given beam under bending vibration,
the zeros and the time delay approximations of them are indeed physically generated by the time delay needed for the bending wave propagation. Therefore,
Eq. (40) can be used as an approximation in determining the upper limit of the
CLS bandwidth imposed by the physical laws.

C. Accuracy of Truncated Models in Predicting the Closed-Loop System Behavior

Three different models under the same controller are studied to predict the
closed-loop behavior of the flexible beam. These include model 1, a finite-
dimensional model with pinned-free mode shapes [Eqs. (32), (36a), and (36b)];
model 2, a finite-dimensional model with clamped-free mode shapes [Eqs. (32),
(34a), and (34b)]; and model 3, an infinite-dimensional model [transcendental
transfer function, Eqs. (20), (23), and (25)]. A normalized numerical example
($\rho a = l = EI = 1.0$) will be used for comparison. The closed-loop behavior predicted by truncated and nontruncated models under the same controllers will be
discussed. Comparisons will be made for the controller parameters which will provide a high-CLS bandwidth compared to the beam flexibility, since the effect of
flexibility and the modeling of it will be of importance in the high-CLS bandwidth
region.

To observe the effect of the number of mode shapes, models 1 and 2 are studied
for two and 10 mode shapes in addition to the rigid-body mode. The purpose is
to determine the accuracy of model 1 and 2 compared to that of the ideal case
(model 3) in terms of controller design and the prediction of closed-loop behavior.
The closed-loop pole-zero locations will be compared as a function of feedback
controller parameters. Notice that although a finite number of the exact model
eigenvalues as well as eigenfunctions are solved for model 3, the solutions are
free from truncation errors, and exact to the extent of the numerical accuracy of
the computer and root finding algorithm.

A realistic noncollocated controller will involve hub rate feedback, since hub
rate feedback increases the stability margin of the system. Without hub rate feedback, the CLS would go unstable immediately as tip position and/or velocity
feedback were increased. For a tip position PD controller with hub rate feedback,
Figs. 6a and 6b show the root locus of the CLS as the tip controller gain (k_{py}) is
increased. Again controller parameters are selected such that the CLS bandwidth
is sufficiently high while all other modes are stable. Simulation is repeated for
finite-dimensional models with 2 and 10 mode shapes, in addition to the rigid-body
mode.

From these figures one can recognize first that the clamped-free model root
locus is much closer to the infinite-dimensional model root locus (Figs. 6a and
6b), even when the number of modes used is very small, such as two. As expected,
as the number of modes is increased, the root locus of the dominant poles of
the pinned-free model converge to those of the infinite-dimensional model too.
If the loop gain is increased even after the CLS is unstable, the root locus of the

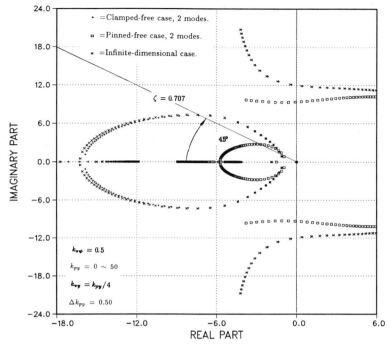

Fig. 6a Root locus of CLS as the tip controller gain (k_{py}) is increased for J_h = 0.00333.

finite- and infinite-dimensional models will converge to their open-loop zero locations, which are complex conjugate for the finite-dimensional model and real for the infinite-dimensional model, as can be verified by Eqs. (32) and (25). Therefore, in the limit (as loop gain goes to infinity) the closed-loop poles will converge to different locations for the finite- and infinite-dimensional models. However, from the view point of designing a practical feedback controller, the asymptotic locations of the CLS unstable poles are not important. If the CLS is already known to be unstable, it really is not important exactly where the CLS poles end up in the right half s plane. What is important is the accurate prediction of the controller parameter values for which the CLS is in the stability boundary, which is discussed next. For inverse dynamics feedforward control techniques this issue is not clear at this point in time.

The second result observed in these figures concerns the stability limit, i.e., at what value of feedback gain does the closed-loop system reach the stability boundary. For two-mode shape approximations the clamped-free model and infinite-dimensional model results agree almost perfectly ($k_{py}^* \cong 6.75$) (Fig. 6a). The pinned-free model predicts that the CLS would go unstable at a value of $k_{py}^* \cong 4.5$ for the two-mode shape approximation model. Again, as the number of modes is increased, the prediction concerning the stability margin of all models converges to the same value, which is about $k_{py}^* \cong 6.75$. It is very interesting

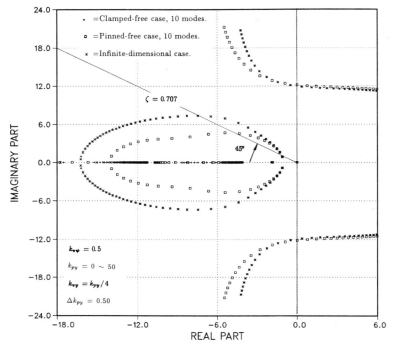

Fig. 6b   Root locus of CLS as the tip controller gain (k_{py}) is increased for J_h = 0.00333.

to note that the clamped-free model prediction was very accurate even with a two-mode approximation. When judging these numerical results concerning the limiting value of feedback gain, the reader should note that the results are for normalized flexible beam parameters ($\rho a = l = EI = 1.0$, and $J_h = 0.00333$). When J_h (hub inertia) is increased, the CLS behavior predictions of the dominant poles by the three different models make no significant difference as shown in Figs. 7a and 7b (for $J_h = 0.333$).

The CLS bandwidth can be characterized as the magnitude of the smallest eigenvalue of the closed-loop system with sufficient damping with all higher modes stable. The third numerical results extracted from Figs. 6a and 6b concerning the CLS bandwidth are such that clamped-free mode shape models (2 modes and 10 modes) give almost perfect numerical accuracy compared to the infinite-dimensional model ($-7.8 \pm 7.8j$), whereas the bandwidth predicted by the pinned-free model with 2 and 10 mode shapes are about $-3.2 \pm 3.2j$ and $-5.0 \pm 5.0j$, respectively, as can be seen from Figs. 6a and 6b.

All of our results consistently indicate that models with clamped-free mode shapes predict the closed-loop behavior more accurately than models with pinned-free modes when the feedback control attempts to drive the arm at a high-CLS bandwidth. However, as the hub inertia increases relative to the beam inertia $[J_h/(\frac{1}{3}\rho al^3) \to 1$ or higher values], the difference between the closed-loop performance predictions of different models becomes insignificant.

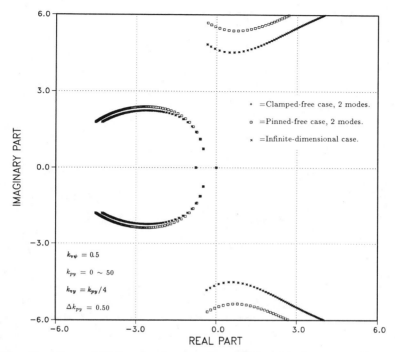

Fig. 7a Root locus of CLS as the tip controller gain (k_{py}) is increased for J_h = 0.333.

V. Summary

The pole-zero structure of transfer functions between joint torque and various output sensor positions (collocated and noncollocated) of a flexible beam is discussed. Particular attention was given to the nonminimum phase zeros and their physical meaning, as well as their significance in terms of the limitations they impose on the CLS bandwidth. The physical source of such behavior is, of course, the presence of an elastic medium between actuation and the sensing point and the fact that there is an unavoidable time delay for the deformation waves to travel through the elastic medium. Truncated and nontruncated models are developed and their accuracy in predicting the CLS dynamics is compared.

The CLS stability under an output feedback control algorithm depends on the locations of the actuators/sensors. If the actuators and sensors are collocated, the CLS will stay stable as loop gain increases. However, if they are noncollocated, the CLS will become unstable as loop gain increases. Collocated output feedback control can be implemented with completely passive components (i.e., springs and dampers) and hence results in a stable CLS. The CLS stability under a finite-dimensional state feedback control algorithm depends on the robustness of the overall system under this particular controller against the residual dynamics uncertainty. The CLS may become unstable even if the actuators and sensors are collocated. The spillover between controlled and residual dynamics is the mechanism affecting the CLS stability.

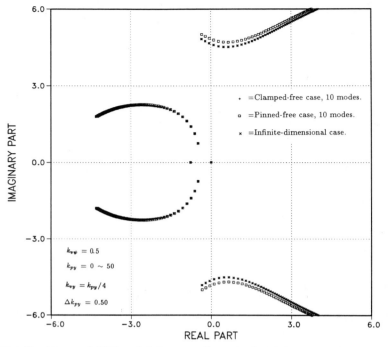

Fig. 7b   Root locus of CLS as the tip controller gain (k_{py}) is increased for J_h = 0.333.

References

[1]Gevarter, W. B., "Basic Relations for Control of Flexible Vehicles," *AIAA Journal,* Vol. 8, No. 4, 1970, pp. 666–672.

[2]Balas, M. J., "Active Control of Flexible Systems," *Journal of Optimization Theory and Applications,* Vol. 25, No. 3, 1978, pp. 415–436.

[3]Cannon, R. J., Jr., and Schmitz, E., "Initial Experiments on the End-Point Control of a Flexible One-Link Robot," *International Journal of Robotics Research,* Vol. 3, No. 3, 1984, pp. 62–75.

[4]Hastings, G. G., and Book, W. J., "Verification of a Linear Dynamic Model for Flexible Robotic Manipulators," *IEEE Control Systems Magazine,* IEEE Control Systems Society, Vol. 7, No. 1, April 1987, pp. 61–64.

[5]Rovner, D. M., and Franklin, G. F., "Experiments Towards On-Line Identification and Control of a Very Flexible One-Link Manipulator," *International Journal of Robotics Research,* Vol. 6, No. 4, 1987, pp. 3–19.

[6]Book, W. J., Maizza-Netto, O., and Whitney, D. E., "Feedback Control of Two Beam, Two Joint Systems with Distributed Flexibility," *Journal of Dynamic Systems, Measurement, and Control,* Vol. 97G, Dec. 1975.

[7]Barbieri, E., and Özgüner, Ü., "Unconstrained and Constrained Mode Expansions for a Flexible Slewing Link," *ASME Journal of Dynamic Systems, Measurement, and Control,* Vol. 110, No. 4, Dec. 1988, pp. 416–421.

[8]Book, W. J., "Modeling, Design and Control of Flexible Manipulator Arms," Ph.D. Thesis, Massachusetts Inst. of Technology, Cambridge, MA, 1974.

[9]Spector, V., and Flasher, H., "Modeling and Design Implications of Non-collocated Control in Flexible Systems," *ASME Journal of Dynamic Systems, Measurement, and Control,* Vol. 112, June 1990, pp. 186–193.

[10]Wei, B., "Active Vibration Control Synthesis for the Control of Flexible Structures Mast Flight System," *Journal of Guidance, Control and Dynamics,* Vol. 11, No. 3, 1988, pp. 271–277.

[11]Miu, D. K., "Physical Interpretation of Transfer Function Zeros for Simple Control Systems with Mechanical Flexibilities," *ASME Journal of Dynamic Systems, Measurement, and Control,* Vol. 113, 1991, pp. 419–424.

[12]Kolsky, H., "Stress Waves in Solids," Dover Pub. Inc., 1963.

Chapter 13

Stability and Control of Robotic Space Manipulators

John L. Junkins[*]
Texas A&M University, College Station, Texas 77843
and
Youdan Kim[†]
Seoul National University, Seoul, Korea

I. Introduction

I N the present chapter, we present some elegant concepts from stability theory, and consider their applicability to the problem of designing control laws for multiple-degree-of-freedom robotic space manipulators. While the spirit of our presentation is classical, we include some novel stability results and a methodology for designing globally stable control laws for nonlinear finite-dimensional and distributed-parameter systems. We find that the Lyapunov approach is attractive not only because it plays an important role in stability and control theory, but also because it provides the only broadly applicable approach to design guaranteed stable control laws for nonlinear, time-varying, and distributed-parameter systems. Especially significant is the fact that the Lyapunov approach leads to a unified stability and control perspective for both linear and nonlinear systems, as well as systems described by ordinary, partial, and hybrid differential equations. The first half of this chapter is an efficient summary of the main features of Lyapunov stability theory, however a few examples are considered to help illustrate this material. The second half of the chapter is addressed to significant studies wherein we apply the analytical developments to formulate stabilizing feedback control laws for idealized manipulator systems undergoing large generally nonlinear motions.

In Sec. II, several important definitions and concepts fundamental to stability analysis are stated. Section III reviews three basic stability and instability theorems stemming from the so-called direct method of Lyapunov. In Sec. IV, a Lyapunov stability analysis for systems described by linear first order systems of differential

---

Copyright © 1994 by the author. Published by the American Institute of Aeronautics and Astronautics, Inc., with permission. Released to AIAA to publish in all forms.
[*]George Eppright Professor, Department of Aerospace Engineering.
[†]Assistant Professor, Department of Aerospace Engineering.

equations is presented. In Sec. V, the indirect method of Lyapunov is presented to characterize the local stability of motion of a nonlinear system near an equilibrium state. Section IV presents a method of designing guaranteed stable controllers for two or more multilink robots which are cooperatively manipulating a payload through large nonlinear motions.

II. Basic Definitions

Consider any member of the family of continuous, finite-dimensional dynamical systems which can be described by a first-order nonlinear vector differential equation of the form

$$\dot{x} = f(x, t), \qquad x \in R^n \tag{1}$$

where $x(t)$ is the state vector at time t, and the dot denotes time differentiation.

A. Definition: Equilibrium State

A vector $x_e \in R^n$ is said to be an equilibrium state of the system described by Eq. (1) at time t_0 if

$$f(x_e, t) = 0 \qquad \forall\, t \geq t_0 \tag{2}$$

If x_e is an equilibrium state of Eq. (1) at time t_0, then x_e is also an equilibrium state of Eq. (1) at all times $t_1 \geq t_0$. In other words, motion initiating at x_e remains there for all time.

B. Definition: Stability of an Equilibrium State

The equilibrium state x_e, or the equilibrium solution $x(t) = x_e$, is said to be stable, if for any given t_0 and positive ϵ, there exists a positive $\delta(\epsilon, t_0)$ such that every time-varying trajectory (or solution) $x(t)$ initiating (time t_0) at a point x_0 which lies within a δ neighborhood of $x_e\{\|x_0 - x_e\| < \delta, x_0 \equiv x(t_0)\}$ remains for all time within an ϵ neighborhood of $x_e\{\|x(t) - xe\| \; \forall\, t \geq t_0\}$. The equilibrium state is said to be unstable if it is not stable.

C. Definition: Asymptotic Stability of an Equilibrium State

The equilibrium state x_e is said to be asymptotically stable if 1) it is stable (Sec. II.B) and if, in addition, 2) for any t_0, there exist a $\delta_1(t_0)$, such that

$$\|x_0 - x_e\| < \delta_1 \qquad \text{implies that} \qquad \lim_{t \to \infty} x(t) \to x_e$$

If δ and δ_1 are not functions of t_0, then the equilibrium state is said to be uniformly stable and uniformly asymptotically stable, respectively. The definition in Secs. II.B and II.C constitute the two basic definitions of stability of an equilibrium state (a fixed point in the state space) for an unforced continuous time system. More generally, we need to consider the stability of a trajectory or a motion. Qualitatively, stability of a trajectory is concerned with whether or not a perturbed motion remains near the unperturbed trajectory, or diverges from it. Stability of a motion is of central interest in many practical feedback control situations whereby a system is designed to execute a large nominal motion, and control inputs must be developed not only to generate the nominal motion but also closed-loop feedback is required to stabilize neighboring motions, with respect to the nominal motion, so that the actual system will behave in a near-nominal fashion.

D.   Definition: Stability of a Motion

The motion $x(t)$ is said to be stable if for all initial times t_0 and prescribed positive ϵ, there exists a positive $\delta(\epsilon, t_0)$, such that

$$\|x(t) - \bar{x}(t)\| < \epsilon \qquad \forall \, t \geq t_0 \text{ if } \|x_0 - \bar{x}_0\| < \delta$$

where $x(t)$ and $\bar{x}(t)$ are neighboring trajectories with the given initial conditions x_0 and $\bar{x}_0$, respectively, at time t_0.

This bounded motion stability property is sometimes referred to as path stability. Qualitatively, path stability means that if the perturbed initial state $\bar{x}(t_0)$ is near $x(t_0)$, then the ensuing perturbed trajectory $\bar{x}(t)$ will remain near $x(t)$ for all time t.

E.   Definition: Asymptotic Stability of a Motion

The motion $x(t)$ is said to be asymptotically stable if 1) it is stable (see the definition in Sec. II.D), and if, in addition; 2) for any t_0, there exist a positive $\delta_1(t_0)$, such that

$$\|x_0 - \bar{x}_0\| < \delta_1 \quad \text{implies that} \quad \lim_{t \to \infty} \|x(t) - \bar{x}(t)\| = 0 \qquad (3)$$

Note that $\bar{x}(t)$ is any member of the set of it all neighboring (perturbed) trajectories satisfying Eq. (3), and all members of this set asymptotically approach $x(t)$.

The preceding definitions are not directly concerned with the global properties of systems, but of the motion in a finite local neighborhood of an equilibrium state or a motion of the system of differential equations. If a system has a globally asymptotically stable equilibrium state, then it is obviously the only equilibrium state, and every motion converges to that unique equilibrium. An analogous global stability property can be defined for the stability of a motion.

The simplest class of Lyapunov stability analysis methods arises in the context of systems described by linear unforced differential equations. We now summarize some of the central ideas.

Consider the linear system

$$\dot{x}(t) = A(t)x(t)$$

which obviously has one equilibrium state at the origin. This linear system can be classified as stable, asymptotically stable, or unstable, depending on the stability of the origin.[1,2]

Now, we introduce two definitions associated with the concept of positive definite functions; these are very important to understand when applying Lyapunov stability theory.

F.   Definition: Positive Definite Function

A single-valued function $U(x)$, which is continuous and has continuous partial derivatives with respect to the components of the vector x, is said to be positive definite in some region Ω about the origin if it vanishes at the origin and is positive elsewhere, i.e.,

$$\text{i) } U(0) = 0$$
$$\text{ii) } U(x) > 0 \quad \text{for all nonzero } x \in \Omega$$

If the positivity condition ii) is relaxed to simply the nonnegative condition $U(x) \geq 0$ for all $x \in \Omega$, then $U(x)$ is said to be positive semidefinite. If the inequality sign in ii) is reversed, then the condition for a negative definite function is obtained. If a function is neither positive nor negative definite, then it is indefinite.

G. Definition: Positive Definite Quadratic Forms

In the analysis of linear dynamical systems, quadratic functions of the state vector arise often in the context of energy, stability, and control analyses. Especially important are symmetric quadratic forms. The quadratic form $U(x) = x^T Q x$ is said to be positive definite if

$$U(x) = x^T Q x > 0 \quad \text{for all nonzero } x \in R^n$$

where Q is a real symmetric matrix.

The definition in Sec. II.G is equivalent to requiring that all the eigenvalues of Q are strictly positive, such a matrix is naturally called a positive definite matrix.

For further discussion of background material, the reader is referred to Vidyasagar[1] and Willems.[2] The following example is useful to illustrate the ideas underlying the preceding discussion.

H. Example

Consider the functions $U_1(x) = x_1^2 + x_2^2$ and $U_2(x) = (x_1 + x_2)^2$. Clearly U_1 satisfies the condition of the definition in Sec. II.C, therefore it is a positive definite function in a two-dimensional space; but U_1 is only positive semidefinite if the underlying space has more than two dimensions. U_2 is only positive semidefinite in two-dimensional space, since it is zero everywhere along the line $x_1 + x_2 = 0$.

III. Lyapunov Stability Theory (Lyapunov's Direct Method)

The central ideas of the Lyapunov stability theorem are now introduced. For a given general dissipative, forced mechanical system, it is often useful to consider a conservative idealized approximation of system without the dissipative or non-conservative external forces acting. For this idealized system, suppose that there exists one equilibrium state x_e of the system at the desired target postion. Also suppose that the total mechanical energy or Hamiltonian of this idealized system is a positive definite function and is an exact integral of the idealized system. For a broad class of practical applications, the total energy or Hamiltonian of an idealized conservative system is a suitable Lyapunov function for studying the stability of the system, including dissipative internal and external forces. More generally, a candidate Lyapunov function must belong to a class of admissible energy functions which have as the most fundamental property the fact that they are zero at the desired equilibrium state and positive everywhere else.

Now suppose that the system is initially perturbed to a state neighboring the desired equilibrium point where the energy level is positive by assumption, and we consider the time evolution of the distance to the desired equilibrium as measured by the energy function. Depending on the nature of the selected energy function (Lyapunov function), the stability of the motion may be described qualitatively as follows:

1) If the system dynamics evolve such that the initial energy of the system is not increasing with time for all starting points in a finite neighborhood, we can conclude that the equilibrium state is stable.

2) If the system dynamics evolve such that the energy of the system is monotonically decreasing with time for all initial conditions in the neighborhood (and thus eventually approaches zero), the equilibrium state is asymptotically stable.

3) If the energy of the system is increasing with time, for any initial condition in the neighborhood, then the equilibrium state is unstable.

4) If the chosen energy measure is indefinite (i.e., it is neither strictly decreasing nor increasing), then no conclusion can be drawn on the stability of the system.

The following theorem, which is a rigorous statement of the preceding remarks, is the basic stability concept underlying Lyapunov's direct (second) method.

A.   Theorem: Stability Theorem

The equilibrium state x_e is stable if there exists a continuously differentiable function $U(x)$ such that

$$\text{i) } U(x_e) = 0$$
$$\text{ii) } U(x) > 0 \quad \text{for all } x \neq x_e, x \in \Omega$$
$$\text{iii) } \dot{U}(x) \leq 0 \quad \text{for all } x \neq x_e, x \in \Omega$$

where $\dot{U}(x)$ denotes the time derivative of the function $U(x)$, and Ω is some region containing x_e. Notice that the energy rate $\dot{U}(x)$ is evaluated along a typical trajectory $x(t)$, and the conditions ii) and iii) must hold along all infinity of trajectories of the dynamical system, which ensue from initial states in Ω.

A modest perturbation of the above theorem (making the final inequality strict) results in the following theorem which provides necessary and sufficient conditions for asymptotic stability.

B.   Theorem: Asymptotic Stability Theorem

The equilibrium state x_e is asymptotically stable if there exists a continuously differentiable function U such that

$$\text{i) } U(x_e) = 0$$
$$\text{ii) } U(x) > 0 \quad \text{for all } x \notin x_e, x \in \Omega$$
$$\text{iii) } \dot{U}(x) < 0 \quad \text{for all } x \notin x_e, x \in \Omega$$

Both of the previous theorems relate to local stability in the vicinity of the equilibrium state. A system has global asymptotic stability with respect to a unique equilibrium point if the following theorem is satisfied.

C.   Theorem: Global Asymptotic Stability Theorem

The equilibrium state x_e is globally asymptotically stable if there exists a continuously differentiable function U with the following properties:

i) $U(x_e) = 0$

ii) $U(x) > 0$ for all $x \neq x_e$

iii) $\dot{U}(x) < 0$ for all $x \neq x_e$

iv) $U(x) \to 0$ as $\|x\| \to \infty$

Note that the stable region Ω extends to infinity in the theorem in Sec. III.C. The reader is referred to Ref. 1 for further discussion, including the complete proofs of the preceding theorems. Observe that there is no one unique Lyapunov function for a given system; some may be better than others. This is especially important when we seek the least conservative stability information when, for example, we seek to determine the size of the Ω region in which we have stability. If a poor choice of $U(x)$ results in a pessimistic conclusion that the Ω is much smaller than it actually is, then this is an obvious (and very frequently occurring) concern. It should also be noted that if a Lyapunov function cannot be found, nothing can be concluded about the stability of the system, since the Lyapunov stability theorem provides only sufficient conditions for stability. Therefore the conditions required to prove stability, based on an arbitrary choice of Lyapunov function, may be very conservative.

Unfortunately, the Lyapunov theorems are not constructive; these theorems do not reveal a process to find a candidate Lyapunov function. It is often difficult to find a suitable Lyapunov function for a given nonlinear system. The physical and mathematical insights of the analyst has historically played an important role in most successful applications of this approach, however more systematic methods have recently emerged[3-5] for certain classes of control design problems. In particular, when the stability analysis and the control design analysis are merged, one is often able to exploit the additional freedom to simultaneously design control laws and select a Lyapunov function which guarantees stability of the closed-loop (controlled) system.

D.   Example

Consider the system described by the nonlinear ordinary differential equation

$$\ddot{x}(t) - \epsilon x^2(t)\dot{x}(t) + x(t) = 0$$

The objective is to use Lyapunov analysis to investigate the stability of motion near the origin for this system.

Introducing the state variable representation of this system with the definitions $x_1 = x$, $x_2 = \dot{x}$, we write the equivalent first-order system

$$\dot{x}_1 = x_2, \quad \dot{x}_2 = -x_1 + \epsilon x_1^2 \dot{x}_2$$

It is easy to verify that the preceding oscillator with quadratic damping has an equilibrium state at the origin $(x_1, x_2) = (0, 0)$. Our goal is to determine if this state is stable. For this purpose, let us choose the simplest candidate Lyapunov function

$$U(x_1, x_2) = (x_1^2 + x_2^2)/2$$

We note that a physical motivation for choosing this positive definite function as a candidate Lyapunov function that it is an exact (total mechanical energy) integral of the motion, for $\epsilon = 0$. Clearly $U(0, 0) = 0$ and $U(x_1, x_2) > 0$ in any neighborhood of $(0, 0)$, and investigating the energy rate, we find

$$\dot{U}(x_1, x_2) = x_1\dot{x}_1 + x_2\dot{x}_2 = x_1x_2 + x_2(-x_1 + \epsilon x_1^2 x_2) = \epsilon x_1^2 x_2^2$$

Thus U is a positive definite function which is strictly decreasing along all system trajectories if $\epsilon < 0$. Therefore, by the preceding theorems, $(0, 0)$ is a globally stable equilibrium point for $\epsilon = 0$, is globally asymptotically stable for $\epsilon < 0$, and is globally unstable for $\epsilon > 0$. Thus Lyapunov analysis was completely successful in this case of establishing the global stability characteristics of this system.

E. Example

Investigate the stability of the system of nonlinear differential equations

$$\dot{x}_1 = x_1(x_1^2 + x_2^2 - 1) - x_2, \quad \dot{x}_2 = x_1 + x_2(x_1^2 + x_2^2 - 1)$$

We try the candidate Lyapunov function

$$U(x_1, x_2) = x_1^2 + x_2^2$$

which is an exact integral of the simplified system $\dot{x}_1 = -x_2, \dot{x}_2 = x_1$. This choice for U is obviously a positive definite function having its global minimun at the origin. It is also obvious after inspection that the origin is the only equilibrium point of the nonlinear system. Investigating the energy rate, we find

$$\dot{U}(x_1, x_2) = 2(x_1^2 + x_2^2)(x_1^2 + x_2^2 - 1)$$

It is evident that $\dot{U}$ is negative definite over the finite circular region $\{(x_1, x_2) \mid x_1^2 + x_2^2 < 1\}$ which includes the equilibrium point at the origin. Hence, the origin $(0, 0)$ is an asymptotically stable equilibrium state of this system. Note that all points within the unit circle are asymptotically attracted to the origin. However, because $\dot{U}$ is not a negative definite function over all of R^2, we cannot conclude global asymptotic stability without more information. Although we are certain we have stability within the unit circle, this conclusion results from a particular choice of $U(x_1, x_2)$, and without further analysis we cannot conclude that the stable region is not actually larger than the unit circle. However, since $\dot{U}$ is positive everywhere outside the unit circle, we can finally conclude, using the following theorem (in Sec. III.F) that we have instability for all trajectories which initiate outside the unit circle and asymptotic stability for all trajectories initiating inside the unit circle. Thus we are able to use both the stability and instability insights simultaneously to establish the complete story vis-a-vis the global stability properties of this system, since the stable and unstable regions have a mutual boundary, and together the stable and unstable regions span all of state space R^2.

The following theorem is sometimes useful in avoiding a fruitless search for Lyapunov functions for systems which can be proven unstable. This theorem is also useful in obtaining theoretical closure of the stability analysis, in the sense that it is sometimes possible to simultaneously apply the instability theorem with the stability theorems to conclusively establish a particular system's global stability properties. For example, in Sec. III.E, we concluded that our simple choice on U gave us all of the stability information (i.e., the system is stable only within the unit circle).

F.   Theorem: Instability Theorem

The equilibrium state x_e is unstable in Ω if there exists a continuously differentiable function U such that

i) $U(x_e) = 0$ and $\dot{U}(x_e) = 0$

ii) $\dot{U}(x) > 0$ for all $x \neq x_e, x \in \Omega$

iii) and there exists points x arbitrarily close to x_e such that $U(x_e) > 0$

Some versions of the instability theorem require a function U such that both U and $\dot{U}$ are positive definite, and this unnecessarily more restrictive version is somewhat less efficient[1] than the preceding theorem, which makes no requirement on the definiteness of U. Actually, if one can find any such function U satisfying the preceding conditions, then x_e is a completely unstable equilibrium point in Ω, and the quest for Lyapunov functions can be halted. In the example shown in Sec. III.E, the Ω for the instability theorem is clearly the compliment of the Ω for the asymptotically stable region, and it is apparent that the stable and unstable regions being complimentary is the key to establishing global stability/instability information.

IV.   Stability of Linear Systems

A.   Lyapunov Theorem for Linear Systems

Lyapunov's method is easily applied to test the stability of a linear system. Consider an autonomous system described by the linear vector differential equation

$$\dot{x}(t) = Ax(t) \qquad (4)$$

The preceding system is said to be stable in the sense of Lyapunov, if the solution of Eq. (4) tends toward zero (which is obviously the only equilibrium state if A is of full rank) as $t \to \infty$ for arbitrary initial condition.

Consider the case of a constant A matrix. If all eigenvalues of A are distinct, the response of system (4), given initial condition x_0, can be written as

$$x(t) = \sum_{i=1}^{n} \underline{\psi}_i^T x_0 e^{\lambda_i t} \underline{\phi}_i \qquad (5)$$

where λ_i are the eigenvalues of A, and $\underline{\phi}_i$ and $\underline{\psi}_i$ are, respectively, the right and left eigenvectors of A associated with λ_i. For the repeated eigenvalue case, the situation is much more complicated (i.e., we should solve for the generalized eigenvectors of A). The generalization of Eq. (5) for the case of generalized eigenvectors has a similar form but will not be discussed here.[8]

From Eq. (5), we can see by inspection that the system will be asymptotically stable if and only if all the eigenvalues of A have negative real parts, i.e.,

$$\Re[\lambda_i(A)] < 0 \tag{6}$$

Thus, we have the well-known result that the stability of a linear constant-coefficient dynamical system can be completely characterized by the eigenvalues of the system. This approach to stability analysis yields both necessary and sufficient conditions. However, calculating all the eigenvalues of the system matrix is not always desirable. As will be evident later, other stability viewpoints lead to important insights and generalized methods, especially vis-a-vis stability analysis for time-varying, distributed-parameter, and nonlinear systems.

For the linear dynamical system of Eq. (4), we choose a symmetric quadratic form as a candidate Lyapunov function

$$2U(x) = x^T P x \tag{7}$$

where P is a positive definite, real symmetric matrix. Thus U is positive definite with its global minimum at the origin, which is obviously an equilibrium state. Differentiating Eq. (7) and substituting Eq. (4) into the result gives the energy rate

$$\dot{U}(x) = x^T(A^T P + P A)x \tag{8}$$

Using the Lyapunov stability theorem in Sec. III.B, we require $\dot{U}(x)$ to be negative definite. We can rewrite the energy rate of Eq. (8) as

$$\dot{U}(x) = -x^T Q x \tag{9}$$

So we see that, for asymptotic stability, P and Q must be positive definite matrices which satisfy the condition

$$A^T P + P A = -Q \tag{10}$$

Equation (10) is commonly known as the algebraic Lyapunov equation.

To examine the stability of a linear system via the preceding Lyapunov approach we can proceed as follows: Choose Q to be any positive definite matrix for a given A, and check the eigenvalues of the resulting P which we obtain by solving Eq. (10), if P is positive definite (all positive eigenvalues), the given system is asymptotically stable, whereas if P has any negative eigenvalues, the system is unstable. One of the potential difficulties with selecting Q and solving the Lyapunov equation (which, of course, depends on the system matrix A) is the uniqueness of the resulting solution for P. The following theorem gives the necessary and sufficient conditions for the Lyapunov Eq. (10) to have a unique solution.

B.   Theorem

If $\{\lambda_1, \ldots, \lambda_n\}$ are the eigenvalues of the system matrix A, then the Lyapunov equation [Eq. (10)] has a unique solution P if and only if

$$\lambda_i + \lambda_j^H \neq 0, \qquad i, j = 1, \ldots, n$$

where $(\ )^H$ denotes complex conjugate.

The reader is referred to Chen[8] for the proof. One cannot solve the Lyapunov equation for undamped second-order systems having pairs of eigenvalues on the imaginary axis (including rigid-body modes, whose eigenvalues reside at the origin of the complex plane), and so we see that stability analysis for systems having a neutrally stable subspace cannot be completed via solution of an algebraic Lyapunov equation.

C.   Theorem: Lyapunov Stability Theorem for Linear Systems

A linear system is asymptotically stable or, equivalently, all the eigenvalues of A have negative real parts, if and only if for any given positive definite symmetric matrix Q there exists a positive definite (symmetric) matrix P that satisfies the Lyapunov equation

$$A^T P + PA = -Q$$

D.   Proof of Theorem in Section IV.C

1.   Necessity

The uniqueness condition of the solution (theorem of Sec. IV.B) is automatically met in the theorem Sec. IV.C, since $\Re[\lambda_i(A)] < 0$ for all i, for asymptotically stable systems. The unique solution P of the Lyapunov equation can be expressed as[9]

$$P = \int_0^\infty e^{A^T t} Q e^{At} dt \tag{11}$$

The integrand is an infinite summation of terms of the form $t^k e^{\lambda t}$, where λ is an eigenvalue of A, and since $\Re[\lambda_i(A)] < 0$, the integrand will exist for all values of t. Next, by direct substitution of Eq. (11) into the left hand side of Eq. (10), we can verify that the combined integrand is a perfect differential, so we see

$$A^T P + PA = \int_0^\infty \frac{d}{dt} \left(e^{A^T t} Q e^{At} \right) dt$$
$$= e^{A^T t} Q e^{At} \Big|_0^\infty = -Q$$

Note that e^{At} (and $e^{A^T t}$) $\to 0$ as $t \to \infty$, since all eigenvalues of A have negative real parts, and since e^{At} (and $e^{A^T t}$) equal 1 at $t = 0$.

2.   Sufficiency

Consider

$$U(\mathbf{x}) = \mathbf{x}^T P \mathbf{x}$$

From Eqs. (7–10), we see that $\dot{U}(\mathbf{x}) = -\mathbf{x}^T Q \mathbf{x}$ along all trajectories of $\dot{\mathbf{x}} = A\mathbf{x}$. From the assumptions, P and Q are positive definite matrices, therefore we see that for any $\mathbf{x} \neq 0$ that $U(\mathbf{x}) > 0$, and $\dot{U}(\mathbf{x}) < 0$. We can apply the theorem in Sec. III.C and conclude that the equilibrium state [the state space origin $(\mathbf{x} = 0)$] is globally asymptotically stable, and of course, all eigenvalues of A must have negative real parts.   Q.E.D.

Note that the Lyapunov equation is equivalent to a set of $n(n + 1)/2$ linear equations in $n(n + 1)/2$ unknowns for an nth-order system. The Lyapunov equation can be solved by using numerical algorithms utilizing QR factorization, Schur decomposition, or spectral decomposition, however our experience indicates that the most efficient and robust algorithms utilize the QR factorization.[32]

E.   Example

Consider the system matrix

$$A = \begin{bmatrix} -2 & 1 \\ -1 & 1 \end{bmatrix}$$

The simplest choice of Q is the identity matrix or some other diagonal matrix; we take $Q = I$ for this example, and let the three distinct elements in P be denoted

$$P = \begin{bmatrix} p_1 & p_2 \\ p_2 & p_3 \end{bmatrix}$$

Substituting this A and P into the Lyapunov equation [Eq. (10)] yields the following three linear algebraic equations

$$-4p_1 - 2p_2 = -1$$
$$p_1 - p_2 - p_3 = 0$$
$$2p_2 + 2p_3 = -1$$

The solution of these three equations is straightforward, we find

$$p_1 = -\frac{1}{2}, \quad p_2 = \frac{3}{2}, \quad p_3 = -2$$
$$\Rightarrow P = \begin{bmatrix} -1/2 & 3/2 \\ 3/2 & -2 \end{bmatrix}$$

Even though we have a unique solution, the resulting matrix P is not positive definite. Hence, we conclude that the system is unstable, and implicitly, that not all of the eigenvalues of A have negative real parts. We would have to calculate the eigenvalues to make further assessments of eigenvalue placement.

In the case of a linear time varying system $\dot{x}(t) = A(t)x(t)$, the sufficient conditions for the stability of the equilibrium state can be discussed based on the concept of matrix measure,[1] and if the system is asymptotically stable, then a quadratic Lyapunov function exists for this system. Of course, eigenvalue analysis is not applicable to the time-varying case and therefore the more general Lyapunov approach provides one possible avenue to characterize the stability of nonautonomous systems. The reader is referred to Ref. 1 for more information on this generalization of the preceding discussion.

F.   Linear Dynamic Systems Subject to Arbitrary Disturbances

To make the Lyapunov stability analysis in this section more complete, we will briefly discuss stability in the presence of disturbances. We consider the class of systems described by the matrix differential equation

$$\dot{x}(t) = Ax(t) + f[t, x(t)] \tag{12}$$

where the uncertainty and/or perturbations of the system are assumed representable by an arbitrary nonlinear function $f[t, x(t)]$ [except we require $f(t, 0) = 0$, so that the origin of the state space remains an equilibrium state for this class of model errors or disturbances]. Furthermore, we assume that exact expressions for $f[t, x(t)]$ are unknown and only bounds on $f[t, x(t)]$ are known. A central question is the following: Given that A is asymptotically stable, and without using specific knowledge of $f[t, x(t)]$, is it possible to obtain a bound on all $f[t, x(t)]$ such that the system maintains its stability? Put another way, can we determine some measure of how large $f[t, x(t)]$ can be without destabilizing a given stable linear system? Some insight on these issues are embodied in the following theorem.

G.   Theorem[10, 37]

Suppose that the system of Eq. (12) is asymptotically stable for $f[t, x(t)] = 0$, then the system remains asymptotically stable for all nonzero perturbations $f[t, x(t)]$ which are sufficiently small that they satisfy the following inequality:

$$\frac{\|f\|}{\|x\|} \leq \frac{\min \ \lambda(Q)}{\max \ \lambda(P)} \equiv \mu_{PT} \tag{13}$$

where P and Q satisfy the following Lyapunov equation

$$A^T P + PA = -2Q$$

and where the otherwise arbitrary $f[t, x(t)]$ vanishes at the origin $f(t, 0) = 0$.

The Proof of this theorem is given in Refs. 10 and 32. Since P is a positive definite matrix, the maximum eigenvalue of P is same as the largest singular value of P. It has been also shown in Refs. 10 and 32 that when the identity matrix is chosen for Q, μ_{PT} in Eq. (13) is a maximum and for this choice, μ_{PT} can be expressed as

$$\mu_{PT} = \frac{1}{\max \lambda(P)} = \frac{1}{\sigma_{max}(P)} \tag{14}$$

The preceding bound is often very conservative, since it is only a sufficient condition for the stability of the system, and this stringent bound is not usually necessary.

An important special case is for the class of perturbations having the linear structure

$$f[t, x(t)] = Ex(t) \tag{15}$$

Clearly this corresponds to an additive error in the A matrix (i.e. $A \rightarrow A + E$). We can apply the theorem Sec. IV.G to arrive at the desired result. That is, the system remains stable if E is bounded by the following modified stability margin:

$$\|E\| \leq \frac{\min\left[-\Re\{\lambda_i(A)\}\right]}{\mathcal{K}(\Phi)} \tag{16}$$

where $\mathcal{K}(\Phi)$ is the condition number of Φ, and Φ is the normalized eigenvector (modal) matrix of A. The condition number can be conveniently computed as the ratio of the largest and least singular values of Φ,

$$\mathcal{K}(\Phi) = \frac{\sigma_{max}(\Phi)}{\sigma_{min}(\Phi)}$$

As is evident in the preceding discussion, the stability margin is closely related to the Patel-Toda[10] robustness margin; the more stable the nominal system is, the larger the bound on the allowable perturbation E becomes. However, the important ingredient evident in Eq. (16) is the fact that a large condition number $\mathcal{K}(\Phi)$ degrades the effective stability margin. The intimate connection of the Patel-Toda robustness measure (for stability of linear dynamical systems in the presence of additive perturbations) to the Bauer-Fike Theorem (original result in Ref. 37; for conditioning of the algebraic eigenvalue problem see Ref. 32) is clear.

Note that the condition number $\mathcal{K}(\Phi)$ approaches its smallest possible value of unity if Φ is any unitary matrix (one for which $\Phi^H \Phi = I$), and the upper bound on the condition number is infinity which occurs if Φ is any singular matrix. Observe that an infinity of unitary matrices exist, some of them are closer to Φ than others. When one has the freedom to modify A (and therefore Φ), a natural question arises: For a given class of A modifications, how can we make Φ as nearly unitary as possible? Of course, one way to modify the A matrix is through design of a feedback controller, and one avenue toward designing gains in linear robust control laws is to maximize the right-hand side of Eq. (16) by minimizing $\mathcal{K}(\Phi)$. It is also of significance that the choice of actuator locations considered simultaneously with the design of control gains can often significantly reduce the condition number $\mathcal{K}(\Phi)$. These ideas provide some of the motivation for the robust eigenstructure algorithms and actuator placement optimization approaches presented in Ref. 32.

V. Nonlinear and Time Varying Dynamical Systems

In this section, we present stability analysis methods for nonlinear systems. In Sec. V.B, we consider a method known as Lyapunov's indirect (or first) method, whereby we can determine partial stability information for nonlinear systems by examining the behavior of locally linearized systems. In Sec. V.C, we develop an important result which provides easy-to-test sufficient conditions to determine if we have asymptotic stability in spite of the common situation that the energy function's time derivative is only a negative semidefinite function of the state variables. In addition to the classical stability analysis for which the Lyapunov methods were developed, these ideas can be used to motivate design methods which yield control laws for the control of large maneuvers for distributed parameter systems.

This approach is used throughout the remainder of this chapter. In Sec. VI, we consider a nonlinear multibody idealization of two robots cooperatively manipulating a payload. Both open-loop and feedback control designs are studied, and Lyapunov methods are used to ensure stability of the resulting tracking control law.

A. Local Stability of Linearized Systems

The stability analysis of linear motion arises often in practical analysis of nonlinear systems when one is concerned with motion near an equilibrium state. The results presented in Secs. IV.A–IV.E enable us to obtain necessary and sufficient conditions for the stability of linear systems but also provide us with a method for determining the local stability of a nonlinear system by linearization, which is called Lyapunov's indirect method.

Consider the autonomous system

$$\dot{x}(t) = f[x(t)] \quad \text{with} \quad f(x_e) = 0 \tag{17}$$

Let $z(t)$ be the perturbation (departure motion) from the equilibrium state as

$$x(t) = x_e + z(t) \tag{18}$$

Using Taylor's series expansion of $f(\cdot)$ around the equilibrium state x_e, we can write

$$f[z(t) + x_e] = f(x_e) + \left[\frac{\partial f}{\partial x}\right]_{x=x_e} z(t) + O[z(t)]^2 \tag{19}$$

Using Eq. (19) in Eq. (17) gives the perturbation equation

$$\dot{z}(t) = Az(t) + O[z(t)]^2 \tag{20}$$

where A denotes the Jacobian matrix of f evaluated at $x = x_e$, $A = [\partial f/\partial x]_{x=x_e}$ and so we find the linear, constant coefficient matrix differential equation

$$\dot{z}(t) = Az(t) \tag{21}$$

The following theorem is given here (without proof); this is the main stability result of Lyapunov's indirect method.

B.   Theorem: Lyapunov's Indirect Method

If the linearized system [Eq. (21)] is asymptotically stable, then the original nonlinear system [Eq. (17)] is also asymptotically stable if the motion initiates in a sufficiently small neighborhood containing the equilibrium state.

The preceding theorem is useful since we can analyze the local stability of an equilibrium state of a given nonlinear system by examining a linear system. However, the conclusions based on linearizations are local, and therefore to study global stability we should rely on Lyapunov's direct method. On the other hand, if one can find all equilibrium points and investigate their local stability, a fairly complete picture of the overall global stability characteristics can often be derived. Note that one key shortcoming (of the indirect approach) is the absence of information on the size or boundary of the domain of attraction of each locally stable equilibrium point; this is precisely the information which a completely successful application of the direct approach determines. Finally, we note the most important point: If the linear motion is critical (e.g., zero damping, some eigenvalues have zero real parts), then the stability of the locally linearized analysis should be considered inconclusive and nonlinear effects must be included to conclude local stability or instability.

C.   What to Do When $\dot{U}$ is Negative Semidefinite

Several subtle possibilities arise if the function derived for $\dot{U}$ is not negative definite. For a significant fraction of the practical occurences of this condition, including several applications considered subsequently in this chapter, we can prove global asymptotic stability in spite of the fact that the function derived for $\dot{U}$ is negative semidefinite. The main results from the traditional literature for dealing with this problem are embodied in a theorem due to LaSalle and Lefschetz[29]; this theorem sometimes allows us to conclude that we have local asymptotic stability for the case that $U > 0$ and $\dot{U} \leq 0$, provided we can prove that the equilibrium point is contained in a region of state space known as the *maximum invariant subspace* M. It is usually easy to identify the subset Z of points in the state space for which $\dot{U} = 0$, but LaSalle and Lefschetz's maximum invariant subspace M is in general a subset of Z. The main challenge of applying LaSalle and Lefschetz's theorem then reduces to the quest to identify or approximate M; this is difficult when the differential equations are complicated nonlinear functions. While these ideas are elegant, we elect not to discuss this approach in greater detail. Instead we present a recently developed result[30, 31, 36] which is easier to apply.

D.   Theorem

Prior to stating the theorem, we need some notations: Let $x = 0$ be an equilibrium state of the nonlinear system $\dot{x} = f(t, x)$, where f is a smooth, twice differentiable n vector function of t and x. Note that the trajectories of the nonlinear differential equation $\dot{x} = f(t, x)$ generates a smooth vector field in the region Ω which includes $x = 0$. Let $U(t, x)$ be a scalar analytic function in Ω, which is locally positive definite. Suppose $\dot{U}(t, x)$ is only negative semidefinite. Let Z denote the set of points for which $\dot{U}(t, x)$ vanishes. We will be concerned with the first k derivatives $d^k U / dt^k$, evaluated on the set Z. We are now prepared to state the theorem:

A sufficient condition for asymptotic stability, when $U > 0$ and $\dot{U} \leq 0$ for all $x \in \Omega$ is that the first $(k - 1)$ derivatives of U vanish on Z, up through some even order $(k - 1)$ as

$$\frac{d^j U}{dt^j} = 0, \quad \forall x \in Z, \quad \text{for} \quad j = 1, 2, \ldots, \quad k - 1 \tag{22}$$

and the first (the kth) nonzero derivative of U (evaluated on Z) is of odd order and is negative definite for all points on Z:

$$\frac{d^k U}{dt^k} < 0, \quad \forall x \in Z, \quad \text{for } k \text{ odd} \tag{23}$$

In the event that all infinity of U derivatives vanish on Z, sufficient conditions for stability are that U is positive definite and that $x = 0$ is the only equilibrium point.

The proof of this theorem is given in Ref. 30 and further discussed in Ref. 31. This theorem is fairly easy to apply and useful results can often be obtained efficiently, as will be evident in the several nonlinear and distributed parameter applications considered in the remainder of this chapter. As is evident in the following example, this theorem is also useful for determining the stability of time varying systems.

E. Example[30]

Consider the damped Mathieu equation: $\dot{x}_1 = x_2, -\dot{x}_2 = -x_2 - (2 + \sin t)x_1$. We select the candidate Lyapunov function: $U(t, x_1, x_2) = x_1^2 + x_2^2/(2 + \sin t)$, which we observe is positive definite and analytic for all (t, x_1, x_2). On differentiation of U, and substitution of the equations of motion, we find that

$$\dot{U}(x) = -x_2^2 g(t), \quad \text{where} \quad g(t) = \frac{4 + 2\sin t + \cos t}{(2 + \sin t)^2}$$

Since $\dot{U}(x)$ does not depend on x_1, it is obviously not positive definite and without further analysis, we can only conclude mere stability, however, we'd like to make a stronger statement and conclude asymptotic stability. This can be done by considering the applicability of the Sec. V.D theorem. Note that the set Z of points for which $\dot{U}(x)$ vanishes is the set of all real values for x_1, and zero values for x_2. On taking the second and third derivatives of U, and evaluating them on Z, we find that

$$\frac{d^2 U}{dt^2} = 0, \quad \text{and} \quad \frac{d^3 U}{dt^3} = -2(2 + \sin t)^2 g(t)x_1^2, \quad \forall x \in Z$$

Since the second derivative vanishes on Z and the third derivative is negative on Z, except at the origin, we conclude that all of the conditions of the Sec. V.D theorem are satisfied; indeed this system is globally asymptotically stable.

F. Lyapunov Control Law Design Method

Here, we present a method for generating globally stable feedback control laws for maneuvers of nonlinear systems and distributed parameter systems. A Lyapunov function is selected which is conserved for the uncontrolled system. Then when the control $u(t) \neq 0$ is considered, $\dot{U}(x)$ depends on $u(t)$ through the equations of motion. One strategy is to select the control function $u(t, x)$ (from a set of admissible controls) to make $\dot{U}(x)$ as negative as possible; this Lyapunov optimal control strategy ensures that $U(x)$ will locally approach zero as fast as possible. On the other hand, any control law which makes $\dot{U}(x)$ negative is asymptotically stabilizing, and in many instances it will be seen that very simple, yet globally stable control laws can be determined which are attractive for applications.

We will use specific dynamical systems to introduce Lyapunov control design methods for nonlinear and distributed-parameter systems using Lyapunov's direct method in this section. The broadest and most productive viewpoint is to simultaneously consider $U(x)$ and $u(t, x)$ to be determined in the design process; as will be seen, the class of problems for which globally stable feedback laws can be obtained is surprisingly large. As will be evident, we place the initial emphasis on using work/energy methods together with stability theory to determine the structure of a stabilizing feedback law and thereby parameterize an infinite family of stable controllers. Conventional nonlinear programming algorithms can then be invoked to optimize a specified closed-loop performance criterion over the linearly stable set. Although we subsequently develop methods for controlling multibody manipulators, and for distributed parameter systems governed by hybrid coupled sets of ordinary and partial differential equations, we first consider a system described by a sixth-order set of nonlinear, ordinary differential equations.

G. Example: Large Angle Rigid-Body Maneuvers

Some key ideas are easily introduced by considering general three-dimensional nonlinear maneuvers of a single rigid body. The equations governing large motion can be written as[17]

$$
\begin{aligned}
I_1 \dot{\omega}_1 &= (I_2 - I_3)\omega_2\omega_3 + u_1 \\
I_2 \dot{\omega}_2 &= (I_3 - I_1)\omega_3\omega_1 + u_2 \\
I_3 \dot{\omega}_3 &= (I_1 - I_2)\omega_1\omega_2 + u_3 \\
2\dot{q}_1 &= \omega_1 - \omega_2 q_3 + \omega_3 q_2 + q_1(q_1\omega_1 + q_2\omega_2 + q_3\omega_3) \\
2\dot{q}_2 &= \omega_2 - \omega_3 q_1 + \omega_1 q_3 + q_2(q_1\omega_1 + q_2\omega_2 + q_3\omega_3) \\
2\dot{q}_3 &= \omega_3 - \omega_1 q_2 + \omega_2 q_1 + q_3(q_1\omega_1 + q_2\omega_2 + q_3\omega_3)
\end{aligned}
\tag{24}
$$

where $(\omega_1, \omega_2, \omega_3)$ and (q_1, q_2, q_3) are the principal-axis components of angular velocity and the Euler-Rodriguez parameters ("Gibbs vector"), respectively. Note that (I_1, I_2, I_3) and (u_1, u_2, u_3) are the principal moments of inertia and the principal-axis components of the external control torque, respectively.

For the case of zero control torque, it can be readily verified that total rotational kinetic energy is an exact integral of the motion described by differential Eq. (24), viz., $2T = (I_1\omega_1^2 + I_2\omega_2^2 + I_3\omega_3^2)$. Motivated by total system energy integral, we investigate the trial Lyapunov function

$$
\begin{aligned}
U &= \frac{1}{2}\left(I_1\omega_1^2 + I_2\omega_2^2 + I_3\omega_3^2\right) + k_0\left(q_1^2 + q_2^2 + q_3^2\right) \\
&\equiv \text{kinetic energy} + k_0 \tan^2\left(\frac{\phi}{2}\right)
\end{aligned}
\tag{25}
$$

where ϕ is the instantaneous principal rotation angle (about the instantaneous Eulerian principal rotation axis, from the current angular position to the desired final angular position of the body[17]). It is apparent that the additive term $k_0(q_1^2 + q_2^2 + q_3^2)$ can be viewed as the potential energy stored in a conservative spring. We can therefore anticipate that the system dynamics will evolve such that U is constant if the only external torque is the associated conservative moment. Of course, we are not interested in preserving U as a constant, but rather we seek to drive it to zero, because it measures the departure of the system from the desired equilibrium state at the origin. We therefore anticipate the necessity of determining an additional judicious control moment to guarantee that U is a decreasing function of time. It is obvious by inspection that U is positive definite and vanishes only at the desired state $q_i = \omega_i = 0$. Differentiation of Eq. (25) and substitution of Eq. (24) lead directly to the following (power) expression for $\dot{U}$:

$$
\dot{U} = \sum_{i=1}^{3} \omega_i[u_i + k_0 q_i(1 + q_1^2 + q_2^2 + q_3^2)]
\tag{26}
$$

Of all of the infinity of possible control laws, we can see that any control u_i that reduces the bracketed terms to a function whose sign is opposite to ω_i will guarantee that $\dot{U}$ is globally negative semi definite. The simplest choice consists of the following: Select u_i so that the ith bracketed term becomes $-k_i\omega_i$. This gives the control law

$$u_i = -[k_i\omega_i + k_0q_i(1 + q_1^2 + q_2^2 + q_3^2)], \qquad i = 1, 2, 3 \qquad (27)$$

The closed-loop equations of motion are obtained by substitution of the control law of Eq. (27) into Eq. (24) to establish

$$\begin{aligned}
I_1\dot{\omega}_1 &= (I_2 - I_3)\omega_2\omega_3 - [k_1\omega_1 + k_0q_1(1 + q_1^2 + q_2^2 + q_3^2)] \\
I_2\dot{\omega}_2 &= (I_3 - I_1)\omega_3\omega_1 - [k_2\omega_2 + k_0q_2(1 + q_1^2 + q_2^2 + q_3^2)] \qquad (28) \\
I_3\dot{\omega}_3 &= (I_1 - I_2)\omega_1\omega_2 - [k_3\omega_3 + k_0q_3(1 + q_1^2 + q_2^2 + q_3^2)]
\end{aligned}$$

Since $\dot{U} = -(k_1\omega_1^2 + k_2\omega_2^2 + k_3\omega_3^2)$ does not depend on the qs, it is only a negative semidefinite function, and while we have stability; if we choose all $k_i > 0$, we cannot immediately conclude that we have asymptotic stability. We can prove that we do indeed have asymptotic stability; for illumination we establish this truth by two logical paths.

1. Path 1

This analysis is physically motivated. We try to see if there is some equilibrium point or trajectory other than the target state where the system can get stuck with $\dot{U}(x) = 0$. We directly investigate the preceding three closed-loop equations of motion [Eq. (28)] for the existence of equilibrium points in these nonlinear closed-loop equations of motion. It can be verified that $(q_1, q_2, q_3, \omega_1, \omega_2, \omega_3) = (0, 0, 0, 0, 0, 0)$ is the only equilibrium state where all velocity and acceleration coordinates vanish. In fact, imposing the conditions $(\dot{\omega}_1, \dot{\omega}_2, \dot{\omega}_3) = (0, 0, 0)$ and $(\omega_1, \omega_2, \omega_3) = (0, 0, 0)$ on the preceding closed-loop equations of motion immediately gives the requirement that the qs satisfy the three equations

$$0 = -[k_0q_i(1 + q_1^2 + q_2^2 + q_3^2)], \qquad \text{for} \quad i = 1, 2, 3$$

and it is obvious by inspection that these three nonlinear equations are simultaneously satisfied only at the origin.

Since we have shown that with the body at rest, $(\dot{\omega}_1 \neq 0, \dot{\omega}_2 \neq 0, \dot{\omega}_3 \neq 0)$, for $(q_1 \neq 0, q_2 \neq 0, q_3 \neq 0)$, everywhere except the origin $x = (q_1, q_2, q_3, \omega_1, \omega_2, \omega_3)^T = (0, 0, 0, 0, 0, 0)^T$, we conclude that $\dot{U}(x) = 0$ can only be encountered for $(q_1 \neq 0, q_2 \neq 0, q_3 \neq 0)$ at (possibly) apogee-like points in the behavior of U [$\dot{U}$ instantaneously vanishes but these points cannot be equilibrium states because $(\dot{\omega}_1 \neq 0, \dot{\omega}_2 \neq 0, \dot{\omega}_3 \neq 0)$]. Therefore we are guaranteed that $\dot{U}(x) < 0$ almost everywhere (thus, we have the ideal situation that the largest invariant subspace is all of state space). We asymptotically approach the origin from all finite initial states and therefore have global asymptotic stability.

2. Path 2

This analysis is more formal (exactly analogous to Sec. V.E); we simply apply the Sec. V.D theorem. First notice that the set Z where $\dot{U}(x)$ vanishes is the set of arbitrary real values for the qs and zero values for the ωs. It can be verified by direct differentiation of U that for general motion

$$\frac{d^2 U}{dt^2} = -2\sum_{i=1}^{3} k_i \omega_i \dot{\omega}_i, \quad \text{and} \quad \frac{d^3 U}{dt^3} = -2\sum_{i=1}^{3} k_i(\dot{\omega}_i^2 + \omega_i \ddot{\omega}_i) \qquad (29)$$

On evaluation of these derivatives for motion on Z where angular velocity vanishes $(\omega_1, \omega_2, \omega_3) = (0, 0, 0)$; but from the closed-loop equations of motion, the nonzero acceleration components are $\dot{\omega}_i = -k_0(1 + q_1^2 + q_2^2 + q_3^2)(q_i/I_i)$, we find that

$$\frac{d^2 U}{dt^2} = 0, \quad \text{and}$$

$$\frac{d^3 U}{dt^3} = -2k_0^2(1 + q_1^2 + q_2^2 + q_3^2)^2 \sum_{i=1}^{3} k_i\left(\frac{q_i}{I_i}\right)^2, \quad \forall x \in Z \qquad (30)$$

Since the second derivative of U vanishes everywhere on Z, the third derivative is negative definite everywhere on Z, the conditions of the theorem Sec. V.D are fully satisfied, and we again conclude that the nonlinear control law of Eq. (27) gives us globally asymptotically stable attitude control.

Since we have shown U to be a positive definite, decreasing function of time along all trajectories, and since it vanishes at the origin, then the necessary and sufficient conditions are satisfied for global Lyapunov stability. We have implicitly excluded the geometric singularity $(q_i \to \infty)$ associated with this parameterization of rotational motion as $\phi \to n\pi$; we can use the quaternion or Euler parameter description of motion and avoid all geometric singularities as well. This path has been successfully pursued in Refs. 3 and 20.

The nonlinear feedback control law of Eq. (27) guarantees stability of the nonlinear closed-loop system under the assumption of zero model errors. In practice, of course, guaranteed stability in the presence of zero model error is not a sufficient condition to guarantee stability of the actual plant having arbitrary model errors and disturbances. On the other hand, rigorously defining a region in gain space, guaranteeing global stability for our best model of the nonlinear system is an important step; it is reasonable to restrict the optimization of gains to this stable family of designs. The determination of the particular gain values, selected from the space of globally stabilizing gains, is usually based on performance optimization criteria specified in consideration of the disturbance environment, sensitivity to model errors, desired system time constants, actuator saturation, and sensor/actuator bandwidth limitations.

Before generalizing the methodology to consider multibody and partial differential equation systems, it is important to reflect on the selection of the Lyapunov function previously given. Notice that, if a system has no inherent stiffness with respect to rigid-body displacement, it is necessary to augment the open-loop energy integral by a pseudopotential energy term [such as $k_0(q_1^2 + q_2^2 + q_3^2)$ in the preceding example]; generally speaking, the pseudoenergy term should be defined, if possible, such that the resulting candidate Lyapunov function (U) is a positive definite measure of departure motion that has its global minimum at the desired target state. Then the still-to-be-determined controls are usually selected as simply as possible (from an implementation point of view) to force pervasive dissipation

($\dot{U} < 0$) of the modified energy (Lyapunov) function along all trajectories of the closed-loop system, and thereby guarantee closed-loop stability.

Although the preceding insights are useful, the Lyapunov function is generally not unique. However, these ideas lead to an attractive strategy that defines the Lyapunov function with relative weights on the portions of total mechanical energy associated with structural subsystems,[5] and leads to a systematic work/energy method to bypass much of the algebra and calculus leading to the power equations, analogous to Eq. (26), for each particular physical system.[3] The lack of uniqueness of the Lyapunov function is not necessarily a disadvantage in practice because it is a source of user flexibility providing control design freedom that is qualitatively comparable to the freedom one has in selecting performance indices when applying optimal control theory. Indeed, formulating the Lyapunov function as a weighted error energy to be dissipated by the controller is qualitatively attractive for both linear and nonlinear systems, since this gives intuitive and physical meaning to the Lyapunov function and the control gains. Other stability analysis approaches, e.g., Refs. 7 and 11–18, as illustrated in the application studies (Refs. 19 and 21–28), remain useful alternatives and supplement the present discussion.

VI.   Cooperative Control of Multibody Manipulators

A.   Mechanics

Consider the class of dynamical systems whose behavior is governed by the discrete coordinate version of Lagrange's equations

$$\frac{d}{dt}\left(\frac{\partial \mathcal{L}}{\partial \dot{q}_i}\right) - \frac{\partial \mathcal{L}}{\partial q_i} = Q_i, \qquad i = 1, 2, \ldots, N \tag{31}$$

or, in matrix form

$$\frac{d}{dt}\left(\frac{\partial \mathcal{L}}{\partial \dot{q}}\right) - \frac{\partial \mathcal{L}}{\partial q} = Q \tag{32}$$

where the Lagrangian $\mathcal{L}$ is defined in the classical form $\mathcal{L} = T - V$. Restrictions imposed in deriving Eq. (32) are such that the coordinates q_i are independent functions of time only and that the potential and kinetic energies have the functional forms $T = T(q, \dot{q}, t)$, $V = V(q)$, and the nonconservative virtual work has the form $\delta W_{nc} = \Sigma_{i=1}^{N} Q_i \delta q_i = Q^T \delta q$. Thus, Eq. (32) are valid for nonlinear, nonconservative systems as well as linear, conservative systems.

A modest generalization allows Eq. (32) to be applied to a significant class of redundant coordinate or constrained systems (i.e., the coordinates q_i are not independent). To accommodate kinematic constraints which depend on the qs and their time derivatives, Lagrange multipliers can be introduced to generate additive generalized constraint forces on the right-hand side of Eq. (32).[17] In particular, for m Pfaffian (linear in the generalized velocities) constraints of the matrix form

$$A\dot{q} + a_o = 0 \tag{33}$$

The generalized constraint force that needs to be added to the right-hand side of Eq. (32) is the vector $A^T \lambda$, where q is an $N \times 1$ vector containing the generalized coordinates, $A = A(q)$ is an $m \times n$ continuous, differentiable matrix function,

$a_o(q)$ is a smooth, $m \times 1$ vector function, and λ is an $m \times 1$ vector of Lagrange multipliers. One standard solution process is to differentiate the kinematic constraint of Eq. (33) to obtain

$$A\ddot{q} + \dot{A}\dot{q} + \dot{a}_o = 0 \qquad (34)$$

Note that the N differential equations of Eqs. (32) must be solved simultaneously with the m kinematic equations [Eq. (34)] to determine the $N + m$ unknowns in the vectors $q(t)$ and $\lambda(t)$. During recent years, significant methodology has evolved for effecting numerical solutions for differential/algebraic systems of equations; see Refs. 33 and 34 for discussion of the recent literature.

For a specific class of systems, the algebra and calculus required in a straightforward application of Lagrange's equations can be dramatically reduced. For the most common case of natural systems in which kinetic energy is a symmetric quadratic form in the generalized-coordinate time derivatives, one finds:

$$T = \frac{1}{2} \sum_{i=1}^{N} \sum_{j=1}^{N} m_{ij}(q)\dot{q}_i\dot{q}_j = \frac{1}{2}\dot{q}^T M \dot{q} \qquad (35)$$

Note that q is an $N \times 1$ configuration vector of generalized coordinates. It is convenient to collect the mass matrix $M = M(q)$ before the differentiations implied by Lagrange's equations are carried out. Including the possibility of Pfaffian nonholonomic constraints, the equations of motion follow from Eq. (32) as the following $N + m$ system of differential and algebraic equations:

$$M\ddot{q} + G + \frac{\partial V}{\partial q} = Q + A^T\lambda, \qquad A\dot{q} + a_o = 0 \qquad (36)$$

where $\partial V/\partial q$ is the $N \times 1$ vector gradient of the potential energy function, and $G = G(q, \dot{q})$ is the $N \times 1$ vector:

$$G = [\dot{q}^T C^1 \dot{q} \cdots \dot{q}^T C^N \dot{q}]^T, \qquad c_{jk}^{(i)} = \frac{1}{2}\left(\frac{\partial m_{ij}}{\partial q_k} + \frac{\partial m_{ik}}{\partial q_j} - \frac{\partial m_{jk}}{\partial q_i}\right) \qquad (37)$$

and where the last equation that generates the typical element $C_{jk}^{(i)}$ of the $N \times N$ symmetric matrix $C^{(i)} = C^{(i)}(q)$ is the Christoffel operator. It is apparent that deriving the equations of motion, for natural systems subject to Pfaffian nonholonomic constraints, has been reduced to formation of the kinetic energy to identify the mass matrix, then carrying out the indicated gradient operations on the mass matrix elements m_{ik} and the potential energy to form the vectors $G = G(q, \dot{q})$ and $\partial V/\partial q$.

For the case wherein the nonconservative forces are generated by an $m_c \times 1$ vector $\mathbf{u}$ of control inputs, we have $Q = Bu$ and Eq. (36) assume the form

$$
\begin{aligned}
M(q)\ddot{q} + \frac{\partial V}{\partial q} + G(q, \dot{q}) &= Bu + [A(q)]^T\lambda \\
A(q)\dot{q} + a_o(q) &= 0
\end{aligned}
\qquad (38)
$$

To appreciate some of the issues of cooperation associated with control design for redundantly actuated robotic systems, we consider a specific example in the following discussion.

B. Prototype Cooperative Manipulation Example

1. Equations of Motion

Consider the pair of robot arms shown in Fig. 1. We assume that there are four active joints, namely, the shoulder and elbow joints on the left and right robots, for simplicity; the wrist torques are neglected. The objective is to design a feedback controller to command the four torques so as to stabilize the payload with respect to a prescribed trajectory of the payload moving from an arbitrary, reachable state A to an arbitrary, reachable state B. It is desired that the control law have the following attributes: 1) accommodate an arbitrary feasible reference trajectory, 2) be of a simple feedforward/output output error feedback form, 3) guarantee global asymptotic stability, including nonlinear kinematics, and 4) handoff smoothly between large trajectory-tracking motion and terminal error suppression, without gain scheduling.

We present a control strategy possessing these four desirable attributes.

Under the assumption that each manipulator is composed of two rigid links, that the payload is a rigid body, and that the entire system undergoes only planar motion, but retaining all nonlinear kinematic effects, the kinetic energy of the system has the natural form

$$T = \frac{1}{2}\dot{q}^T[M(q)]\dot{q}$$

$$= \frac{1}{2}\dot{q}_L^T[M_L(q_L)]\dot{q}_L + \frac{1}{2}\dot{q}_R^T[M_R(q_R)]\dot{q}_R + \frac{1}{2}\dot{q}_P^T[M_P(q_P)]\dot{q}_P \qquad (39)$$

where the configuration coordinate vector naturally partitions into left (L), right (R), and payload (P) configuration coordinate subsets as

$$q_L = \begin{Bmatrix} \theta_1 \\ \theta_2 \end{Bmatrix}, \quad q_R = \begin{Bmatrix} \theta_6 \\ \theta_5 \end{Bmatrix}, \quad q_P = \begin{Bmatrix} \theta_3 \\ x_{c_3} \\ y_{c_3} \end{Bmatrix}$$

$$q = \begin{Bmatrix} q_L \\ q_R \\ q_P \end{Bmatrix} = \{\theta_1 \quad \theta_2 \ \vdots \ \theta_6 \quad \theta_5 \ \vdots \ \theta_3 \quad x_{c_3} \quad y_{c_3}\}^T$$

The 7×7 system mass matrix has the block diagonal structure

$$M(q) = \begin{bmatrix} M_L & & \\ & M_R & \\ & & M_P \end{bmatrix} \qquad (40)$$

where, introducing the elbow angles $\theta_{ij} = \theta_j - \theta_i$, the substructure mass matrices are compactly written as

$$M_L = \begin{bmatrix} I_1 + \frac{1}{4}m_1 l_1^2 + m_2 l_1^2 & \frac{1}{2}m_2 l_1 l_2 \cos\theta_{12} \\ \frac{1}{2}m_2 l_1 l_2 \cos\theta_{12} & I_2 + \frac{1}{4}m_2 l_2^2 \end{bmatrix} \qquad (41)$$

$$M_R = \begin{bmatrix} I_5 + \frac{1}{4}m_5 l_5^2 + m_4 l_5^2 & \frac{1}{2}m_4 l_5 l_4 \cos\theta_{65} \\ \frac{1}{2}m_4 l_5 l_4 \cos\theta_{65} & I_4 + \frac{1}{4}m_4 l_4^2 \end{bmatrix} \qquad (42)$$

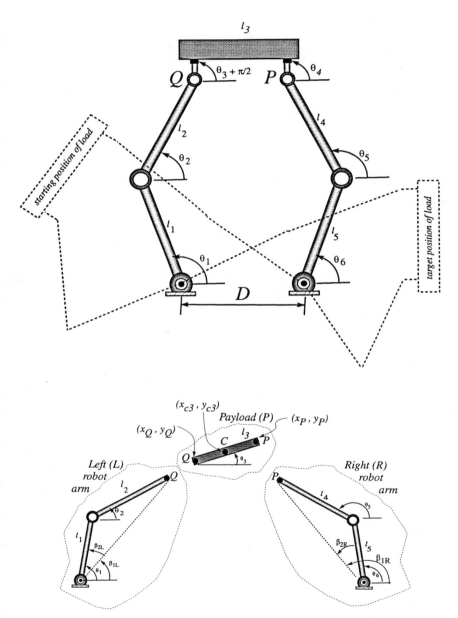

Fig. 1   Dual robot cooperative manipulation example.

and

$$M_P = \begin{bmatrix} I_3 & & \\ & m_3 & \\ & & m_3 \end{bmatrix} \tag{43}$$

The equations of motion follow in the form of Eq. (38), where, using Eq. (37), the nonlinear vector $G(q, \dot{q})$ has the following specific form

$$G(q, \dot{q}) = \left\{ \begin{matrix} G_L \\ G_R \\ 0 \end{matrix} \right\}, \quad \left\{ \begin{matrix} G_L \\ G_R \end{matrix} \right\} = \frac{1}{2} \left\{ \begin{matrix} -m_2 \dot{\theta}_2^2 l_1 l_2 \sin \theta_{12} \\ m_2 \dot{\theta}_1^2 l_1 l_2 \sin \theta_{12} \\ \cdots\cdots\cdots \\ -m_4 \dot{\theta}_5^2 l_4 l_5 \sin \theta_{65} \\ m_4 \dot{\theta}_6^2 l_4 l_5 \sin \theta_{65} \end{matrix} \right\} \tag{44}$$

The control vector (containing the four shoulder and elbow torques) is

$$u = \{u_1 \quad u_2 \quad u_6 \quad u_5\}^T \tag{45}$$

and, using the virtual work principle, we can establish that the control influence matrices are

$$B = \begin{bmatrix} B_L & O \\ O & B_R \\ O & O \end{bmatrix}, \quad B_L = B_R = \begin{bmatrix} 1 & -1 \\ 0 & 1 \end{bmatrix} \tag{46}$$

Taking the origin for a nonrotating (x, y) coordinate system in the joint at the base of the left arm, the geometric constraint arising from the pinning of the left and right robot wrists to the payload at points Q and P are captured by the four holonomic constraints:

$$\left. \begin{matrix} l_1 \cos \theta_1 + l_2 \cos \theta_2 + \frac{1}{2} l_3 \cos \theta_3 - x_{c_3} = 0 \\ l_1 \sin \theta_1 + l_2 \sin \theta_2 + \frac{1}{2} l_3 \sin \theta_3 - y_{c_3} = 0 \\ l_5 \cos \theta_6 + l_4 \cos \theta_5 - \frac{1}{2} l_3 \cos \theta_3 - x_{c_3} + D = 0 \\ l_5 \sin \theta_6 + l_4 \sin \theta_5 - \frac{1}{2} l_3 \sin \theta_3 - y_{c_3} = 0 \end{matrix} \right\} \tag{47}$$

On differentiation with respect to time, Eqs. (47) yield a kinematic constraint of the Pffafian form [the second equation of Eqs. (38)], with $a_o = 0$ and with

$$A(q) = \begin{bmatrix} -l_1 \sin \theta_1 & -l_2 \sin \theta_2 & 0 & 0 & -\frac{1}{2} l_3 \sin \theta_3 & -1 & 0 \\ l_1 \cos \theta_1 & l_2 \cos \theta_2 & 0 & 0 & \frac{1}{2} l_3 \cos \theta_3 & 0 & -1 \\ 0 & 0 & -l_5 \sin \theta_6 & -l_4 \sin \theta_5 & \frac{1}{2} l_3 \sin \theta_3 & -1 & 0 \\ 0 & 0 & l_5 \cos \theta_6 & l_4 \cos \theta_5 & -\frac{1}{2} l_3 \cos \theta_3 & 0 & -1 \end{bmatrix} \tag{48}$$

and also, for subsequent use, we record the time derivative of A as

$$A(q, \dot{q}) = \begin{bmatrix} -l_1 \dot{\theta}_1 \cos \theta_1 & -l_2 \dot{\theta}_2 \cos \theta_2 & 0 & 0 & -\frac{1}{2} l_3 \dot{\theta}_3 \cos \theta_3 & 0 & 0 \\ -l_1 \dot{\theta}_1 \sin \theta_1 & -l_2 \dot{\theta}_2 \sin \theta_2 & 0 & 0 & -\frac{1}{2} l_3 \dot{\theta}_3 \sin \theta_3 & 0 & 0 \\ 0 & 0 & -l_5 \dot{\theta}_6 \cos \theta_6 & -l_4 \dot{\theta}_5 \cos \theta_5 & \frac{1}{2} l_3 \dot{\theta}_3 \cos \theta_3 & 0 & 0 \\ 0 & 0 & -l_5 \dot{\theta}_6 \sin \theta_6 & -l_4 \dot{\theta}_5 \sin \theta_5 & \frac{1}{2} l_3 \dot{\theta}_3 \sin \theta_3 & 0 & 0 \end{bmatrix} \tag{49}$$

Now, solving the first of Eqs. (38) and Eq. (34) simultaneously for the generalized constraint force $Q_c = A^T \lambda$ and $M\ddot{q}$, we obtain

$$Q_c = A^T \lambda = F_1 + F_2 u$$
$$F_1 = A^T (AM^{-1}A^T)^{-1}(AM^{-1}G - \dot{A}\dot{q}) \tag{50}$$
$$F_2 = -A^T (AM^{-1}A^T)^{-1}AM^{-1}B$$

and

$$M\ddot{q} + \tilde{G} = \tilde{B}u$$
$$\tilde{G} = G - A^T \left(AM^{-1}A^T\right)^{-1} \{AM^{-1}G - \dot{A}\dot{q}\} \tag{51}$$
$$\tilde{B} = [I - A^T(AM^{-1}A^T)^{-1}AM^{-1}]B$$

It is natural to introduce the consistent partitions

$$M = \begin{bmatrix} M_L & & \\ & M_R & \\ & & M_P \end{bmatrix}, \quad \tilde{G} = \begin{Bmatrix} \tilde{G}_L \\ \tilde{G}_R \\ \tilde{G}_P \end{Bmatrix}, \quad \tilde{B} = \begin{bmatrix} \tilde{B}_L \\ \tilde{B}_R \\ \tilde{B}_P \end{bmatrix} \tag{52}$$

and rewrite the first of Eqs. (51) as three equations

$$M_L \ddot{q}_L + \tilde{G}_L (q, \dot{q}) = \tilde{B}_L (q, \dot{q}) u$$
$$M_R \ddot{q}_R + \tilde{G}_R (q, \dot{q}) = \tilde{B}_R (q, \dot{q}) u \tag{53}$$
$$M_P \ddot{q}_P + \tilde{G}_P (q, \dot{q}) = \tilde{B}_P (q, \dot{q}) u$$

This constraint-free form of the equations of motion implicitly reflects the constraints; the third of Eqs. (53) is sufficient to describe the dynamics of the system, since all other coordinates can be determined as a function of $(q_P, \dot{q}_P)$ through use of the constraint equations.

Prior to a discussion of control law design approaches, it is useful to consider the inverse kinematics problem: Given a smooth desired (prescribed) payload motion $q_P(t)$, determine feasible/desirable corresponding control inputs. Inverse kinematics for the case of redundant coordinates involves some subtle issues which are captured in the following sections.

2. Inverse Kinematics

Notice that the four holonomic constraints of Eqs. (47) reduce the number of degrees of freedom from seven to three. Thus, in principle, we could derive all coordinates and their time derivatives history from a given trajectory of the payload coordinates $q_P(t) = [\theta_3(t) x_{c_3}(t) y_{c_3}(t)]^T$. Obviously, if we know all of the coordinates and their first two time derivatives, then the differential equations of motion [Eqs. (51) or (53)] can be considered algebraic equations for determination of the corresponding control torques. Since there are only three degrees of freedom and four control torques, there is obviously an issue of uniqueness, and it is through the exploitation of the lack of uniqueness that we can seek an optimal control by which the robot arms may cooperate in carrying out the manipulation. It is also important to anticipate geometric singularities on the boundary of the reachable region (the maximum feasible workspace). First let us consider some geometric issues.

With reference to Fig. 1, observe that a given motion $q_P(t)$ of the payload dictates the motion of points P and Q through the four geometric formulas:

$$
\left.
\begin{aligned}
x_Q &= x_{c_3} - \left(\frac{l_3}{2}\right)\cos\theta_3 \\[4pt]
y_Q &= y_{c_3} - \left(\frac{l_3}{2}\right)\sin\theta_3 \\[4pt]
x_P &= x_{c_3} + \left(\frac{l_3}{2}\right)\cos\theta_3 \\[4pt]
y_P &= y_{c_3} + \left(\frac{l_3}{2}\right)\sin\theta_3
\end{aligned}
\right\}
\tag{54}
$$

and obviously, the companion equations can be obtained to determine the first two time derivatives of the grapple point coordinates (x_P, y_P, x_Q, y_Q) as a function of the payload motion

$$
(\theta_3, x_{c_3}, y_{c_3}, \dot\theta_3, \dot x_{c_3}, \dot y_{c_3}, \ddot\theta_3, \ddot x_{c_3}, \ddot y_{c_3})
$$

These straightforward equations are not recorded for the sake of brevity. However, given the payload motion, we can obviously determine the grapple point's velocity and acceleration coordinates

$$
(\dot x_P, \dot y_P, \dot x_Q, \dot y_Q, \ddot x_P, \ddot y_P, \ddot x_Q, \ddot y_Q)
$$

from the time derivatives of Eqs. (54). We consider how to determine the motion of the left and right robot arms. Considering the geometry of the left robot arm, from Fig. 1, it is evident that the left shoulder and elbow angles θ_1 and θ_2 are related to the instantaneous position of the grapple point (x_Q, y_Q) by

$$
\left.
\begin{aligned}
\theta_1 &= \beta_{1L} + \beta_{2L} \\[4pt]
\beta_{1L} &= \tan^{-1}(y_Q/x_Q) \\[4pt]
\beta_{2L} &= \cos^{-1}\left(\frac{l_1^2 - l_2^2 + (x_Q^2 + y_Q^2)}{2l_1(x_Q^2 + y_Q^2)^{1/2}}\right), \quad \text{two roots, take } \beta_{2L} > 0 \\[4pt]
\theta_2 &= \tan^{-1}\left(\frac{y_Q - l_1\sin\theta_1}{x_Q - l_1\cos\theta_1}\right)
\end{aligned}
\right\}
\tag{55}
$$

Similarly, considering the right robot, it is evident that the right robot angles θ_6 and θ_5 are related to (x_P, y_P) by

$$
\left.
\begin{aligned}
\theta_6 &= \beta_{1R} - \beta_{2R} \\[4pt]
\beta_{1R} &= \tan^{-1}(y_P/x_P) \\[4pt]
\beta_{2R} &= \cos^{-1}\left[\frac{l_5^2 - l_4^2 + (x_P^2 + y_P^2)}{2l_5\left[(D - x_P^2)^2 + y_P^2\right]^{1/2}}\right], \quad \text{two roots, take } \beta_{2R} > 0 \\[4pt]
\theta_5 &= \tan^{-1}\left(\frac{y_P - l_5\sin\theta_6}{D - x_P - l_5\cos\theta_6}\right)
\end{aligned}
\right\}
\tag{56}
$$

It can be verified that taking β_{2L} and β_{2R} as positive corresponds to the elbows out configuration shown in Fig. 1. Obviously, the elbows in configuration results from choosing the negative signs for β_{2L} and β_{2R}, and two other asymmetric configurations are possible if opposite signs are selected. The lack of uniqueness is a consequence of redundancy and the choice of manipulation modes is dictated by practical configurations. Except near certain singular configurations discussed later, it is possible to manipulate smoothly through an infinite family of neighboring configurations for any one of the four choices of signs for $\beta_{2L}(t)$ and $\beta_{2R}(t)$. Straightforward differentiation yields the following kinematic equations which determine the first two time derivatives of the left and right shoulder and elbow angles:

$$\begin{Bmatrix} \dot\theta_1 \\ \dot\theta_2 \end{Bmatrix} = A_L^{-1} \begin{Bmatrix} \dot x_Q \\ \dot y_Q \end{Bmatrix}, \quad \begin{Bmatrix} \ddot\theta_1 \\ \ddot\theta_2 \end{Bmatrix} = A_L^{-1} \left[\begin{Bmatrix} \ddot x_Q \\ \ddot y_Q \end{Bmatrix} - \dot A_L \begin{Bmatrix} \dot\theta_1 \\ \dot\theta_2 \end{Bmatrix} \right]$$

$$\begin{Bmatrix} \dot\theta_6 \\ \dot\theta_5 \end{Bmatrix} = A_R^{-1} \begin{Bmatrix} \dot x_P \\ \dot y_P \end{Bmatrix}, \quad \begin{Bmatrix} \ddot\theta_6 \\ \ddot\theta_5 \end{Bmatrix} = A_R^{-1} \left[\begin{Bmatrix} \ddot x_P \\ \ddot y_P \end{Bmatrix} - \dot A_R \begin{Bmatrix} \dot\theta_6 \\ \dot\theta_5 \end{Bmatrix} \right]$$

$$(57)$$

where we have introduced the matrices

$$A_L = \begin{bmatrix} -l_1 \sin\theta_1 & -l_2 \sin\theta_2 \\ l_1 \cos\theta_1 & l_2 \cos\theta_2 \end{bmatrix}, \quad A_R = \begin{bmatrix} -l_5 \sin\theta_6 & -l_4 \sin\theta_5 \\ l_5 \cos\theta_6 & l_4 \cos\theta_5 \end{bmatrix} \quad (58)$$

It is easy to verify that the preceding matrices are singular if $\theta_1 = \theta_2$, and $\theta_6 = \theta_5$, respectively. It is obvious that these singularities corresponded to the left and right arms being fully extended, and it is clear that these boundaries of the workspace are to be avoided [the reachable set of points interior to the workspace must be taken into account in the trajectory planning for the payload, leading to the nominal trajectory $q_P(t)$ of the payload].

3. Cooperative Actuation

Given the inverse kinematic solution for all system coordinates and time derivatives, as a function of a prescribed payload trajectory $q_P(t)$, the corresponding control torque vector $u(t)$ is not unique for the case of more actuators and degrees of freedom. In our particular example, since we have four actuators and three degrees of freedom, we expect an infinity of torque vectors for the nominal maneuver. As in the case of human beings jointly manipulating a heavy object, we desire to exploit the redundancy of actuation to cooperate in the sense that large, nonworking constraint forces are avoided.

To capture these considerations as a control strategy, we introduce the following cooperation criterion to be minimized

$$J = \frac{1}{2} u^T W_u u + \frac{1}{2} Q_c^T W_c Q_c \quad (59)$$

subject to satisfying the third of Eqs. (53). Notice that the weight matrix selection permits us the flexibility of emphasizing small torques (u), or small constraint forces ($Q_c = A^T \lambda$), or a compromise between these two competing objectives.

Using the Lagrange multiplier rule, we introduce the $m \times 1$ Lagrange multiplier vector γ and the augmented function $\tilde{J}$, and use Eqs. (50) and (53) to write

$$\tilde{J} = \frac{1}{2}u^T W_u u + \frac{1}{2}(F_1 + F_2 u)^T W_c (F_1 + F_2 u) + \gamma^T (M_P \ddot{q}_P + \tilde{G}_P - \tilde{B}_P u)$$
(60)

Requiring that the gradients $\nabla_u \tilde{J}$ and $\nabla_\gamma \tilde{J}$ both vanish as a necessary condition for minimizing J leads to the solution

$$u = H\{\tilde{B}_P^T \gamma - F_2^T W_c F_1\}$$
$$\gamma = (\tilde{B}_P H \tilde{B}_P^T)^{-1}\{M_P \ddot{q}_P + \tilde{G}_P + \tilde{B}_P H F_2^T W_c F_1\}$$
(61)
$$H = (W_u + F_2^T W_c F_2)^{-1}$$

Some simple calculations with example payload motions reveal the utility of this formulation of the inverse kinematics and cooperative actuation strategy.

4. Example Nominal Payload Trajectory

Perhaps the simplest and easiest to motivate scheme for prescribing a nominal motion $q_P(t)$ for the payload is to adopt a smooth polynomial spline from the initial state $q_P(t_0)$ to the target final state $q_P(t_f)$ of the form

$$q_P(t) = f(\tau)\{q_P(t_f) - q_P(t_0)\} + q_P(t_0), \quad \tau = \frac{(t - t_0)}{(t_f - t_0)}$$

$$\dot{q}_P(t) = \dot{f}(\tau)\{q_P(t_f) - q_P(t_0)\}, \quad \dot{f}(\tau) = \frac{1}{(t_f - t_0)}\frac{df}{d\tau}$$
(62)

$$\ddot{q}_P(t) = \ddot{f}(\tau)\{q_P(t_f) - q_P(t_0)\}, \quad \ddot{f}(\tau) = \frac{1}{(t_f - t_0)^2}\frac{d^2 f}{d\tau^2}$$

where we choose the particular shape function

$$f(\tau) = \tau^3(10 - 15\tau + 6\tau^2)$$
$$\frac{df}{d\tau} = \tau^2(30 - 60\tau + 30\tau^2)$$
(63)
$$\frac{d^2 f}{d\tau^2} = \tau(60 - 180\tau + 120\tau^2)$$

This trajectory can be shown to be optimal for the idealized case where we consider only the payload trajectory and the vector sums (F, M) of the forces and moments applied to the payload, without regard to how these are generated. Equations (62) and (63) can be shown[17] to simultaneously minimize the translational and rotational jerk integrals

$$J_1 = \int_{t_0}^{t_f} \dot{F}^T \dot{F} dt, \quad \text{and} \quad J_2 = \int_{t_0}^{t_f} \dot{M}^T \dot{M} dt$$

subject to satisfaction of the third of Eqs. (53), and the boundary conditions:

$q_P(t_0) =$ specified initial position $q_P(t_f) =$ specified final position

$$\dot{q}_P(t_0) = 0 \qquad\qquad\qquad \dot{q}_P(t_f) = 0 \qquad\qquad (64)$$

$$\ddot{q}_P(t_0) = 0 \qquad\qquad\qquad \ddot{q}_P(t_f) = 0$$

Since the idealized optimal trajectory [Eqs. (62) and (63)] does not explicitly consider workspace constraints, this nominal motion must be checked to make sure it remains feasible throughout the motion, and of course, optimality with respect to the entire system's dynamics and minimization of other performance measures cannot be claimed. These smooth, easy-to-compute, motions usually represent excellent starting solutions, however, and we elect to use this family of solutions to generate the nominal trajectories throughout the remainder of this chapter. A typical example motion of the system is shown in Fig. 2.

5. Lyapunov Stable Tracking Control Law

A smooth nominal (reference) trajectory for the entire system can be computed using Eqs. (62) and (63), and via inverse kinematics; the left and right robot joint coordinates are determined from Eqs. (54–57), while the nominal (cooperative) shoulder and elbow torques are determined from Eqs. (61). This is a possible way to determine the reference trajectory, and can be replaced by a more appropriate path-planning method in particular applications. However the reference trajectory satisfying the boundary conditions of Eqs. (64) is determined, we denote all state and control variables along the reference trajectory with a subscript reference. Of course, in actual applications, we can expect that the system will not follow the reference trajectory $q_{\text{ref}}(t)$ exactly when we command the control $u_{\text{ref}}(t)$, because of model errors, external disturbances, and nonideal actuation. We seek a control law $\delta u = \text{function}[\delta q(t), \delta\dot{q}(t)]$ which will guarantee that an initially disturbed motion will asymptotically return to the reference trajectory in the absence of model or implementation errors. Actually, it is preferable that the control perturbation δu is in output feedback form where it depends only on a measurable subset of the coordinates and their time derivatives.

In view of the four kinematic constraints, we know that a minimal coordinate description requires only three generalized coordinates. By considering $(q, \dot{q})$ to be functions of $(q_P, \dot{q}_P)$, in the third of Eqs. (53), we are motivated to investigate the kinetic energy

$$T_P = \frac{1}{2}\dot{q}_P^T M_P \dot{q}_P \qquad\qquad (65)$$

and observe that

$$\dot{T}_P = \dot{q}_P^T \tilde{B}_P u \qquad\qquad (66)$$

This motivates the Lyapunov function

$$U = \frac{1}{2}\delta\dot{q}_P^T M_P \delta\dot{q}_P + \frac{1}{2}\delta q_P^T K_1 \delta q_P \qquad\qquad (67)$$

where $\delta q_P = q_P - q_{P\text{ref}}(t)$. For the simplest case where $q_{P\text{ref}}(t)$ is a constant vector, then it is easy to verify that the Lyapunov function derivative is

$$\dot{U} = \delta\dot{q}_P^T[\tilde{B}_P u + K_1\delta q_P] \tag{68}$$

and selecting the bracketed term to equal $-K_2\delta\dot{q}_P$ (so that $\dot{U}$ is never positive), we are led to the global stability condition

$$\tilde{B}_P u = -[K_1\delta q_P + K_2\delta\dot{q}_P] \tag{69}$$

Since $\tilde{B}_P$ is a 3×4 matrix, it is evident that u is underdetermined and we are free to introduce an optimization criterion to select a particular control satisfying Eq. (69). One attractive possibility is to minimize $u^T u$; this gives the minimum actuator torque controller

$$u = -\tilde{B}_P^T(\tilde{B}_P\tilde{B}_P^T)^{-1}[K_1\delta q_P + K_2\delta\dot{q}_P] \tag{70}$$

For the trajectory tracking case, in which we desire to stabilize the motion with respect to a prescribed reference motion, the situation is more complicated. Suppose that the reference trajectory $q_{P\text{ref}}(t)$ and an associated control $u_{\text{ref}}(t)$ are determined consistent with the system dynamics [for example, using Eqs. (54–64)]. Then it follows that the payload dynamics at every instant on the actual and reference trajectories satisfy

$$M_P\ddot{q}_P = \tilde{G}_P + \tilde{B}_P u$$
$$M_{P_{\text{ref}}}\ddot{q}_{P_{\text{ref}}} = \tilde{G}_{P_{\text{ref}}} + \tilde{B}_{P_{\text{ref}}}u_{\text{ref}} \tag{71}$$

and it also follows that the Lyapunov function [Eq. (67)] has the time derivative

$$\dot{U} = \delta\dot{q}_P^T\left[\tilde{B}_P u - \tilde{B}_{P_{\text{ref}}}u_{\text{ref}} + K_1\delta q_P - \delta\tilde{G}_P - \delta M_P\ddot{q}_{P_{\text{ref}}} + \frac{1}{2}\dot{M}_P\delta\dot{q}_P\right] \tag{72}$$

Setting the bracketed term to $-K_2\delta\dot{q}_P$ gives the stabilizing control condition

$$\tilde{B}_P u = \tilde{B}_{P_{\text{ref}}}u_{\text{ref}} - [K_1\delta q_P + K_2\delta\dot{q}_P] + \left[\delta\tilde{G}_P + \delta M_P\ddot{q}_{P_{\text{ref}}} - \frac{1}{2}\dot{M}_P\delta\dot{q}_P\right] \tag{73}$$

and for the case of minimum control torque, a particular solution of Eq. (73) gives the nonlinear feedback law

$$u = \tilde{B}_P^T\left(\tilde{B}_P\tilde{B}_P^T\right)^{-1}\left\{\tilde{B}_{P_{\text{ref}}}u_{\text{ref}} - [K_1\delta q_P + K_2\delta\dot{q}_P]\right.$$
$$\left. + \left[\delta\tilde{G}_P + \delta M_P\ddot{q}_{P_{\text{ref}}} - \frac{1}{2}\dot{M}_P\delta\dot{q}_P\right]\right\} \tag{74}$$

This law, while guaranteeing stability (neglecting model errors), is cumbersome to implement because of the detailed computation required to produce all of the nonlinear terms. Note that the payload coordinates $q_P = [\theta_3 \quad x_{c3} \quad y_{c3}]^T$ may not be directly measurable. For example, assume that the measurable quantities are $q_L = [\theta_1 \quad \theta_2]^T$ and $q_R = [\theta_6 \quad \theta_5]^T$, and the time derivatives thereof; then it is easy to verify from geometry that the payload coordinates are computable as follows

$$\theta_3 = \tan^{-1}\left[\frac{y_Q - y_P}{x_Q - x_P}\right]$$

$$= \tan^{-1}\left[\frac{(l_5 \sin\theta_6 + l_4 \sin\theta_5) - (l_1 \sin\theta_1 + l_2 \sin\theta_2)}{(D + l_5 \cos\theta_6 + l_4 \cos\theta_5) - (l_1 \cos\theta_1 + l_2 \cos\theta_2)}\right]$$

(75)

$$x_{c_3} = \frac{1}{2}(x_Q + x_P) = \frac{1}{2}[(D + l_5 \cos\theta_6 + l_4 \cos\theta_5) + (l_1 \cos\theta_1 + l_2\cos\theta_2)]$$

$$y_{c_3} = \frac{1}{2}(y_Q + y_P) = \frac{1}{2}[(l_5 \sin\theta_6 + l_4 \sin\theta_5) + (l_1 \sin\theta_1 + l_2 \sin\theta_2)]$$

and the time derivative $\dot{q}_P = [\dot{\theta}_3 \quad \dot{x}_{c_3} \quad \dot{y}_{c_3}]^T$ follows from differentiation of Eq. (75).

As an alternative to the preceding developments, and to obtain a direct output error feedback form for the control law, we can observe the following kinematic form for the work rate of the control torques

$$\dot{T} = u_1\dot{\theta}_1 + u_2\dot{\theta}_2 + u_6\dot{\theta}_6 + u_5\dot{\theta}_5 = \left\{\begin{matrix} \dot{q}_L \\ \dot{q}_R \end{matrix}\right\}^T u = \dot{q}_L^T u_L + \dot{q}_R^T u_R$$

(76)

and it is obvious by inspection that setting $u_L = -K_{2L}q_L$, $u_R = -K_{2R}\dot{q}_R$ will decrease T for all nonzero motion of the system. This energy dissipative control suggests the following output error feedback law for controlling the departure motion relative to the reference trajectory

$$u = u_{ref}(t) - \left\{K_1\left(\begin{matrix} \delta q_L \\ \delta q_R \end{matrix}\right) + K_2\left(\begin{matrix} \delta\dot{q}_L \\ \delta\dot{q}_R \end{matrix}\right)\right\}$$

(77)

where the 4×4 positive definite gain matrices have the structure $K_i = \begin{bmatrix} K_{iL} & 0 \\ 0 & K_{iR} \end{bmatrix}$.

It can be verified that the control law of Eq. (77) is globally stabilizing only for the case that $q_{ref} = $ constant. Although global asymptotic stability is not guaranteed during the time interval $\{t_0 < t < t_f\}$, it is guaranteed during the interval $\{t > t_f\}$, for all reference maneuvers satisfying the boundary conditions of Eq. (64). These developments can be better appreciated in the light of some illustrative numerical examples, as provided in the next section.

6. Cooperative Robotic Manipulation: A Numerical Example

To illustrate the preceding discussion, we consider each link of the robots to be 1 m long and to have a mass of 1 kg. The distance D between the shoulder joints is taken as 0.75 m, and the nominal initial and desired target values of five angles are listed in Table 1. The inverse kinematic process of Eqs. (54–64) was used to compute the solution shown in Fig. 2. All the initial conditions were then perturbed by moderately large angles (order of 10 deg), and the feedback control law of Eq. (77) was used. A typical controlled response from large initial disturbances is shown in Fig. 3. Notice that the order of 10-deg initial errors are less than 0.5 deg by the nominal final time of 10 s, however a few more seconds of terminal control are required to effectively null the errors. The weight

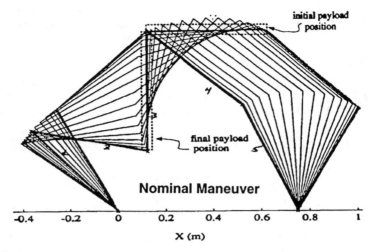

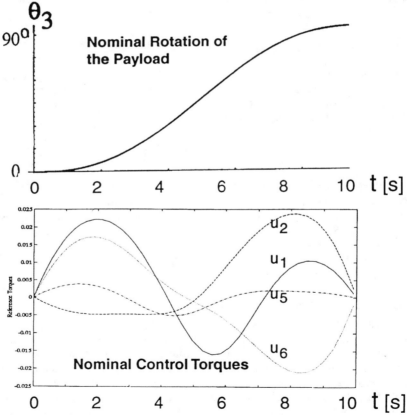

Fig. 2   Cooperative manipulation example: open-loop reference motion.

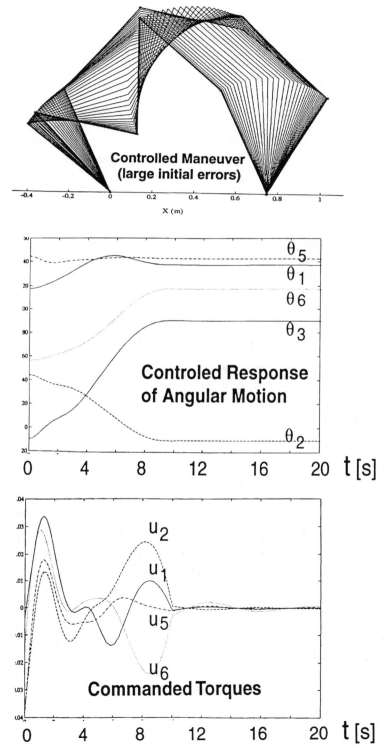

Fig. 3 Cooperative manipulation example: closed-loop tracking controller.

Table 1   Initial and final angles for the nominal maneuver

	θ_1, deg	θ_2, deg	θ_3	θ_6, deg	θ_5, deg	Time, s
Initial	121.0430	40.0323	00.0000	58.9570	139.9677	0
Target	137.2041	−10.3342	90.0000	117.3017	142.3095	10

matrices [in Eq. (59)] were $W_u = I$, $W_c = 0$, and the control gains [in Eq. (77)] were $K_1 = 0.5I$, $K_2 = 0.2I$; these affect the controlled response, however, we found a large family of feasible values. From evaluating the response using several other initial conditions and variations in the selections of the control gains and weight matrices, we confirmed that a wide range of choices give excellent tracking stability over a large domain of initital condition errors. Thus the control law of Eq. (77) seems to be an attractive candidate for practical applications.

The preceding results have been extended, including a successful experimental study in Ref. 35.

VII.   Concluding Remarks

In this chapter, we have summarized the central aspects of Lyapunov stability theory with particular emphasis on the role that it can play in designing stable controllers for nonlinear multibody robotic manipulation. Several elementary analytical and numerical examples are provided to illustrate the ideas and to provide some basis for extrapolating the practical implications of the methods presented. A more extensive example is offered to introduce some ideas on cooperative robotic manipulation, in which two or more manipulators are manipulating a payload while cooperating with each other to minimize a measure of the associated control and constraint forces and moments. The chapter concludes with a numerical example which supports the practical value of these developments.

Acknowledgments

This work has been supported by the Air Force of Scientific Research under Grant F49620-92-J-0496, the administrative and technical support of Spencer Wu is appreciated. The following individuals contributed directly to these developments: Brig Agrawal, Hyochoong Bang, Johnny Hurtado, and Gary Yale. These individuals all made significant contributions to the development and validation of the ideas, as well as the numerical implementations underlying Figs. 1–3. These contributions are appreciated and are warmly acknowledged. A portion of this effort was completed while the first author was on sabbatical at the Naval Postgraduate School, the administrative and technical support of Professors Brij Agrawal and Rudy Panzholzer are warmly acknowledged. The interactions with Rao Vadali and Ranjan Mukherjee has made significant indirect contributions which are most appreciated.

References

[1]Vidyasagar, M., *Nonlinear System Analysis,* Prentice-Hall, Englewood Cliffs, NJ, 1978.

[2]Willems, J. L.,   *Stability Theory of Dynamical Systems,*   John Wiley and Sons, New York, 1970.

[3]Oh, H. S., Vadali, S. R., and Junkins, J. L.,   "On the Use of the Work-Energy Rate Principle for Designing Feedback Control Laws,"   *Journal of Guidance, Control, and Dynamics,* Vol. 15, No. 1, 1992, pp. 275–277.

[4]Junkins, J. L., Rahman, Z. H., and Bang, H.,   "Near-Minimum-Time Control of Distributed Parameter Systems: Analytical and Experimental Results,"   *Journal of Guidance, Control, and Dynamics,*   Vol. 14, No. 2, 1991, pp. 406–415.

[5]Junkins, J. L., Rahman, Z. H., and Bang, H.,   "Near-Minimum Time Maneuvers of Flexible Vehicles: A Lyapunov Control Law Design Method,"   *Mechanics and Control of Large Flexible Structures,*   edited by J. L. Junkins, Vol. 129, Progress in Astronautics and Aeronautics, AIAA, Washington, DC, 1990, Chap. 22.

[6]Kalman, R. E.,   "Lyapunov Function for the problems of Lure in Automatic Control,"   *Proceedings of the National Academy Science,* Vol. 49, 1963, pp. 201–205.

[7]Popov, V. M.,   "Hyperstability and Optimality of Automatic Systems with Several Control Functions,"   *Review of Romanian Science and Technology,*   Vol. 9, 1964, pp. 629–690.

[8]Chen, C. T.,   *Linear System Theory and Design,*   Holt, Rinehart, and Winston, New York, 1984.

[9]Kalman, R. E., and Bertram, J. E.,   "Control System Analysis and Design via the Second Method of Lyapunov,"   *Transactions of the ASME, Journal of Basic Engineering,* Vol. 82, 1960, pp. 371–392.

[10]Patel, R. V., and Toda, M.,   "Quantitative Measures of Robustness for Multivariable Systems," Proceedings of the JACC, TP-8A, San Francisco, CA, 1980.

[11]Inman, D. J.,   *Vibration with Control, Measurement, and Stability,*   Prentice-Hall, Englewood Cliffs, NJ, 1989.

[12]Creamer, N. G., and Junkins, J. L.,   "A Pole Placement Technique for Vibration Suppression of Flexible Structures," AIAA/AAS Astrodynamics Conference, Minneapolis, MN, Aug. 15–17, 1988.

[13]Junkins, J. L., and Kim, Y.,   "A Minimum Sensitivity Design Method for Output Feedback Controllers,"   *Mechanics and Control of Large Flexible Structures,*   edited by J. L. Junkins, Vol. 129, Progress in Astronautics and Aeronautics, AIAA, Washington, DC, 1990, Chap. 25.

[14]Canavin, J. R.,   "The Control of Spacecraft Vibrations Using Multivariable Output Feedback," AIAA/AAS Astrodynamics Conference, Palo Alto, CA, Aug. 7–9, 1978.

[15]Goh, C. J., and Caughey, T. K.,   "On the Stability Problem Caused by Finite Actuator Dynamics in the Collocated Control of Large Space Structures,"   *International Journal of Control,* Vol. 41, No. 3, 1985, pp. 787–802.

[16]Joshi, S. M.,   *Control of Large Flexible Space Structures,*   Lecture Notes in Control and Information Sciences, Vol. 131, Springer-Verlag, New York, 1989.

[17]Junkins, J. L., and Turner, J. D.,   *Optimal Spacecraft Rotational Maneuvers,*   Elsevier, Amsterdam, The Netherlands, 1986.

[18]Juang, J. N., Horta, L. G., and Robertshaw, N. H.,   "A Slewing Control Experiment for Flexible Structures,"   *Proceedings of the 5th VPI and SU Symposium on Dynamics and Control of Large Structures,*   Virginia Polytechnic Inst. and State Univ., Blacksburg, VA, 1985, pp. 547–551.

[19]Fujii, H., Ohtsuka, T., and Udou, S.,   "Mission Function Control for Slew Maneuver Experiment,"   *Journal of Guidance, Control, and Dynamics,*   Vol. 12, No. 6, 1989, pp. 858–865.

[20]Wie, B., Weiss, H., and Araposhathis, A., "Quaternion Feedback for Spacecraft Eigenaxis Rotations," *Journal of Guidance, Control, and Dynamics,* Vol. 12, No. 3, 1989, pp. 375–380.

[21]Thompson, R. C., Junkins, J. L., and Vadali, S. R., "Near-Minimum Time Open-Loop Slewing of Flexible Vehicles," *Journal of Guidance, Control, and Dynamics,* Vol. 12, No. 1, 1989, pp. 82–88.

[22]Vadali, S. R., "Feedback Control of Space Structures: A Lyapunov Approach," *Mechanics and Control of Large Flexible Structures,* edited by J. L. Junkins, Vol. 129, Progress in Astronautics and Aeronautics, AIAA, Washington, DC, 1990, Chap. 24.

[23]Meirovitch, L., and Quinn, R., "Maneuvering and Vibration Control of Flexible Spacecraft," *Journal of Astronautical Sciences,* Vol. 35, No. 3, 1987, pp. 301–328.

[24]Singh, G., Kabamba, P., and McClamroch, N., "Planar Time Optimal Slewing Maneuvers of Flexible Spacecraft," *Journal of Guidance, Control, and Dynamics,* Vol. 12, No. 1, 1989, pp. 71–81.

[25]Breakwell, J. A., "Optimal Feedback Control for Flexible Spacecraft," *Journal of Guidance, Control, and Dynamics,* Vol. 4, No. 5, 1981, pp. 427–479.

[26]Slotine, J. E., and Weiping, L., *Applied Nonlinear Control,* Prentice-Hall, Englewood Cliffs, NJ, 1991.

[27]VanderVelde, W., and He, J., "Design of Space Structure Control Systems Using On-Off Thrusters," *Journal of Guidance, Control, and Dynamics,* Vol. 6, No. 1, 1983, pp. 759–775.

[28]Byers, R. M., Vadali, S. R., and Junkins, J. L., "Near-Minimum-Time Closed-Loop Slewing of Flexible Spacecraft," *Journal of Guidance, Control, and Dynamics,* Vol. 13, No. 1, 1990, pp. 57–65.

[29]LaSalle, J., and Lefschetz, S., *Stability by Lyapunov's Direct Method with Applications,* Academic Press, New York, 1961.

[30]Mukherjee, R., and Chen, D., "Stabilization of Free-Flying Under-Actuated Mechanisms in Space," Proceedings of the 1992 American Control Conference, 1992.

[31]Mukherjee, R., and Chen, D., "An Asymptotic Stability Theorem for Nonautonomous Systems," *Journal of Guidance, Control, and Dynamics,* Vol. 16, No. 6, Nov.–Dec. 1993, pp. 1191–1194.

[32]Junkins, J. L., and Kim, Y,, *An Introduction to Dynamics and Control of Flexible Structures,* AIAA Education Series, AIAA, Washington, DC, 1993.

[33]Ahmad, S., and Zribi, M., "Lyapunov Based Control Design for Multiple Robots Handling a Common Object," *Lecture Notes in Control and Information Sciences,* edited by J. M. Skowronski, H. Flashner, and R. S. Guttalu, Vol. 170, Springer-Verlag, New York, 1991, Chap. 1.

[34]Krishnan, H., "Control of Nonlinear Systems with Applications to Constrained Robots and Spacecraft Attitude Stabilization," Ph.D. Dissertation, Univ. of Michigan, Ann Arbor, MI, 1992.

[35]Yale, G., "Cooperative Control of Multiple Space Manipulators," Naval Postgraduate School, Ph.D. Dissertation, Montercy, CA, Sept. 1993.

[36]Mukherjee, R., and Junkins, J. L., "An Invariant Set Analysis of the Hub-Appendage Problem," *Journal of Guidance, Control, and Dynamics,* Vol. 16, No. 6, 1993, pp. 1191–1193.

[37]Bauer, F. L., and Fike, C. T., "Norms and Exclusion Theorems," *Numerische Mathematique,* Vol. 2, 1960, pp. 137–141.

**Part 5.   Telerobot System Design
and Applications**

Chapter 14

Teleoperation: From the Space Shuttle to the Space Station

Phung K. Nguyen*

Spar Aerospace Ltd., Brampton, Ontario, Canada

and

Peter C. Hughes†

University of Toronto, North York, Ontario, Canada

I. Introduction

T HE U.S. Space Shuttle program was not only the start of a new era in space transportation, it also brought forth a new technology: teleoperation in space. The Shuttle remote manipulator system (SRMS), also known as the Canadarm (Fig. 1), made its inaugural appearance in space onboard the Shuttle Orbiter Columbia in the second space transportation system mission on November 4, 1981. Almost twenty years later, a second-generation space manipulator system, the Space Station remote manipulator system (SSRMS), shown in Fig. 2, will be on board the U.S. Space Station providing services such as payload transporting and maneuvering, Shuttle Orbiter berthing and deberthing, etc. Another teleoperated remote manipulator system will also be on board the Space Station to perform more delicate, dexterous assembly tasks: the special purpose dexterous manipulator (SPDM). The SSRMS and the SPDM are part of the mobile servicing system (MSS), which is the Canadian contribution to the U.S. Space Station program. Another remote manipulator system will be contributed by Japan, another international partner to the U.S. Space Station program, to support the Japanese experiment module (JEM). Over the years, the field of space teleoperation has evolved a great deal helping man to perform more and more complex and difficult tasks in space.

Space remote manipulator systems (srms) help extend human capability in terms of reach and object handling in the space hostile environment, and eliminate boredom by performing repetitive tasks with precision. However, unlike their terrestrial counterparts, srms's design and operation are subjected to a number

Copyright © 1994 by the American Institute of Aeronautics and Astronautics, Inc. All rights reserved.

*Formerly, Senior Staff Engineer, Advanced Technology Systems Group.
†Professor, Institute for Aerospace Studies.

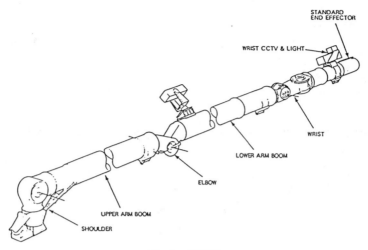

Fig. 1   SRMS.

of constraints which exist only in space programs. For example, an srms must satisfy the transportation constraints associated with launch. The launch cost is proportional to its weight and its lengthwise occupancy of the Orbiter cargo bay. The launch cost must be minimized, which in turn leads to the optimization of srms mass/stiffness subject to the constraint of packaging for its installation in the Orbiter. In addition, the srms needs to survive the storage and launch environment, including salt fog, liftoff vibrations, and pyrotechnic shock during the Orbiter separation from the external tank. Next, while in orbit, an srms must be thermally protected from the heat induced by solar radiation, reflected radiation from Earth, and from cold when it is in Earth's shadow. It also depends on the mother spacecraft for power and communications, which means that the srms system architecture must conform with the architecture of the mother spacecraft. The srms must then be designed within its mass, stiffness, power and data budgets. Before being qualified as a space system element, an srms must go through extensive testings to demonstrate survival under harsh transportation and storage conditions, launch conditions, and the extreme thermal-pressure space environment. Also, as part of the design verification, prior to launch, some system tests must be done on the ground, complemented by computer simulations, leaving the final system verification to be performed in space.

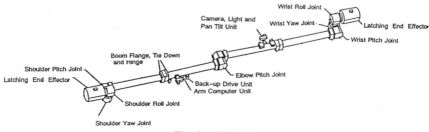

Fig. 2   SSRMS.

From an operation and logistics viewpoint, an srms is quite different from its terrestrial counterpart. First of all, as the srms's operators, astronauts, are thoroughly instructed on the hardware and software details of the system, and are well trained for its operation. Highest priority is placed on the safety of the crew, the spacecraft, and its equipment, while preparing operation plans and procedures. Undesired contacts between the srms elements (including attached payload) with themselves or with other structures must be avoided. In this regard, potential malfunctions must be thoroughly studied during the design phase to have them designed out if possible; otherwise, adequate malfunction detection, trouble shooting, and handling procedures must be provided. In terms of operation supports, premission computer simulations are carried out to verify the feasibility of the planned mission; to ensure the safety of the crew, the srms, and the spacecraft; and to establish time lines for on-orbit tasks. During the mission, monitoring the srms operation is performed in real time by the on-board astronauts, with assistance provided by ground support personnel via data telemetry. In the event of an abnormal observation, the ground personnel would analyze the telemetered data immediately to determine its cause for a go/no-go decision. If possible, they would provide the crew with a fix which has been verified normally by simulations. Spares of srms elements must therefore be carried onboard; some faulty components can be replaced via a robotic task, others by an astronaut performing an extravehicular activity (EVA). Redundancy provides an alternative design solution: when a failure occurs in the primary system, the redundant system will be automatically activated to prevent interruption in the manipulator operation. Backup system design provides the last resort in the event the redundant system fails.

To facilitate the tasks of payload deployment, handling, and retrieval, srms's are often required to have a large reach envelope and to be able to place and orient the manipulator tip practically anywhere within the envelope. In a payload deployment using an srms with its base fixed to the mother spacecraft like the SRMS, the payload can be placed at a location that is safely cleared of structures. In retrieving a payload, the spacecraft does not need to be brought too close to a payload, thus avoiding plume impingement on the payload by the spacecraft thrusters. For an srms with a moving base like the SSRMS, the reach capability helps simplify the design of its base structure: the base structure can move along a straight line (e.g. on rails). In any case, the required reach envelope has six dimensions (three in translation and three in rotation); hence, an srms must have at least six joints. The joint type selection (i.e. prismatic or revolute, with or without offset), the joint placement, and the manipulator segmentation into shoulder, elbow, and wrist depend on the intended use of the srms.

The srms base is not fixed to inertial space. Control of the srms is relative to the mother spacecraft, which always has some motion while the srms is maneuvered. The magnitude of the spacecraft motion depends on its mass properties, srms payload mass properties, srms tip speed, srms configuration, and location of the srms base on the spacecraft. To maintain the spacecraft attitude with respect to a certain control reference frame, the spacecraft attitude control system might need to be active during an srms maneuver. Controller-controller interactions and controller-structures interactions become concerns: the controller stability must be ensured and the structural integrity of the srms must be protected. Often, flight rules and operation procedures are developed to restrict the spacecraft thruster activities while the srms is in operation.

Although they have a lot in common, the srms's are in fact different from each other as illustrated by the SRMS and SSRMS examples. In the launch configuration, the SRMS is stowed in a straight-out configuration along the Orbiter cargo bay door hinges, avoiding occupancy of the payload-carrying area in the Orbiter cargo bay. It is treated as an element of the Orbiter, not as a payload. The SSRMS, as an Orbiter payload, must fit into a launch package that requires the SSRMS to be segmented, folded, and secured across the cargo bay. Different load paths and load levels in these two srms launch configurations necessitate different designs of their support structures. Topologically, the SRMS arm is anthropomorphic with six in-line joints: two at the shoulder, one at the elbow, and three at the wrist. The SSRMS is designed for self-locatability which can be achieved using an arm with seven offset joints: three at the shoulder, one at the elbow, and three at the wrist. Furthermore, both ends of the SSRMS are equipped with identical latching end effectors (LEE) to allow either end to act as its base or its tip. The latching mechanism is needed to further secure its base attachment to the Station. In contrast, the SRMS has only one end effector at its tip. Logistically, different SRMS are installed in different Shuttle Orbiters on a rotary basis to maximize their 10-yr lifetimes in space. They are inspected and serviced on the ground in between space missions. Designed for a 30-yr lifetime, the SSRMS is launched in one launch package; servicing is to be done on the Station. As a result, SSRMS elements need to be designed with modularity and commonality to allow for easy on-orbit replacement. Other differences will be discussed later.

The preceding highlights will be discussed in more detail based on the SRMS and the SSRMS designs and tests. The limited space in one chapter does not allow a full treatise of the entire srms design and tests. Instead, discussion will be focused on some selected topics only: kinematics, dynamics, control and tests of the srms. An overall system description will be given briefly in Sec. II to illustrate the srms's subsystems and their inter-relationships. Typical system parameters and characteristics will also be presented.

From a kinematics viewpoint, an srms arm is simply a mechanism. The kinematic relationship between joint movement and arm tip movement will be analyzed in Sec. III. Important issues will be addressed such as reference frame selection, joint placement, forward and inverse kinematics, arm singularities, etc. Arm tip positioning error estimates based on kinematic relations will be presented. In passing, a similar statics relationship between joint loads and arm tip loads will also be discussed.

Sec. IV begins with a description of typical srms mass and stiffness distributions. SRMS and SSRMS data will be used for illustration purposes. The discussion then continues with various srms dynamical problems associated with different srms operational scenarios. Different srms topologies will be included, ranging from open chain to closed loop. Typical srms dynamics characteristics will be presented using the equation of motion for an open-chain srms. In particular, inertial coupling between different degrees of freedom will be emphasized. A brief discussion will also be given to address the issues of dynamics analysis and computer simulation.

The srms's control system is presented in Sec. V. The control system hierarchy and its design philosophy will be described. At the arm control level, the description includes control laws, control features, and operation modes. The controls/displays device, system health monitor, and malfunction detection hardware and software will be described as well. At the joint control level, the description will cover the servo design and implementation including motors, joint encoders, tachometers, and resolvers.

Sec. VI deals with the ground tests and on-orbit tests required for the verification and certification of the srms's design. Emphasis will be placed on SRMS tests simply because the SSRMS testing is only in its planning phase at the time of writing. Also included are tests that are routinely performed on-orbit to demonstrate normal arm behavior. Test equipment, flight instrumentation, flight data analysis, and their role in the validation of srms simulators are some topics discussed in this section.

II.   Space Remote Manipulator System Description

An srms can be viewed as part of a three-body chain with the space vehicle at one end, a payload at the other, and the srms in between. Consequently, the srms must interface with the space vehicle system, and with the payload system. Compatibility across these interfaces must be established and maintained in srms external interface control documents (ICD). (See Ref. 1 for example.) Within the system design architecture, the functions of communications with ground control and power distribution are assigned to the space vehicle system. The srms can then be designed separately provided, of course, that its external interface requirements are satisfied. Typically, an srms is made up by the following subsystems: 1) mechanical subsystem, 2) electrical subsystem, 3) thermal subsystem, 4) displays and controls subsystem, 5) software subsystem, 6) computer subsystem, and 7) vision subsystem. These subsystems are connected to one another as illustrated in Fig. 3. In addition, ground support equipment (GSE) and flight support equipment (FSE) are also required for maintenance, preparation for launch, logistics, etc.

Some key srms design parameters are presented in Table 1 based on the requirements of the SRMS and SSRMS. For example, the SSRMS fail-operational requirement leads to the dual-channel (prime and redundant) design for most elements in the SSRMS, whereas the fail-safe requirement in the SRMS allows all elements to be designed on a single string basis. The SSRMS self-relocatable requirement leads to its unique configuration which is symmetrical with respect to the elbow located at the middle of the arm.

A.   Mechanical Subsystem

The mechanical subsystem consists of all structures within the srms's arm and its interfaces: joints, electronics housings, arm booms, and end effector(s). The joints, joint housings, and electronics housings are normally constructed of titanium alloy, aluminum alloy, and steel. Graphite/epoxy composite material is used for the booms to meet their stiffness budget and high stiffness per mass requirement. The end effectors are made of light gauge aluminum.

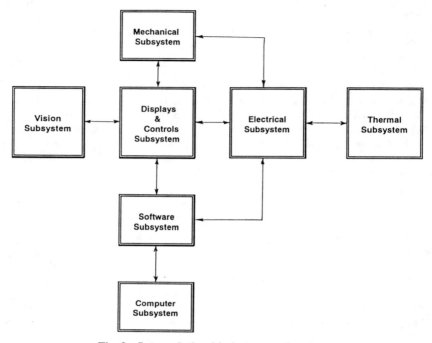

Fig. 3   Inter-relationship between subsystems.

Table 1   Design parameters

Design parameter	SRMS	SSRMS
Reliability requirement	Fail-safe	Fail-operational
Mobility	No requirement	Self-relocatable
Maximum payload (kg)	14,515	116,000
Stopping distance (m)	0.61	Payload and operation dependent (≤ 1.25)
Power budget (W)	1000 (operation)	915 (keep-alive)
	1050 (heaters)	2000 (operation)
On-orbit mass budget (kg)	450	1334
Stiffness budget based on straight-out arm configuration (N/mm)	1.35	0.67
Positioning accuracy budget (relative to a reference frame at arm base) (m and deg)	0.0508 and 1	0.065 and 0.7
Data transfer rate (ms)	once every 80	once every 50
Nominal bus DC voltage (V)	28	120

The joints are often identical thanks to their commonality design. The SRMS's joints have identical design, except its gear train, as shown in Table 2. The SSRMS joints are functionally identical with the same performance characteristics; however, its yaw joints are not interchangeable with the pitch and roll joints because of a slight difference in the yaw joint dimension. Figures 4–6 illustrate the SRMS shoulder, elbow and wrist joints; Fig. 7 shows the cross-section of an SSRMS joint. Each SRMS joint consists of a brushless dc motor, a pair of commutators, a high-speed gear train, an epicyclic low-speed output gear train, an optical encoder on the gearbox output shaft, a tachometer on the motor output shaft, and an electromechanical brake located on the motor output shaft. The motor has a fly wheel attached to its shaft; the inertia of the wheel is selected to meet the servo stability requirement. The brake is a spring-engaged friction type which releases when the dc coil is energised. The brake is automatically released when the motor is powered up, and engaged when the motor is powered down; it can also be engaged via a brake-on command.

To satisfy its fail-operational requirement, each SSRMS joint, however, has two sets of motors, brakes, joint resolvers and motor resolvers in its prime and redundant channels. Typical values for important joint parameters are shown in Table 2.

The arm booms are thin-walled tubular sections that are connected to the joint structure via end flanges.[2] The end flanges are made of aluminum alloy, bonded and bolted to the arm booms. Internal stabilization rings are provided in the arm booms to maintain ovalization frequencies above a design limit (180 Hz for the SRMS's booms). The rings also resist local shell buckling. The SSRMS launch packaging requirement leads to a rather unique configuration: each SSRMS boom

Table 2   Typical values for joint parameters

Parameter	SRMS			SSRMS
	Shoulder	Elbow	Wrist	All joints
Gear ratio	1842	1260	738	1845
Software travel limits (deg)	±177.4 (Yaw) 0.6, 142.4 (Pitch)	−157.6 −0.4	±116.4 (Yaw and pitch) ±442 (Roll)	±270
Torque capability (Nm)	1047 (min) 1570 (max)	716 (min) 1074 (max)	313 (min) 470 (max)	1044 (min) 2332 (max)
Brake slip torque (min) (Nm)	1113	780	403	1044
Backdrive threshold (Nm)	144	73	73	200
Max joint rate Unloaded arm (deg/s)	2.29	3.21	4.76	4 (software limit can be lower)
Arm + max payload	0.229	0.321	0.476	0.03

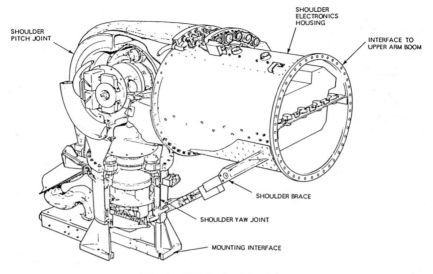

Fig. 4   SRMS shoulder joint.

is made of two segments hinged together. The two segments are folded and tied down to an integrated truss assembly during launch. When in orbit, the segments are untied and bolted together by an EVA astronaut.

The end effector is a motor-driven grappling device which can capture a payload by first snaring a grapple fixture on the payload, pulling it into the end effector, and finally rigidizing to secure the mating. The SRMS end effector and a grapple fixture are shown in Fig. 6. The end effector has three cutouts on its opening face to accommodate the grapple fixture cam during mating. This feature is designed to allow misalignment errors (up to 15 deg about the end effector longitudinal axis and 10 deg about a lateral axis) to be self-corrected during payload capture. The SRMS end effector comprises a motor, a brake and clutch associated with the actuation of the snare system, a brake and clutch associated with the actuation of the rigidization system, cable snares and drive train, and a rigidization carriage and drive train,

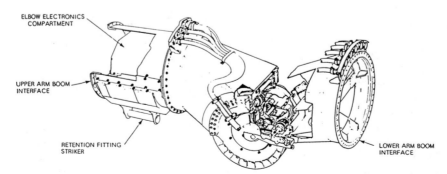

Fig. 5   SRMS elbow joint.

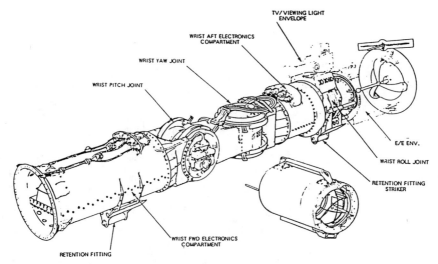

Fig. 6   SRMS wrist joint.

and a spring return device and clutch as a payload backup release mechanism. The SSRMS end effector as shown in Fig. 8 is mechanically equivalent to the SRMS end effector, i.e., compatible with the grapple fixtures used in the Shuttle program. It also has additional features: power and data transmission to its power and data grapple fixture (PDGF), a latching mechanism to further secure its mating to the PDGF, a force and moment sensor, and a redundant motor and control circuit. The SSRMS end effector is also known as the LEE.

B.   Electrical Subsystem

The electrical subsystem is responsible for distributing and managing power and data from the space vehicle to the srms's joints and end effector. In particular, the SSRMS is also capable of transferring up to 2500 W to a grappled payload via a PDGF if required.

Electrical power is supplied to the SRMS from the Orbiter Power system, fed from the Orbiter fuel cells through the SRMS/Orbiter interface. The SRMS electrical subsystem then distributes it along the arm by means of two main dc power buses, primary and backup, selectable from a panel on the Orbiter aft station. When backup is selected, power is supplied to the backup channels of the displays and controls subsystem and the backup arm based electronics (ABE). Similarly, electrical power is supplied to the SSRMS from the Space Station power system through a PDGF connected to its base LEE. It is then distributed along the arm via two power buses: prime and redundant, as illustrated in Fig. 9. A backup drive power and data bus is dedicated to the operation of the SSRMS in the backup drive mode. Power conditioning is performed locally within each SRMS and SSRMS unit.

The SRMS electrical subsystem consists of six units: the manipulator controller interface unit (MCIU), servo power amplifier (SPA), signal conditioning unit

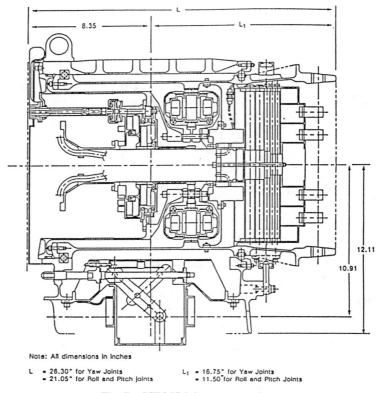

Note: All dimensions in inches

L = 26.30" for Yaw Joints   L₁ = 16.75" for Yaw Joints
   = 21.05" for Roll and Pitch Joints   = 11.50" for Roll and Pitch Joints

Fig. 7   SSRMS joint cross section.

(SCU), end effector electronics unit (EEEU), joint power conditioner (JPC), and backup drive amplifier (BDA). The MCIU handles information exchange between the Orbiter general purpose computer (GPC) and the SRMS. The SPA provides the signal to drive the joint motor, releases the brake when commanded, processes signals produced by the tachometer, encoder and commutator, transmits digital data to the MCIU, and performs self-testing using built-in test equipment (BITE). The SCU processes tachometer output for use in the SPA. Located within the SRMS end effector, the EEEU controls and monitors the end effector operation. It consists of a power conditioner, a command and commutator logic, a motor driver, and brake/clutch drive circuits. The EEEU also processes end effector status signals and transmits them to the MCIU. The JPC main function is to convert the +28 V dc bus to provide secondary regulated supply voltages to the SPA, motor commutators, and position encoders. The BDA is a backup unit to any of the SPAs in the event of a failure of an SPA or its associated JPC. Its main functions are to provide drive to any joint motor, as selected from the displays and controls panel; and to provide power conditioning from the backup +28 V bus. The SRMS has one MCIU, six SPAs, six SCUs, two JPCs, one EEEU, and one BDA.

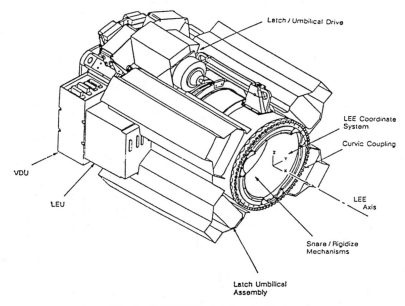

Latch / Umbilical Drive

LEE Coordinate System

Curvic Coupling

VDU

LEE Axis

LEU

Snare / Rigidize Mechanisms

Latch Umbilical Assembly

Fig. 8   SSRMS latching end effector.

In the SSRMS, the joint electronic unit (JEU), the LEE electronics units (LEU), and the backup drive unit (BDU) play the same roles as the SRMS SPA, SCU, JPC; EEEU; and BDA, respectively, as shown below:

SSRMS	←	SRMS
JEU	←	SPA + SCU + JPC
LEU	←	EEEU
BDU	←	BDA

The SSRMS arm computer unit (ACU) plays the same role as the Orbiter GPC plus MCIU in the SRMS. The SSRMS has two ACUs (1 prime and 1 redundant), two LEUs, one BDU and seven JEUs.

C. Thermal Subsystem

The srms thermal protection is passive in nature, supplemented by active devices. The passive devices are multilayer thermal blankets, radiators, white paint, thermal isolators, and thermal interface fillers. The active devices consist of heaters, thermostats, and thermistors.

The SRMS heater system is an active redundant system capable of supplying up to 1050 W to the SRMS arm; if a failure occurs on one power bus, the other is capable of supplying 525 W. The heaters can be in two states: off or auto, controlled by a switch on Panel A8A2 in the Orbiter aft station. In auto, the individual heaters are turned on and off by quad-redundant thermostats, designed to operate within the 10°C to +6°C temperature range. Fourteen thermistors capable of sensing

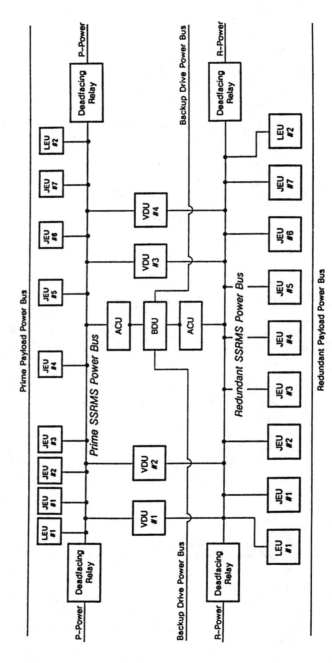

Fig. 9   SSRMS power distribution.

temperatures in the range of −40°C to +100°C are placed at its six joints and at the end effector. Caution annunciators are issued by the Orbiter GPC when the sensed temperatures lie outside their allowable limits which can be specified from a keyboard while on orbit. The normal limits are −30°C and 62°C; the upper limit is based on the worst hot case in which the SRMS has continuous operation for two hours. After the caution has been annunciated, the SRMS arm can still be operated for another 15 min before it has to be shut down.

The SSRMS heater system is similar with the following exceptions:

1) Prime and redundant power buses supply power to prime and redundant heaters respectively.

2) Both prime and redundant power buses will be on in the SSRMS keep-alive state, whereas in the SSRMS operational state only one power bus will be on.

3) There are units external to the arm structure which need to be thermally protected, e.g., the ACU, the BDU, and the LEU.

4) Each heater has its own set point.

5) The minimum operating temperature for heaters, thermistors, and the thermostat are lower (−50°C).

D.   Displays and Controls Subsystem

The displays and controls subsystem provides the essential interface between the srms's operator and other subsystems. The SRMS's displays and controls subsystem consists of a 3-DOF translational hand controller, a 3-DOF rotational hand controller, a keyboard, a controls monitor, two closed circuit television (CCTV) monitors, and a displays and controls panel. Figure 10 shows the SRMS's displays and controls panel, which is located in the Orbiter aft station.

The control functions associated with the displays and controls panel are arm operation mode selection, end effector mode selection, mode entry control, braking, safing, single joint rate command, and backup mode commands. When the automatic mode is selected, the SRMS's operator can enter a desired position and orientation of the point of resolution (POR) or call up a prestored set of POR position and orientation via the keyboard. The POR is a point fixed to, but not necessarily physically on, a payload selected for control and display purposes. When the manual augmented mode is selected, the SRMS operator commands the POR rate using the hand controllers. A coarse/Vernier switch and a rate hold switch are built into the rotational hand controller. In either single or direct drive mode, the joint command is effected by a three-way switch on the displays and controls panel. The end effector operation can be commanded in either the auto or manual mode via a switch on the panel. In its manual mode, the end effector can be rigidized or derigidized manually via a three-way switch. Payload capture can be commanded using a switch built into the rotational hand controller.

The displays and controls panel provides the following displays: arm mode status indication; automatic sequence status indications; end effector meter indicating commanded rate vs actual rate; end effector rate flags; brake status indication; end effector status indication; three numeric readouts which can be selected for POR position, orientation, linear velocity, angular velocity, or joint angle; as well as caution/warning annunciators and master alarm.

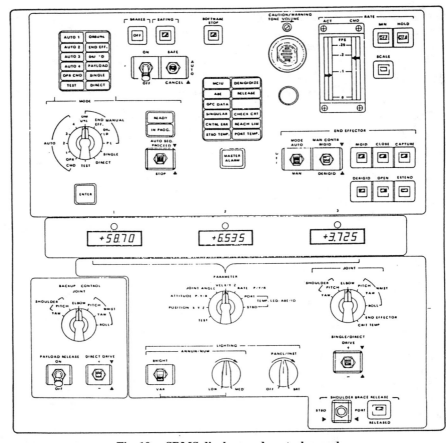

Fig. 10   SRMS displays and controls panel.

The SSRMS's displays and controls subsystem has three CCTV cathode ray tubes (CRTs), a displays and controls monitor screen, a mouse, a keyboard, and a 6-DOF hand controller. Different pages of displays and controls can be called up on the monitor screen. Commands are effected either via a contact to the touch screen or by selecting a soft switch displayed on the screen. It is to be noted that at the time of writing, the SSRMS human-machine interface function is under development and some features mentioned in this paragraph might be different in the final design. Unlike its SRMS counterpart, the SSRMS displays and controls subsystem does not have a fixed number of components and physical location. For example, the SSRMS can be commanded from the ground control station in which case hand controller operations are prohibited; hence the hand controller is not needed for the ground control work station. In orbit, the SSRMS can be commanded from a Space Station intravehicular activity (IVA) multipurpose application console (MPAC). The SSRMS's soft displays and controls are designed to accommodate the many SSRMS design requirements which did not exist in the

SRMS, e.g., graphic displays, more control features and system feedback flags, arm control from different work stations, etc.

E. Software Subsystem

The software subsystem consists of a number of principal functions which perform the required mathematical and logical operations to support the srms operation. These functions are usually described in detail in a functional subsystem software requirements (FSSR) document (see Gift et al.[3] for example). They can be modified from time to time to introduce new features or improvements provided that the software change request is approved by a NASA-controlled software control board. Detailed function descriptions in the FSSR document are supplemented by flow charts, tables of design constants, initialization (I)-load parameters, and table-maintenance (T-M) specable parameters.

At the highest level, an executive (EXEC) function specifies srms software modes and permits mode transitions provided that prerequisites are met. Depending on the software mode, the EXEC calls a specific set of principal functions in a particular sequence to control and monitor arm motion. Some typical principal functions are

1) coordinate transformation function (to transform data in different coordinate systems), 2) resolved rate function (to resolve an arm tip rate command into appropriate joint rate commands), 3) kinematic data generator function (to compute kinematic parameters for controls and displays purposes), 4) automatic principal function (to control the arm along an automatic trajectory), 5) single control function (to control the arm motion in a joint-by-joint basis), 6) manual augmented function (to process hand controller outputs for arm tip rate commands), and 7) system health monitoring function.

Figure 11 shows the SRMS's principal functions and the data flow between them. Although the SSRMS software subsystem is under development, it is expected to be similar to its SRMS counterpart, but with more complicated algorithms and logics to handle SSRMS-specific design features.

F. Computer Subsystem

The SRMS does not have a computer of its own. All data processing is carried out by one of the Orbiter GPCs. The MCIU, which is part of the electrical subsystem, however, contains a microprocessor that handles the data communication between the MCIU and the GPC, the ABE and the displays and controls panel.

The SSRMS has a computer subsystem distributed over three different units (ACU, LEU, and JEU), each of which is dedicated to special processing functions. At the heart of the subsystem is the ACU, which is responsible for arm mode control, arm tip position and rate control, arm health monitoring/safing, force/moment accommodation, data collection, local bus control, and video distribution unit (VDU) control. The LEU performs force moment sensor processing, monitors LEE health and status, and controls the LEE operations. The JEU provides joint position and rate control, and monitors joint health and status. The ACU communicates with the LEU, the JEU, as well as its higher level controller, the MSS computing and control facility.

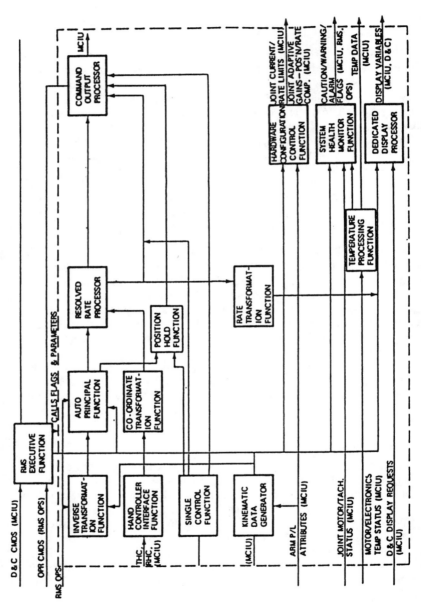

Fig. 11   SRMS principal functions.

G. Vision Subsystem

The SRMS vision subsystem consists of a CCTV camera and lighting unit mounted on the roll section of its wrist roll joint, and a CCTV camera with a pan and tilt unit mounted on the lower arm boom near its elbow joint. The cameras are identical monochrome cameras featuring a remotely controlled zoom lens. At minimum zoom, the effective field of view as seen on the monitor display is 30 deg vertical, and 42.6 deg horizontal. The pan and tilt unit can be controlled within ±70 deg in elevation (tilt) and ±170 deg in azimuth (pan). In addition, Orbiter CCTV cameras and lights located in the cargo bay can also be used to assist SRMS operations such as cradling/uncradling, and payload berthing/deberthing.

Similar to the SRMS, the SSRMS vision subsystem consists of two camera/light units mounted on each LEE, two camera/light units mounted on pan and tilt units located on the two arm booms as shown in Fig. 2 and four VDUs mounted in the vicinity of each camera/light unit. All cameras are identical color cameras with maximum zoom ratio 8.5:1, and maximum wide field of view of 52 deg. The pan and tilt unit can be controlled within ±90 deg in tilt and 0–350 deg in pan. The light unit is capable of providing a minimum illumination level of 380 ft-c at a distance of 0.68 m, and 3.5 ft-c at a distance of 10 m. Each VDU transfers power and command from the ACU to its associated light, camera, pan, and tilt unit. It receives video signals and status, e.g. camera setting, pan and tilt angles, prior to distributing them onto SSRMS and Space Station video lines.

In addition, the mobile servicing system also has an artificial vision unit (AVU), which can perform image processing and camera control functions to support SSRMS and SPDM operations such as payload tracking, capturing, and berthing. Located in the Space Station pressurised environment, the AVU generates image processed output video for display to the IVA operator. It also processes the video images from a camera to provide payload position, attitude, and rates to the SS-RMS, the IVA operators, and camera controls to the Space Station communications and tracking.

III. Kinematics

Topologically, the srms is always an open chain in all operation scenarios. There are instances in which an srms can form a closed chain with other objects, for example, the SRMS grappling a payload latched to the Orbiter cargo bay, or the SRMS handing over its payload to the SSRMS while the Orbiter is attached to the Station. In these instances, the srms itself retains its open-chain topology. Consequently, an open-chain model such as in Fig. 12 is appropriate for describing the trajectory of the srms tip relative to its base.

A. Forward Kinematics

From a kinematics viewpoint, the srms chain can be represented by n, joints which are connected to one another by rigid links. The SRMS has six active revolute joints and one passive joint, which is located between the first SRMS active joint and the Orbiter longeron. Known as the swing-out joint, the passive joint, is designed to keep the stowed SRMS in the swung-in position allowing the Orbiter cargo bay door to be closed for launch and landing. It is locked in the

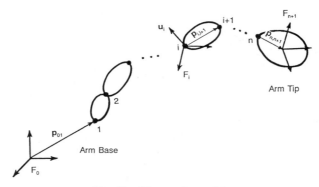

Fig. 12   Kinematic model.

swing-out position during SRMS operations to reduce interference of the Orbiter structure with the SRMS's reach envelope, thus gaining more operation space for the SRMS. The SSRMS has seven active revolute joints, any one of which can be selected to have its position maintained by its (joint) control system.

Let the joints be numbered from 1 to n, starting from the srms base as shown in Fig. 12. At each joint i, an outboard reference frame F_i is selected such that it is fixed to the link connecting joint i to joint $i + 1$, and one of its axes coincides with the joint drive axis. On the other side of joint i, an inboard reference frame F_i is selected such that it is fixed to the link connecting joint i to joint $i - 1$, and one of its axes coincides with the joint drive axis. For revolute joints, the origins of F_i and F_i coincide; the two axes in F_i and F_i, that coincide with the joint drive axis, also coincide. The joint angle α_i is defined as the angle by which F_i is rotated about the joint drive axis to coincide with F_i. For srms controls and displays purposes, two reference frames F_0 and F_{n+1} are selected: the base frame F_0 being fixed to the space vehicle and the tip frame F_{n+1} being fixed to the POR. The srms's motion can then be described by the translation and rotation of F_{n+1} relative to F_0. The position and orientation of F_{n+1} relative to F_0 are defined by

$$p_{0,n+1} = \sum_{i=1}^{n+1} R_{0,i-1} \, p_{i-1,i} \tag{1}$$

$$R_{0,n+1} = \prod_{i=1}^{n+1} R_{i-1,\hat{i}} \, R_{\hat{i},i} \tag{2}$$

$$R_{\hat{i},i} = \mathbf{1} \cos\alpha_i + (1 - \cos\alpha_i)u_i u_i^T + u_i^\times \sin\alpha_i \tag{3}$$

where $p_{0,n+1}$ is the position vector of the POR relative to F_0, expressed in F_0; $p_{i-1,i}$ is the components in F_{i-1} of the vector joining the origin of F_{i-1} to the origin of F_i; $R_{0,i-1}$ is the rotation matrix that transforms the components of a vector in F_i into those in F_0; $R_{i-1,i}$ is the rotation matrix that transforms the components of a vector in F_i into those in F_{i-1}; $R_{\hat{i},i}$ is the rotation matrix that transforms the components of a vector in F_i into those in $F_{\hat{i}}$; u_i is the unit vector defining the drive axis of joint i, expressed in F_i; $\mathbf{1}$ is the 3×3 unit matrix; and $(\cdot)^\times$ is the matrix representation of the vector cross product. The SRMS and SSRMS POR orientations are actually represented by Euler angles based on a pitch, yaw, roll Euler sequence. The orientation angles are extracted from $R_{0,n+1}$.

When all joint angles are zero, the srms is said to be in its reference configuration. Furthermore, an srms joint is called a roll, or pitch, or yaw joint depending on whether its drive axis u_i is aligned with the roll, or pitch, or yaw axis of the space vehicle when the srms is in its reference configuration. The joint placement in the SRMS's and the SSRMS's arms is similar: two roll joints at the two ends of the chain (the SRMS's swing-out joint is a roll joint), followed by two yaw joints in the shoulder and in the wrist, and three pitch joints in between. The three wrist joints are designed mainly for reach in orientation; the shoulder yaw and pitch joints provide reach in azimuth and elevation; and the elbow pitch joint allows the arm to reach the work space between its shoulder and wrist. The two links connecting the three pitch joints define the so-called arm pitch plane . It is interesting to note that except for the wrist roll joint, all other SRMS's joints always remain in the SRMS's arm pitch plane. The SRMS's elbow joint has a dog-leg design to allow for its lower angular travel limit of -160 deg (cf, software stop limit of -157.6 deg in Table 2). In contrast to the SRMS's in-line joint design, the SSRMS's features an offset joint design which is essential to provide wide angular travels from -270 deg to $+270$ deg as shown in Table 2.

In Fig. 12, a joint is represented as a point which can arbitrarily lie anywhere along the joint drive axis within the joint housing structure. This representation allows analysts to conveniently select the SRMS and SSRMS joint locations such that the link vector $p_{i-1,i}$ has one or two zero elements, leading to a simple expression for $p_{0,n+1}$ when Eq. (1) is expanded. Similarly, the orientation of the F_i axes can be selected in different ways. Since the drive axes of the roll, pitch, and yaw joints are always normal to each other, the simplest choice is to keep F_i parallel to F_{i-1}; i.e., $R_{i-1,\hat{i}} = 1$.

Kinematically, the link vector $p_{i-1,i}$ is time invariant; so is the drive axis unit vector u_i, leaving the joint angles α_i to be the only time variant parameters. Let v_{n+1} and ω_{n+1} be the POR linear and angular velocities as seen in F_0, then differentiating Eqs. (1) and (2) with respect to time and realizing

$$\omega_i^\times = -\dot{R}_{0,i} R_{0,i}^T \qquad (4)$$

one finds

$$\begin{pmatrix} v_{n+1} \\ \omega_{n+1} \end{pmatrix} = J(\alpha)\dot{\alpha} \qquad (5)$$

where J is the $6 \times n$ Jacobian matrix; and α is the $n \times 1$ is the column matrix of joint angles. The set of equations (1), (2) and (5) represents the srms's forward kinematics. They are used in the SRMS's and SSRMS's kinematic data generator functions to compute the POR position, orientation, and rates for displays purposes. The Jacobian matrix J has the simplest form when it is expressed in the shoulder yaw joint frame instead of F_0. The shoulder yaw frame has two of its axes parallel to the arm pitch plane; thus the POR velocities v_{n+1} and ω_{n+1} as seen in the shoulder yaw frame would have components parallel and normal to the arm pitch plane.

At this point it is interesting to note a special relationship which has an expression similar to Eq. (5) although it is not part of the srms's kinematics. Consider the srms in static equilibrium.

Let τ = the $n \times 1$ column matrix representing the drive-axis torques at the n joints, expressed in the local joint reference frames F_i. Let f = the 3×1 column matrix representing a force applied at the POR, expressed in the base frame F_0; and let t = the 3×1 column matrix representing a torque applied at the POR, expressed in the base frame F_0.

Then the relationship between τ and f, t can be derived based on the position vectors from the joints to the POR and the rotation matrices R_{0i}; or it can be obtained using the method of virtual work: the virtual work done by the joints ($\tau^T \delta \alpha$) must be equal to the virtual work done by the applied force f and torque t ($f^T \delta p_{0,n+1} + t^T \delta \theta$ where θ are the attitude angles defined by ($R_{0,n+1}$). It then follows from Eq. (5) that

$$\tau = J^T \begin{pmatrix} f \\ t \end{pmatrix} \tag{6}$$

B.   Inverse Kinematics

Prior to placing the POR at some desired position and orientation, it is desirable to know whether such a position and orientation can be reached by the srms. This can be posed as the inverse kinematics (resolved position) problem: Find the joint angles α, that satisfy Eqs. (1) and (2) subject to the constraint that each joint angle α_i is within its travel limits. Since Eqs. (1) and (2) are transcendental equations, finding the solution to the preceding problem is not a trivial task. Furthermore, the problem can have more than one solution even for a six-jointed srms. Different numerical techniques have been proposed ranging from least-squares solution to numerical simulation which simulates a straight-line trajectory between the initial and the desired POR positions and orientations.

The solution to the SRMS's inverse kinematics has an interesting history of development. The limited computation capability of the Orbiter GPC precludes iterative and other time consuming numerical techniques (e.g., matrix inversion) in the SRMS kinematic algorithms. Such a constraint thus became a challenge to find an analytical solution to the SRMS's inverse kinematics problem. At first glance, finding an analytical solution to the resolved position problem did not seem to be possible because the SRMS wrist joints do not have the characteristics of a spherical joint; i.e., the wrist pitch joint drive axis does not pass through the intersection of the wrist yaw and wrist roll drive axes. However, thanks to the observation that five SRMS joints always lie in the arm pitch plane, a simple analytical solution was discovered: the POR position and orientation is first transformed into the position of the wrist yaw joint and the orientation of the arm pitch plane; the latter provides the solution for the shoulder yaw joint angle, and the former helps determine the three pitch joint angles. The wrist yaw and wrist roll joint angles are determined next. The conditions for zero, one, and more than one solution(s) are identified analytically. Details of this analytical solution can be found in the inverse transformation function, see Fig. 11 and Gift et al.[3] No attempt has been made to find a corresponding analytical solution for the SSRMS mainly because it would be more difficult to find an analytical solution for a seven-jointed srms, and also because there is no similar constraint on the ACU computation capability.

The srms POR rates can be determined from its joint rates as shown in Eq. (5). Conversely, the joint rates can be manipulated to provide a desired set of POR rates, which poses the inverse kinematics (resolved rate) problem: for a given arm configuration defined by α, find the joint rates $\dot{\alpha}$ such that the desired POR rates v_{n+1} and ω_{n+1} can be achieved. The solution to this problem is simple:

$$\dot{\alpha} = J^{\oplus} \begin{pmatrix} v_{n+1} \\ \omega_{n+1} \end{pmatrix} + \mu \dot{\alpha}_N \tag{7}$$

where $J^{\oplus}$ is the right inverse of J, $\dot{\alpha}_N$ is a vector in the null space of J, and μ is a scalar.

The SRMS Jacobian is a 6×6 matrix and the second term on the right-hand side of Eq. (7) vanishes. Because of the computation constraint of the Orbiter GPC mentioned earlier, the SRMS resolved rate solution is obtained analytically via a simple transformation which reduces the original six-dimensional problem to two uncoupled three-dimensional problems. More precisely, one reduced problem involves three pitch joint rates (which provide the arm pitch in-plane motion) and the other involves the roll and yaw joint rates (for out-of-plane motion). The SRMS's Jacobian matrix has three singularities:

1) Shoulder yaw singularity: the wrist yaw joint is along the shoulder yaw joint drive axis; no POR rotation about the shoulder yaw joint drive axis is possible.

2) Elbow singularity: the three pitch joints are collinear; the arm is stretched out to its limit, no further stretching is possible.

3) Wrist yaw singularity: wrist yaw joint angle is ± 90 deg, a gimbal lock situation; the wrist roll joint then behaves as a pitch joint.

When these singularities occur, no joint rates can be found to satisfy the desired POR rates. Special algorithms known as the singularity management algorithms are designed to keep the SRMS operation away from its singularities. Details of the preceding analytical solution can be found in the resolved rate processor function, see Gift et al.[3] and Fig. 11.

Using a similar transformation, the SSRMS resolved rate solution is also obtained analytically. The original seven-dimensional problem is reduced to a three-dimensional problem and a four-dimensional problem. The null space solution is associated with the reduced 3×4 Jacobian matrix, which is converted to its equivalent in the seven-dimensional space in the final solution. The null space solution can be used to provide the proper solution for special augmented cases. For example, an SSRMS joint can be selected to have 0 rate; the scalar μ can then be selected such that the null space solution for that joint would cancel the corresponding solution from the right inverse of J. As another example, in an arm pitch plane change maneuver, both ends of the SSRMS are held fixed by specifying $v_{n+1} = \omega_{n+1} = \mathbf{0}$; the elbow joint can still translate along a direction normal to the arm pitch plane. The translational rate of the elbow joint can be regarded as an augmented linear constraint to Eq. (5) to form seven linear equations for seven joint rates; or it can be used to determine the scalar μ in Eq. (7). In other words, Eq. (7) is the general solution which can be used to accommodate additional linear constraints to the SSRMS's kinematics.

The SSRMS's Jacobian has five singularities which correspond to det $J^T = 0$:

a) The wrist yaw joint angle is ± 90 deg *and* the shoulder yaw joint is in a plane normal to the arm pitch plane and containing the wrist yaw joint drive axis.

b) The shoulder yaw joint angle is ± 90 deg *and* the wrist roll joint is in a plane normal to the arm pitch plane and containing the shoulder yaw joint drive axis.

c) The elbow joint angle is 0 deg, or ± 180 deg corresponding to straight-out and folded arm configurations;

d) *Both* shoulder yaw and wrist yaw joint angles are ±90 deg.

e) The wrist roll joint is along an axis going through the shoulder yaw joint and normal to the arm pitch plane.

The similarity between the SRMS and SSRMS singularities can be readily recognized: a and b correspond to 1, c corresponds to 2; and d corresponds to 3. The two singularities e and c (with elbow joint angle $= \pm 180$ deg) have no correspondence in the SRMS because of the SRMS's joint travel limits.

The srms's singularities have some interesting characteristics. As the arm approaches its singularity, the joints need to speed up to maintain the same speed at its POR. In a similar manner some joint torques also need to be high to balance the applied force and torque at the POR, as a direct consequence of Eq. (6). In the joint (angle) space, the singularity behaves as a boundary separating different families of joint (angle) trajectory. A joint (angle) trajectory is simply the solution to the nonlinear ordinary differential equation (5) for a given initial condition $\alpha(0)$ with the POR rates being some functions of time. If the initial arm configuration is not near a singularity, then for slightly perturbed $\alpha(0)$, all perturbation trajectories converge to a unique joint trajectory. However, when $\alpha(0)$ corresponds to a singularity, then a perturbation on $\alpha(0)$ can lead to completely different joint trajectories. Repeatability of joint trajectory therefore can be ensured only when the starting point of the trajectory is away from a singularity.

The preceding forward and inverse kinematic relationships discussed so far are general enough for applications other than the SRMS and SSRMS. In fact, several general purpose kinematics software programs have been developed together with a user-friendly graphics package to allow users to quickly test different manipulator designs. One popular graphics package is the commercially available Interactive Graphics Robot Instruction Program (IGRIP). However, all general purpose programs have a common drawback: the algorithms coded are either not transferrable from the program to an application software, or not optimised for a particular application (for example, numerical pseudoinverse $J^{\oplus}$ solution in a general purpose software vs analytical inverse solution).

C. Error Analysis

Up to this point, all kinematics parameters $p_{i-1,i}$, u_i, α_i have been treated as srms design parameters and their design (i.e., specification) values have been implicitly assumed. In reality, these parameters might have values different from their specification values for a number of reasons such as manufacture imperfections, assembly errors, thermal effects, etc. The joint angle α_i might also contain an error which is associated either with the joint encoder accuracy or with the joint control system error. The contribution of these errors to the POR position and orientation can be determined easily by partial differentiations of Eqs. (1) and (2) with respect to each parameter. If the errors are known, e.g., by measurements, then their total contribution to the POR position and orientation errors can be determined by summing the individual contributions algebraically.

At the beginning of an srms's development, such possible errors can only be estimated based on current technologies and past experience. At the same time, it is desirable to establish the srms's positioning accuracy which is used to drive the system element designs. If these errors are assumed to have the same sign, or to have signs that would add up their contributions to the POR errors, then the algebraic sum of POR errors will be unrealistically large. However, if they are considered to be statistically independent, then their contributions to the POR errors can be added in an rss sense to yield a more realistic POR positioning error. In this regard, errors due to thermal effects cannot be assumed to be statistically independent, whereas it is legitimate to consider mechanical errors and control system errors to be statistically independent. On the other hand, error estimates are often carried out for the worst case conditions. In this connection, the straight arm configuration is chosen for estimating the arm positioning accuracy, for it can be demonstrated that the arm tip position error is at its maximum when the arm is straight out.

IV. Dynamics

The dynamics of the srms is excited either internally by its control system actions, or externally by the space vehicle propulsion system/attitude control system activities. The environmental disturbance forces, such as aerodynamics, gravity gradient, etc., are much less significant than the control system forces. Thus, there are mainly two concerns that require analyses: controller-arm structural dynamics interactions and controller-controller interactions. Only the former is discussed in this section with emphasis on the effect of control system activities on the arm structural dynamics.

A.   Inertial and Structural Properties

The SRMS and SSRMS arms do not have uniform mass and stiffness distributions. In fact, most of their masses are concentrated at the joints, and the joint structural flexibilities contribute a major portion to the overall arm flexibility. From a global dynamics viewpoint, the srms arm can be represented by a number of structural bodies, B_i, each of which represents the structure between two consecutive joints. For an n-jointed arm, there are $n + 1$ bodies. Except the base and tip bodies, each body B_i has two connections to its adjacent bodies B_{i-1} and B_{i+1}: one at the inboard joint and the other at the outboard joint. The inboard joint is defined as the joint which is closer to the arm base when the arm is straight out. The dynamical characteristics of a body are defined by its inertial (also known as rigid-body mass) and structural properties in joint dynamics reference frames. Similar to the reference frame F_i defined in Sec. III.A, the joint dynamics reference frame is body-fixed, with its origin located at the body inboard joint which may or may not coincide with the origin of F_i defined in Sec. III.A. Figures 13 and 14 show the SRMS's and SSRMS's joint dynamics reference frames (shown offset from the joints for clarity); they also identify the bodies B_i for the SRMS and SSRMS models.

The inertial properties of each body are defined in terms of its mass m_i, mass center location c_i, and second mass moments about its mass center I_i, expressed in each body's joint dynamics reference frame. The inertial properties of the SRMS and SSRMS are presented in Table 3. Also included in Table 3 are the position vectors $p_{i,i+1}$, which define the location of the outboard joint in each body relative to the joint dynamics reference frame.

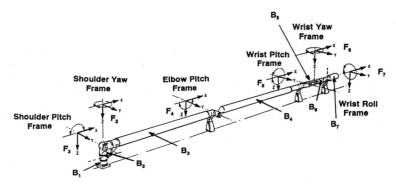

Fig. 13   SRMS joint dynamics reference frames.

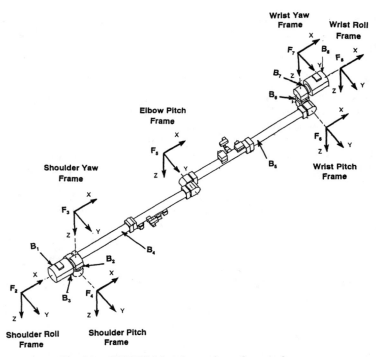

Fig. 14   SSRMS joint dynamics reference frames.

Table 3   SRMS and SSRMS inertial properties

Body	SRMS m_i(kg)	SRMS c_i(m)	SRMS I_i(kg-m^2)	SRMS $p_{i,i+1}$(m)	SSRMS m_i(kg)	SSRMS c_i(m)	SSRMS I_i(kg-m^2)	SSRMS $p_{i,i+1}$(m)
1	63.24	0	$I_{xx}=3.664$	0	243.66	0.669	$I_{xx}=9.336$	1.3589
		0	$I_{yy}=3.803$	0		0	$I_{yy}=44.413$	0
		0.2856	$I_{zz}=0.686$	0.5712		0	$I_{zz}=44.413$	0
			$I_{xy}=0.0044$				$I_{xy}=0$	
			$I_{xz}=0.0047$				$I_{xz}=0$	
			$I_{yz}=0.506$				$I_{yz}=0$	
2	32.12	0	$I_{xx}=1.115$	0	105.98	0	$I_{xx}=12.19$	0
		0	$I_{yy}=1.417$	0		0	$I_{yy}=12.13$	0
		0.1524	$I_{zz}=0.595$	0.3048		0.2983	$I_{zz}=3.061$	0.7026
			$I_{xy}=-0.019$				$I_{xy}=0$	
			$I_{xz}=-0.080$				$I_{xz}=0$	
			$I_{yz}=-0.037$				$I_{yz}=0$	
3	139.35	3.1883	$I_{xx}=3.225$	6.377	105.98	0	$I_{xx}=8.088$	0
		0	$I_{yy}=998.63$	0		-0.2316	$I_{yy}=3.061$	-0.5692
		0.0282	$I_{zz}=997.18$	0.1524		0	$I_{zz}=8.446$	0
			$I_{xy}=-1.928$				$I_{xy}=0$	
			$I_{xz}=9.185$				$I_{xz}=0$	
			$I_{yz}=-0.0302$				$I_{yz}=0$	
4	87.42	3.53	$I_{xx}=1.590$	7.06	314.88	3.55	$I_{xx}=15.41$	7.11
		0	$I_{yy}=632.90$	0		-0.08	$I_{yy}=2094.71$	-0.475
		-0.0066	$I_{zz}=632.51$	–	0	$I_{zz}=2103.19$	0
			$I_{xy}=0.8021$	0.1524			$I_{xy}=49.52$	
			$I_{xz}=-3.786$				$I_{xz}=0$	
			$I_{yz}=-0.080$				$I_{yz}=0$	

(Table 3 continued on next page.)

Table 3   (continued) SRMS and SSRMS inertial properties

Body	SRMS				SSRMS			
	m_i (kg)	c_i (m)	I_i (kg·m²)	$p_{i,i+1}$ (m)	m_i (kg)	c_i (m)	I_i (kg·m²)	$p_{i,i+1}$ (m)
5	8.48	0.2286	$I_{xx} = 0.0893$	0.4572	279.2	3.55	$I_{xx} = 9.522$	7.11
		0	$I_{yy} = 0.402$	0		0	$I_{yy} = 1966.28$	0
		0	$I_{zz} = 0.402$	0		0	$I_{zz} = 1966.28$	0
			$I_{xy} = 0$				$I_{xy} = -39.95$	
			$I_{xz} = 0$				$I_{xz} = 0$	
			$I_{yz} = 0$				$I_{yz} = 0$	
6	45.94	0.381	$I_{xx} = 0.408$	0.762	105.98	0	$I_{xx} = 8.305$	0
		0	$I_{yy} = 4.68$	0		-0.337	$I_{yy} = 3.061$	-0.5692
		0	$I_{zz} = 4.68$	0		0	$I_{zz} = 8.386$	0
			$I_{xy} = 0$				$I_{xy} = 0$	
			$I_{xz} = 0$				$I_{xz} = 0$	
			$I_{yz} = 0$				$I_{yz} = 0$	
7	45.21	0.3302	$I_{xx} = 1.903$	0.6604	105.98	0	$I_{xx} = 12.13$	0
		0	$I_{yy} = 3.816$	0		0	$I_{yy} = 12.13$	0
		0	$I_{zz} = 2.823$	0		-0.404	$I_{zz} = 3.061$	-0.7026
			$I_{xy} = 0.385$				$I_{xy} = 0$	
			$I_{xz} = 0.966$				$I_{xz} = 0$	
			$I_{yz} = 0.353$				$I_{yz} = 0$	
8	N/A	N/A	N/A	N/A	243.66	0.669	$I_{xx} = 9.336$	1.3589
						0	$I_{yy} = 44.41$	0
						0	$I_{zz} = 44.41$	0
							$I_{xy} = 0$	
							$I_{xz} = 0$	
							$I_{yz} = 0$	

The srms's structural properties are defined primarily by the bending and torsional stiffnesses of its structural elements; other secondary properties, such as
elongation, shear and rotary inertia effects, can be neglected. The srms's structural
elements consist of the base support structure (mechanical positioning mechanism
for the SRMS; the mobile transporter plus mobile remote servicer base system
plus PGDF plus shoulder LEE for the SSRMS), the joints, the gearboxes, the arm
booms, and the end effector (plus grapple fixture). Except the gearboxes, other elements can be represented by their linear stiffnesses as shown in Table 4. Based on
ground tests and flight data, and neglecting hysteresis, the gearbox characteristics
can be represented by the following nonlinear function $g(\delta)$:

$$t_g = g(\delta) \, \text{sign} \, (\delta)$$

$$g(\delta) = \begin{array}{ll} (t_\Delta/\Delta^2)\delta^2 & |\delta| \leq \Delta \\ k_g(|\delta| - \Delta) + t_\Delta & \text{otherwise} \end{array}$$

where t_g is the gearbox output torque; δ the gearbox twist angle defined to be the
motor shaft angle—joint angle × gear ratio; Δ the boundary on δ, beyond which
the gearbox stiffness becomes linear; t_Δ the gearbox output torque at Δ; and k_g
the gearbox linear stiffness.

For a given arm configuration, linearizing the gearbox stiffness at a value of
δ, the srms's base-to-tip stiffness can be determined from the flexibility matrix F
given here. Let m represents the number of structural elements in the srms;

F_i is the 6 × 6 matrix representing the structural flexibility between the tip and
the base of element i, expressed in a reference frame F_i fixed in element i;

F is the 6 × 6 matrix representing the structural flexibility between the tip and
the base of the srms, expressed in a reference frame F_0 located at the srms base;

p_i represents components in F_i of the vector from the tip of element i to the tip
of the srms;

R_{i0} is the 3 × 3 rotation matrix that transforms the components of a vector in
F_0 into F_i; and

T_i is the 6 × 6 transformation matrix.

$$T_i = \begin{pmatrix} R_{i0} & O \\ p_i^\times R_{i0} & R_{i0} \end{pmatrix}$$

where $(.)^\times$ is the matrix representation of the vector cross-product. Then since the
elements are in series, it can be shown that

$$F = \sum_{i=1}^{m} T_i^T F_i T_i \qquad (8)$$

B.   Dynamics Problems

The simplest dynamics problem involves a passive arm subjected to external
loads through its interface(s) with the space vehicle. By passive, we mean the
condition where the srms control system is inactive, e.g., an arm having brakes
applied. This problem occurs when the srms is launched, or stowed, or parked
on orbit. These scenarios are characterised by the number of interfaces between
the arm and the space vehicle: one for the srms in parking configuration (srms in

Table 4   SRMS and SSRMS structural properties

Element	SRMS	SSRMS
Base structure (Nm/rad)/m	(Mechanical positioning mechanism)	(Mobile transporter plus mobile remote servicer base system plus PDGF plus shoulder LEE)
	$K_{\theta,x} = 3.28 \times 10^5$	$K_{\theta,x} = 8 \times 10^5$
	$K_{\theta,y} = 1.29 \times 10^6$	$K_{\theta,y} = 8 \times 10^5$
	$K_{\theta,z} = 2.35 \times 10^5$	$K_{\theta,x} = 6.93 \times 10^5$
Shoulder joint housing (Nm/rad)	Drive axis[a]: $K_{\theta,z} = 6.5 \times 10^6$	Drive axis[b]: $K_{\theta,z} = 1.08 \times 10^6$
	Cross axis[a]: $K_{\theta,x} = 3.51 \times 10^6$	Cross axis[b]: $K_{\theta,x} = 1.32 \times 10^6$
	$K_{\theta,y} = 2.93 \times 10^6$	$K_{\theta,y} = 10^6$
Shoulder joint gearbox		
(Nm/rad)	$K_g = 1.6 \times 10^6$	$K_g = 1.33 \times 10^6$
(rad)	$\Delta = 9.16 \times 10^{-4}$	$\Delta = 9.17 \times 10^{-4}$
(Nm)	$t_\Delta = 424$	$t_\Delta = 424.5$
Upper arm boom	EI $= 4.2 \times 10^6$	EI $= 3.8 \times 10^6$
(Nm²)	GJ $= 2.1 \times 10^6$	GJ $= 1.93 \times 10^6$
Elbow joint housing (Nm/rad)	Drive axis: $K_{\theta,y} = 1.56 \times 10^6$	Drive axis: $K_{\theta,y} = 1.08 \times 10^6$
	Cross axis: $K_{\theta,x} = 8 \times 10^5$	Cross axis: $K_{\theta,y} = 1.32 \times 10^6$
	$K_{\theta,z} = 1.22 \times 10^6$	$K_{\theta,x} = 10^6$
Elbow joint gearbox		
(Nm/rad)	$K_g = 2.58 \times 10^6$	$K_g = 1.33 \times 10^6$
(rad)	$\Delta = 7.14 \times 10^{-4}$	$\Delta = 9.17 \times 10^{-4}$
(Nm)	$t_\Delta = 343$	$t_\Delta = 424.5$
Lower arm boom	EI $= 2.82 \times 10^6$	EI $= 3.8 \times 10^6$
(Nm²)	GJ $= 1.48 \times 10^6$	GJ $= 1.93 \times 10^6$
Wrist joint housing (Nm/rad)	Drive axis[c]: $K_{\theta,y} = 4.47 \times 10^5$	Drive axis[c]: $K_{\theta,z} = 1.08 \times 10^6$
	Cross axis[c]: $K_{\theta,x} = 2.56 \times 10^5$	Cross axis[c]: $K_{\theta,x} = 10^6$
	$K_{\theta,z} = 4 \times 10^5$	$K_{\theta,y} = 10^6$
Wrist joint gearbox		
(Nm/rad)	$K_g = 1.42 \times 10^6$	$K_g = 1.33 \times 10^6$
(rad)	$\Delta = 1.25 \times 10^{-3}$	$\Delta = 9.17 \times 10^{-4}$
(Nm)	$t_\Delta = 321$	$t_\Delta = 424.5$
End effector	Axial: 1.95×10^6	Axial: 2.85×10^6
(Nm/rad)	Lateral: 5.1×10^5	Lateral: 3.79×10^6

[a]Based on SRMS shoulder yaw joint data.
[b]Shown for SSRMS shoulder and wrist yaw joints. Data for the pitch and roll joints are the same except that their drive and cross axes have different orientation.
[c]Shown for SRMS wrist pitch joint. Data for the other two wrist joints are the same except that their drive and cross axes have different orientation.

open-chain topology), and more than one for stowing and launch configurations (srms in closed-loop topology with the space vehicle). In this regard, the simplest problem is actually associated with the srms in its parking configuration on orbit.

To maintain the arm's structural integrity, loads induced in the arm must be kept below design limits, which requires operational constraints to be placed on the space vehicle propulsion/attitude control systems. Brake holding torques at the joints, however, are lower than load limits in neighbouring structures. Consequently, when the induced loads become excessively high, some brakes would slip to relieve the arm stress unless the arm were in a singularity configuration. If brake slippage is to be avoided (e.g., to maintain arm configuration), then it is necessary to constrain the space vehicle propulsion/control system activities such that the loads induced in the joints are below their minimum brake torque values. In this case, a simple model is possible: the arm structure can be represented by a beam with a tip body. The beam length is the distance between the arm base and its tip. The beam structural properties can be determined from the element stiffnesses given in Table 4 and the overall arm flexibility using Eq. (8). The external excitation is the linear and angular accelerations applied at the root of the beam. The beam response can then be derived for various acceleration profiles such as those resulted from a single or repetitive thruster firings. The loads induced in the joints can be computed from the beam response and analytical expressions can therefore be derived to define the constraint on the base accelerations to keep the induced loads below critical levels (e.g., braking slip torques). The arm dynamics in the launch and stow configurations are more complex, and finite element analyses are required. These results are often verified by extensive simulations before they become operation constraints.

When the srms's control system is active, the arm dynamics becomes more complex and simple analysis is often not possible. However, a particular case does exist where the arm configuration is maintained in position hold mode. In this case, the joint control system can be approximated by a proportional controller; in other words, the joint servo behaves as an electronic spring. The preceding simple model and analysis are still appropriate provided that the joint drive-axis stiffness is modified to represent two springs in series: one spring modeling the joint structure and the other possessing the servo electronic stiffness. Alternatively, the arm dynamics can be analyzed using a model developed for the more general case in which the srms's control system action is represented by joint control torques as discussed hereafter.

The motions of interest in the srms's dynamics consist of the translation and rotation of the space vehicle relative to its orbit, joint angular motions, and structural deformations within the srms, its payload, and the space vehicle. These motions can be studied using different formulations available for the dynamics of elastic multibody structures, of which the Newton-Euler formulation by Hughes,[5] and Sincarsin and Hughes[6] has been selected by Spar Aerospace Ltd. for their SRMS and SSRMS dynamics simulators. The selection was made based on the following considerations:

1) The srms is in an open-chain topology unless it berths, or deberths a payload to its space vehicle.

2) Loads generated within the srms are to be determined as part of the analysis and/or simulation.

3) Understanding of the dynamical characteristics can be gained from the motion equations.

When payload berthing or deberthing is to be analyzed, additional dynamical constraints can be imposed on the motion equation to specify either contact conditions (transient topology) or closure of the open-chain topology (closed loop). The basic motion equation for the open-chain srms dynamics is of the following form

$$M(x)\ddot{x} + Kx = f_i + f_d + f_c + f_e \qquad (9)$$

where M is the system mass matrix; x the column matrix of space vehicle translational and angular degrees of freedom, joint angles, and elastic degrees of freedom in the system; K the system stiffness matrix; f_I the column matrix of nonlinear inertial forces; f_d the column matrix of damping/friction forces; f_c the column matrix of control forces; and f_e the column matrix of environmental forces.

The preceding forces are actually combinations of forces and torques. Detailed derivation of Eq. (9) can be found in Sincarsin and Hughes[6] and Hughes and Sincarsin.[7] Without going through complicated expressions for the preceding terms, the following points can still be made:

1) Equation (9) is a nonlinear ordinary differential equation. Closed-form solution is not possible.

2) The mass matrix M is a function of x; more precisely, a function of joint angles. The system's natural frequencies therefore vary with arm configuration.

3) M is a full, positive-definite matrix. The nonzero off diagonal elements in the matrix represent inertial coupling between different degrees of freedom in the system: rigid-body and elastic degrees of freedom alike. In particular, the degree of coupling between the joint rigid-body degrees of freedom depends on the angles between the joint drive-axis unit vectors. The coupling is strongest when these vectors are parallel (e.g., in three pitch joints in the SRMS and the SSRMS), and weakest when they are normal to each other.

4) The elements of M and K corresponding to the elastic degrees of freedom depend on the model used to represent the elastic deformations in the system. They can be computed using finite-element models, models with assumed shape functions, or from lumped stiffness models such as those discussed in Sec. IV.A.

5) The expression for the nonlinear inertial forces f_I is very complicated. Its numerical evaluation requires a large amount of CPU time and becomes a burden in computer simulations. However, thanks to the rather low operational speeds of the SRMS and SSRMS, f_I is a slowly varying function of time. A periodic update of f_I in simulations has been found to be adequate for the treatment of nonlinear inertial forces.

6) The friction/damping forces f_d mainly represent friction at the joints and structural damping in the joint housing and in the arm booms. When the joint friction is represented by Coulomb friction model, a special algorithm will be required to handle the effect of inertial coupling on the joint friction.

7) The control forces f_c represent the space vehicle control forces and the joint control torques. Therefore, to completely describe the srms dynamics, two more sets of equations are required: the space vehicle propulsion/attitude control system equation, and the joint control system equation. It is to be noted that f_c is a function

of x because the joint control torque is the gearbox output torque t_g discussed in Sec. IV.A.

8) As mentioned earlier, the component of the environmental forces f_e exerted on the srms bodies is negligible in comparison with the corresponding components in f_I, f_d and f_c. However, the component of f_e exerted on the space vehicle can be significant and may need to be retained in the motion Eq. (9).

From a simulation viewpoint, Eq. (9) must be solved as efficiently as possible. On the one hand, algorithms have been developed to speed up the computation of the mass matrix M, the nonlinear inertial forces f_I, etc. as in Golla et al.[8] On the other hand, instead of solving Eq. (9), recursive algorithms have also been developed to solve the body equations of motion without assembling them to form an equation as in Eq. (9), see D'Eleutario.[9] In addition to these algorithm developments, other approximations can also be made to further save computing time. For example, the coefficients in Eq. (9) are updated periodically with the update rate increasing with the simulated srms speed. In between updates, the coefficients M and K are considered to be constant and quasistatic modal analysis can be performed allowing the displacement vector x to be expressed in terms of the system modal coordinates. In this connection, different modal coordinates have been used involving a free, fixed space vehicle, and free, locked joints. Model order reduction is then possible when certain system modes are discarded from the model.

Simulations based on Eq. (9) and control system equations are often used to support srms operation procedures development and mission planning. For a given srms operation scenario, it is desired to establish the POR trajectory, the time line, and the load levels within the srms's structure and at its interfaces with the payload and the space vehicle. The POR trajectory can be constructed from the joint angles and the elastic deformations in the system, both of which are provided by the elements of x. If only the joint angles are used, then the resultant POR position/orientation is called the kinematic POR position/orientation although the elements in x are derived from a dynamics simulation [The algorithms for the POR position and orientation in this case are identical to the kinematic relationships (1) and (2)]. When the elastic degrees of freedom in x are also used, the resultant POR position/orientation then becomes the flex POR position/orientation referring to the flexible effect contributed by the elastic degrees of freedom. Normally the difference between these two kinds of POR position/orientation is not significant because most of the SRMS's and SSRMS's flexible motion is contributed by the joint gearbox flexibility which is contained in the joint angle response. The POR flex trajectory is analyzed to determine the effect of srms structural flexibility on the arm motion, the capability of the arm to follow its command, and other performance parameters such as stopping distance, settling time, etc. The srms loads are computed based on the stiffness matrix K and the displacement vector x; they are used mainly to determine whether design load limits have been exceeded. If need be, constraints will be placed on the control systems of the srms and/or of the space vehicle to protect the structural integrity of the srms and its payload; for example: to reduce srms operation speed limits, or to prohibit thruster firing when the arm maneuvers a heavy payload. Both normal operation and contingency scenarios with srms malfunctions are considered.

In addition to payload handling scenarios in which the srms and its payload form an open-chain topology, there are other scenarios which are of interest to the srms's design and operation as well. For example, consider a scenario in which the SRMS grapples a payload which is latched to the Orbiter cargo bay. If the latching mechanism fails to release while the SRMS attempts to lift the payload out of the bay, then sooner or later, the SRMS motors would stall with high loads generated in the SRMS. This scenario is referred to as a constrained motion situation. Topologically, the SRMS is in a closed loop with the Orbiter and its payload. To analyze this case, a closure constraint must be applied to Eq. (9) and a special algorithm is required (e.g., cut-and-try, singular decomposition). A model-order-reduction technique is especially needed to speed up the simulation of this closed-loop system. Unfortunately, this problem has not received much attention from the analysts yet. Without a proper closed-loop solution, the preceding problem can still be solved using some approximation: the constrained payload is modeled by a very heavy payload and a very heavy Orbiter, for which Eq. (9) is still valid. In other situations, such an approximation is not valid and proper solution is required; for example when the SRMS hands over its payload to the SSRMS while the Orbiter is docked to the Station.

Next, consider a payload capture or release scenario. During the process of capture, the snaring mechanism in the end effector pulls the payload grapple fixture shaft into the end effector. Sooner or later, the end effector would contact with the payload grapple fixture while the srms is limped. The contact force would help the end effector to align itself with the grapple fixture, thereby correcting misalignment errors between the end effector and the grapple fixture which may exist at the time of capture. The limped srms allows the end effector to move freely relative to its base. This portion of the capture process requires a contact dynamics analysis, for which a formulation and modeling are required. Following the misalignment correction, the snaring mechanism continues pulling the grapple fixture into the end effector until the end effector has been completely rigidized. In this portion of the capture process, a pulling force exerts on the end effector causing torques to develop at the joints. If the joint torque is higher than its friction torque, the joint will backdrive. Otherwise, the joint will not rotate causing loads to build up and possibly jeopardizing the completion of the capture process. To analyze this situation, the joint friction must be correctly represented and implemented, keeping in mind the difficulty alluded to in the earlier discussion of f_d.

Finally, in the context of contact dynamics, a similar situation arises when a payload is released. When the end effector is withdrawn to release a payload, small uncommanded motion of the end effector could cause the end effector to contact with the payload temporarily (the dynamical effect in the arm motion in response to the arm kinematic commands causes uncoordinated motion at the arm tip). The short-duration contact acts as an impulse on the payload causing it to tip off. This was in fact a major concern in the long-duration exposure facility (LDEF) mission: if the LDEF was tipped off with an unfavorable impulse, it could end up with a wrong orientation and its experiments would be ruined. The LDEF is gravity-gradient stabilized. There were no contact dynamics simulations available to analyze this scenario. Instead, a worst-case analysis was performed in which a free-flyer LDEF was assumed to be disturbed by an impulse of maximum magnitude (defined in the SRMS specifications) applied in the most unfavorable direction. The result was positive: the LDEF would be in the correct orientation after release provided that the SRMS placed the LDEF in the correct orientation at release. The mission went ahead.

V.   Control System

The srms control system is designed to handle payloads with respect to its space vehicle, for example, to grapple a stowed payload, then transport it to another location for release or restorage, or to capture a payload, then berth it to the space vehicle. Therefore, to capture a payload which is not stowed in the space vehicle, either the space vehicle must be maneuvered towards the payload or the payload needs to be positioned such that it is within the srms's reach envelope. Similarly, the srms must place the payload in a desired attitude relative to the space vehicle prior to releasing it. Following the release, the payload's control system will take over for orbital maneuvering or for attitude control. In this fashion, the srms and the space vehicle control systems are active sequentially most of the time, allowing the two control systems to be designed independently of each other. Controller-controller interactions might occur when both control systems are active simultaneously, in which case operational constraints would be required to maintain their stability.

The srms control system design is based on a hierarchical organization in which members of higher levels control several members of lower levels in parallel, and members of lower levels are designed in isolation based on requirements placed on the next level. The hierarchy consists of the operator as the task controller, the arm control algorithms, and the joint servos/end effector actuator as depicted in Fig. 15.

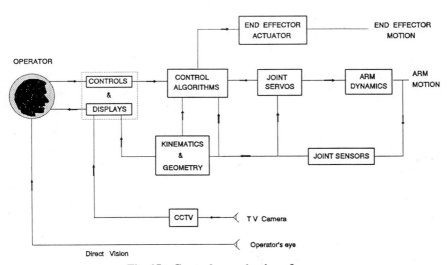

Fig. 15   Control organization of srms.

A.   Control Modes

The human operator interfaces with the srms via the displays and controls subsystem described in Sec. II.D. Three types of srms control modes are available: 1) Human-in-the-loop modes, including the manual augmented mode, and the single joint rate mode; 2) automatic trajectory modes, including the operator commanded POR mode, the prestored POR autosequence mode, and the following modes which are available in the SSRMS only: operator commanded joint position mode, prestored joint autosequence mode, and artificial vision function (AVF)—supported tracking mode; and 3) contingency modes, including the backup drive mode, and the direct drive mode, which is available in the SRMS only.

In the manual augmented mode, operator commanded POR mode, prestored POR autosequence mode, and AVF-supported tracking mode, the srms control is effected by POR linear and angular rate commands relative to a reference frame fixed to the space vehicle, i.e. end-point control. In the operator commanded joint position mode, and the prestored joint autosequence mode, all joints are controlled simultaneously via joint position commands. In the remaining modes, the srms is controlled in a joint-by-joint basis.

In the manual augmented mode, the operator commands the POR velocities using the hand controller(s). The commands are resolved into joint rate commands using the resolved rate algorithm which is based on Eq. (7). Joint rate commands are then fed to joint servos whose primary function is to provide joint motions as commanded. In this fashion, the joints are in closed-loop control while the arm control loop is closed at the operator. In other words, when the commands are constant, the POR has an open-loop control. Since the resolved rate commands are based on the arm kinematics only, actual arm trajectory often deviates from its command due to dynamics effects; consequently, the operator is required to adjust his/her POR commands to counteract the arm's uncommanded motion. To relieve the operator from such a control burden, which might be annoyance while berthing a payload, a position/orientation hold submode (POHS) is introduced as a selectable control feature. In this submode, the POR rate commands are integrated forward to determine the desired POR trajectory, starting from the time this submode is entered. The desired trajectory is maintained along those directions in which the POR commands remain unchanged. Thus, the POR control is assisted by the srms's control software and the control loop is closed at the arm control level. In the manual augmented mode, the POR can be controlled at variable rates which are proportional to the deflections of the hand controllers. In addition, coarse and Vernier POR rate limits are selectable via the displays and controls subsystem. Stopping an srms's motion can be achieved either by returning the hand controllers to their null positions (i.e., servo stopping), or by applying brakes to the joints (i.e., braking).

In the operator commanded POR mode, the operator specifies the POR destination in terms of its position and orientation using the keyboard in the displays and controls subsystem. In the SRMS application, before the autosequence gets started, the specified POR destination is verified for its reachability using the resolved position algorithm discussed in Sec. III.B. Coarse or Vernier POR rate limits can be selected. Once the sequence begins, the POR is commanded to travel along straight-line trajectories based on the instantaneous POR position/orientation and its destined position/orientation. Synchronization between POR linear and angular rate commands is performed to ensure that both the POR commanded position and orientation will be satisfied simultaneously at the end of the trajectory. The SS-RMS's control algorithms, however, have an added feature to maintain the initial straight-line trajectories throughout the sequence by adjusting the POR commands mentioned earlier with a command to drive the POR toward the initial straight-line trajectories. The POR is normally commanded at maximum rate until it approaches its destination. Within a predetermined wash-out distance from the destination, the POR rate commands are reduced in preparation for smooth stopping. When the POR is within the (predetermined) position-hold distance from its destination, the sequence is terminated and the srms's control mode is automatically switched to the position hold submode. The srms will be held in the same configuration that exists when the position hold submode is entered. The srms can be stopped at any point along the sequence and the sequence can be restarted from where the srms stops.

The prestored POR autosequence mode is the same as the operator commanded POR mode, with the exception of the following differences. The trajectory is made up by a series of straight-line segments. The data for POR position and orientation in these segments are prestored in the srms's computer; they are called up when the prestored POR autosequence mode is entered. The intermediate POR data points can be fly-by points or pause points; the end point in the trajectory is always a pause point. The POR is commanded towards the next POR data point in the trajectory. When the POR is within a fly-by distance from a fly-by point, the srms's motion continues with the POR commanded to the next POR data point. When the POR approaches a pause point, the arm is commanded to stop in exactly the same manner as it is in the operator commanded POR mode.

The AVF in the SSRMS provides the position and orientation of a photogram-metric target relative to a camera, based on which the SSRMS can be controlled in an automatic mode to track the target or to help berth a payload. In the former case, the SSRMS LEE tip camera tracks a target, and in the latter case a camera on the Space Station views a target which is fixed to the payload attached to the SSRMS. In either case, a POR can be defined for the purpose of controlling the SSRMS, and a target position and orientation can be defined as the final POR destination. In the AVF-supported tracking mode, the POR is commanded along an automatic trajectory toward its destination in a fashion similar to that of the operator commanded POR mode. The target must be in the field of view of the camera before this mode can be entered and in case the photogrammetric data from the AVF is lost, the AVF-supported tracking mode is automatically terminated and the manual augmented mode will be selected automatically.

The operator commanded joint position mode and the prestored joint autose-quence mode are similar to the operator commanded POR mode and the prestored POR autosequence mode, respectively, the difference being that joint angles are defined in place of the POR position and orientation. Prior to the execution of the operator commanded joint position mode, the commanded joint angles are verified for their validity by checking them against the joint angular soft(ware) limits. Along the trajectory, joint rate commands are based on the angular distances from their destinations and the joint rate limits. The joint rate commands are synchronized to allow all joints to reach their destinations simultaneously.

In the single joint rate mode, only one joint in the srms is selected to be driven; the remaining nondriven joints are in the position hold submode where they are maintained at the same positions that exist when the position hold submode is entered. In the SRMS, control in the single rate mode is effected using a switch on the displays and controls panel as shown in Fig. 10. The driven joint is commanded with a ramp which is held constant at the joint rate limit, either in a clockwise or counterclockwise direction. Stopping the joint motion is achieved by returning the switch to its null position, which is equivalent to a zero joint rate command, or by applying brakes (to all joints). The SSRMS single joint rate control is similar, except that the command is effected by using the hand controller thereby allowing variable joint rate commands.

The contingency modes are human-in-the-loop modes designed mainly for safety reasons. In the event of an srms malfunction wherein the preceding computer-supported control modes are not available, the operator can use these modes to maneuver the srms to a safe position (e.g., stowing). In the backup drive mode, one joint is selected to be driven, the remaining joints have brakes applied. There is no joint rate control; only joint motion direction can be selected. The direct drive mode, which is available in the SRMS only, is similar to the backup drive mode with one exception: the motor of the driven joint receives the input voltage through its own motor drive amplifier whereas in the backup drive mode, a common motor drive amplifier is used with output switchable to different joints.

In the position hold submode, the arm configuration is maintained by holding the joints to their desired positions. The desired positions are defined at the time the position hold submode is entered, except that in the case of the single rate mode the desired position of a driven joint will be redefined after the joint has been deselected. Joint angular deviations from the desired positions are converted to joint rate commands to the joint servos. Joint rate limiting is applied to the joint rate commands to keep the arm's stopping distance within its requirement.

A number of control features are designed to facilitate the srms's operation as shown here:

1) The rate hold feature holds POR rate commands constant. This feature is activated by switching the rate hold switch on the hand controller and returning it to the null position. Any subsequent commands via hand controllers are treated as biases to the rate hold commands. This feature is available only in those modes where hand controllers are used as command-input devices.

2) The coarse/Vernier rate feature offers selectable coarse or Vernier rate limits.

3) The pause feature allows an automatic trajectory to pause until the operator commands it to resume.

4) The position/orientation hold feature holds POR trajectory to the initially commanded trajectory by compensating for uncommanded motion. This feature is available in the manual augmented mode only.

5) The position hold submode feature holds the joints to the positions existing when the position hold submode is entered. Joint rate commands are proportional to joint angular deviations from their reference (hold) values and limited by joint rate limits. This feature is available in all computer-supported control modes.

The following control features are available for the SSRMS control only:

6) The bidirectional control feature allows the SSRMS to be controlled with either end as its base. This feature is available in all SSRMS control modes.

7) The arm pitch plane change feature allows the arm pitch plane to rotate while maintaining the position and orientation of the SSRMS's tip. This feature is possible because the SSRMS's tip position and orientation can be reached by different sets of joint angles; it is available in the manual augmented mode only.

8) The force-moment accommodation feature assists the arm in moving along a desired direction when the arm and/or its payload come in contact with another object (e.g. in payload berthing). The outputs of the force-moment sensors in the SSRMS's tip LEE are first transformed into the same command reference frame as the POR, then scaled and subtracted from the POR rate commands before they are resolved into joint rate commands. This feature is available only in the manual augmented mode and in automatic POR modes. The arm and the joint servos are operated in Vernier when this feature is activated.

9) The coordinate rereferencing feature allows the operator to select the SSRMS's control reference frame and display reference frame. This feature is available in all computer-supported control modes.

While controlling the srms, the SRMS operator can observe the arm and payload motions through the Orbiter aft and overhead windows by direct viewing or by looking at the two CCTV monitors mentioned in Sec. II.D. The SSRMS IVA operator, however, cannot observe the arm and payload motion all the time because of the restricted view from the MPAC cupola and because the SSRMS base is mobile along the Space Station. The srms operator can also obtain from the displays and controls panel (or screen) the following digital readouts: 1) POR position and orientation, 2) POR linear/angular velocities, and 3) joint angles. Caution and warning annunciations, system health status, and various flags are also available by calling appropriate pages on the controls monitor; for example, arm singularities, joint soft stops, invalid POR position and orientation commands, ABE failure, brakes-on request, brakes-on indicator, etc.

In the SRMS, the system health is monitored by the hardware BITE which is supplemented by the system health monitoring function (SHMF) in the control software. The SHMF monitors the following critical SRMS operational parameters to detect anomalies in the SRMS performance:

1) With the encoder check, the encoder output is compared with the time integral of the tachometer output; within a small tolerance the two should be equal. The SRMS encoder in each joint does not have any hardware BITE, but the tachometer does.

2) With the consistency check, the tachometer output is integrated and compared with the change in encoder output. If they differ by more than their tolerance and if the difference persists for some predetermined length of time, an inconsistency has been detected. In addition, a rate envelope check is also performed where the tachometer output is checked to determine whether it stays within an envelope around its joint rate command time history. If it stays outside its envelope for more than 3 GPC cycles (= 0.240 s), an inconsistency has been detected. When an inconsistency is detected, brakes are applied automatically to stop the arm motion. The consistency check is performed only when there is no failure in the

tachometer, and when the joint is not backdriven (backdrive is detected by the arm control software based on motor current and motor rate).

3) If the encoder check has not detected a failed encoder, then a reach limit check is performed where the encoder output is checked against the joint lower and upper reach limits. The reach limits are slightly lower than the software travel limits shown in Table 2. When a reach limit is detected, an annunciator will be issued to warn the operator that a joint is approaching its software limits.

4) With the Software stop, if a reach limit has been detected, the encoder output will be checked against its software travel limits. If the encoder output exceeds the joint software limits, a software stop flag will be set and all joint rate commands will be set to zero to stop the arm motion. The single rate mode must be used next to bring the joint out of its software stop regions.

5) The three SRMS arm singularities are monitored in every GPC cycle. If the arm is in the vicinity of a singularity within a tolerance, a singularity warning is issued.

6) End effector flag checks occur when the EEEU BITE detects a drive electronics failure. A flag is sent to the MCIU and the GPC where 3-N filtering is performed before the MCIU EE flag is set. The MCIU EE flag indicates that the end effector status flags are in an invalid state.

7) The release and derigidize flags are monitored against the release and derigidize commands, respectively. If they change their states without being commanded, then warnings will be sent to the displays and controls panel.

8) The thermistor outputs are checked against their lower and upper limits with a temperature limit check. A warning is issued to the displays and controls panel if the temperature exceeds its limits for more than two GPC cycles ($= 0.16$ s).

The SRMS's hardware BITE are designed to detect and display critical failures. The BITE consists of ABE BITE and MCIU BITE. The ABE BITE detects 28 V power failure in the SPA (main bus voltage below 17 ± 2 V), SPA commutator failure (invalid transition on the input lines to the optical commutator for more than 120 μs), tachometer failure (phase lock loop, inductosyn drive and output circuit malfunctions), MDA failure (overcurrent in the supply to output power switches), JPC failure (output voltage outside operational limits), and EEEU failure (motor driven without being commanded, commutator signals in invalid state). The MCIU BITE investigates the integrity of the communication links between itself and the ABE, GPC, and the displays and controls subsystem. It also monitors end effector functioning, thermistor circuit operation, and its own internal consistency.

B. End Effector Control Modes

The end effector can be controlled in either the automatic mode or manual mode to capture and release a payload. The sequence of events in an end effector operation is the same in both modes, but their execution is slightly different from each other.

In the SRMS end effector automatic mode, the payload capture sequence begins with the actuation of the *capture* switch on the rotational hand controller when the payload grapple fixture is within the *capture envelope* inside the end effector. The two snare cables will close on the grapple fixture, then center and capture the payload, as illustrated in Fig. 16. The rigidization will then start

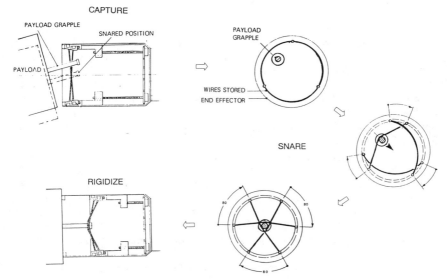

Fig. 16   SRMS capture and rigidize sequence.

automatically to pull the grapple fixture all the way to the end of the rigidization carriage. During the rigidization, misalignments between the end effector and the payload will be self-corrected. At the end of the rigidization process, the end effector rigidization motor applies a pulling force on the order of 1200 N to the grapple fixture to secure the payload/end effector mating. The end effector operation status is fed back to the displays and controls panel via four flags: CAPTURE = ON indicating the snare cables wrapping around the grapple fixture, CLOSE = ON indicating the completion of the snaring, EXTEND = OFF indicating the starting of the rigidization, and rigid = on indicating the completion of rigidization.

The SRMS's payload capture operation in the manual mode is similar, with the following exception: the operation of snaring or rigidizing is possible regardless of the end effector status; the rigidization does not start until the *rigidize* switch is actuated, and both the *capture* and *rigidize* switches must be held continuously during the capture and rigidization operations. The preceding features allow for complete overriding capability in the manual mode.

In the SRMS end effector automatic mode, the payload release sequence begins with the actuation of the *derigidize* switch on the displays and controls panel. The rigidization carriage will be driven towards the end effector opening end until it crosses the zero tension point. The snare cables will then be opened to release the payload. When the cables are fully open, the rigidization carriage will be driven outward until it reaches the end of its travel. The following flag setting indicates the end effector status during a payload release: DERIGID = ON when the rigidization carriage reaches the zero tension point, CLOSE = OFF when the snare cables start to open, CAPTURE = OFF when the snare cables release the grapple fixture, OPEN = ON when the snare cables are fully open, and EXTEND = ON when the rigidization carriage reaches the end of its travel.

The sequence of events in the SRMS payload release manual mode is essentially the same as the preceding, except that the derigidization and release operations are performed independently. The *release* switch may be actuated when the DERIGID changes to ON; and to complete the release, it must be held until the OPEN flag changes to ON. The *derigidize* switch must be held to operate the carriage, but it must be in the OFF position during the release operation. Following the release operation, the *derigidize* switch must be reactivated and held until the EXTEND flag changes to ON.

The SSRMS's LEE is controlled in automatic and manual modes in a similar fashion. In addition, the SSRMS's LEE control has the following features:

1) When mated with a PDGF, the LEE latching mechanism will be activated. The latching mechanism will be deactivated before the LEE is derigidized.

2) The IVA operator can select either LEE for control purposes. At the time of writing, the base LEE is designed to be unpowered while the SSRMS is operational so that unintentional release of the base LEE resulting in a fly-away SSRMS will not be possible.

3) The IVA operator can terminate an automatic LEE operation at any time.

C. Joint Servos

The srms servo is designed using an independent joint control approach, although dynamic coupling exists between the joints via inertia loading and structural flexibility effects as discussed in Sec. IV.B. Coupling effects are treated as disturbances to the joint servos. In principle it is possible to decouple the joints through a centralized feedback system using gain-adjusting algorithms which are dependent on arm and payload configuration. In the SRMS, such a scheme was precluded by the complexity of the required algorithms and the limited availability of computation power. As a result, the SRMS's servos were designed with fixed gains for all arm loading conditions. shows the SRMS's servo requirements derived from the SRMS's specification.

In the SSRMS, the independent joint control approach is also used. However, taking advantage of the computation power in the JEU, the SSRMS joint servos have settable gains based on the computed moments of inertia about the joint drive axes.

Some principal considerations in SRMS servo design include the following:

a) The servo bandwidth, which should enclose the dominant arm flexible modal frequencies for all arm configurations and all payloads including the design case maximum payload (15000 kg, 18.3 m long, and 4.5 m in diameter). Based on modal analysis, the SRMS servo bandwidth requirement is >120 rad/s for an unloaded arm and 0.3 rad/s for the 15000 kg payload. It is to be noted that the joint gearbox behaves as a mechanical filter so that certain low arm modal frequencies can affect the servo response.

b) The moment of inertia of the motor and its load (about the motor shaft) has the widest variation at the shoulder yaw joint, from 0.00035 to 3 kg-m^2. Thus, the shoulder yaw joint was chosen to be the baseline design joint.

Table 5 SRMS servo requirements

		Rate mode	Position hold submode
Stability	Gain margin (dB)	≥ 6	≥ 6
	Phase margin (deg)	≥ 30	≥ 30
Steady-state performance	Rate accuracy	±1% of maximum unloaded rate	Not applicable
	Limit cycles (deg)	Amplitude ≤ 0.025	Amplitude ≤ 0.025
	Response to disturbances	Rate offset ≤ 2% of maximum unloaded rate for 68 N-m constant torque	Position offset ≤ 0.1 deg for 68 N-m constant torque
Transient performance	Threshold	Rate error ≤ 1% of maximum unloaded rate	Position error ≤ 0.025 deg
	Dynamic range	Threshold to maximum rate	Full range of joint travel
	Deceleration	Joint stopping angular travel corresponding to ≤ 0.6 m end effector stopping distance	Same as in rate mode
	Control of flexible dynamic	Servo shall not act as a destabilizing influence	Same as in rate mode

c) Joint stiction/friction of the mechanical drive train must be overcome. No jerky motion is allowed for low rate commands. The servo must have a high gain at the lowest tracking rate required.

d) The joint output torque must be capable of reaching high levels within the joint stress limits for both forward drive (motoring) and backdrive (generating) conditions to maximize the servo performance.

Based on the preceding design considerations, the SRMS's servo was designed using classical frequency domain techniques. Figure 17 shows the block diagram of the SRMS's servo design. Some features of the SRMS's servo design are discussed hereafter.

The SRMS servo is a rate servo, even when the SRMS is in the position hold submode. Input to the servo is the joint rate command available from the MCIU every 42 ms although the GPC issues joint rate commands every 80 ms. The joint motor rate is sensed by a tachometer, the output of which is processed before being fed back to the servo. A low-pass filter is in series with the digital tachometer to remove undesirable processing delays at high frequencies (a digital tachometer provides excellent low-noise drift-free responses at very low rates, but suffers from undesirable processing delays at high frequencies). A high-pass filter is in series with the analogue tachometer to eliminate the drift and accuracy problem at very low rates (an analogue tachometer provides good high-frequency characteristics and very little phase shift at high acceleration levels, but suffers from drift and accuracy problem at very low rates. Note that the SRMS's motor acceleration can be as high as 3500 rad/s^2). The two filters have overlapping frequency range and constant gains. It is to be noted that in implementation an additional capacitively coupled, high-amplification stage is added to the high-pass filter to bring the extremely low-level inductosyn tachometer signal to required levels. The capacitive coupling prevents passage of any dc offset to the motor drive amplifier (MDA) input.

The integral trim comprising an integrator and a limiter is added to the forward path to provide high gains at low frequencies to break motor and drive-train stiction,

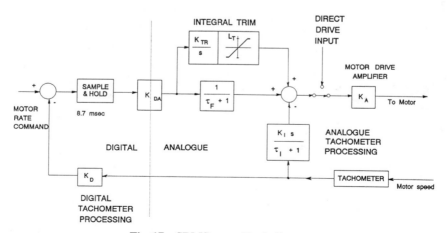

Fig. 17 SRMS servo block diagram.

and to null small errors. The integrator output is limited to avoid undesirable large error offsets. The limiter reduces overshoot in the large signal transient response, its value is selected to drive the motor under worst-case stiction levels. When the integral trim operates within its limits, the rate servo is a type-1 servo. When the integral trim is saturated, it reverts to a type-0 servo which requires an error to produce the steady-state rate as commanded. Consequently, the motor rate commands in the GPC must be shaped using a dual-gain function in the command output processor (see Fig. 11) so that accurate steady-state rates can be achieved regardless of the state of the integral trim.

The SRMS motor can be directly driven via a hard-wired command in direct drive mode. When the direct drive mode is selected, a constant voltage is supplied to the motor drive amplifier resulting in 6.5 V applied to the motor. The motor can be driven in clockwise or counterclockwise directions. Not shown in Fig. 17 is the motor current limiting scheme designed to maintain approximately equal maximum joint torques for the forward and backdrive conditions. Each joint has its own forward and backdrive current limits which are established based on gearbox forward and backdrive efficiencies, gear ratio, and maximum joint torque. A joint is forward driven when its motor current has the same sign as its motor rate and the motor rate itself is outside of a threshold; otherwise, it is backdriven. Using such logic, the joint is considered to be backdriven when its motor is at rest. Different current limit settings are used for different SRMS operations: 15/15 (of current limits) for SAFING (which is one type of servo braking), 13/15 for normal arm operations, and 1/15 for limping the arm (the 0 current limit is not used to avoid the indeterminate condition in the forward/backdrive logic).

The SRMS motors are reversible, three-element, brushless dc types, consisting of a permanent magnet rotor and a wound stator. Attached to the motor shaft is a flywheel designed to raise the moment of inertia about the motor shaft for servo stability purposes. The SRMS's (and SSRMS's) motor characteristics are shown in Table 6.

SRMS joint sensors consist of tachometers and encoders, a pair of which is located in each joint. The tachometer is a rotary inductosyn device consisting of a single-coil rotor and two-coil stator. The tachometer electronics provides ac excitation to the stator windings, and processes the sensor output, following signal preconditioning by the signal conditioning unit, to produce an analog signal and a digital signal. The digital tachometer output is represented by a 12-bit word; the first bit represents sign and the remaining bits represent the motor speed up to 180 rad/s which is more than double the SRMS's maximum motor speed. The encoder is an electro-optical device using a light-emitting diode light source and a single disc providing a 16-bit grey code output which represents an angular position from 0 deg. to 360 deg. The encoder data is converted to binary in the SPA, and to the proper joint angle in the GPC. Encoders are calibrated via ground tests and biases are taken into account in the encoder output processing in the GPC.

The SSRMS servo is similar to the SRMS servo, except for the following differences:

1) The SSRMS's servo can be a position servo or a rate servo depending on a flag set in the ACU.

2) The joint angle is measured by a joint resolver. In the position servo mode, the difference between the joint commanded angle and the joint resolver output is

Table 6   SRMS and SSRMS motor characteristics

Parameter	SRMS	SSRMS
Moment of inertia about motor shaft $(kg\text{-}m^2)$	3.68×10^{-4} (Shoulder) 3.94×10^{-4} (Elbow) 3.48×10^{-4} (Wrist)	2.67×10^{-4}
Friction/stiction (Nm)	0.028	0.1
Forward current limit (A)	4.53 (Shoulder and elbow, 15/15 value) 3.43 (Wrist, 15/15 value)	4.44
Backdrive current limit (A)	2.57 (Shoulder and elbow, 15/15 value) 1.95 (Wrist, 15/15 value)	1.66
Torquing gain (Nm-A)	0.23	0.318
Motor resistance (Ω)	2.85	5.16
Time constant (s)	0.001	0.00114
Back emf gain (V/rad/s)	0.23	0.318

converted in the JEU software to a motor rate command with a lead-lag compensation. The command is then fed to the rate servo at the rate of once every 50 ms. The lead-lag gain and time constant are payload dependent; their values have not been finalized at the time of writing.

3) Shaping of the motor rate command is performed in the JEU software at the rate of once every 50 ms.

4) Motor rate feedback and compensation are performed in the JEU software at the rate of once every 8 ms.

5) The integral trim has a reset logic which allows the integrator output to be reset to 0 when desired.

6) The motor resolver senses the motor shaft position based on which the motor rate is derived. The resolver outputs are motor angle and motor rate in digital format and are available every 0.5 ms.

Other servo and gearbox parameters have values shown in Table 7.

VI.   Tests

The first SRMS flight model—a contribution of the National Research Council of Canada (NRCC) to the Space Shuttle program (also known as the design, development, test and evaluation (DDTE—arm) was delivered in March 1981 after the successful completion of a series of ground tests. The DDTE model was used to verify the SRMS on-orbit performance in the space transportation system (STS) flights 2 (unloaded), 3 (loaded with the 156-kg plasma diagnostics package), 4 (loaded with the 370-kg induced environment contamination monitor), 7 (loaded with the 1439-kg Shuttle Pallet Satellite) and 8 (loaded with the 3384-kg payload flight test article). Three subsequent SRMS flight models built under contract with Johnson Space Center did not go through on-orbit testing although there were flight test objectives (FTOs) that called for SRMS maneuvers. Among all SRMS flight models, the DDTE arm is the only one that is instrumented with strain

Table 7   SRMS and SSRMS servo and gearbox parameters

Parameter	SRMS	SSRMS
Gearbox forward efficiency (%)	84.5 (Shoulder) 83.5 (Elbow) 82.5 (Wrist)	84.5
Gearbox backdrive efficiency (%)	78.5 (Shoulder) 76.5 (Elbow) 77.3 (Wrist)	78
Digital-analog converter gain, K_{DA} (V/count)	0.1615	N/A
Integrator gain, K_{TR} (s^{-1})	0.05	Payload dependent[a]
Integrator limiter, L_I (V)	1.5	56
Motor drive amplifier gain, K_A (V/V)	1.92	1
Digital tachometer quantization gain, K_D (counts/rad/s)	11.3778	N/A
High-pass filter gain, K_I (V/rad/s)	0.12	6.5
Low-pass filter time constant, θ_1 (s)	0.1	0.1
High-pass filter time constant, θ_F (s)	0.1	0.1

[a]Varying between 0.1 and 3 volts/rad/s. Values have not been finalized at the time of writing.

gauges to measure structural loads in the SRMS's shoulder and wrist. The DDTE arm was also used in flight 41-C to deploy the 8768-kg LDEF.

A.   Ground Tests

The DDTE arm ground testing consists of hardware tests and tests by computer simulations; other SRMS flight models underwent hardware tests only. The hardware tests comprise subsystem tests and system tests.

In subsystem tests, the shoulder joint, the elbow joint, the wrist joint, the end effector, the interface electronics, the control panel, and the hand controllers were tested separately in a sequence of ambient, vibration, and thermal/vacuum cycling tests. The test sequence was the same for each subsystem where ambient performance tests were performed at the beginning and at the end of the sequence to verify all test performance requirements. A shortened version of ambient performance test, known as the baseline performance test, was performed before and after the vibration test to ensure that the subsystem performance was not degraded by the vibrations. The thermal/vacuum cycling tests consist of a series of cold-hot-cold-hot baseline performance tests performed in a thermal chamber 9 m high and 3 m in diameter. Figure 18 illustrates the thermal vacuum cycle used in the tests.

Subsystem tests of the joints were performed with the joints supported by an exercise fixture and by the system test rig. Figure 19 illustrates the wrist exercise fixture configuration and Fig. 20 shows the SRMS's arm mounted on its system test rig. The rig supports the arm on air bearings through a mechanism which automatically compensates for changes in the floor flatness which is less than

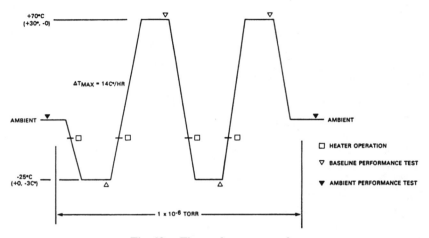

Fig. 18   Thermal vacuum cycle.

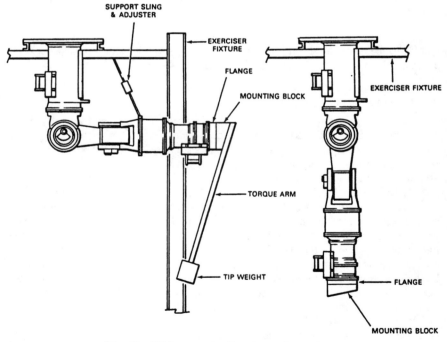

Fig. 19   Wrist exercise fixture configuration.

Fig. 20   SRMS system test rig.

1/8 in. in variation per 3 ft. The additional inertia of the test rig was accounted for in the arm control test software. With the joint on the system test rig, tests were performed to verify angular travel between hardware limits, steady-state and overshoot responses for motor rate step commands up to 60 rad/s, zero rate drift, threshold rate response, joint acceleration and deceleration in the direct drive mode, direction of joint motion in the backup mode, and JPC BITE performance (for shoulder and wrist joints only, elbow does not have a JPC). With the joint on an exercise fixture, tests were performed to verify joint backdrive/stiction capability and joint stall torque. With the end effector mounted on a support structure, the snaring/unsnaring, rigidizing/derigidizing capabilities of the end effector were verified. In particular, the pulling force by the end effector motor at the end of the rigidisation was measured, and the misalignment self-correction capability was tested for misalignments within the allowed payload capture envelope (10 deg for pitch and yaw, 15 deg for roll).

An interesting result was found during the thermal/vacuum test of the joint brakes: the brake slip torque varied somewhat unexpectedly. In some conditions the brake slip torque was below its (minimum value) specification as reported by Trenouth and MacKenzie.[10] The same phenomenon was observed later in on-orbit tests.

Following successful testing of all the subsystems, the subsystems were integrated together with the upper and lower arm booms and associated harnesses. Standard verification of thorough power, continuity, and joint operation were performed in successive stages. The thermal protection system and the camera subsystem functions were also verified. Prior to performing the system tests, the integrated arm was split into two halves (upper and lower arms) for functional testing where all six joints were driven simultaneously. The split arm was supported on a ground handling device known as the strongback. In this test configuration, the wrist pitch and yaw joints were counterbalanced. On completing this intermediate integration and verification test, the arm was ready for its system tests where only ambient performance tests were performed. The system environmental tests were to be carried out in orbit.

The system tests were performed with the arm mounted on the system test rig and with the cabin equipment subsystem integrated. The shoulder end of the test rig was attached to a concrete structure fixed to the floor, leaving the end effector free to travel in a plane parallel to the floor. There were two phases of the tests known as the pitch coupled mode (PCM) tests and the yaw coupled mode (YCM) tests. In the former, the three pitch joints, the wrist roll joint, and the end effector could be commanded, whereas in the latter, only the shoulder yaw joint, the wrist yaw joint, the wrist roll joint and the end effector could be operational. In the PCM tests, thanks to the three pitch joints, the end effector could travel practically anywhere within the arm planar reach envelope, allowing more tests to be done than in the YCM modes. The arm control test software was modified from the actual SRMS control software to allow for planar motion of the arm in these system tests.

The following performance tests were performed during the PCM test phase:

1) Single drive, direct drive and backup drive mode tests were performed, in which the wrist roll joint and each pitch joint were driven one at a time. Motor rates and joint angles were monitored. Safety limits were checked.

2) Manual Augmented mode tests were performed, in which the capability to drive the arm using the hand controllers was verified. System operation features were also verified, e.g. coarse/vernier rate commands, rate hold, position hold, and stopping performance.

3) Automatic mode tests were performed, in which four automatic trajectories with fly-by and pause points were tested. Positioning accuracy was verified.

4) Payload maneuvering test was performed, in which the end effector's ability to capture and release a payload in the manual and automatic modes was verified. The test payload mass is approximately 900 kg. The arm's ability to maneuver a payload was verified by moving the loaded arm.

5) Constrained motion test was performed. With the payload constrained to the floor, the loaded arm was commanded to move. Stall conditions of the elbow and wrist joints were obtained.

6) System safety tests was performed, in which auto braking as in the consistency check was verified.

7) Arm characterization tests measured positioning accuracy, arm stiffness, and maximum tip force were measured with the arm in the straight out configuration. The arm's ability to respond to a very small rate command given by the integral trim was also verified.

In addition, the SRMS performance requirements were verified using non-real-time and real-time computer simulations to make up for those tests that could not be done under the 1-g environment. The simulations were also set up to demonstrate the arm performance under worst-case conditions (e.g., using worst-case combinations of maximum and minimum system parameter values) and in the presence of malfunctions. While the non-real-time simulations using detailed models of the arm provided high-fidelity results for engineering analyses, the real-time simulations with the participation of NASA astronauts and test pilots from NRCC were necessary to test the arm performance with human-in-the-loop control. The real-time simulator used in these tests is the simulator facility (SIMFAC) which was jointly developed by Spar Aerospace Ltd. and CAE Ltd., featuring a mockup of the Shuttle Orbiter which houses a functionally equivalent displays and control panel, hand controllers, keyboard, CCTV monitors, and aft-view windows. The main objective of the simulations was to verify SRMS compliance with specific performance requirements and the simulations were designed to address the following: 1) payload end-to-end maneuvering capability, including capture and retrieval of up to 14500-kg free-flying payload and deployment of up to 29500-kg payload; 2) loaded and unloaded arm performance in the manual augmented mode, including POR rate limiting, joint rate limiting, and arm stopping performance from maximum rates; 3) loaded and unloaded arm performance in the automatic mode, including trajectory tracking performance and positioning accuracy; 4) load exertion capability at the arm tip; 5) SRMS fail-safe performance, including malfunction detection and arm commands in the contingency modes; 6) berthing payload to Orbiter cargo bay; 7) arm stowing; and 8) SRMS performance with the Orbiter reaction control system jet firings.

The SRMS control software was implemented by IBM. The implementation was checked out using the Shuttle Avionics Integration Laboratory (SAIL) at Johnson Space Center. In SAIL simulations, the arm and servo dynamics are simulated using math models which account for structural flexibility in the SRMS and in its interface with the Orbiter. The GPC which hosts the SRMS's control software is a flight-like computer; the arm interface unit MCIU, the displays and control panel, and the hand controllers are actually flight-like hardware.

Following successful testing at Spar, the complete SRMS was shipped to Kennedy Space Center for installation in the Orbiter. The cabin equipment subsystem was installed in the Orbiter crew compartment while the arm was kept on its strongback over the Orbiter cargo bay. The arm and its cabin equipment were then electrically interfaced with the Orbiter wiring, followed by suspended arm tests before the arm was mechanically mated to the Orbiter. The suspended arm tests were similar to the system tests described earlier, but with test procedures modified to account for the suspension of the arm. The heater operation was also tested before the thermal protection system was installed. Following the successful mechanical mating of the arm to the Orbiter, post mechanical mate tests were performed to verify correct joint data, end effector operation, brake operation, shoulder brace release, and latch operation. The CCTV subsystem was then installed and verified. The deployment and stowing of the arm were verified by commands to the manipulator positioning mechanism. The opening and closing of the Orbiter cargo bay door was performed and the door/arm clearance was measured. At that point, the arm ground testing was complete.

The SSRMS ground tests will be similar to those of the SRMS, but with some differences. The SSRMS system (ground) tests will be performed in MSS Avionics Integration Facility (MAIF) where, unlike the SRMS, there will be no system test rig and no complete SSRMS mechanical arm will be involved. At the time of writing, some subsystem tests, e.g., joint performance tests, have been performed; the SSRMS ground tests and on-orbit tests are being planned.

B.   On-Orbit Tests

SRMS on-orbit tests were designed to verify the SRMS's performance and to provide data for the SRMS's model validation while maintaining the safety of the SRMS, Orbiter, and its crew. The feasibility of the tests were determined in premission simulations in terms of time lines, crew workload, malfunctions, and collisions using SIMFAC; and in terms of arm loads, SRMS/Orbiter configurations, and control activities using non-real-time simulators. Details of the tests were defined in flight test objectives (FTOs) and test procedures were incorporated into the crew activity plan.

During on-orbit testing, SRMS performance was monitored using some or all of the following instrumentation described below.

1) Operational instrumentation provides SRMS command data such as hand controller outputs, switch inputs, and joint rate commands; SRMS status such as end effector flags, joint angles, motor rates; SRMS health monitoring such as BITE annunciations; SRMS temperature; and Orbiter data such as attitude angles, and reaction control system (RCS) thruster firings. Data was recorded on board and downlisted at 12.5-Hz and 1-Hz rates.

2) Development flight instrumentation consists of strain gauges located on the shoulder pitch and wrist pitch electronics compartments to measure bending and torsional moments, and thermistors to measure temperatures at the strain gauge locations to compensate thermal effects. Strain gauges data was downlisted at a 12.5-Hz rate.

3) Closed circuit television provides the clear views of the SRMS. Four cameras were mounted in the Orbiter cargo bay: two on the forward bulkhead and two on the aft bulkhead, on both port and starboard sides.

4) Data acquisition cameras record the SRMS's motion using six 16-mm movie cameras mounted in groups of three on the forward and aft bulkheads. Postmission photogrammetric analyses were to be done to provide accurate arm position and orientation relative to the Orbiter, arm oscillation frequencies, etc. Unfortunately, the films failed to provide the desired data either because of unexpected mishaps onboard, or because of the poor photogrammetric results.

All SRMS operation modes, except the backup drive mode, were used in on-orbit tests. In direct drive, single drive, and manual augmented mode tests the arm was placed in one of the three preselected initial configurations with and without payloads attached. The arm was commanded to move in different directions for a preselected duration, then was brought to rest by removing the commands, by applying brakes, or by commanding safing. Flight data was analyzed for arm performance related to command tracking, transient response, POR rate limiting, stopping distance, etc. In automatic mode tests, the arm was first brought to preselected initial configurations with and without payloads attached, and commanded towards desired POR positions and orientations. Flight data was analyzed to determine positioning accuracy and the arm's ability to track an automatic trajectory. Data from the preceding tests were also used to validate simulation models.

In addition, there were SRMS thermal tests and primary reaction control system (PRCS) tests. The thermal tests comprised a cold case evaluation and a hot soak test in which the Orbiter was maneuvered belly and top to the Sun, thus placing the arm in a cold and a hot environment respectively. The cold test was designed to study the performance of the SRMS's heaters with the arm unpowered to keep the arm temperature as low as possible. In the hot test the arm was maneuvered continuously for two hours; the heat generated by the arm electrical subsystem helped maximize the arm temperature. The PRCS tests were designed to study the interactions between the PRCS's thruster firings and the arm dynamics. The arm was placed in different configurations and held in place by applying brakes, or via selecting the single drive mode without a joint command (all joints were in the position hold submode), or via safing. The PRCS's roll and pitch thrusters were fired in single, double, and doublet pulses for predetermined lengths of time. Data were analyzed to determine induced loads in the arm, arm frequencies, and arm damping. They were also used for simulation validation. Table 8 summarizes the primary objectives of the SRMS on-orbit tests. For a detailed description of the SRMS flight tests, see Ref. 11.

All SRMS on-orbit tests were completed successfully at the end of the STS-8 mission. Some examples of the results are presented below. Figure 21 shows the shoulder pitch joint rate responses to commands in the direct drive and single drive modes when the SRMS was loaded with the induced environmental contamination monitor (IECM) and the SPAS-01 payloads, respectively. The joint rates were actually computed from the downlisted motor rates as the SRMS does not have any sensory device to measure joint rates. In the direct drive mode, joint motion is controlled by direction command only; steady- state joint rates differ from joint to joint by their gear ratios. The initial transient response in the direct drive test corresponds to the gearbox winding up. In the single drive mode, the joint rate command is ramped up in 1.2 s (for coarse rate commands) to the joint rate limit and held constant until a change in command occurs. A similar initial transient response was found at the beginning of the single drive test. The overshoot in the joint rate response was within the joint servo design specification. In both examples in Fig. 21, the joint was stopped by applying brakes, which resulted in large amplitude oscillations.

Figure 22 shows the POR motion in response to a $+Z$ command in the manual augmented (Orbiter loaded) mode when the arm was loaded with the payload flight test article (PFTA) payload. The $+Z$ command corresponds to the payload motion along the same direction as the Z axis of the Orbiter body axis reference frame. The command was constant for 15 s, then a safing command was given to stop the arm motion. The translation of the POR along the X and Y axes (of the Orbiter body axis reference frame) and the three attitude motions (pitch, yaw, and roll) were small uncommanded motions since the arm was driven in open loop. Oscillations near the end of the test were the result of the safing command (to stop the motion of the arm while it was traveling at high speed). Figure 23 shows the relationship between the POR commanded speed and its response in the manual augmented mode when the arm was unloaded and when it was loaded with the small payload plasma diagnostic package (PDP). Although the POR commands were constant in these tests, the effective POR speed commands varied slightly

Table 8 Primary objective of SRMS on-orbit test

| Test | SRMS component | | | | | | Engineering data | | | |
| | Motor | Brake | Gearbox | Servo | Software | Coupling | Arm dynamics | | | Orbiter |
							Frequency	Damping	Loads	Mass properties
Direct drive	x	x	x				x	x		
Single drive		x	x	x	x		x	x	x	
Manual augmented		x	x	x	x	x	x	x	x	
Automatic			x	x	x	x	x	x	x	
PRCS		x	x	x		x	x	x	x	x

[a] Arm mode shapes were not measured because of inadequate development flight instrumentation.

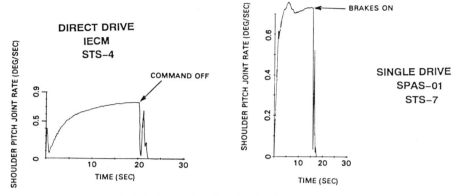

Fig. 21   Shoulder pitch joint rate responses.

because of the joint rate limiting in the resolved rate algorithm which is arm configuration dependent.

Figure 24 shows a typical comparison of flight data vs simulation results. The simulation results were obtained from simulations performed by Spar Aerospace Ltd. using the nonreal-time simulator all singing all dancing (ASAD). ASAD was developed for the design, development, and verification of the SRMS. Figure 24 also shows the arm loads in response to a PRCS roll thruster firing with the arm in its straight configuration and loaded with the IECM payload. The U and V bending moments in Fig. 24 are the components of the induced moments at the shoulder pitch joint, measured by the strain gauges. These components are at 45 deg from the shoulder pitch drive axis. In this test, the joint dynamics about its drive axis was strongly excited; as a result, the nonlinear gearbox characteristics are clearly evident in the bending moment time histories. The arm's structural damping can be determined from results such as those in Fig. 24. The damping was found to be dependent on arm configuration and varied approximately from 3% to 9%.

Not discussed so far were the results of the payload capture and release which were performed to verify the end effector's performance. The test payloads were grappled successfully when they were in the Orbiter cargo bay or when they were in high hover positions over the bay. In STS-7, the SRMS successfully captured the SPAS-01 which spun up to 0.1 deg/s. Also, in STS-7, tip-off rates (< 0.36 deg/s) were measured thanks to the gyro on board the SPAS-01. The gyro data could be correlated with the end effector rates computed by the kinematic data generator (KDG) function in the SRMS control software, and the angular momentum imparted on the payload at release could be estimated as reported by Nguyen et al.[12] Figure 25 shows the payload angular momentum relative to the Orbiter based on the SPAS-01 gyro data and on the end effector estimated rates.

When in orbit, the SRMS is checked out before its first use. Following the power up and software initialization, the manipulator positioning mechanism is deployed and the arm retention mechanism is unlatched. The six joints in the unloaded arm are then driven in the direct drive mode, one at a time, for approximately 10 s each, in a sequence starting with the shoulder yaw joint, then the shoulder pitch

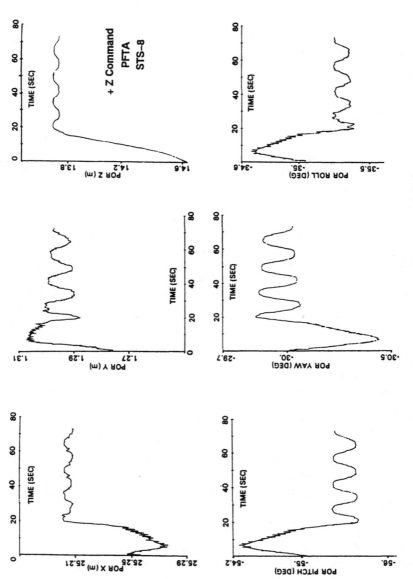

Fig. 22 POR response in manual augmented mode.

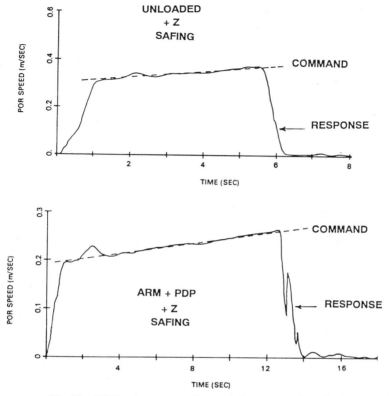

Fig. 23   POR rate response in manual augmented mode.

joint, etc. Joint angles and motor rate responses are recorded and downlisted for analysis and for comparison with data from previous missions. Anomalies in joint and motor behaviors can be detected by the checkout. It is interesting to note that in the preceding sequence of direct drive checkout tests, the moment of inertia as seen at a joint about its drive axis is independent of the joint angles of the inboard joints. As a result, the motor rate response at a joint is expected to be the same in all flights under normal conditions, regardless of the duration of the commands to drive other inboard joints. Of course, data from the same arm must be used in the comparison.

VII.   Concluding Remarks

Some key features of space remote manipulator systems have been presented and discussed in terms of the design, development, test and operation of the SRMS and SSRMS. A trend of improvements associated with new requirements and their implementation in the SSRMS can be readily observed. To cite a few, the capabilities of the SSRMS to self-relocate, to sense and accommodate loads in the end effector, and to have computer-aided vision, pose technical challenges in

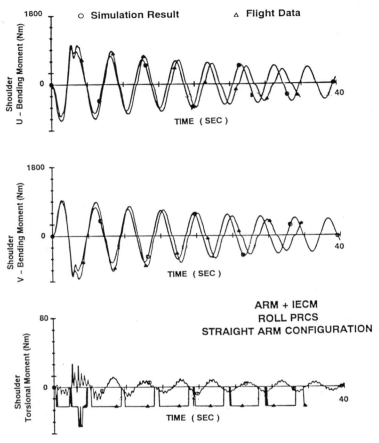

Fig. 24   Simulation results vs flight data.

the SSRMS's design and development. On the other hand, from an operation point of view, more complex operation procedures are anticipated and more extensive training will be required for the SSRMS as it will handle more complex and difficult robotic tasks, sometimes under less favorable conditions in comparison with the SRMS. However, with the success of the SRMS, it is hoped that the SSRMS will be successful as well.

The SPDM, which is part of the MSS, has not been discussed so far. Although it is similar to the SSRMS in many respects, it is a different kind of space remote manipulator system. Designed to handle much lighter and smaller payloads and to perform payload capture/release, berthing/deberthing more frequently, the SPDM requires different attention and considerations to be focused on the fine handling quality of the manipulator, although the reach and load handling capability deserves no less attention than it does in the design of larger manipulators such as the SRMS and the SSRMS. At the time of writing, the SPDM's design has gone past its preliminary phase and no significant problems have been encountered.

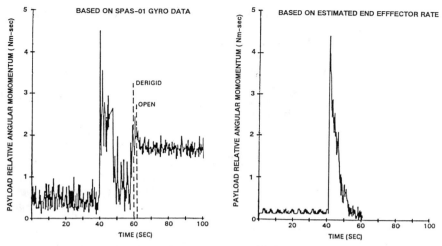

Fig. 25   Payload angular momentum during release.

Acknowledgments

The authors would like to thank Spar Aerospace Ltd. for permission to publish the information in this chapter.

References

[1]Anon., "Space Station Freedom Program Interface Control Document U.S. Space Station Manned Base to Mobile Servicing System," NASA SSP 42003C, Oct. 15, 1992.

[2]Dunbar, D. R., Robertson, A. R., and Kerrison, R., "Graphite/Epoxy Booms for the Space Shuttle Remote Manipulator," *International Conference for Composite Materials,* 1978.

[3]Gift, K., Gray, C., Reasor, G., and Turnbull, J., "Space Shuttle Operation Level C Functional Subsystem Software Requirements Document Remote Manipulator System (RMS)," *The Charles Stark Draper Lab. Inc.,* OI20, STS 87-0017D, Cambridge MA, March, 1991.

[4]Anon, *IGRIP Version 2.2 User Manual,* Deneb Robotics Inc., Auburn Hills, MI, 1992.

[5]Hughes, P. C., "Dynamics of a Chain of Flexible Bodies," *Journal of Astronautical Sciences,* Vol. 27, 1979, pp. 359–380.

[6]Sincarsin, G. B., and Hughes, P. C., "Dynamics of an Elastic Multibody Chain: Part A—Body Motion Equations," *Dynamics and Stability of Systems,* Vol. 4, Nos. 3 and 4, 1989, pp. 209–226.

[7]Hughes, P. C., and Sincarsin, G. B., "Dynamics of an Elastic Multibody Chain: Part B—Global Dynamics," *Dynamics and Stability of Systems,* Vol. 4, Nos. 3 and 4, 1989, pp. 227–244.

[8]Golla, D. F., Buhariwala, K., Hughes, P. C., and D'Eleutario, G. M. T., "Efficient Algorithms for the Dynamical Simulation of Structurally Flexible Manipulators," 9th Symposium on Engineering Applications of Mechanics, London, Ontario, Canada, May 29–31, 1988.

[9]D'Eleutario, G. M. T., "Dynamics of an Elastic Multibody Chain: Part C—Recursive Dynamics," *Dynamics and Stability of Systems,* Vol. 7, No. 2, 1992, pp. 61–89.

[10]Trenouth, J. M., and MacKenzie, C. W., "Investigation of the Torque Degradation Observed During Qualification Testing of Canadarm Joint Brakes," *National Research Council of Canada,* Rept. No. 26161, Ottawa, Canada, July, 1986.

[11]Middleton, J. A., Ashworth, K. L., and Aikenhead, B., "Flight tests of the Shuttle Remote Manipulator System (Canadarm)," 2nd Canadian Conference on Astronautics, Ottawa, Canada, Nov. 30–Dec. 1, 1982.

[12]Nguyen, P. K., Assaf, S. A., and Ravindran, R., "The Payload Deployment/Retrieval Performance of the Space Shuttle Remote Manipulator System," *Rendezvous and Proximity Operations Workshop,* Houston, TX, NASA Johnson Space Center, Feb. 20, 1985.

Bibliography

Beyer, G., Diebold, B., Brimley, W., and Kleinberg, H., "The Development of the Canadian Mobile Servicing System Kinematic Simulation Facility," Graphics Technology in Space Applications Conference, NASA Johnson Space Center, Houston, TX, April 12–14, 1989.

Burns, G., Ravindran, R., Aikenhead, B., and Windler, M., "Shuttle Remote Manipulator System Orbital Flight Tests and Results," CASI Conference, Toronto, Canada, 1985.

Butt, C., "Space Shuttle Remote Manipulator System Design Definition Report," Spar Aerospace Ltd., SPAR-R.776 E, Brompton, Ontario, Canada, March 1980.

Gossain, D. M., and Smith, P. J., "Structural Design and Test of the Shuttle RMS," AGARD-NATO Structures and Materials Panel 55th Meeting, Toronto, Canada, Sept. 22–24, 1982.

Hunter, J. A., Ussher, T. H., and Gossain, D. M., "Structural Dynamic Design Considerations of the Shuttle Remote Manipulator System," AIAA/ASME/ASCE 23rd Conference of Structures, Structural Dynamics, and Materials, May 10–12, 1982.

Kumar, P., Truss, P., and Wagner-Bartak, C. G., "System Design Features of the Space Shuttle Remote Manipulator," Proceedings of the Fifth World Congress on Theory of Machines and Mechanisms, American Society of Mechanical Engineers, 1979, pp. 839–842.

Kumar, R., "Space Station Remote Manipulator System Specification," Spar Aerospace Ltd., SPAR-SS-SG-0379 E, Brampton, Ontario, Canada, Aug. 1992.

Anon, "Payload Deployment and Retrieval System Simulation Database, Version 1.0," Lockheed Engineering and Sciences Co., NASA. Johnson Space Center, JSC/25134, Houston, TX, 1991.

Longman, R. W., "The Kinetics and Workspace of a Satellite-Mounted Robot," *The Journal of the Astronautical Sciences,* Vol. 38, No. 4, 1990.

Nguyen, P. K., Ravindran, R., Carr, R., Gossain, D. M., and Doestch, K. H., "Structural Flexibility of the Shuttle Remote Manipulator System Mechanical Arm," AIAA Guidance and Control Conference, AIAA Paper 82-1536-CP, San Diego, CA, Aug. 9–11, 1982.

Quittner, E., Borduas, H., Chumak, S., Mobrem, M., Mullins, M., and Piatkowski, M., "The Structural Design of the Space Station Mobile Servicing System Under Multiple Constraints," 4th Canadian Symposium on Aerospace Structures and Materials, 1988.

Ravindran, R., and Doetsch, K. H., "Design Aspects of the Shuttle Remote Manipulator Control," AIAA Guidance and Control Conference, AIAA Paper 82-1581-CP, San Diego, CA, Aug. 9–11, 1982.

Young, T. R., "Shuttle Remote Manipulator Testing and Performance Verification," CASI Flight Test Symposium, Montreal, Québec, Canada, March 16–17, 1982.

Chapter 15

Overview of International Robot Design for Space Station Freedom

W. Brimley,[*] D. Brown,[†] and B. Cox[‡]
Spar Aerospace Limited, Brampton, Ontario, Canada

I. Introduction

T O assemble and maintain the Space Station Freedom (SSF), a number of robotic designs have been developed. Large-scale manipulators that will be used include the Shuttle remote manipulator system (SRMS) and the Space Station remote manipulator system (SSRMS). Both the SRMS and SSRMS are Canadian designs required for SSF assembly. The complete on-orbit Canadian contribution at the man-tended SSF is the mobile servicing system (MSS), which comprises the SSRMS, a mobile base system (MBS), a special purpose dexterous manipulator (SPDM) and a MSS maintenance depot. A U.S.-supplied mobile transporter provides translation capability along the Space Station truss for the Canadian robotic equipment.

Japan is also providing a large-scale manipulator on the Japanese experimental module called the JEM RMS, as well as a dexterous manipulator or JEM small fine arm. These robots will be provided before the Space Station reaches permanently manned configuration.

The United States has proposed a flight telerobotic servicer (FTS) with capabilities similar to those of the SPDM.

The European Space Agency has also proposed the use of the HERMES spaceplane to rendevous with and berth to the SSF. The concept for the manned HERMES also had a RMS included, not necessarily for SSF operations, but for operations with a Columbus free-flying module.

Copyright © 1994 by the American Institute of Aeronautics and Astronautics, Inc. All rights reserved.

[*]Assistant Program Manager, Mobile Servicing System Operations, Advanced Technology Systems Group.
[†]Senior Member Technical Staff, Mobile Servicing System Operations, Advanced Technology Systems Group.
[‡]Senior Systems Engineer, Mobile Servicing System Operations, Advanced Technology Systems Group.

All of the aforementioned robot designs and concepts are presented in this chapter. Overall system configuration descriptions are provided, and the functions allocated to each robot are described. These functions include free flyer capture and berthing, payload manipulation, payload transportation, SSF assembly, maintenance and servicing, and crew extravehicular activity (EVA) support.

Missions are allocated to each robot. Representative design reference missions (DRMs) are described for each robot. These DRMs include Space Shuttle and Hermes berthing, Space Station assembly, Space Station logistics and resupply, and Space Station and MSS servicing and maintenance missions.

II.   Space Shuttle Remote Manipulator System

A.   Shuttle Remote Manipulator System Historical Perspective

As early as 1969, during the Apollo program, Canada was invited to participate in the Shuttle program. Under the guidance of the National Research Council (NRC), Canada evaluated the possible systems that Canadian technology could provide.

While NASA continued to develop the Shuttle concept, DSMA ATCOM Ltd. of Toronto and Spar Aerospace Ltd. were relentlessly pursuing the concept of a robotic arm, namely the remote manipulator system (RMS), which later became popularly known as the "Canadarm."

Following an evaluation of Canada's technological capabilities by NASA in 1973 and 1974, Canada and the U.S. signed an agreement in July 1975 to build the Canadarm. Canada was required to fund the design, development, and testing of the first arm and to deliver it by 1980. In return three additional robot arms were to be delivered in 1982, 1983, and 1984. The NRC was assigned to manage the project, acting as the focal point between NASA and Spar, the prime contractor. Portions of the development went to DSMA ATCOM, CAE Electronics of Montreal and RCA as subcontractors.

Development of the arm took nearly seven years, and the official signing over ceremony took place at Spar in Toronto in February 1981. Following the formalities the arm was packaged for road transport to the launch site at Kennedy Space Centre. The arm was installed into the Orbiter Columbia and flew in November 1981. During this second launch of Columbia, four hours of intensive checkout and performance verification proved to the operators, and NASA, that the arm met or exceeded specifications and expectations. During the third mission in March 1982 the RMS received much more extensive and exhaustive testing including thermal, end effector and payload retrieval testing. Braking and runaway performance were also examined in detail. Following the fourth shuttle flight in June of 1982 the SRMS was declared ready for operational use, and an agreement was signed in November of that year between NRC and NASA.

Since that time the SRMS has been used extensively for payload deployment and retrieval, with missions including the retrieval and subsequent repair of Solarmax (a solar observation satellite) on Flight 41-C in April of 1984, deployment of the long-duration exposure facility (LDEF) in April 1984, and the retrieval and subsequent return to Earth of communications satellites Westar IV and Palapa. A complete list of SRMS missions is given in Table 1.

Table 1   Shuttle Remote Manipulator System mission details

Flight	Date	Orbiter	Mission
STS-2	04 Nov. 81	Columbia	Unloaded arm and thermal tests
STS-3	22 March 82	Columbia	Loaded arm and thermal tests
STS-4	27 June 82	Columbia	Loaded arm and thermal tests
STS-7	18 June 83	Challenger	Shuttle pallet satellite unberthing/berthing tests
STS-8	30 Aug. 83	Challenger	Medium loaded arm tests
41-B	03 Feb. 84	Challenger	Astronaut on end of arm
41-C	06 April 84	Challenger	Deployment of Long Duration Exposure Facility, Solarmax
41-D	30 Aug. 84	Discovery	Removal of ice from shuttle
41-G	08 Oct. 84	Challenger	Deployment and release of Earth radiation budget satellite
51-A	06 Nov. 84	Discovery	EVA activities and Palapa satellite maneuvers
51-C	24 Jan. 85	Discovery	Department of Defense (classified)
51-D	12 April 85	Discovery	Attempted repair of Syncom satellite
51-G	17 June 85	Discovery	Release and retrieve astronomy (SPARTAN) payload
51-F	29 July 85	Challenger	Deploy and release plama diagnostic package payload
51-I	27 Aug. 85	Discovery	Repair and deploy Syncom satellite
61-A	30 Oct. 85	Challenger	Observation of water dumps
61-B	26 Nov. 85	Atlantis	EASE/ACCESS (Space Station construction activities)
51-L	28 Jan. 86	Challenger	Orbiter and crew lost during launch
STS-27	02 Dec. 88	Atlantis	Department of Defense (classified)
STS-32	09 Jan. 90	Columbia	Retrieval of Long Duration Exposure Facility
STS-31	24 April 90	Discovery	Deployment of Hubble space telescope
STS-41	06 Oct. 90	Atlantis	SRMS used as experimental platform, no payload
STS-37	05 April 91	Atlantis	Deploy gamma ray observatory and EVA development
STS-39	28 April 91	Discovery	Deployment and retrieval of infrared background signature satellite
STS-48	12 Sept. 91	Discovery	Deployment of upper-atmosphere research satellite
STS-49	07 May 92	Endeavour	Retrieval of Intelsat VI and EVA
STS-46	31 July 92	Atlantis	Deploy EURECA experiment
STS-52	22 Oct. 92	Columbia	Supported Canadian experiments (CANNEX)

B.   Shuttle Remote Manipulator System Description

The SRMS shown in Figs. 1 and 2 is the payload deployment and retrieval system fitted to the Space Shuttle Orbiter. The Orbiter was originally designed to accommodate two manipulator arms, one to be fitted on either side of the payload bay, although only one arm has ever been used on any mission. The arm is operated from the port side longeron, and there is a provision for operating a SRMS starboard side for specific missions.

The manipulator is 50 ft. long, consists of six joints connected together with structural boom sections, and is terminated at the free end with an end effector which forms the interface to the payload. The arm joints are driven by brushless dc motors connected through epicyclic gear trains which provide the required torque and speed characteristics.

The SRMS is controlled from the aft flight deck of the Orbiter by one of the crew members at a control station consisting of a display and control (D and C) panel and two, 3-degree-of-freedom (DOF) hand controllers. The manipulator arm comprises shoulder pitch, roll, and elbow pitch joints (providing end point translation) plus wrist pitch, yaw, and roll joints (providing rotations of the end effector). Limited shoulder yaw is provided.

The Orbiter general purpose computer is used via its software to convert the end point commands to individual joint commands. The operator may command joint movement using a number of reference frames, and may also choose automatic modes whereby automatic point of resolution (POR) trajectories may be executed. The SRMS systems are built to stop motion (fail-safe) in the event of a failure, and there is also the provision of a secondary back up drive which will allow the completion of a mission or movement of the SRMS to a safe position.

Built-in test equipment monitors the system for hardware and software faults, and allows diagnosis of system malfunctions. In the event of a major malfunction, the system is provided with an automatic means of bringing the system to rest. Each motor is provided with a friction brake which holds the joints in position when they are not actively commanded, the secondary function being emergency braking. An active control mode called position hold also holds the joints in position with no brakes on. In the event of a major malfunction which prevents the restowage of the arm to its cradle on the longeron the arm may be jettisoned at the shoulder using a pryotechnic cutting device.

C.   Shuttle Remote Manipulator Systems Operations on Space Station Freedom

The SRMS is the principal assembly tool for the Space Station until the delivery of the SSRMS. The SSRMS is launched on flight mission build 3 (MB-3), and will be checked out and available for nominal operations on MB-6A. The SRMS will continue to be fitted to the Orbiter after MB-6A to provide an additional means of removing SSF elements from the payload bay, and also to provide the capability to berth the Orbiter to the Space Station. The SRMS will have enhanced capabilities and upgrades to allow Orbiter berthing, which is a function beyond its original specification.

On MB-1, MB-2, and MB-3 the SRMS is used to deploy the truss sections to form the initial structure of the Space Station. The first section (S4/S3) consists of

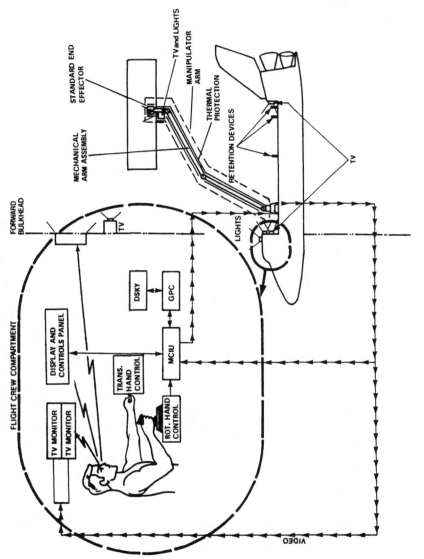

Fig. 1   SRMS.

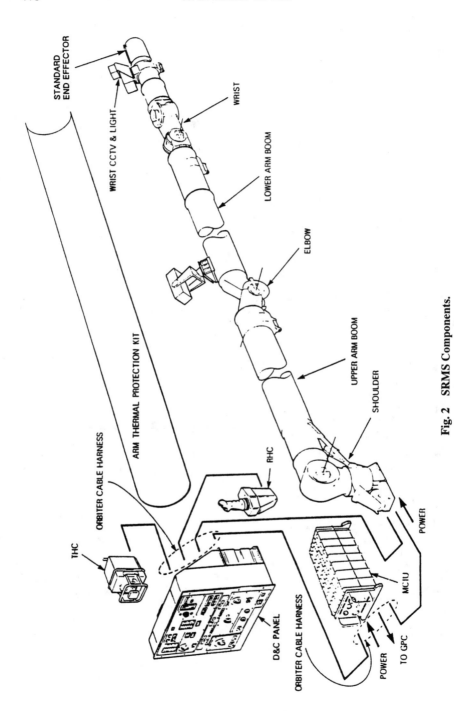

Fig. 2   SRMS Components.

the inboard photo voltaic system, the alpha joint, propulsion module platforms, the mobile transporter and the unpressurized berthing adapter. The first truss section remains inactive following its deployment and flies in a gravity gradient mode at an altitude of 221 n.mi. until the Orbiter rendezvous on MB-2.

The second mission consists once again of assembling a truss section containing communications and propulsion systems, and early avionics assemblies. The SRMS is used to deploy the truss section, and EVA is required to mate the truss section to the existing passive station. During the mission the station is powered via the Orbiter. Following activation of the power and propulsion systems, the station is actively stabilized, and the electronic systems can be kept in a keep-alive state. On MB-3 the SSRMS is delivered and manually deployed as outlined in Sec. III. Once again the SRMS is used to deploy the truss section carrying the SSRMS from the payload bay.

During the continuing build phases of 4, 5, and 6 the SRMS capabilities are utilized to build the Station to a configuration capable of supporting man tended capability. The spacecraft is fully operational under both automatic and ground control during periods when the Orbiter is not present. The habitable volumes of the lab and node are pressurized, and there is thermal and humidity control and CO_2 removal. There is an active power generation and management system, and active stabilization via gyros and cold gas thrusters. At the end of flight MB-6A the full operational capabilities of the SSRMS are available including berthing the Orbiter and unloading the payload bay. From flight MB-6A onwards the Shuttle arm is used to reach into the payload bay, but because of the configuration of the Station from this stage onward the majority of operations require a handoff maneuver to the SSRMS. Handoff maneuvers are used extensively both during the assembly and mature operations of the SSF.

The maneuver itself presents many difficulties with mechanical compatibility, dynamic performance of the respective arms at the point of handoff, and control source organization. A typical maneuver would consist of the following tasks.

1) The SRMS maneuvers to payload in the Payload Bay.

2) The SRMS grapples payload.

3) The Payload latches are released.

4) The SRMS maneuvers payload to position within reach of the SSRMS for handoff.

5) The SRMS holds position with brakes on.

6) The SSRMS maneuvers from a safe position, approaches the payload and positions itself ready to grapple the payload.

7) The SSRMS grapples the payload (during the grappling the SSRMS is in a limp condition to aleviate stored energy loads).

8) The SSRMS assumes control authority of the payload.

9) The SRMS releases the payload and moves away to a safe distance.

10) The SSRMS maneuvers the payload to its final location.

Handoffs can be broadly categorized into two groups.

1) There are handoffs from the SRMS to the SSRMS during which the payload is deposited directly into the Station. This category of maneuvers are principaly for SSF assembly operations in the early phases of the Station, when the reach of the SSRMS from its base on the MBS or lab/hab is adequate to deposit the payload directly into its berthing location. An example of such a maneuver is the

installation of the MBS onto the MT on MB-6A. Figure 7 illustrates the MBS handoff from the SRMS to the SSRMS.

2) There are handoffs from the SRMS to the SSRMS after which the payload is deposited onto the MBS. During the later assembly phases and during mature operations it is necessary to deposit the payload onto the MSC, perform a transportation along the truss and then deposit the payload using the SSRMS. An example of such a maneuver would be the change out of propulsion modules.

III.   Space Station Mobile Servicing System

A.   Space Station Freedom Introduction

The international Space Station Freedom is being designed to be constructed and occupied by international crews for an estimated 30-yr time period. Scientific tasks including astronomy, observation of the Earth's resources and environment, microgravity, and life-science experiments which will be performed by the crew on the Space Station. The Space Station will be assembled in a circular orbit approximately 220 n.mi. above the Earth, with an orbital period of approximately 90 min over the equator.

Components for assembling the 355-ft-long Space Station into it's final configuration will be lifted into orbit via the NASA fleet of four Space Shuttles. Launches from Cape Canaveral in Florida will be scheduled for every three months for a total of 18 flights. The initial flight, or first element launch (FEL), is scheduled for late 1995, and will carry the first part of the power generating system and truss structure. Additional flights will carry the SSF components including the European laboratory module, the American-supplied laboratory and habitation module, and the Japanese laboratory with two robotic arms. The Canadian contribution consisting of a large manipulator and a smaller dexterous robot is scheduled to be launched on the third and sixth Shuttle flights. Figure 3 shows the SSF in its permanently manned configuration, flight MB-17.

B.   Space Station Freedom Configuration

The Space Station is made up of preintegrated truss (PIT) segments of a long hexagonal structure called the integrated truss assembly (ITA), supplied by the U.S., to which all of the physical components required to support life and conduct the various planned experiments will be attached. Each ITA segment as well as the three laboratory modules and the habitation module are designed to fit into the cargo bay of the Space Shuttle, to be deployed using the Canadarm or SRMS.

Attached to both ends of the Space Station are large solar array panels, resembling wings, which will deliver a maximum power output of 58 kW to the users by tracking the sun with gimbaled motors as the Space Station orbits the Earth. In addition to the physical modules for the crew members, there are miles of pipes, electrical wires, and TV video cables which will run the 355-ft length of the Space Station connect to the various subcomponents required to keep Space Station functional.

In order for Freedom to be assembled and maintained in orbit for a 30-yr period it will be serviced by either astronauts performing EVA or by teleoperated extravehicular robots (EVR). Canada will be building and delivering these robots

a)

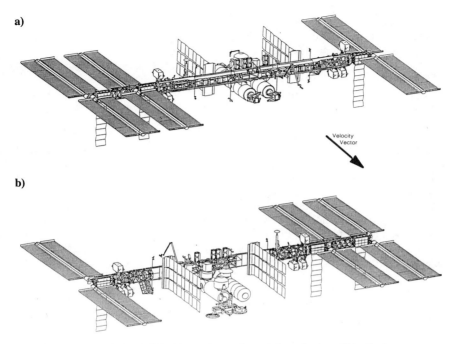

b)

Fig. 3   MB-17 MTC-PR configuration: a) foreview, and b) aft view.

as its contribution to the Space Station program. In return Canada will obtain 3% of the SSF usage (resources), and Canadian astronauts and experiments will be assigned to missions for on-orbit experiments and research. The following presents a general physical description of the Canadian robotic equipment supplied to the Space Station, and details some of the operating characteristics of those robotic components. Some of the components of the Space Station, as well as the Canadian equipment itself, are being designed to be serviced by either EVA crews or by telerobots operated by intravehicular activity (IVA) crews from IVA workstations within the pressurized environment.

C.   Mobile Servicing System System Configuration

Because of the complexity, cost, and time required to design and build the Space Station, the different physical elements that make it up have been apportioned between the U.S., various European nations, Canada, and Japan. The Canadian Space Agency (CSA) is providing, as the Canadian contribution to Space Station Freedom, the MSS. Canada previously developed and gave the first of the Space Shuttle robot arms to NASA, known as the Canadarm or SRMS. For SSF Canada will be supplying a new and longer manipulator, as well as a smaller, more dexterous two-armed version. These two robotic manipulators make up the major part of what is known as the Canadian MSS, and are known, respectively, as the

SSRMS and the SPDM. An introductory description of the MSS elements and subelements is described as part of this section.

The SSRMS is being designed to maneuver large payloads around the Space Station while the SPDM is being designed to perform maintenance tasks on the MSS, as well as on the Space Station Freedom.

To perform both the assembly and maintenance operations required to have an operational Space Station, either astronauts working together in pairs using EVA in procedures, or the use of teleoperated robots, or a combination thereof, will be necessary. EVA is potentially dangerous, time consuming, and requires the participation of internal and external crew members (who could be performing other tasks onboard the Space Station). The use of teleoperated robots is therefore attractive to mission operations planners. Time, and the division of it, is a key resource. In addition, because of the limitation of the amount of oxygen that can be carried within an appropriate safety margin, EVA is limited to a maximum of 6 h, and is always conducted by pairs of astronauts working in a buddy-system. Variations in resources available such as power crew member scheduling for task completion and consideration hazards associated with maintenance tasks must also be incorporated into any mission operations analysis. The mission scenarios may be analyzed using a computerized simultation which is a low-cost initial method of developing the planned scenario. EVAs must then be developed and tested in NASA's weightless environment training facility (WETF). EVA crews selected for the specific mission must then be trained using the WETF training mock ups.

Telerobots such as the SSRMS and the SPDM are not under the same constraints as EVA, although other variables such as dynamic lighting conditions as the Space Station orbits the Earth, power resources, and IVA crew time will still have to be accounted and budgeted for.

D.   Mobile Servicing System

The MSS is comprised of the mobile servicing center (MSC), the SPDM and the mobile servicing system maintenance depot (MMD). The MSC is the mobile portion of the MSS and consists of the mobile remote servicer (MRS) mounted on the U.S.-supplied mobile transporter (MT) element (see Fig. 4). The MRS includes the MRS base system (MBS), and the SSRMS. The MRS along with the SPDM will traverse from one end of the Space Station to the other on the MT. The MBS serves as the fully integrated structural carrier for the SSRMS and the SPDM. The MBS also provides the mechanisms for the support of and attachment of the SSRMS and the SPDM, which includes power, data linkage, and video signal routing via the MSC utility ports and the power data grapple fixtures (PDGFs). In addition to serving as the primary operations platform for both robotic devices the MBS provides payload orbit replaceable units (ORU) accommodation (POA), and other mechanisms for the attachment of Space Shuttle payloads, ORU pallets, SSRMS tools, portable work platforms (PWP), and the unpressurized logistics carrier (ULC). The MSC will also transport MSS robotic ORUs via the MMD which can also be attached via a suitable interface to the MBS. All of these systems are being designed to reflect the new, restructured Space Station Freedom requirements for assembly and maintainability.

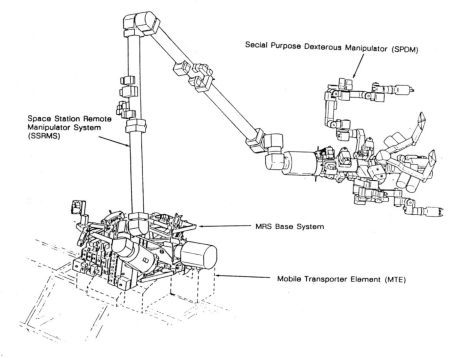

Fig. 4   MSC with SPDM.

E.   Mobile Transporter

The U.S.-supplied MT element provides the physical interfaces between the MRS and the Space Station truss, and provides translational mobility along the front face of the truss including the ability to cross over the alpha joints at either end of the Space Station. Electrical power for translational mobility, keep-alive, and safing is supplied by batteries which are part of the transporter energy storage system (TESS). A trailing umbilical system provides the hardwired link for data communications as well as power for fixed-base operations.

F.   Mobile Servicing Center (Space Station Remote Manipulator System and Mobile Remote Servicer Base System) Overview

The SSRMS is the capable of manipulating large payloads up to the mass of a fully loaded Orbiter (116,000 kg). The system consists of a seven-jointed arm which is terminated at each end with an end effector. The SSRMS is functionally symmetrical about the central (elbow joint) and is capable of relocating itself from its normal operating location on the MBS to SSF interfaces which are on the modules and truss. The PDGF interfaces provided for SSRMS operations on the lab or hab modules allow the SSRMS sufficient reach to berth and unload the Orbiter.

The MBS is delivered to orbit on Flight 6A as an independent package in the payload bay, upon which the SPDM and MMD are mounted on the MBS for launch. On-orbit, the MBS forms the primary support structure for the SSRMS and SPDM, and provides attachments for payloads through two different mechanical interfaces. These interfaces are an active snare-type POA (which is identical to an end effector), and a propulsion module attachment system (PMAS) which has active latches specifically designed for carrying the SSF propulsion modules. The MBS is also the primary location for MSS maintenance and provides the physical platform for many of the electrical, electronic, and video systems.

1. Space Station Remote Manipulator System

A unique feature of the SSRMS (see Fig. 5) is its ability to be relocatable by virtue of its being symmetrical and having identical latching end effectors (LEE) either of which can act as a shoulder or wrist to the SSRMS. The SSRMS can relocate onto PDGFs either on the MBS or strategically located on SSF. Video cameras, pan, and tilt units (PTUs), and lighting will be provided to monitor SSRMS operations. As with the SSF, the SSRMS is being designed to remain on orbit for the projected 30-yr operational time period.

2. Space Station Remote Manipulator System Performance

Table 2 shows the payload handling characteristics including maximum tip velocities for various Mass payloads and stopping distances.

3. Space Station Remote Manipulator System Control

The SSRMS is controlled from an IVA workstation located in either the cupola or the node. Each workstation consists of a multipurpose application console (MPAC), capable of displaying control and command data, display monitors for video views, and two 3-DOF hand controllers similar to the SRMS configuration. Using display pages the IVA operator may select suitable operating modes for the SSRMS in both manual and automatic modes. In manual mode the operator may

Table 2   Performance requirement: payload handling (tip velocity and stopping distances)

Function	Reference payload mass (kg)	Maximum velocity		Maximum tip stopping distance	
		m/s (ft/s)	deg/s	m(ft)	deg
Capture	20,900	0.024 (0.079)	0.08	1.46 (4.79)	5.4
	116,000	0.024 (0.079)	0.08	2.85 (9.35)	16.1
	0	0.37 (1.214)	4.0	0.61 (2.00)	3.0
Maneuvering	1000	0.15 (0.492)	1.2	0.61 (2.00)	2.62
	20,900	0.02 (0.066)	0.15	0.61 (2.00)	3.16
	116,000	0.012 (0.039)	0.04	1.25 (4.10)	7.2
Berthing/	1000	0.10 (0.328)	0.8	0.305 (1.00)	1.82
deberthing	20,900	0.011 (0.036)	0.085	0.305 (1.00)	1.69
	116,000	0.006 (0.020)	0.02	0.61 (2.00)	3.19

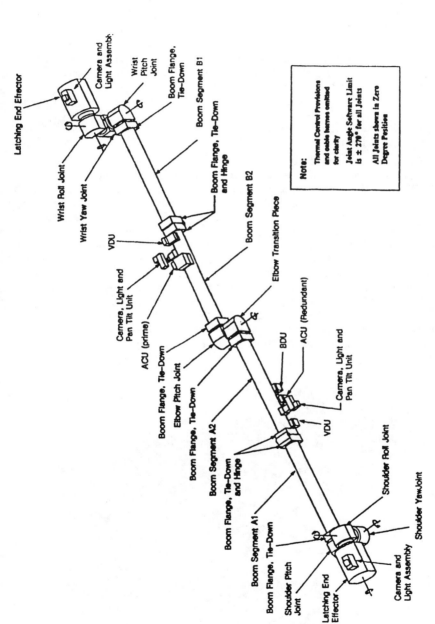

Fig. 5 SSRMS physical configuration.

command the point of resolution (POR) of the arm to any allowable position in the work space as well as command individual joints at various rates. In automatic mode the operator may choose to initiate prestored sequences for joint or POR movement, command the joints or POR to predetermined points or allow the SSRMS to be semiautonomous with the assistance of the artificial vision system (AVS). The AVS is capable of automatically tracking targets to reduce operator workload.

4. Space Station Remote Manipulator System Launch Configuration, Delivery to Orbit, and Unstowing

The SSRMS is delivered to orbit on the third Space Station assembly flight as part of a highly integrated launch package based on part of the PIT. As well as the SSRMS, the PIT structure contains radiator panels to dissipate heat from the power system, antennae, and utilities. The SSRMS is integrated onto the PIT in a tightly folded configuration to sustain launch loads. Both latching end effectors are mechanically interfaced to the PIT, and tie down bolts secure the booms. On flight three of the assembly sequence the Orbiter docks to the existing Space Station, and the PIT segment including the folded SSRMS is deployed from the payload bay using the Canadarm (SRMS). The PIT segment is attached to the existing truss and utilities are connected using a combination of automatic fastening devices and EVA. The SSRMS is then deployed from its launch configuration.

Deployment of the SSRMS is an EVA intensive activity, and present estimates indicate two astronauts will be engaged in the task for 5.5 h (see Fig. 6). The deployment sequence consists of the following major activities:

Fig. 6 Astronants unfolding SSRMS booms.

1) Eight flight support equipment bolts pass through the SSRMS, four at either end of the folded configuration. Each of them is removed with an EVA power tool, temporarily stored in a quiver, and then stored in a rack in the truss itself.

2) The light end of the SSRMS is raised manually to clear the PIT structure.

3) The astronauts position themselves on either side of the hinge and manually rotate the boom segments 180 deg so the SSRMS is unfolded. The hinge is then bolted.

4) The LEE are also held down firmly with a support structure to survive launch loads, and the two astronauts must remove the tie downs using power tools.

5) The astronauts elevate the booms manually so as to clear a grapple fixture, which is used by the Orbiter for deberthing on flight 3 and reberthing on flight 4. The grapple Fixture is positioned close to the SSRMS launch package.

Once the deployment is complete the SSRMS is left in a keep-alive state, checkout testing and verification of the performance of the SSRMS has to be completed prior to the first nominal operation of the SSRMS. Full control of the arm is established on flight 6 when the cupola workstation is available for IVA control of the SSRMS.

5. Space Station Remote Manipulator System First Nominal Operation

On flight MB-5 the cupola is installed using the SRMS, and on flight MB-6 the lab is installed and the cupola is pressurized and may be entered. The SSRMS is checked out and performance is verified on MB-6.

The first nominal operation of the SSRMS is the installation of the MBS onto the MT on MB-6A. The SSRMS operates from its initial location on the PIT 3 segment. The SRMS reaches into the cargo bay, grapples the MBS and then positions itself so that the SSRMS can take this payload in a handoff maneuver. Once the SSRMS has grappled the MBS payload, the SRMS releases it and moves clear so that the SSRMS can berth the MBS to its final location on the MT. Once again EVA is required to mate the interface, deploy cameras and lights, and to connect utilities. On completion of the MBS installation the SSRMS relocates onto the MBS. After a full checkout of the SSRMS and MBS on flight MB-6A, the SPDM/MMD payload is also installed onto the MBS in a handoff maneuver using the SRMS and the SSRMS (see Fig. 7) The SSRMS is left with both end effectors down on the MBS until its next mission.

6. Space Station Remote Manipulator System Operation from Lab Power Data Grapple Fixture

As was previously explained, the SSRMS has the ability to relocate itself to another PDGF and to use that as its base, and perform nominal operations. An example of such a maneuver is the changeout of the pressurized logistics module (PLM). The PLM is a module carried in the Orbiter (every other flight after assembly flight 7) which contains consumable supplies for the crew, and spares and equipment for the scientific experiments. PLMs are exchanged on a one-for-one basis, the spent module returns waste products, equipment for repair, and returned scientific experiments.

Once the MSC (SSRMS and MBS) is powered up and checked out, the SSRMS steps off from the MBS to the underside of the lab module, latches onto the PDGF

Fig. 7   Handoff of MBS from SRMS to SSRMS.

and releases from the MBS. The arm then maneuvers its free end to the Orbiter Payload bay, aligns with the target on the PLM and grapples the PLM. Control for this maneuver is maintained by one of the crew in the cupola workstation where direct viewing is available.

Having established that the PLM is grappled correctly, the Orbiter payload latches are released and the PLM is taken from the Orbiter and positioned beneath the vacant node. The PLM is berthed to the node using an active berthing interface, so EVA is not required. After the release of the new PLM, the SSRMS maneuvers over to the spent PLM, grapples it and then returns it to the Orbiter payload bay. To complete the mission, the SSRMS steps back onto the MBS, and back to its keep-alive and storage position. The complete sequence is represented by Figs. 8 and 9.

G.   Special Purpose Dexterous Manipulator

1.   Special Purpose Dexterous Manipulator Functions

The SPDM will feature capabilities that will allow it to successfuly complete SSF servicing and MSS maintenance tasks. A major goal of the SPDM is to reduce EVA task hours, which are potentially hazardous and time consuming, since EVA requires two suited crew members cooperating in a buddy system. A third IVA crew member monitors their activity visually either directly or indirectly via closed circuit television (CCTV).

The SPDM which is being designed for external assembly, maintenance, and inspection operations for SSF, performs missions similar to EVA operations. The

Fig. 8 SSRMS relocating to module.

Fig. 9 PLM in its final berthing position beneath node.

SPDM will perform the task of removal and replacement of both SSF and MSS ORUs. The selected design of the SPDM provides the required operational capabilities.

The SPDM is a teleoperated robotic device which will conduct both nominal and contingency operations in the unpressurized environment onboard Space Station Freedom. Primary missions and tasks will involve removing and replacing ORUs on the Station on both a scheduled and unscheduled basis. The SPDM is being designed to berth ORUs classified as small (24 in. × 24 in. × 24 in.), medium (39 in. × 39 in. × 19 in.), and large (39 in. × 39 in. × 39 in.), respectively. There are currently almost 300 SSF ORUs fulfilling various functions which are being designed to be compatible with the SPDM for removal and replacement. In addition to the on-orbit removal and replacement of ORUs, other maintenance actions which SPDM is expected to fulfil include adjustment and/or maintenance of ORUs in situ, support to EVA operations, and inspection of SSF components and hardware over the 30-yr operational life of SSF.

2. Special Purpose Dexterous Manipulator Physical Description, Configuration, and Capabilities

The SPDM consists of a single-DOF base body roll segment incorporating a PDGF for SSRMS attachment, and a LEE to provide for attachment to the MBS. The SPDM will be capable of operating from any one of the four MBS PDGFs, from the end of the SSRMS, or from a PDGF located on the Space Station itself (i.e. lab, JEM). The physical mass allocation for the SPDM and tools is 815 kg (2118 lb). The SPDM body section is attached to the base by a roll joint (a modified SSRMS yaw joint), and incorporates the onboard avionics, has provisions for the stowing of four tools as well as two ORU temporary storage locations. The body will have three cameras, two mounted on PTUs and the third attached to the LEE for attaching the SPDM's LEE to a PDGF when transported by the SSRMS. AVF is a planned feature of the SPDM which will allow it to automatically lock onto and track a target, and orient the SPDM's arm for final grappling by the IVA teleoperator.

Part of the coaxial designed body is a shoulder bar structure to which two 7-DOF arms are attached. Both arms are composed of two equal boom segments housing their joints ORU tool change out mechanism (OTCM) associated electronics to yield an overall arm length of 136.34 in. and 644 lbs (Fig. 10). Each arm has seven offset joints seven joint electronics units (JEUs) which provide the electronics control. Each joint is housed in an 8-in.-diam structure and each joint has a rated torque of 150 ft lbs. Because the two arms are mounted above the roll joint they will be able to rotate 360 deg with respect to the LEE and ORU accommodations platforms.

Each 7-DOF dexterous arm will incorporate an OTCM. The OTCM incorporates a camera and dual lamp assemblies to visually acquire ORU targets. The OTCM will also incorporate a force moment sensing (FMS) device which, following calibration, will accurately sense the amount of force being generated on the arm. Each OTCM will also have an electrical connector capable of delivering a total of 500 W of power via a separate interface connector to proposed tools and inspection equipment.

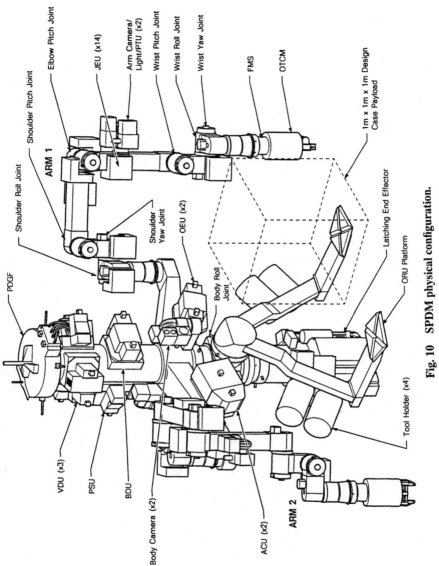

Fig. 10   SPDM physical configuration.

SSF power consumption allocation for the SPDM is 1900 W peak with an average power usage of 1300 W. Every operation which will utilize the SPDM will be examined for power budget allocation requirements prior to commencement of the operation.

3. Special Purpose Dexterous Manipulator Launch Configuration

The design concept at the time of writing, to manifest and launch the SPDM, is a four-trunnion pin design, which would have the SPDM bolted down and latched to the MMD in the Shuttle payload bay. Strict launch criteria must be observed with respect to the Shuttle physical volume constraints and center of gravity requirements. The MMD must also 1) be able to be carried by the MBS without obstructing the SSF forward truss face transport corridor when it passes in front of the berthed Space Shuttle, 2) provide adequate MSS ORU storage capabilities, and 3) provide the launch carrier for SPDM.

4. On-Orbit Operations of the Special Purpose Dexterous Manipulator

The SPDM's maintenance role will include servicing of the SSRMS and MBS, as well as maintaining the Space Station itself. Operating from MBS PDGF # 2, in a stand-alone mode for SSRMS maintenance, the SPDM will be able to access all five robotically compatible SSRMS ORUs. These include the two SSRMS Visual display units (VDUs), two arm computer units (ACUs) and one backup drive unit (BDU).

Operational tip translational velocities for the SPDM have been proposed to be as follows; 5.0 cm/s for maximum unloaded vernier mode, and 1 cm/s in vernier mode when fully loaded with 600-kg payloads. Rotational velocities for tip vernier velocities are proposed to be 1.67 deg/s, when loaded and 0.33 deg/s when fully loaded. Dynamic simulations to be conducted at Spar will confirm these preliminary values.

In addition to MSS servicing operations, nominal SSF maintenance operations are scheduled to be performed by the SPDM, including the routine exchange of the oxygen and nitrogen cryo carriers which will be changed on a scheduled basis. Special tie down bolts which are SPDM compatible are being incorporated as part of the design to these cryo carriers. Other tasks include the removal and replacement of battery units which will not be able to be recharged after some time, and small electronics units of varying functions onboard SSF.

5. Special Purpose Dexterous Manipulator Control Characteristics

The SPDM will be controlled teleroboticaly from a workstation inside the IVA environment known as the IVA control station (IVA CS). The SPDM is being designed with specific capabilities. FMS and accommodation will be implemented with SPDM's initial launch. The SPDM will be capable of delivering a tip force of 111 N (25 lbf) (which would be useful for inserting and removing a box type ORU) and of generating a torque of 54 Nm (40 lbf) required for fastening and unfastening a bolt. The SPDM has a positioning repeatability of 0.125 cm (0.05 in.) to get onto that bolt, and a payload handling capability of up to 600 kg (1320 lbs) to maneuver the ORU to and from the Space Station. It is projected that most ORUs onboard SSF will be in the 100-300 kg range.

There will be three distinct modes of operating the SPDM. The first is defined as the manual augmented mode, where the IVA operator would input commands via two 3-DOF hand controllers which would cause the selected arm to move to a particular POR within the task space area. The second is the single joint mode in which each of the SPDM's 19 individual joints could be commanded, (for example, to reposition the base body joints for better access to the ACU ORU). The third mode, the automated trajectory mode, commands the SPDM along prescribed trajectories generated either from previously stored information, or as determined from POR target coordinates, or vision system tracking signals.

6. Special Purpose Dexterous Manipulator Operating Locations

The SPDM will operate either from the end of the SSRMS or attached to a PDGF on the MBS. The PDGF serves as the connector outlet which would deliver all required power, robot command signals, and TV video links to/from the SPDM at appropriately located utility sites. There are four PDGFs located on the MBS as well as at additional sites located elsewhere on the SSF as determined by access requirements for performing the Station maintenance tasks.

7. Special Purpose Dexterous Manipulator Kinematic Simulations

As a first step, to verify the task scenarios, kinematic simulations of the current SPDM configuration have been developed on a computer workstation with special software tailored to robotic motions. This SPDM simulation is then used in representative tasks, such as an ACU changeout scenario, which helps to derive and define the operation of the SPDM performing that task.

8. Special Purpose Dexterous Manipulator Design Reference Missions

The MSS Operations Group at Spar, in conjunction with CSA and NASA, has developed realistic and representative assembly and maintenance scenarios. These scenarios are used to investigate and help derive requirements which will determine such SPDM design factors as physical configuration, joint angle limits, and rates, as well as working envelopes and fault tolerance requirements. These scenarios are commonly referred to as design reference missions (DRMs) and encompass both MSS specific maintenance tasks as well as the Space Station servicing and maintenance. The MSS is a unique system on the Space Station; it is a robotic system that is largely autonomous and capable of self-maintenance. The MSS is one of the most complex systems on the Space Station, so if the SPDM can maintain the MSS, it will likely be able to maintain most other Space Station equipment.

SPDM times are derived by using joint angle travel rates which are on the order of a maximum of 5 deg/s for an unloaded SPDM but are reduced to 1 deg/s for a 600-kg payload. Some common routines have standardized times. As an example, for the SPDM arm to release from a microinterface, a time of 30 s (0:30) is allocated.

Special purpose dexterous manipulator example. The dexterous manipulator's task of replacing the SSRMS upper or lower boom segment ACU can be summarized as follows. The spare ACU is checked for readiness through BIT. The SSRMS and the SPDM are individually powered up and checked out. The SSRMS is maneuvered to the SPDM base PDGF, where it attaches and latches itself to the SPDM. Together they are manipulated to the spare ACU location where they acquire the spare ACU and temporarily stow it on the SPDM base ORU 1 location. The SSRMS deposits the SPDM back on the MBS and then SSRMS is stowed in its shoulder roll ACU maintenance configuration. The failed ACU is disconnected by the SPDM, then removed and temporarily stowed on the SPDM base 2 ORU location. The replacement ACU is removed from its temporary stowage at base location 1, fitted to the SSRMS, electrically connected and checked out by BIT. The failed ACU is removed from its temporary stowage and placed in long-term storage with keep-alive power connected, using the same SSRMS and SPDM combination. The manipulators are stowed in their normal operating configurations and powered down.

As a robotic task, removal and replacement of the ACU requires additional trajectory planning over and above that of an EVA activity. The manipulator arms could arrive at a singularity configuration at some point within the task scenario. A singularity configuration occurs when no definite solution for a particular point in space can be achieved, or if a position is not reachable by either SPDM arm. Validated kinematic and dynamic simulations performed a priori on the ground will ensure consideration of this situation. In addition, this mission takes longer than one orbital period of 94 min, so that lighting must be arranged to provide adequate camera views for the IVA operator during all phases of the task. Robotic activities may cause some vibrations, which may require some SSF scientific experiments to be rescheduled or, alternatively, SPDM maintenance to be postponed to avoid disrupting these experiments. Power consumption also needs to be planned to ensure availability and compatibility with any on-going science experiments and other SSF operations.

Special purpose dexterous manipulator robotic arm computer unit design reference mission example. The following task outline in DRM format lists the procedure and its associated task times (min:s) for removing and replacing the SSRMS shoulder roll ACU ORU. IV1 is used to designate the SPDM intravehicular teleoperator crewmember 1 operating the SPDM from a SSF control station.

The operator would first verify that the spare ORU was operational, probably while it was still in its storage configuration. This would be followed by a MSS system power up and test of the MBS, the SSRMS, and the SPDM which would last approximately 1/2 h. After moving down the forward truss face from its parked location, the MBS would be stopped and its brakes applied adjacent to the MMD. The time necessary for this activity to be completed depends on the distance to be moved along the truss, although a rate of 1 ft/s is planned. The operator would then acquire the SPDM with the free end of the SSRMS, and, with the SPDM attached, would transfer the SSRMS/SPDM to the location where the new ACU is stored. It is planned that each of the ORUs would ideally have one bolt passing through the centerline which would attach the ORU to the Space Station. Electrical connectors would have a similar bolt incorporated as part of the connector which would be attached prior to the ORU being undone and after it was reattached to the Space Station. The IVA astronaut operator would maneuver the SPDM's arm over the bolt using a camera attached to the end of the arm and then close the robot's fingers on a doorknob-like device called an H handle because of its shape. In the middle of the H would be the bolt which would then be undone by a tool built into the end of the arm. A typical step-by-step breakdown of the sequences involved is as follows. Examples are given in min:s.

1) SSRMS move SPDM to new ACU storage location.

2) SPDM acquire replacement ACU.

3) IV1 align SPDM arm #1 and attach it to electrical connection mate/demate mechanism of replacement ACU (2:40).

4) IV1 disconnect electrical connector using SPDM arm #1 (1:40).

5) IV1 detach SPDM arm #1 from electrical connection mate/demate mechanism then record arm #1 position as position 2 (0:45).

6) IV1 align SPDM arm #1 and attach it to single tie down mechanism of replacement ACU (2:40).

7) IV1 unfasten single tie down screw using SPDM arm #1, then record position as Position 3 (1:55).

8) SPDM remove replacement ACU from MMD stowage location using SPDM arm #1, then record SPDM position as position 4.

9) SPDM store ACU on body stanchions.

10) SSRMS move SPDM to worksite.

A similar sequence of events would take place at the location of the failed ORU where the SPDM would first remove the failed article, stow it, and would then pick up the new ORU and place it in the original's location. A diagnostic check test would then be run to verify that the new unit is functioning properly. The maintenance activity could take place from either the end of the SSRMS or from the MBS itself with the SPDM reaching over to acquire the failed ORU.

H.   Mobile Servicing System Maintenance Depot

The MMD is situated at a fixed location on the truss where MSS-specific spares and tools are stored, and can be accessed for MSC maintenance. This location is tentatively on P1, face 4, situated on a preassigned module attach system (PMAS) interface. The MMD could be removed onto the MBS for transport by the SSRMS if needed for MSS maintenance operations.

IV.   Japanese Experimental Module Remote Manipulator System/Small Fine Arm

A.   Introduction

As part of their contribution to SSF the Japanese Space Agency NASDA are providing a Japanese Experimental Module (JEM). The JEM is a multipurpose research and experimental facility attached to the Space Station, which consists of a pressurized microgravity laboratory, two external exposed facilities for attaching scientific experiments, a pressurized logistics module, a large manipulator known as the main arm, and a small fine arm (SFA) for dexterous activities around the exposed facilities. The complete robotic system with the control equipment and

computing facilities in the pressurized module is known as the JEM remote manipulator system (JEM RMS). The configuration of the JEM is shown in Fig. 11.The
assemblies are launched on MB flights 12 and 15.

B.   Main Arm and Small Fine Arm

The main arm is attached to the aft end cone of the pressurized module (PM)
and has the capability to reach out over the exposed facilities for payload and
experiment placement and retrieval. The main arm is terminated in an end effector
similar to the Canadarm (SRMS) interface, which can attach itself to the small
fine arm. In fact, Spar has a contract to provide the end effector for the JEM arm.

The SFA would normally be stored on the exposed facility and picked up when
needed. The SFA is terminated at one end with a grapple fixture that can be
interfaced with the main arm end effector, and at the other end by a gripper for
dexterous tasks around the exposed facility workspace.

An airlock is built into the aft face of the laboratory pressurized module allowing
experiments to be taken out to the space environment. Direct viewing of the
exposed facilities is available from the JEM RMS operator's console inside the
pressurized module. The JEM RMS also has the reach capability to accept a
handoff of a payload to and from the SSRMS, which allows payload retrieval
from the Orbiter payload bay, and allows the return to Earth of experiments on
their completion. The general arrangement and performance characteristics of the
JEM RMS as shown in Figs. 12 and 13 and Table 3.

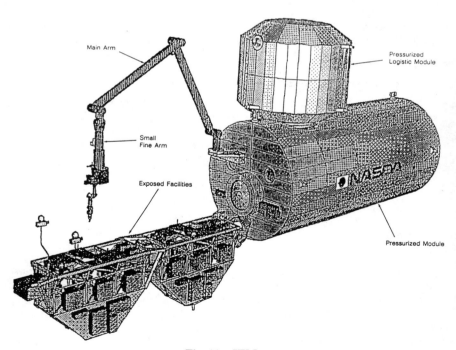

Fig. 11   JEM.

Table 3   Japanese experiment module remote manipulator system characteristics

		Main arm	Small arm
Arm length (m)		9.7	1.6
Arm boom (cm o.d.)		31	11
Weight (kg)		375.5	85.5
	Shear force (N)	70	30
Tip force	Bending moment (Nm)	140	4.5
	Role torque (Nm)	200	4.5
Tip speed	No load (cm/s)	45	49
	Full load (cm/s)	4.5	4.9
Payload	Mass (kg)	7000	TBD[a] (700)
	Volume (m ϕ × mL)	4 × 4	TBD
Positioning accuracy		±50/ ± 1°	±10
(automatic) (mm/deg)	(mm)		
Control		Position/velocity control	Bilateral
Man machine		Hand control	Master-slave

[a]TBD = to be determined.

C.   Control System

The control console is situated close to the airlock and window at the aft end of the JEM PM. At the console the operator may input commands to the RMS using a 6-DOF hand controller. Video viewing is also available. Manual control is a master-slave arrangement, and automatic trajectories may be selected.

The SFA is also equipped with force moment sensing and force moment accommodation which allows feedback to return to the operator regarding the forces and moments developed at the tip of the SFA. FM sensing and accommodation allows faster berthing of payloads and prevents the potentially damaging buildup of forces in the arm structure. FMS allows the SFA to feel its way into passive berthing mechanisms.

D.   Launch Package and Delivery to Orbit

The main arm is folded onto the end of the JEM PM for launch, and is attached to the end of the pressurized structure by flight support equipment. EVA is required to deploy the main arm. In addition to the PM core, flight MB-12 includes outfitting and payload racks for the module.

The JEM PM is assembled to the SSF cluster in a complex handoff maneuver. The SRMS unloads the JEM payload from the Orbiter bay and positions it beneath the modules, so that the SSRMS positioned on its base can accept the handoff and take control of the payload. The SSRMS then positions the JEM PM close to the active interface on the node, and the final berthing stage takes place. EVA is required to connect utilities and to unstow the JEM RMS. On flight MB-15

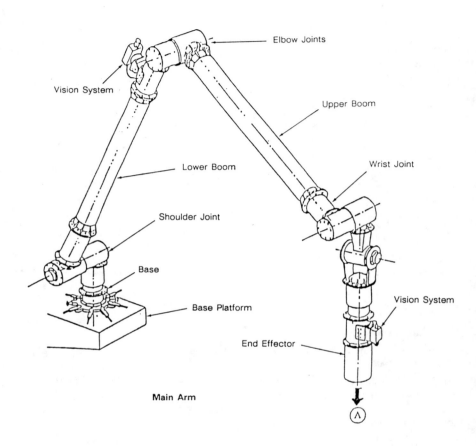

Elbow Joints

Vision System

Upper Boom

Lower Boom

Wrist Joint

Shoulder Joint

Base

Base Platform

Vision System

End Effector

Main Arm

Replaceable in Orbit

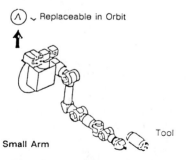

Small Arm

Tool

Fig. 12   JEM RMS subsystem.

the exposed facilities are launched and both are handed off to the SSRMS which positions the exposed facilities for EVA attachment.

V. Flight Telerobotic Servicer

The flight telerobotic servicer, or FTS (Fig. 13), was conceived as a means of incorporating U.S. robotics technology on Space Station Freedom. The U.S. Congress was interested in advancing both robotics and automation technology for the benefit of the Station, as well as directing spin-offs to the U.S. economy. In addition to ensuring technology transfer between various U.S. industries, the FTS would also serve to provide telerobotics assistance to early Station assembly tasks, service attached scientific payloads, and serve as a telerobotic assistant to EVA crewmembers.

The prime contract to manufacture the FTS was allocated to Martin Marietta Corp. with NASA's Goddard Spaceflight Center (GSFC) serving as the technical/managerial lead for integration onto the Space Station. The FTS was scheduled to fly as part of the Station's first element launch package in 1995, but the FTS program was terminated in 1991 because of SSF budgetary cutbacks. The FTS is included here for completeness. Figure 13 shows the final FTS concept.

A. Flight Telerobotic Servicer Elements/General Requirements and Capabilities

The FTS consisted of a short body torso section into which were incorporated the batteries, battery regulator/charger and communication module, which would allow independent operation from a worksite following the FTS being relocated

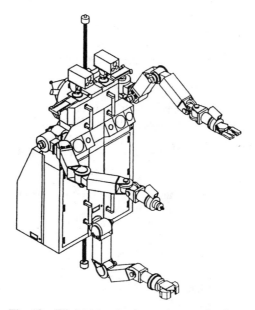

Fig. 13   Flight telerobotic services—telerobot.

there by the SSRMS. These systems were located in the aft section of the FTS body. On the front side were two system data processors (SDP), prime and redundant telerobot computer controllers.

Operational duty cycles were based on a projected 30-h/week utilization of the FTS for operational tasks. System power was rated at 1000 W average and 2000 W peak, with 350 W as standby power for keep-alive requirements.

Two 7-DOF manipulator arms, each 7-ft long with forearm cameras with a 2:1 zoom ratio were attached to the upper torso, above which was the camera positioning assembly from which two additional head cameras with a 6:1 zoom ratio were attached. These cameras would be used for the IVA teleoperator to acquire or position a robotic tool or payload. The FTS was designed to exert a manipulator tip force of 20 lbf and a torque of 20 ft-lbf, and achieve a repeatability of 0.005 in.

An attachment stabilization positioning system (ASPS) was fixed below the body, and was intended to provide a stabilization path for reducing dynamic oscillation when the FTS operated from the end of the SSRMS, or when it was attached independently to the Station. The FTS, which was allocated a launch weight budget of 1500 lbs was scheduled for launch on MB-1, the first flight for the assembly of the Space Station.

VI. HERMES Manipulator

The HERMES spaceplane proposed by the European Space Agency (ESA), was to carry crew into space and berth with the SSF and European freefly-ers. The concept of the HERMES Space Shuttle incorporates a manipulator arm called the HERMES robot arm (HERA). The HERA functions are capture, berthing/unberthing, payload transfer (manipulation) and release, spaceplane in-spection, ORU exchange, tool manipulation and actuation, EVA support, and jettison capability. The HERA system breakdown is into three main areas: 1) the manipulator arm 2) the controls 3) the manipulator ground facilities.

A. HERMES Robot Arm

The configuration of HERA is a 7-DOF manipulator symmetrical about its elbow, and stowed on the back of the spaceplane's expendable resource/propulsion modules. The HERA has two carbon composite booms joined by the elbow joint. Each wrist has 3-DOF, and each end effector is identical. The HERA weight is 200 kg, and the arm is 11.2 m long. Positioning accuracy is specified as 5 mm for a design case payload of 3500 kg.

The HERA is a relocatable manipulator that could be walked off the HERMES spaceplane and onto the Columbus man-tended free-flying laboratory (MTFF). The design life is for 10 yr on-orbit attached to the MTFF.

B. Future of HERMES/HERMES Robot Arm Buran

Unfortunately, European support for the HERMES spaceplane, and with it the HERA, has declined, and late in 1992 ESA support for HERMES was dropped. Early in 1993, France was still showing some determination to proceed with HERMES itself. Reasons for the loss of European interest in HERMES include the program cost of the European community, and some interest in pursuing a joint program with Russia to develop the Buran orbiter.

It is of interest to note that the Russians are reported to be developing a manipulator for Buran. The arm is 15.3 m long, with a mass of 350 kg. It is reported that the arm is capable of deploying 30 tons from the cargo bay of the Buran. The tip speed of the arm is 4 cm/s, and the arm can be controlled manually or automatically.

VII.   Future Evolution of Space Station Robotics

Once the space station has been completed, there will be possibilities for further growth of the SSF and its associated robotics.

A.   Growth of the Space Station

The original concepts for the SSF included dual keels and upper and lower transverse booms. A large enclosed servicing facility was also considered. Although the present SSF has been considerably descoped through numerous scrubs and restructurings, it may be possible to build the Space Station into a dual-keel configuration. Enclosed servicing bays for lunar transfer vehicles, orbital maneuvering vehicles, etc., may be provided. The growth SSF may provide an assembly and servicing platform for assembly of manned interplanetary vehicles.

B.   Space Station Freedom Robotics Evolution

If the Space Station grows significantly, a second SSRMS and SPDM would be desirable to provide robust servicing capability. Redundant manipulators would allow continued robotic capability, even when a manipulator is withdrawn from operational use for maintenance. The first operational requirement to be met before any manipulator upgrades are initiated, would be to replace the manipulator to be withdrawn from service. Therefore a second SSRMS and SPDM would first be upgraded on the ground before launch with the latest improvements. After installation and checkout on orbit, the original SSRMS and SPDM could be withdrawn from operations for upgrading and refurbishing. This could be performed on orbit, but may be more easily performed on the ground.

Aside from the possible upgrade/refurbishment of the existing manipulators, additional capabilities could be added to the MSS. These capabilities include ground-controlled telerobotics, collision detection using on-orbit sensors, advanced vision systems, and tactile sensing. The potential of using two SSRMSs or two SPDMs simultaneously also exists, with the ramifications of coordinated control of two manipulators, and the increased need for collision prediction and avoidance.

C.   Uses of Robots on Interplanetary Manned Space Vehicles

1.   On-Orbit Assembly

The MSS has the capability to assist in the on-orbit assembly and maintenance at the SSF of the large space vehicles required for manned interplanetary flights. It has already been proposed by a Rockwell/Spar team to use a copy of the MSS to perform a self-build of a nuclear Mars transfer system (MTS). The MTS servicing

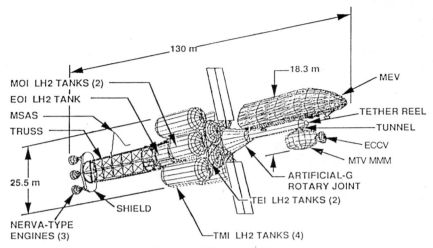

MOI LH2 TANKS (2)
EOI LH2 TANK
MSAS
TRUSS

130 m

18.3 m

MEV

TETHER REEL
TUNNEL
ECCV
MTV MMM
ARTIFICIAL-G
ROTARY JOINT
TEI LH2 TANKS (2)

25.5 m

SHIELD

NERVA-TYPE
ENGINES (3)

TMI LH2 TANKS (4)

Fig. 14   MTS description.

Fig. 15   SPDM evolution mounted.

and assembly system (MSAS) would be launched on the second flight of the MTS assembly sequence to perform all subsequent berthing, manipulation, and assembly tasks. Six to nine flights using a heavy lift launch vehicle would be required for the assembly of the MTS using the MSAS. Figure 14 shows the MTS and the MSAS during the on-orbit assembly phase. A rail system and transporter provides access by the MSAS to all locations on the MTS.

2. In-Flight Maintenance

Rockwell/Spar also proposed that the MSAS remain on the MTS for the round trip to Mars. The MTS has all the requirements for maintenance and servicing on its mission that the Space Station has. In fact, the MTS requirements are more stringent since once it departs from low Earth orbit, no spares or maintenace can be provided other than what the MTS carries along. The MSAS can provide all the functions for the MTS that the MSS provides for the SSF. Additional functions that the MSAS provides include servicing nuclear thermal rocket engines, berthing/deberthing the Mars excursion vehicle/ascent stage, and providing a backup capability to jettison spent fuel tanks.

3. Support to Landers

A further possibility proposed was that certain robotics, if properly designed, could accompany the Mars lander and be used on Mars. Possibilities include a version of the SPDM to be used on a rover vehicle as shown in Fig. 15. Also considered were large manipulators with thermal control systems suitable for supporting landing site operations such as unloading the lander of its cargo, constructing habitation modules, and housing a nuclear power reactor.

Chapter 16

Use of Manipulators in Assembly of Space Station Freedom

Patrick L. Swaim,[*] Jeffrey J. Arend,[†] Pat J. Bevill,[‡] Roy J. Decker,[§]
John C. Dunn,[§] David A. Read,[‡] Robert E. Reiher,[‡]
Brian J. Richard,[‡] Ken J. Ruta,[§] and Scott Teplitz[§]
McDonnell Douglas Space Systems, Houston, Texas 77062

Introductory Note

T HIS chapter introduces the requirements, systems, and techniques developed for robotic assembly of the Space Station Freedom. Although the Freedom design has since been superceded by that of the International Space Station Alpha, the information provides a broad overview of the use of robotic systems in on-orbit spacecraft assembly, and is presented in the context of earlier Space Station redesign efforts.

I. Introduction

Large robotic systems will be used during the assembly of Space Station Freedom (SSF) to augment and enhance man's capabilities in tasks requiring the movement and positioning of large masses. They will also be used as transport mechanisms to position astronauts for precise assembly tasks. Such transport has the benefit of reducing the time and effort required of the crew in assembly operations.

Smaller dexterous robots also are under development to reduce the need for extravehicular activity (EVA), thereby reducing the exposure of the crew to extravehicular hazards, and providing increased flexibility in contingency situations. Further, it is expected components designed for servicing with dexterous robotics will be more easily serviced by man as well.

Copyright © 1994 by the American Institute of Aeronautics and Astronautics, Inc. All rights reserved.

*Senior Manager, Robotic Systems, 13100 Space Center Blvd.
†Manager, Robotic Systems, 13100 Space Center Blvd.
‡Engineer, Robotic Systems, 13100 Space Center Blvd.
§Senior Engineer, Robotic Systems, 13100 Space Center Blvd.

The use of robotic systems in the SSF program will result in a significantly enhanced understanding of the capabilities and limitations of space robotic systems. Application of this knowledge to advanced space robotic designs will be critical to the success of future space initiatives. The first step on this path will be the assembly of Space Station Freedom.

II. Space Station Freedom Assembly Design Requirements

The Space Station Freedom (SSF) is assembled through the incremental delivery of its component elements to space and their on-orbit integration by astronauts using extravehicular and robotic methods. In planning the order in which the various SSF components are launched, NASA considers the functionality required to sustain an orbiting laboratory and the interdependence of its support systems. For example, all systems depend on electrical power; similarly, most require thermal control to reject heat. Both the power and thermal systems require attitude and pointing control to orient their array elements. These interdependencies dictate that subsystems be delivered and activated in the proper sequence. The interdependencies among systems are most critical early in the build sequence when NASA must deliver to orbit components of the SSF that provide the required functionality and are robust enough to survive on the limited resources available.

In selecting and manifesting the elements to launch, NASA is constrained by the launch system and its capabilities. Currently, the Space Shuttle is the only launch system used for SSF assembly. Therefore, the components delivered on orbit must be configured so that they can be approached by and mated to the Space Shuttle orbiter on subsequent flights. All SSF stages must exhibit stability and structural integrity when subjected to orbiter reaction control jet plume impingement during approach and mating operations. Additionally, each stage must afford acceptable clearances to the orbiter during berthing or docking operations.

Components must be allocated to specific flights of the assembly process in accordance with the Space Shuttle orbiter's payload mass and volume constraints. The orbiter's payload bay (which is essentially a cylinder about 61 ft long with a diameter of 14.5 ft) defines the maximum size of payloads that can be delivered. Unfortunately, not all of this volume is available for manifesting SSF components. Some space must be reserved for support equipment that will be used on each flight, such as the orbiter to SSF mating system. Shuttle delivery capability is further limited by operational considerations, such as how the mass is distributed in the payload bay.

The mass that can be carried on each flight is limited by the Space Shuttle's performance capabilities. At the SSF's planned altitude and inclination, that mass is approximately 36,000 lb. Additionally, the center of gravity of the manifested payloads must be kept within a well-defined volume for flight stability reasons. The payload retention mechanisms further restrict payload placement because they provide only specific, discrete attach points in the payload bay. Furthermore, payloads to be deployed by the Shuttle remote manipulator system (SRMS) must be positioned with at least 24 in. of clearance to any other hardware.

These restrictions, together with a preferred buildup of functionality, must be considered in the design and implementation of the SSF assembly sequence.

Once the SSF elements are delivered to orbit, robotic arms will be used to assemble them. Initially, only the SRMS is available. The Space Station remote manipulator system (SSRMS) must be delivered with the appropriate support systems, assembled, and tested before it can be used. Both of these robotic devices grapple payloads using their end effectors. Specialized interface hardware (grapple fixtures) must be attached to the payloads. Careful analysis must be performed to verify that these grapple fixtures are placed in locations which are structurally compatible with the payloads, fit into the payload bay, and are compatible with the required remote manipulator system (RMS) operations. Kinematic operational constraints include arm reach, manipulator joint limits and singularities, and clearance.

Additionally, the grapple fixture locations must provide sufficient distance for the manipulator end effector to approach and back away from the grapple fixture when grappling or releasing a payload. Adequate clearance must be provided to preclude interfering with other hardware, particularly in the event of certain failure conditions. And operations must be planned to minimize the time required, yet still satisfy load and joint rate limits.

Furthermore, all robotic operations require adequate lighting and viewing. The viewing may be as simple as looking out the window or may require multiple cameras to discern the payload's position and orientation relative to the installation interface. Through viewing analyses, the need for targets is identified that may be affixed to many payloads to assist in operations.

It is these complex requirements and design interactions which NASA has responded to in the development of a viable station design and assembly plan.

III. Evolution of the Space Station Freedom Assembly Plan

Since the initial award of the Space Station Freedom development contracts, both the design and assembly concepts have continually evolved. The changes have been in response to budget reductions, changes in the scientific objectives, and the technical challenges.

The early SSF configurations used a large erectable truss superstructure. This superstructure was to be assembled on orbit, using extravehicular activity and robotic methods (Fig. 1). Pressurized laboratory and habitat modules were attached to this structure. Avionics equipment, propulsion modules, external experiment pallets, and other components would have been brought to the station as required and integrated into the truss work. This approach had the advantage of providing a large degree of flexibility in both assembly (by allowing relatively small components to be brought to the station as required) and growth (by easily supporting attachment of additional truss structure, modules, and support equipment).

Unfortunately, the flexibility of the on-orbit integration of components complicated system integration and verification, since many system components could not be fully assembled and tested before construction on orbit. There were also weight and volume penalties associated with the structural, mechanical, electrical, and fluid interfaces required for easy on-orbit integration.

Compounding these difficulties were the long and complex EVA tasks required for on-orbit integration of the components. Dexterous robots could do little to reduce these demands on crew time, as intravehicular activity (IVA) required to configure, checkout, and operate the dexterous systems also proved excessive. To compress timelines and simplify extravehicular astronaut positioning during

Fig. 1   Flexibility provided by 5-m erectable truss, but significant EVA assembly time required; also shown is a neutral bouyancy test of truss assembly operations.

assembly operations, specialized robotic devices, such as the astronaut positioning system (APS), were designed (Fig. 2).

In spite of these efforts, the required assembly activities consumed so much EVA and IVA time that serial operation of robotic devices could no longer meet mission timeline constraints. Simultaneous EVA and robotic activity and coordinated operation of the robotic systems was necessary. This complicated assembly planning, execution, and monitoring, and threatened mission success and safety factors.

IV.   New Assembly Concept

The difficulty of SSF on-orbit assembly, integration and verification, and associated resource requirements were key technical issues that led NASA and its contractors to restructure the Space Station Freedom in January 1991. Primary objectives of this activity included[1] a Congressional mandate to reduce costs, divide the program into discrete phases, emphasize microgravity sciences, and achieve early man-tended capability; additional recommendations from the Augustine commission to accommodate materials and life sciences, reduce transportation requirements, and reduce SSF size and complexity; and desires by NASA to simplify on-orbit assembly, reduce EVA, increase ground integration and testing, and provide flexibility for changing priorities and budget variations.

The central feature of the new design replaced the on-orbit erectable truss with one in which elements were preintegrated into large payloads prior to flight (Fig. 3). The baselining of the preintegrated truss (PIT) concept significantly reduced the

Fig. 2   Simulated assembly operation of truss elements; use of astronaut positioning devices facilitated astronaut assembly of the truss assembly.

on-orbit assembly time devoted to truss assembly and the installation of orbital replacement units (ORUs), without a reliance on dexterous robots early in the assembly process. The PIT elements formed near full volume Shuttle payloads, which could be assembled through a combination of SRMS, SSRMS and EVA operations.

Other major changes from an assembly viewpoint included the redesign of the mobile transporter and baselining of manipulator berthing as the primary means of mating the orbiter to the SSF. The mobile transporter was radically simplified, changing it from a crawler mechanism to a rail system. All of these actions resulted in increased emphasis on the use of robotic devices capable of positioning large masses during the assembly sequence.

V.   Space Station Freedom Assembly Today

Today, the SSF assembly sequence is well defined. To facilitate a phased construction plan, the SSF program has defined two major milestones, the man tended configuration (MTC) and the permanently manned configuration (PMC). Six assembly flights over approximately 15 months are required to achieve MTC. The incremental growth of SSF through MTC is depicted in Fig. 4. Another 11 flights will be required to reach PMC by the year 2000.

Each phase is attained through a series of Space Shuttle assembly flights, each of which is known as a mission build (MB). After delivering the first SSF elements to orbit, every subsequent Shuttle flight must rendezvous with the previously delivered hardware. Since the Space Station takes on a slightly different appearance at the end of each flight, each new configuration is referred to as a "stage." Thus, the MB-6 flight will rendezvous with the SSF stage 5 configuration.

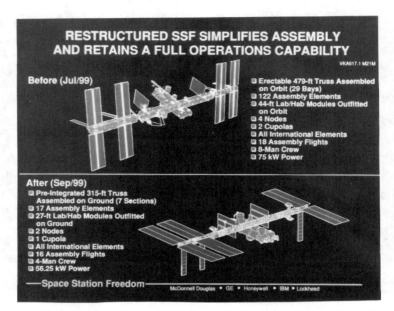

Fig. 3   Preintegrated truss concept packages components prior to flight, reducing EVA and dexterous robot assembly effort; configuration differences and merits of PIT concept.

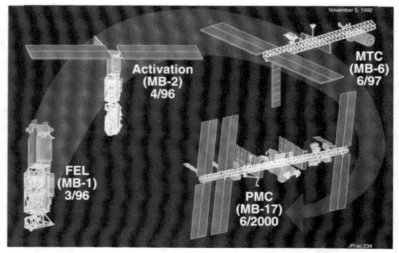

Fig. 4   Space Station Freedom will be developed through a phased assembly plan.

Construction of the station will begin on the starboard side, as described subsequently. Figure 5 presents nomenclature and labeling information essential to this discussion, which includes the SSF coordinate system and integrated truss segment (ITS) labeling convention. Integrated truss segments are identified by their on-orbit position: starboard elements (S), port elements (P), and the middle segment (M). Within these divisions, segments are numbered incrementally from the center of the module pattern outward.

Mission Build 1

Since electrical power is required by most subsystems, the first flight (MB-1) delivers the electrical power system made up of photovoltaic solar arrays, thermal radiators, and the batteries (required for the night pass of the orbit). These resources are packaged within the S3/S4 integrated truss segments. Also manifested on this flight is the mobile transporter (MT), which is attached to rails that are integrally mounted to the integrated truss assembly. Also included on this flight is an unpressurized berthing adapter (UBA) that will be used to berth the SSF to the orbiter on subsequent flights. Manipulator operations begin as the SRMS is used to unberth the S3/S4 segments from the payload bay and mate them to the unpressurized berthing adapter. The attach interface for the unpressurized berthing adapter is on the mobile transporter. This location allows the mobile transporter to translate the integrated truss assembly relative to the orbiter to facilitate SRMS

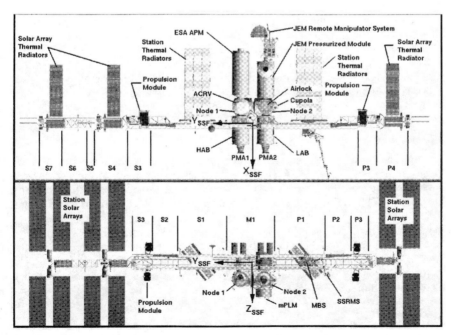

Fig. 5   Space Station Freedom components, coordinate system, and truss segment labeling scheme.

operations and allows the unpressurized berthing adapter to be indexed to the end of the truss for berthing on subsequent flights.

Stage 1, now consisting of both the integrated truss segment and the unpressurized berthing adapter, is unberthed and deployed by the SRMS. The unpressurized berthing adapter remains attached to the stage on orbit and is used for all subsequent orbiter-to-station berthing through MB-5. Since attitude control equipment and central processing units have not yet been made available, stage 1 is left on orbit in a passive mode without power.

Mission Build 2

The second flight delivers attitude control equipment (control moment gyros and propulsion modules), fundamental communications equipment, and two standard data processors. These elements are delivered as an integrated truss segment (S2) and two individual propulsion modules. On orbiter rendezvous with stage 1, the SRMS captures the stage and berths it with the orbiter.

A complex set of SRMS operations are performed to attach the S2 integrated truss segment to stage 1 and then install both propulsion modules on the S3 integrated truss segment of stage 1. This stage now has enough resources to generate electricity, provide communications and attitude control, yet must do so without an active thermal control system to maintain each component's temperature.

Mission Build 3

The third mission build flight, MB-3, delivers the thermal radiators required for the active thermal conditioning, additional communication and tracking equipment (Ku-band and uhf), and the SSRMS. These elements are manifested as an integrated truss segment (S1). The SRMS performs the same basic capture, berthing, and integrated truss segment installation functions as described for MB-2. The additional equipment is not activated until additional avionics equipment is installed on MB-5.

Mission Build 4

Flight MB-4 delivers the M1 integrated truss segment, which contains the gas conditioning assembly (GCA) to maintain the environment in the pressurized modules. The operations begin as stage 3 is captured and berthed to the orbiter by the SRMS. The SRMS is then used to relocate the crew equipment translational aid (CETA) from its transport location on the M1 integrated truss segment to a position on the previously delivered S1 segment. This relocation clears a path for the MT to translate to the end of the M1 segment once it is installed. Next, the SRMS berths the M1 segment to the integrated truss assembly. The final operation is for the MT to translate to the end of the stage so that the stage can be unberthed and deployed by the SRMS.

Mission Build 5

MB-5 delivers a habitable resource node, a cupola (used by the crew to control the SSRMS and provide vital direct viewing of EVA/RMS operations), and a pressurized mating adapter (PMA). The robotic operations performed on this flight begin with the capture and berthing of stage 4. Subsequent to this, the SRMS unberths the node and attaches it to a preassembled module-to-truss structure (MTS). The module-to-truss structure is the attachment interface between the integrated truss assembly and the pressurized module elements. Following the resource node installation, the SRMS installs the pressurized mating adapter on one of the radial ports of the resource node. Finally, the cupola is unberthed by the SRMS and installed on one of the node's axial ports.

Installation of the pressurized mating adapter indicates the space station's transition from a berthing to a docking capability. As a result, all subsequent mission build orbiter flights will dock at the SSF and will only use SRMS berthing as a contingency. The UBA returns to Earth in the orbiter at the end of this flight.

Mission Build 6

Flight MB-6 represents a milestone after which astronauts can begin experiments on station in a shirt sleeve environment while the orbiter is present. Thus, it is called a man tended configuration. It is also at this stage that the orbiter begins docking with the station, as opposed to SRMS berthing of the station to the orbiter. This flight delivers a pressurized laboratory module and the mobile remote servicer base system (MBS). This system is used as an accommodation facility for the SSRMS and to translate SSF elements along the truss.

Robotic assembly operations required on this flight begin with the unberthing and installation of the laboratory by the SRMS. The laboratory is mated to a radial port of the resource node. Because of the complexity involved with installation operations, considerable design has been required to properly locate the laboratory's installation grapple fixture. After the laboratory's installation, the mobile remote servicer base system installation begins by unberthing it from the Shuttle's payload bay with the SRMS. The reach of the SRMS does not allow it to directly install the system, therefore, the assistance of the SSRMS is required. The SRMS performs what is referred to as a hand off, passing the mobile remote manipulator base system to the awaiting SSRMS (Fig. 6).

The cupola, which is now operational, is used by the crew to control and monitor these SSRMS operations. Once the hand off is completed, the SSRMS delivers the mobile remote servicer base system to its final destination on the mobile transporter. The completion of this mission build provides many enhanced SSF operational capabilities and is viewed as a major milestone toward the development of the permanently manned SSF.

Permanently Manned Configuration Mission Build Sequence

After the man-tended configuration is achieved, the Space Station Freedom build continues by installing the port side truss segments and the remaining habitable pressurized modules. As the build progresses, more dependency is placed on the SSRMS to complete the build. For instance, the port side truss segments will be unberthed from the orbiter by the SRMS and subsequently handed off to the SSRMS to perform the installation. Additional flights provide an increase in the available resources and system redundancy. Permanently manned configuration,

Fig. 6   Hand-off operation allows a payload to be passed between arms.

MB-17, occurs once the station has sufficient resources and levels of redundancy to house the astronauts on a permanent basis.

Table 1 summarizes the current plan for manifesting mission build flights.

VI.   Primary Robotic Devices for Space Station Freedom Assembly

To accomplish these assembly tasks, both the existing SRMS and new robotic devices, such as the SSRMS, will be used. Other newly designed robotic devices used primarily for SSF maintenance or specific payload handling will be the special purpose dexterous manipulator (SPDM) and the Japanese experiment module (JEM) main arm (MA) and small fine arm (SFA).

These devices represent major commitments by international partners to the SSF program. Canada, the developer of the SRMS, will provide the SSRMS and the SPDM through their prime contractor, Spar Aerospace. The Japanese Space Agency, NASDA, will provide the JEM manipulators, through an effort led by Toshiba. The SSF manipulators are depicted in Fig. 7.

Shuttle Remote Manipulator System

The SRMS will be the only manipulator available for assembly through MB-5. The geometry and dimensions of the SRMS are depicted in Fig. 8. For SSF assembly, the SRMS will be certified to a higher mass handling capacity and upgrades are planned which will improve the control performance of the system.

The SRMS is a six-jointed serial manipulator system that can be operated in both manual and automatic modes. Manual control is accomplished using two 3-degree-of-freedom (DOF) hand controllers to issue rate commands for rotational and translational motion. These hand controllers are located on the aft flight deck of the orbiter, affording the crew direct viewing of the operations through the overhead and aft windows. Additional views are provided by the orbiter payload bay closed circuit television cameras (CCTV) and CCTV cameras located on the manipulator arm.

Table 1   Space Station Freedom manifest table

Flight	Cargo	Description
MB-1	ITS-S3/S4, UBA	Truss segment with solar arrays and attachment sites for propulsion modules mobile transporter, unpressurized berthing adapter
MB-2	ITS-S2, prop modules	Two propulsion modules and a truss segment with communication and tracking equipment, CMG, and temporary avionics
MB-3	ITS-S1	Truss segment with starboard active thermal control radiators, two antennae, and SSRMS
MB-4	ITS-M1	Truss segment with GCA and attachment sites for nitrogen and oxygen and oxygen; umbilicals for node 2 and U.S. Laboratory, mobile transporter batteries, EVA equip.
MB-5	Node 2, PMA-A cupola	Node 2, PMA (one of two), cupola
MB-6	USL	U.S. laboratory module (USL)
MB-6A	Airlock, MBS, SPDM/MMD	Airlock, mobile base system (a work platform for the SSRMS), SPDM and mobile servicing system (MSS) maintenance depot (MMD)
MB-7	MPLM, ULC, PMA-B	A minipressurized logistics module with racks for the USL, unpressurized logistics module, a second pressurized mating assembly
MB-8	ITS-P1	First port section of truss with more active thermal arrays
MB-9	ITS-P2, prop modules	Segment of truss and two prop modules
MB-10	ITS P3/P4	Port solar arrays and prop module attachment sites
MB-11	Node 1, ITS-S5	Node 1 and a short segment of truss outboard of the starboard solar arrays
MB-12	JEM PM	Japan's pressurized module
MB-13	ESA APM	ESA's Columbus module
MB-14	ITS S6/S7	Third set of solar arrays with batteries and thermal cooling arrays
MB-15	JEM EF, JEM ELM PS and ES	Japanese exposed facility, Japanese experimental logistics module pressurized section and exposed section
MB-16	U.S. hab. module	U.S. habitation module, living quarters for station crew
MB-17	ACRV	Assured crew return vehicle
MB-18	Centrifuge node	Centrifuge for experiments

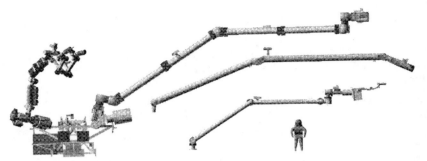

Fig. 7   Relative dimensions of robotic devices available external to SSF.

Both manual and automatic modes allow control of the position and orientation of a point of resolution (POR) that is fixed relative to the manipulator end effector. The point of resolution can be controlled relative to several different coordinate frames, which are selected primarily based on the ease of operating the manipulator.

Four manual operational modes are available. The orbiter loaded and unloaded modes use coordinate frames fixed to the orbiter. The end effector mode uses a coordinate frame fixed to the end effector, and facilitates positioning the arm when viewing operations through the end effector camera. The payload mode uses a predetermined coordinate frame attached to a payload. This mode can simplify the positioning of payloads by providing control frame that corresponds to payload geometry.

Automatic control is accomplished with operator commanded or prestored trajectory autosequences that specify a set of desired end effector positions and attitudes. The two automatic modes use the orbiter fixed coordinate frames.

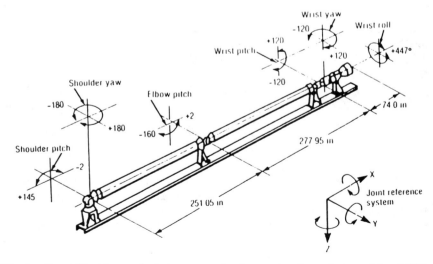

Fig. 8   Space Shuttle manipulator system is a key system in early assembly of SSF.

The operator also may drive the SRMS using a single-joint drive mode in which a rate command can be issued to drive a specific joint whereas the software maintains the other joints at their current position. A direct drive mode can be used to bypass most elements of the SRMS control system and apply current directly to the joint servos in certain contingency situations.

Space Station Remote Manipulator System

The Space Station Remote Manipulator System will be used primarily for handling large objects on the space station. The SSRMS is a symmetric seven-jointed arm (Fig. 9). Each end of the arm will have an end effector which functions as an interface mechanism with external systems. This feature allows operation of the SSRMS with either end effector acting as the reference base of the arm. Thus, the SSRMS can walk along SSF, from one power data grapple fixture (PDGF) to the next, alternately converting the end effector to the operating base.

The SSRMS will provide similar manual and automatic operating modes as provided by the SRMS. In addition, the SSRMS will provide an automatic operator commanded joint mode, prestored joint mode, and greater flexibility for specifying operating coordinate systems. Other control features include an artificial vision function (AVF) tracking mode and a force moment accommodation capability. An arm pitch plane control function (orbit mode) also may be provided to assist in controlling the redundant degree of freedom.

As a redundant manipulator, the SSRMS does not intrinsically have a unique arm configuration for each end effector position and attitude. This provides a great deal of operational flexibility that can be exploited to simplify SSF assembly operations. However, this added flexibility does present the following important control issues.

1) How should arm configurations be defined for each desired end effector position?

2) For each configuration, are the corresponding joint angles continuous functions as the end effector is moved throughout the workspace?

3) How should singularities and joint limits be avoided?

4) Does the geometry of the manipulator itself pose a risk of self collision between its booms and end effectors?

These issues have significant operational impacts, and are discussed in more detail in the Appendix.

In any event, SSRMS operations will involve a significant departure from the current experience base with the SRMS, both in terms of mission operations and planning. Additional tools to assist in planning and monitoring the operation of the SSRMS are already in development.

Mobile Transporter

The primary base from which the SSRMS will operate is the MBS. The system has four strategically oriented power data grapple fixtures which give the operator different options for basing the SSRMS. The MBS is attached to the MT which provides another way to relocate the SSRMS and its payloads. The mobile transporter can translate along the rails located on the front face of the integrated truss assembly (ITA) and stop at any one of the ten designated mobile transporter

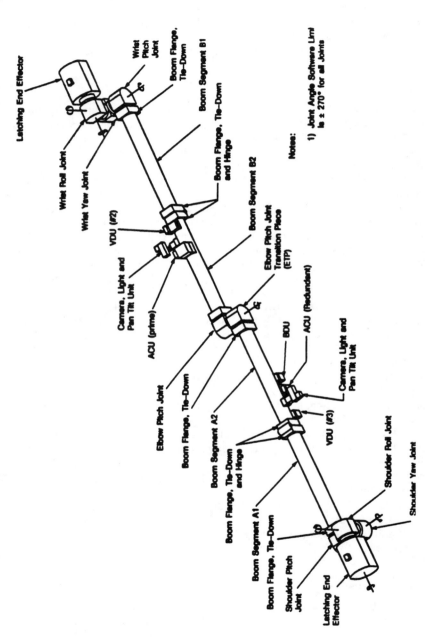

Fig. 9 SSRMS will be available to assist in SSF assembly operations beginning on mission build 6.

worksites. The MBS will also have two zoomable cameras and lights mounted on pan/tilt control units.

Special Purpose Dexterous Manipulator

The SPDM is a dexterous dual arm manipulator designed to handle payloads of approximately 1300 lb or less with positioning repeatability of up to 0.05 in. (Fig. 10). Its two 7-DOF arms attached to a relocatable body provide significant flexibility. It can be mounted and operated from the end of the SSRMS or any power data grapple fixture.

The SPDM will have the dexterity to manipulate hinged panels and doors, mate and demate connectors, operate jackscrews and changeout orbital replacement units. These operations are facilitated by the incorporation of force moment sensing devices in the ORU tool changeout mechanisms which terminate each arm. The SPDM will also have three zoomable cameras, lights, and two pan/tilt control units that make it a key asset for viewing and inspection tasks on SSF.

Other Robotic Devices

Other robotic devices that will be on SSF include the Japanese exposed module main arm and small fine arm (Fig. 11). The JEM main arm and small fine arm are both six-jointed arms designed to manipulate payloads on the JEM exposed facility. The small fine arm is attached to the end effector of the main arm when servicing of small payloads is required. The main arm is based on the Japanese pressurized module and is not relocatable.

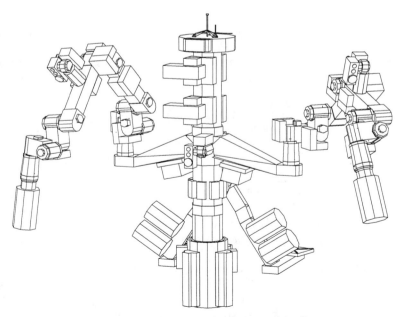

Fig. 10   Special purpose dexterous manipulator.

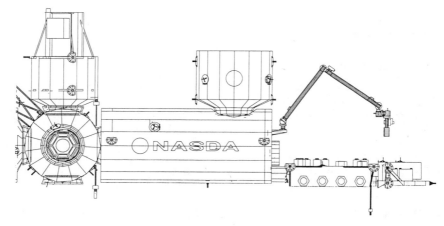

Fig. 11   JEM arms.

VII.   Robotic Interfaces for Assembly Operations

Use of the robotic devices previously described requires a variety of specialized mechanical and electrical interfaces. This section will describe the physical and functional types of robotic interfaces, where those interfaces can be positioned on the payload, and finally how the manipulator attaches itself to this device.

Grapple Fixtures

All current large manipulator devices use a grapple fixture (GF) for interfacing with the payload. There are four different types of grapple fixtures planned for the SSF. Functionally each grapple fixture interface provides a slightly different capability (Table 2). It is important to note that the SRMS, SSRMS, and

Table 2   Grapple fixture types

Type	Description
FRGF	Flight-releasable grapple fixture: can be used by all manipulators as a mechanical interface
RSGF	Rigidize-sensing grapple fixture: provides a single switch which is activated/deactivated with the enhanced rigidity of the end effector when grappling; used to turn on heaters, turn off active attitude control, etc.
EFGF	Electrical flight grapple fixture: provides electrical connectivity only to the SRMS and has the capability to send data back and forth between the payload and the SRMS; interface is used when monitoring is needed
PDGF	Power data grapple fixture: provides electrical connectivity only to the SSRMS and has the capability to send data back and forth between the payload and the SSRMS; interface is used when monitoring is needed

JEM main arm can use any of these grapple fixtures as a mechanical inter-face to the payload, but they do not have the same type of electrical connec-tivity.

Physically each grapple fixture has an abutment plate, three cam lobes, a grapple shaft, and a three-dimensional target (Fig. 12). Because the target rod projects upward from the center of a circle, paralax effects help the manipulator operator align the end effector with the grapple fixture. The proper alignment is achieved when the rod appears centered within the circle when viewed through the end effector camera.

The location of grapple fixtures on the payload is determined by several pa-rameters. First, the grapple fixture location must be structurally compatible with the payload, ensuring an adequate area for mounting and appropriate stiffness and load carrying capacity. A specified volume around the grapple fixture location must be clear of structure to allow both for grappling misalignment and EVA crew access to the grapple fixture release mechanisms in contingency situations. The location must be appropriately positioned relative to the payload center of mass to minimize inertia cross-coupling effects which can degrade remote manipulator system (RMS) control performance. The grapple fixture location must provide acceptable manipulator kinematics, from payload unloading to installation on the SSF; these issues include conformance with arm reach capabilities and provision of adequate margins relative to joint limits and singularity zones. Finally, when the payloads are manifested in the orbiter payload bay, the grapple fixtures must not structurally interfere with the orbiter payload bay or other payload elements, either statically, or due to vibration during Shuttle ascent.

End Effectors

The end effector provides a physical interface between the manipulator arm and its payload. The SRMS, SSRMS, and JEM end effectors are designed for use with payloads or tools with attached grapple fixtures. Mechanically, the end effectors of these manipulators are functionally equivalent and use a snare to achieve a soft capture and a retraction mechanism which aligns and enhances the interface rigidity.

The capture and rigidity sequence has several steps, including placing the grap-ple shaft in the end effector canister, rotating of the snare cables around the grapple fixture shaft to align the interface, and pulling the shaft into the end effector to form a rigid interface. The SSRMS additionally engages a set of collet latches, which allows for the transfer of higher loads, and reduces snare cable lifetime concerns.

In addition to the standard end effectors just described, a dexterous end ef-fector is under development. This end effector incorporates the Johnson Space Center magnetic end effector (MEE), the Jet Propulsion Laboratory force torque sensor (FTS), and the targeting and reflective alignment concept (TRAC). These components are functionally packaged as an extension tool which attaches to the SRMS end effector through an electrical flight grapple fixture. This system could enhance SRMS capability by providing more precise SRMS alignment, force and moment data necessary to perform constrained motion tasks, and improved safety

The FRGF also meets the the following requirements.

Design Approval:
NASA Johnson Space Centre

Cargo Element Interface
Configuration:
per NASA 07700 NSTS Volume XIV

Quality Assurance:
per NASA NHB 5300-4

Reliability:
per NASA NHB 5300-4

Non Operational Temp:
+/-250°F (-156°C to + 121°C)

Operational Temp:
-105°F to + 154°F (-76°C to + 68°C)

Payload Mass:
65,000lbs (143,000kg) maximum

Grapple fixture Mass:
27lbs (59.4kg)

Load Capability:
1200ft lbf Bending Moment
450ft lbf Torsional Moment

(U.S. Patent No. 4105241)

Fig. 12   Grapple fixture and target.

and reliability by eliminating current mechanical components and interfaces. A Shuttle flight experiment tested this concept on STS-62 in February of 1994.

VIII.   Space Station Robotic Assembly Tasks

Although the assembly of the Space Station will involve the installation of many different components at various locations, the tasks themselves can generally be broken down into several generic activities: track and capture, maneuvering, berthing and unberthing, deployment, positioning, and handoffs between manipulators. These activities are briefly described subsequently, along with the operational considerations that are addressed in manipulator trajectory design.

Operational Considerations

The nominal operational uses of the SRMS and the SSRMS are governed by adherence to a number of guidelines and constraints. These constraints can be categorized into several key areas, including reach, clearance, viewing, and lighting.

Reach capability is the first parameter that must be evaluated before an operation is considered possible. From a kinematic aspect, the arm must be able to physically reach all positions along the desired trajectory. One important operational issue in the design of these trajectories is the avoidance of both physical and control singularities, at which one or more degrees of freedom for moving the arm are lost. The RMS software will reconfigure the joints while driving the arm through these singularities in such a way as to return the DOF that was lost; however, uncommanded motion of the arm may occur while in a singularity zone, requiring the operator to monitor the arm closely.

Trajectories must also be planned to provide sufficient additional reach for the end effector to back off the grapple fixture once a payload is in place. To enhance operations, a minimum elbow joint angle is specified to ensure that the arm does not become fully extended. Currently, a minimum angle of 30 deg from the stretched out configuration is used for planning.

After the operation is determined to be kinematically acceptable, then clearances around the manipulator and payload must be evaluated. Each robot has limitations of dexterity and positioning accuracy. These characteristics yield a comfort zone for the operator and a guideline for planning purposes. This zone protects the operator from possible structural contact. Currently, the SRMS uses a 2-ft clearance zone during man-in-the-loop operations. The same 2-ft clearance guideline is being applied to the SSRMS, but ultimately its performance characteristics may permit this clearance zone to decrease.

Trajectory design also must ensure that adequate viewing is available to monitor the operations and meet positioning requirements. Direct viewing is generally preferred; however, for Space Station operations, direct views will often be obscured by intervening structure. Also, SSRMS operations may be conducted at locations far from the operator. Direct views are still important to provide an overall perspective of the operation, but indirect camera views will have to be relied on to a greater degree. Also important will be views available from other crewmembers (either inside or outside the station), which may be verbally transmitted to assist in the operation.

Lighting also is a crucial factor for successful conduct of RMS operations. At the operating altitude of the space station, each orbit will take approximately 90 min. The illumination geometry changes continuously, from orbital sunrise to sunset. Approximately 30 min of each orbit is in darkness due to the Earth's shadow. There are often times when the sun will not illuminate the worksite, and the need for artificial lighting is evident. For either source of lighting, space station components will cast shadows which degrade viewing. These shadows tend to be much more harsh than those on Earth due to the lack of atmospheric scattering and diffusion. The shadows and glare created by sunlight change continuously with the illumination geometry. The relative geometry of cameras, lights, and direct viewing must be carefully analyzed and integrated within operational timelines to ensure that operations can be completed as designed.

Space Station Freedom Capture

One of the first tasks which must be accomplished on each of the early SSF assembly flights is the SRMS berthing of the station to the orbiter. This means of mating the orbiter and the space station will be used as the primary mating method for on-orbit operations at the beginning of mission build flights 2–5 and as a backup method on later flights. In practice, this is achieved by placing the manipulator in an "optimum" configuration, and allowing the orbiter pilot to "fly" the end effector to the near proximity of an SSF grapple fixture. With the orbiter pilot performing stationkeeping between the vehicles, the RMS operator completes the operation by grappling the SSF stage.

To help ensure a successful capture, the arm is positioned in the poised-for-capture position. In this position, the arm reaches over the payload bay to provide the pilot and RMS operator with direct views out the aft flight deck windows. The SRMS elbow joint has an angle of approximately 60 deg from its fully extended configuration to provide good tracking ability and an adequate range of travel to arrest relative motion after capture. Nominally, the SRMS end effector is aligned so that it is normal to the RMS operator's line of sight, and the SRMS is kept in a planar configuration (Fig. 13). This configuration simplifies single-joint operations if coordinated joint control capability is lost, and facilitates grapple in the event of an end effector camera failure.

Station to Orbiter Berthing

On early assembly flights, the orbiter must be berthed to the station using the SRMS. Once the station is grappled, the SRMS must position the station for final mating to the orbiter. On early flights, this mating is accomplished using the UBA. This structure is delivered on orbit as part of the MB-1 hardware and has four trunnion pins and a keel pin which allow solid attachment to the orbiter.

The berthing operation starts after capture of the on-orbit stage. The payload is positioned in a high hover position, where the mating interfaces are aligned such that they can be brought together by a simple downward translation into the payload bay. The operation proceeds by maneuvering the trunnions through V guides which ensure proper positioning for final attachment using the payload retention latch assembly (PRLA).

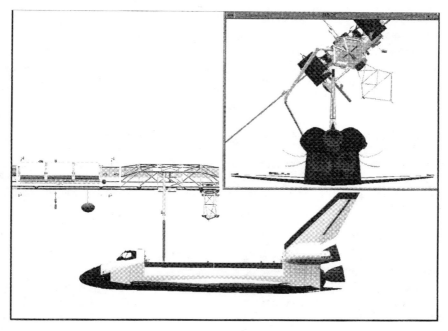

Fig. 13   Space Station Freedom capture.

Since the unpressurized berthing adapter is latched in the forward portion of the payload bay, this berthing task is done with good visual cues from the orbiter aft flight deck windows and forward bulkhead cameras. However, other payload berthing tasks such as the propulsion modules, unpressurized logistics carrier, and mini-payload logistics module (PLM), which are latched in the aft portion of the payload bay, will not have good direct visual cues.

Payload Deployment Tasks

Large station payloads fall into two distinct categories. The first type is a payload that uses the entire volume for a certain length of the orbiter payload bay. All of the pressurized modules are of this type (i.e., their cross section is approximately 175 in. in diameter). The other type of station payload is characterized by its length being at or near the allowable limit for the orbiter payload bay. Several of the truss segments are of this type (i.e., their length is at or near the 540-in. maximum allowed). These two types of payloads special care during unberthing from the payload bay, as well as installing them on orbit.

In the case of either payload, the unberthing task requires the SRMS to translate the payload directly out of the payload bay, while maintaining clearances on all sides. After achieving a position high enough that clearance to the orbiter is no longer a concern, the payload must be moved along a path to its final location on station. There, the operator will align the payload mating interfaces such that a final single axis translation can accomplish mating.

Space Station Freedom Element Berthing Tasks

Berthing tasks require the manipulator arm to maneuver payloads to within the capture envelope of their mating systems. Examples of these tasks are MT berthing to the unpressurized berthing adaptor; truss segment to truss segment berthing; propulsion module to orbiter and to the propulsion module attach structure berthing; unpressurized logistics carrier (ULC) berthing to the propulsion module attach structure on the mobile remote servicer base system; and module to module berthing.

Several different types of capture, mating, and attach mechanisms are used with SSF payloads. The major latching mechanisms that are employed are the propulsion module attach structure (PMAS) and the common berthing mechanism (CBM). In the interests of SSF component commonality, the basic latching mechanism of propulsion module attach structure is used with several different guide systems for mating many of the SSF elements that are handled (or must be replaced) by the large robotic systems. Figure 14 shows the latching mechanism as it is attached to bulkhead of an ITS segment.

Astronaut Positioning Tasks

The EVA community is actively pursuing using the robot arms as a movable base for the astronaut. This combined operation will improve the EVA efficiency and allocated times for assembly activities.

Fig. 14   ITS will be mated using a capture latch which closes about a capture bar, also shown are the alignment cones; power bolts at corners of truss segments will secure the assembly.

Three devices are available that enable an EVA crewmember to be maneuvered by the manipulators. All of these devices use a type of portable foot restraint (PFR), a platform on which the crewmember stands. Two of the three devices use a grapple fixture interface allowing them to be used with either the SRMS or the SSRMS. The other device, which is specific to the SRMS, uses a capture latch on the wrist striker bar instead of a grapple fixture. These devices range from the basics of the standing platform to the full up service station with equipment restraints and tool stanchions (Table 3).

Some example tasks that will utilize this hardware include relocating common berthing mechanism covers, installing trunnion and keel covers, translating antennas, connecting utilities, and servicing station orbital replacement units.

IX. Detailed Example: Mission Build 2

This section presents a detailed example of the operations necessary to successfully conduct a space station assembly flight, using the MB-2 flight as an example. The manifest for this flight includes two of the space station propulsion modules (PMs) and the 280.0-in. Starboard 2 (S2) integrated truss segment with temporary avionics, two S-band antennas, star trackers, and four CMGs.

The orbiter approaches the station along the radius vector from the center of the Earth to the station radial (Earth) direction bar (+R-Bar). The SRMS grapples SSF stage 1 and nulls its relative motion. The final SRMS elbow pitch angle is dependent on vehicle relative rates, poise for capture position, payload mass, relative SRMS configuration, and a number of other factors.

Once the stage is captured, the SRMS begins the berthing operations by pitching, then rolling the stage (relative to the orbiter) and aligning it such that the stowed solar arrays are over the nose of the orbiter and the UBA trunnions are over the V guides in the forward part of the payload bay. The stage is then translated along the orbiter +Z axis in a straight line toward the payload bay until the trunnions are berthed into the PRLAs and the stage is latched.

Table 3   Variety of crew accommodation devices facilitate EVA operations from manipulator systems

Type	Description
PFR	Portable foot restraint: platform for the EVA to stand on plugs into a PFR socket (SSP hardware)
APFR	Articulating PFR platform: platform with joints (PYR) which allows more flexibility than the PFR (SSFP hardware)
PWP	Portable work platform: comprised of an APFR, a tool stanchion, and a temporary equipment restraint aid (TERA) with a PDGF interface (SSFP hardware)
PAD	PFR attachment device: uses the PFR and attaches to the SRMS striker bar to keep the end effector free for a grapple fixture (SSP hardware)
MFR	Manipulator foot restraint: single device which is the Shuttle version of the SSF PWP (SSP hardware)

S2 Integrated Truss Segment Installation

Assembly of the truss segment is relatively simple. The SRMS grapples and un-berths the S2 integrated truss segment from the orbiter payload bay with a straight outward (orbiter Z axis) translation. There is tight clearance (~20 in.) between the S2 integrated truss segment and the ITA during this operation. Once the payload is at the correct height (and there are no pitch, yaw, or roll misalignments) it is mated to the integrated truss assembly by moving it forward (in the orbiter X direction, as shown in Fig. 15). This is the first integrated truss segment to integrated truss assembly mating operation in the station build.

The difficulty with this operation is that the operator cannot easily determine the roll, pitch, and yaw cues either looking directly out the window or using any of the payload bay cameras (one in each corner and one looking straight up from the keel). Alignment aids and targets of some kind will be required to accurately position the payload for mating. Concepts for such aids are being tested and evaluated through crew ground simulations and training.

The attachment of the truss structure is performed as follows.

1) The S2 integrated truss segment is positioned by the SRMS so that its striker bar is within the capture envelope of the latching mechanism on the integrated truss assembly (determined by visual cues and a ready-for-latch indication).

2) The latch is driven and the fine alignment guides brought together.

3) The motorized bolts are driven to complete the connection.

On completion of the attachment operation, the SRMS will release and back off the GF. Once the S2 integrated truss segment is joined with the integrated truss assembly, the keel structure is removed via EVA before the MT translates the integrated truss assembly forward.

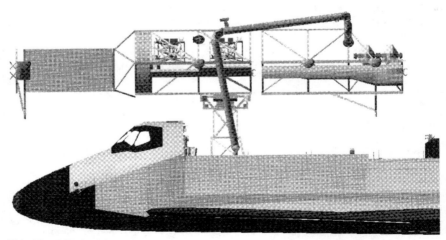

Fig. 15   MB-2 with stage 1 attached showing final alignment phase of ITS berthing.

Forward Propulsion Module Installation

The propulsion module assembly requires the modules to be berthed to the PMAS on the S3 integrated truss segment. These are complicated assembly operations, which require placement of the SSF face 5 propulsion module (which is manifested forward in the payload bay) at a temporary location on face 3, and translation of the mobile transporter in order to complete the required assembly (Fig. 16). In addition, the propulsion module installations are sensitive to the final PM/MT/UBA design because they require the SRMS to operate very near structure while near the ends of its reach limits.

Because the propulsion modules are considered hazardous payloads, they will require continuous monitoring and heater power (provided by the SRMS) during their assembly. This requires an electrical flight grapple fixture (EFGF) on the propulsion module to interface with the SRMS end effector (EE). Also required is one PDGF on each propulsion module to interface with the SSRMS end effector for later propulsion module change out operations.

Before the propulsion module installation operations begin, the SSF solar array alpha joint rotates 90 deg (to allow clearance between the solar arrays and the orbiter nose) and the mobile transporter translates the integrated truss assembly forward, over the nose of the orbiter. The mobile transporter is parked at the utility

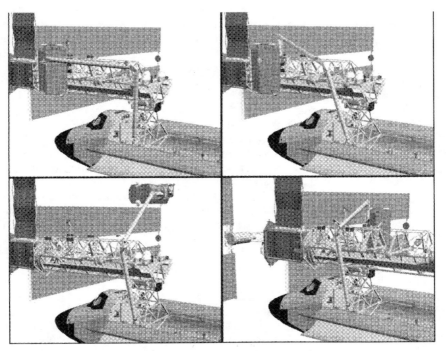

Fig. 16   Time sequence of assembly operations for installing first propulsion module.

port on bay 3 (B3) of the S2 segment. The SRMS unberths the forward propulsion module from the payload bay. The SRMS installs the first propulsion module into the inboard propulsion module attach structure on face 3 of the S3 integrated truss segment (orbiter port side).

The SRMS regrapples the first propulsion module using an additional electrical flight grapple fixture. This regrapple is a difficult SRMS operation due to proximity to the wrist pitch (WRP) limit and the fact that the shoulder yaw (SHY) joint must be driven from one limit to the other. However, there is adequate viewing, as this operation can be seen with the payload bay camera B, the SRMS elbow, and end effector cameras.

The SRMS unberths the first propulsion module from its temporary propulsion module attach structure location on face 3 of the S3 integrated truss segment (orbiter port side). The mobile transporter is driven to index the integrated truss assembly over the payload bay. The mobile transporter is parked on S3-B2, enabling the SRMS to reach over the truss. The propulsion module is installed on the inboard propulsion module attach structure on face 5 of the S3 integrated truss segment (orbiter starboard side).

The electrical flight grapple fixture for the propulsion module is angled on the forward/top face (when the modules are stowed in the payload bay). Because of the physical limitations of the SRMS and the installation location, the propulsion module that is installed on face 5 of the integrated truss assembly (starboard side of the orbiter) requires an additional electrical flight grapple fixture to complete the assembly. The electrical flight grapple fixture is located on support structure attached to the lower starboard side of the module whereas the module is stowed in the payload bay.

Aft Propulsion Module Installation

The truss is again indexed forward, using the mobile transporter, and parked on S2-B3. The SRMS unberths the aft propulsion module from the payload bay in a similar fashion to the forward propulsion module. The installation of the propulsion module is similar to the first few steps of the previous installation (inboard propulsion module attach structure on face 3). The aft propulsion module does not require the additional electrical flight grapple fixture for installation. Therefore, it has one electrical flight grapple fixture for installation and one PDGF for later propulsion module change out operations. The electrical flight grapple fixture for the propulsion module will be angled on the forward/top face of each propulsion module (when the modules are stowed in the payload bay).

Mission Build 2 Issues

As can be seen, the MB-2 flight involves intricate operations. Testing and evaluation of the S2 integrated truss segment positioning aids will need to be performed by the mission operations directorate (MOD) and the flight crew before the segment mating task operation is considered viable with the current hardware.

There is currently no adequate view of the propulsion module base/attach assembly interface. NASA is examining a camera outfitted with the Canadian space vision system (SVS) to provide operator feedback of precise digital position and orientation information to accomplish these installation operations. This system was tested on orbit during STS-52 and proved to be a valuable tool.

Several SRMS joints are driven to within a few degrees of their respective soft stops during the propulsion module installations. This will likely lead to additional crew training and additional on-orbit time (relative to other SRMS operations) to perform this operation safely.

X. Analysis Tools and Techniques

The design iterations required to develop the SSF assembly sequence presented could not have been completed without the aid of sophisticated planning tools. The majority of this robotic analysis was performed using the Manipulator Analysis—Graphic, Interactive, Kinematic (MAGIK) program. MAGIK is the primary kinematic analysis tool used by the robotic systems group at McDonnell Douglas Space Systems and SSF robotic analysts at the Johnson Space Center. It allows users to conduct man-in-the-loop analyses to evaluate space robotic operations, including issues of reach, clearance, viewing, and control. MAGIK can be used to model, specify, simulate, analyze, and modify n-jointed type manipulators and their respective control algorithms. The MAGIK simulation currently implements kinematic models of all robotic devices currently baselined on SSF. The notable features of the MAGIK simulation environment are presented in Table 4.

MAGIK traces its heritage to the manipulator system remote planning software (RPS), which was originally developed for the Space Shuttle program. With RPS engineers were able to analyze payload deployment and retrieval techniques and to generate Space Shuttle flight computer data loads for specific payloads and flights. RPS was designed specifically for the SRMS and does not support other robotic systems.

As plans for Space Station Freedom developed, engineers realized that similar manipulator software would be needed to model the SSF manipulators as they proceeded through design iterations. Thus, in 1985, McDonnell Douglas Space Systems Company began the development of MAGIK as a derivative of RPS, but with the goal of modeling generic manipulators in an interactive graphics simulation. MAGIK was developed in parallel with other software tools, with which it now interacts (Fig. 17).

In 1988, MAGIK was rehosted to a silicon graphics Iris series workstation. At this juncture, the solid surface graphics rendering became fast enough that analysts could interactively control virtual robotic systems. A generic manipulator could be operated in coordinated motion, while viewing the operations from any chosen perspective or orthogonal view. This capability allowed excellent insight into the problems that would be faced by the crew as they attempted to perform space station assembly and maintenance operations.

MAGIK has proven to be an invaluable tool, without which development of the assembly planning work previously presented would have been much more difficult.

XI. Work Remaining

As assembly plans continue to mature, the focus of analysis activity will shift from kinematic assessments to dynamic assessments and performance analysis.

Table 4   Features of the MAGIK analysis tool

Modeling
Models of SRMS flight software control algorithms and SSRMS pseudoflight software
Data-driven modeling that allows new robotic devices to be configured and simulated without rewriting software
Generic modeling capability for n-jointed robotic systems
Manipulator design utility for creating accurate graphic manipulators models

Image generation
High-resolution solid surface color graphics, using silicon graphics Iris workstations, and a variety of graphics rendering options

Control interface
Operator control of manipulators through keyboard, hand-controller devices, or mouse-driven interface
Specification and control of manipulator operator control modes and the operational reference coordinate system
Real-time collision detection among payloads, manipulators, and workspace elements

Planning interface
Scripting capability to specify or record manipulator operations and views for subsequent retrieval or playback
Payload data base management of payload specific mass properties and control constraints
Autosequence editor to specify and modify manipulator trajectories
User-friendly interface with online help

Already, significant studies of SRMS upgrades have been conducted using dynamic simulation tools. These tools include simulators with refined dynamic models of the manipulator system, and man-in-the-loop simulators, such as the systems engineering simulator (SES), which includes a functional orbiter aft flight deck mockup. Spar continues to refine the SSRMS and SPDM designs, and to define and verify their performance capabilities. Other significant work is proceeding to verify the integrated performance of the manipulator systems, which includes space station structure and control systems interactions.

In parallel with this work, hardware testing of the systems also is underway. In addition to the testing required for the development of the new robotic systems, the automation and robotics division at JSC is developing the space systems automated integration and assembly facility (SSAIAF) and the mobile remote manipulator development facility (MRMDF).

The SSAIAF facility provides an integrated simulation capability, combining on-orbit systems simulation with hardware-in-the-loop and man-in-the-loop capabilities. With this facility the crew can evaluate various assembly operations. Hardware elements provide a high degree of realism: forces and torques developed

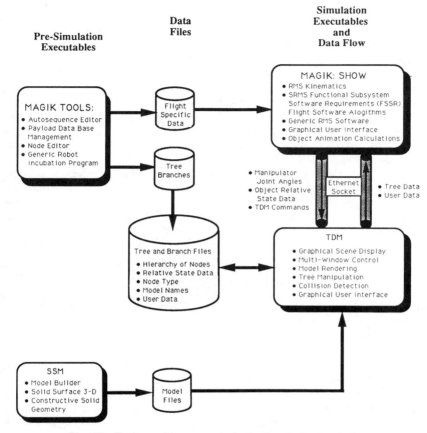

Fig. 17   Tools used in support of robotic mission analysis.

as a result of contact can be fed back into on-orbit environment and system simulations; the equations of motion are integrated to simulate system response; this data is fed to a hydraulic system used to move the hardware and close the simulation loop. The SSAIAF will be a valuable facility for assessing assembly operations.

The MRMDF facility includes a full-scale working model of the SSRMS for crew training. The facility is designed as the SSF program analogue to the manipulator development facility (MDF), which is used for crew training on SRMS operations. The MRMDF will be capable of accurately simulating SSRMS operations with up to 500-lb payload mockups. The facility will substantially preserve SSRMS dimensions and control performance, in spite of the fact that it is a hydraulic system that must operate at surface gravity levels. Attendant requirements for component packaging, structural stiffness, load capacity, sag compensation and modeling following control have presented challenges requiring novel design innovations.

XII. Summary

This chapter has presented an overview of robotic operations necessary for assembly of the Space Station Freedom, approximately current with the design configuration of the June 1993 critical design review. We have also attempted to provide an introduction to the robotic devices, their capabilities and limitations, and the operational issues associated with the assembly operations. The focus of the paper has been primarily on work performed within the Automation and Robotics Division at NASA's Johnson Space Center.

Although a major redesign of the station is in progress, many of the same components will be employed in any new design. The robotic systems described are expected to survive substantially intact. Similar ground rules and constraints are likely to be employed, and similar analyses must also be conducted. We hope and this paper can make a contribution to the understanding of the capabilities and limitation of robotics for space assembly operations, and help provide focus for research and development activities to improve future robotics systems and support software.

Appendix: Redundant Manipulator Control

Control algorithms for the SSRMS are still under development. The resolved rate algorithm is based upon the use of pseudoinverse (Moore-Penrose generalized inverse) to solve the inverse kinematics problem. In this formulation the joint velocities ($\dot{\gamma}$) are related to the workspace (point of resolution) velocities ($\dot{x}$) by the following expression:

$$\dot{\gamma} = J^+(\dot{x}) + (I - J^+J)k$$

where J^+ represents the pseudoinverse of the Jacobian matrix, I the identity matrix, and k an arbitrary vector. The first part of this equation represents a general solution, brought about by the desired workspace motion, which minimizes the Euclidean norm of the joint velocities. Mathematically, the second term is equal to the orthogonal projection of k into the null space of J. Therefore, the addition of this term to the joint velocities does not contribute to point of resolution motion in the workspace and can be used to define the configuration of the manipulator. One way to use this term is to specify that the redundancy in the manipulator should minimize or maximize the value of some objective function $V(q)$. Since k is arbitrary, it can be set to be proportional to the gradient of this objective function. Then the resolved rate law takes the form

$$\dot{\gamma} = J^+(\dot{x}) + (I - J^+J)\xi \nabla V(q)$$

where ξ is the proportionality constant which specifies the rate of convergence, and $\nabla V(q)$ is the gradient of the objective function.

For the SSRMS, several rate laws and objective functions of this form have been proposed. Most algorithms analyzed to date present issues with regard to either multiple local minima for the objective function or discontinous joint space trajectories. In the case of local minima, the configurations of the arm associated with each minima of the objective function are generally noncontiguous, in that it is not possible to drive the arm from one configuration to another using only the redundant degree of freedom (self motion). In the case of discontinuous joint space trajectories, rapid or undesirable reconfiguration of the manipulator will generally result.

Further, it has been demonstrated that, in some cases, the final configuration of the arm after moving from one point to another can be dependent on the path taken by the end effector during that motion. If the end effector returns to the previous position, traversing the same end effector path, the arm may not return to its previous configuration.

The most recent information presented by Spar indicates that the operator will be provided some capability to control the redundancy in the manipulator. An ability to control the manipulator pitch plane (defined by the three contiguous pitch joints) will likely be provided. Operational studies have shown that this mode is intuitive to the operator, as it can be effectively visualized as "orbiting" the elbow joint about a line from the base to the end effector. Specification of a vector k to weight the relative motion of each joint (in some cases, effectively locking a joint) also may be provided.

Because of the offset joints and range of joint travel, another issue with the SSRMS is avoiding self collisions between the end effectors and booms. Several methods to avoid this have been proposed. One way is to define soft stops (software inhibits) that force the SSRMS to stop before a collision could occur. Another solution is to combine soft stops with an objective function that minimizes the opportunity for self-collision configurations. One such algorithm is the end effector opposing control law, which always keeps the elbow joint "behind" the end effector and manipulator base, thus minimizing the opportunity for self collision.

Singularity management will be more difficult with the SSRMS since there are more singularity zones to contend with in a seven-jointed arm. If an objective function is employed, objective function singularity management also must be provided near configurations at which the redundant degree of freedom constraint degenerate. At these points the desired manipulator configuration is undefined, resulting in discontinuities in the joint space trajectory. Single joint operations may be required when operating near these points to ensure specific arm configurations are obtained.

Acknowledgments

The majority of the work described in this paper has been conducted under NASA Contract NAS9-17885, in support the Automation and Robotics Division of the Engineering Directorate at the Johnson Space Center. The authors would like to acknowledge the support, encouragement provided by our branch chiefs, Charles Gott and Edith Taylor, and the technical direction provided by our technical task monitors, Michael Goza, Kevin Lewis, Leslie J. Quiocho, and Larry Walter.

Reference

[1]Cox, J., "Space Station Freedom Status," *Beyond the Baseline 1991,* Proceedings of the Space Station Evolution Symposium, NASA Conf. Pub. 10083, Aug. 6–8, 1991.

Chapter 17

Space Station Robotics Task Validation and Training

Donald Woods,* Mike Kearney,† Dave Crosse,‡ and Mike Massimino§

McDonnell Douglas Aerospace, Houston, Texas 77062

Introduction

A LTHOUGH the challenges of tomorrow will lead in many new directions, the space frontier has and will continue to be one of the most formidable challenges. The Space Station (Fig. 1) will be the first long-duration outpost in that frontier. The harsh environment of space will present countless risks to those who dare to open this frontier. The Space Station program represents both a nationwide and international effort, challenging thousands of engineers with design, development, and integration of space qualified components and systems. In addition, the Canadians are developing a new sophisticated telerobotic manipulator to enable assembly and maintenance of the Freedom Station vehicle. The Japanese and Europeans are providing pressurized modules to expand Freedom's research capabilities. When the assembly phase is completed the station is projected to weigh approximately 400,000 lb and span 400 ft in length. Projected to have a 30-yr life, the Freedom Station will service researchers in life, materials, and physical sciences and also provide a platform for Earth observation and possibly future space exploration missions.

Although the reasons for completing this project are certainly compelling, first the station must be assembled using both existing and new telerobotics systems that will operate well outside the envelope of our current experience base. Over a sequence of Orbiter-based assembly flights, telerobotic systems will maneuver, position, and attach flexible structures with large masses and offset centers of gravity while providing limited visual cues to the operator. These operations will present a significant challenge to the Space Station team.

Two robotics systems will be available for Station assembly; the Shuttle Remote Manipulator System (SRMS) and the Space Station Remote Manipulator System (SSRMS). In addition, two other specialized robotics systems will be resident on

---

Copyright © 1994 by the American Institute of Aeronautics and Astronautics, Inc. All rights reserved.

*Senior Engineer, Space Station Division.
†Director, Space Station Division.
‡Director, S1 P1 Segment IPT.
§Lead Research Engineer, Space Station Division.

Fig. 1   Space Station robotics operations.

the Freedom Station; the Special Purpose Dexterous Manipulator (SPDM) and the Japanese Exposed Module Remote Manipulator System (JEM-RMS). The SPDM will be utilized for several maintenance tasks and while the JEM-RMS will be used in conjunction with experiments in the JEM exposed facility.

Beginning with mission definition, and continuing throughout the preliminary and critical design phases, the system's engineering process is being used to assure that the integrated vehicle and supporting systems will perform their missions as intended. Although simulations have become an essential tool in the verification process, space-based robotics tasks present a new challenge to simulation technology, namely, zero gravity. This challenge becomes critical in the final validation phase, where a determination is made as to whether or not the system meets user needs, and then in the crew training phase required for mission success. With over 100 high-level robotics assembly tasks, new simulation systems were required to perform Station task validation and training.

Simulation facilities can be categorized by the four characteristics as shown in Table 1. Examples of each type of facility are provided. (MAGIK is the manipulator analysis graphic interactive kinematic, SES the systems engineering simulation, MDF the manipulator development facility, MRMDF the mobile remote manipulator development facility, SDTS the six-degree-of-freedom test system,

Table 1   Types of robotics simulations

	Kinematic simulation	Dynamic simulation
Software simulation	MAGIK Cimstation	Trick SES
Hardware simulation	MDF MRMDF	SDTS (SSAIAF)

and SSAIAF the space systems automated integration and assembly facility. Trick is not an acronym but a dynamic software simulation of manipulator systems; it works in tandem with the MAGIK program. Cimstation is a commercial program primarily utilized for kinematic simulations, but also having some dynamic capabilities; the remainder of these programs and facilities are somewhat unique to the NASA robotics community.) Kinematic simulations are used to describe the movement of the robotics device, while dynamic simulations describe the flexibility of the robotics manipulator's structural members and joints. Software simulations present robotics equations of motion in a visual manner for evaluation, but no actual hardware is used. Hardware simulations use representative components for both kinematic and dynamic evaluations. Each of these simulations is appropriate for a different period of the life cycle of a program.

Kinematic simulations (both hardware and software) are used to define requirements pertaining to component dimensions, required visual cues, and grappling locations; this sort of analysis is usually performed during the requirements definition and early design phases. Dynamic software simulations are performed in the precritical design review (CDR) time frame to ensure that no major structural deficiencies exist in the design. Dynamic hardware simulations which represent the highest level of fidelity are used to validate the integrated man-machine's systems capability to perform a task.

This chapter will focus on two new hardware simulation systems currently under development at NASA's Johnson Space Center: the six-degree-of-freedom dynamic test system (SDTS) and the mobile remote manipulator development facility-training system (MRMDF-TS). The MRMDF-TS provides a supporting development capability and is a key element of the STATION crew training program. The SDTS is a real-time dynamic hybrid hardware and software robotics task validation facility SSAIAF which can support STATION flight readiness certification.

Space Systems Automated Integration and Assembly Facility

As part of the SSAIAF definition process an analysis of the planned robotics assembly tasks was completed. Figure 2 summarizes this analysis with a task characterization by two parameters, robotics system utilized and task type. As noted in this figure, the SRMS is the most frequently used robotics device for assembly operations. Once the SSRMS itself is assembled it becomes involved in several joint tasks with the SRMS and several completely by itself. Some tasks are performed in parallel with extravehicular activity (EVA) by crew members. This same task set represents four distinct types of assembly operations. The most frequent task type is mating and demating of the Shuttle and the Station vehicle. Two other frequent activities will be attachment of equipment to the truss structure and attachment of modules to the truss structure and to one another. The last, but far from the least important, task type is the attachment of truss segments to one another.

The SSAIAF provides the capability to validate an integrated man-in-the-loop and robotics system's ability to complete assembly tasks under simulated orbital conditions. Initially these capabilities will consist of the robotics devices identified for the early utilization in the STATION Program: the SRMS, and the SSRMS.

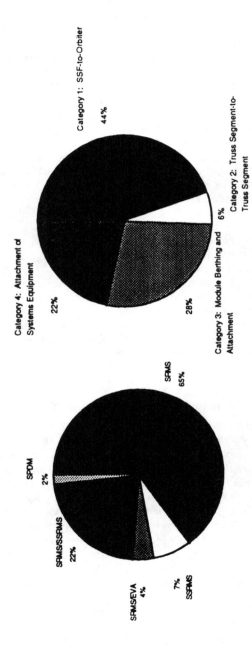

Fig. 2   Manipulator utilization and task types for Space Station robotics assembly tasks.

SSAIAF will provide an integrated dynamic hardware simulation for both of these robotics devices. The test systems are being defined and selected to provide a highly flexible facility that can effectively achieve STATIONP assembly operations validation objectives as well as being used for automation and robotics (A and R) technology development. By developing a modular flexible facility which can easily be adapted to a variety of uses, future space programs' validation and verification costs can be mitigated by using the existing SSAIAF infrastructure.

The detailed layout of the SSAIAF is currently being defined; however, some of the basic support elements have been identified and allocated space within the facility (see Fig. 3). It will include an air-bearing floor(s), 6-DOF table(s), computer(s), and other facility support systems required to complete the planned testing. In addition to these basic capabilities, the SSAIAF will make use of other supporting equipment and capabilities such as the manipulator development facility (MDF), and the MRMDF.

The test support objectives of the SSAIAF are:

1) Validate integrated dynamic performance of hardware/software/ procedures/ crew execution of assembly tasks within planned timelines and acceptable manipulator corridors, under expected variable disturbance conditions such as relative motions caused by flexible body dynamics, arm dynamics, and hardware contact dynamics.

2) Validate assembly hardware design requirements.

3) Validate assembly procedures.

4) Assess selected nominal and off-nominal assembly scenarios.

5) Validate simulations shared with other Station supporting facilities.

6) Enhance crew experience/skill base.

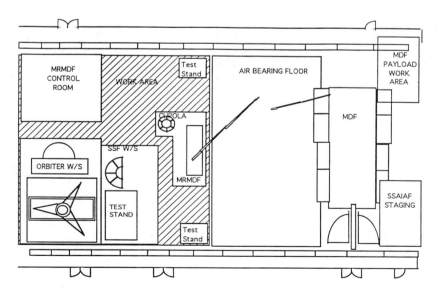

Fig. 3   Preliminary SSAIAF floor plan.

7) Provide for real-time mission support for assembly, as well as flexible rapid response to unanticipated test requirements.

8) Provide a testbed for product/process improvement and for subsequent STA-TION utilization and evolution.

9) Validate the structural operability, and suitability, of the various berthing mechanisms under various contact loading conditions, considering flexible body dynamics, berthing arm dynamics, and hardware contact forces and moments.

10) Validate task completion under off-nominal conditions such as camera failure(s), off-nominal SRMS performance, excessive berthing forces, etc.

The functional architecture for the SSAIAF is shown in Fig. 4. Operator inputs drive the contact dynamics simulation which in turn drive the views which the operator receives. This continuous process creates a set of data which will validate task completion in a realistic dynamic environment. The types of inputs the operator can provide and the visual cues are identified in more detail later. The test system is required to simulate unconstrained manipulator/payload dynamics, and reactions to contact dynamics.

Six-Degree-of-Freedom Test System

SSAIAF will provide all the basic capabilities to support a wide variety of testing scenarios; however, each test will require some engineering analysis to define the test setup (these will be documented in associated test requirements documents and test plans). SSAIAF will simulate contact dynamics and close range movements using the SDTS (Fig. 5). This 6-DOF table approach was used for testing of the Apollo-Soyuz mission docking, and can provide a high level

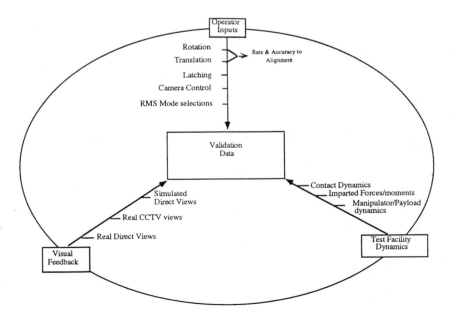

Fig. 4   SSAIAF functional architecture.

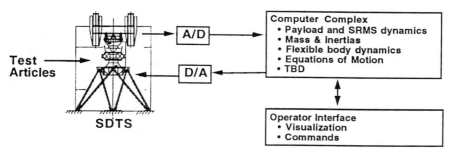

Fig. 5   SDTS functional diagram.

of fidelity for the close in dynamics of Station-related tasks. This system uses a real-time computer system to compute vehicle dynamics, flex body modes, and other system inputs as they affect a chosen assembly process with contact force feedback from the test hardware as it is brought into contact. A test subject or operator controls the simulated robot with hand controllers and interacts with this system based on actual and/or simulated views. This will also simplify the test article design effort since a common test facility interface can be established. The test article fabrication effort will be minimized since passive test hardware can be simulated as opposed to being manufactured.

To better understand how a given operational task is simulated in the SDTS test environment, refer to Fig. 6. The operational scenario shows the berthing of a preintegrated truss (PIT) segment to the berthed Station vehicle. In the test environment a portion of the fixed structure (e.g., the berthed Station) is attached to the super structure, while the portion being maneuvered by the robotics system is mounted on the motion platform. The operator monitors the motion though closed circuit television (CCTV) focused on the test articles and commands the simulated robotics system with hand controllers. The computer systems convert these operator commands into simulated motions of the test article via movement of the motion platform.

As alluded to earlier, the complete set of robotics tasks planned for assembly of the Freedom Station was analyzed and a subset of representative tasks was selected. Some of the criteria used for this selection included both quantitative and qualitative measures such as masses, offset centers of gravity, structural flexibility,

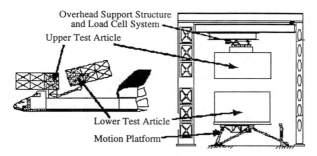

Fig. 6   SSAIAF operational task illustration.

available visual cues and risk to mission success. The chosen "pathfinder test set" for phase I of SSAIAF activation will be a combination of tasks planned for the second assembly flight: berthing the passively damped stage configuration 1 Station vehicle to the orbiter, attaching the PIT segment starboard 2 (S2) to starboard 3 (S3), and the installation of the upper propulsion module to the PIT structure. The successful completion of these tasks is critical to activating the propulsion system which is required to maintain the altitude and attitude of the activated vehicle. (Note: the recent addition of Russian-provided Space Station elements has resulted in significant changes to the planned assembly sequence and robotic utilization. Therefore the planned testing described here will likely undergo some modification.)

Mating the Orbiter with stage 1 is important since this type of maneuver must be performed numerous times over the lifetime of the station. This maneuver constitutes a significant extrapolation of previously demonstrated grappling and berthing capabilities. For example Mission Build 1 (MB1) weight exceeds that of the long-duration exposed facility by 13,000 lb. The stage configuration 1 berthing timeline is also critical (requiring completion in 1 h and 35 min). Attaching the integrated truss segment (ITS) S2 and S3 are critical steps in STATION assembly (basically this step connects the power system with the avionics systems which provides active orbital control) and requires precision control of a large flexible structure with V guides approximately one-third the size of the orbiter V guides and task completion within 2 h and 15 min. Installation of the upper propulsion module is required to facilitate Freedom Station orbit and attitude maintenance. This is also an unusual kinematic task with poor views and less than 2 h to complete. Illustrations of these operational scenarios and how the equivalent SSAIAF test will be setup are shown in Fig. 7.

Both nominal and off-nominal operating conditions will be tested. The tests will validate the human-machine interface during hardware contact phases of assembly operations, emphasizing the human operator's ability to control the arm during critical contact dynamics situations. Therefore the tests will concentrate on the final 3–5 ft of motion through actual hardware contact or positioning within the specified capture envelopes. This pathfinder test set will provide additional confidence in the planned Station assembly process as early as possible in the development process and will provide early testing of the SSAIAF's systems and test equipment.

Recreating the motions of a 6-DOF dynamic test stand so that they appear to the human operator as the SRMS in space is a major challenge for the SSAIAF team. The first step toward creating a realistic environment for testing and training is to develop a mockup that would provide functionality that represented the actual aft flight deck on the Space Shuttle. In addition, a full mockup of the aft flight deck will be used.

To add appropriate functionality to the aft flight deck crew station mockup, a task analysis was conducted to determine which user interfaces were essential for performing the planned assembly operations. The results of this analysis, which identified the required control interfaces are listed here: 1) SRMS mode selection; 2) SRMS joint selection; 3) braking; 4) digital display of rotation/translation parameters; 5) digital display parameter select; 6) camera to monitor and multiplexer (MUX) select; 7) camera pan, tilt, and zoom controls; 8) camera pan, tilt, and zoom

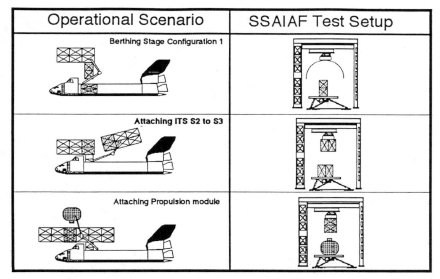

Operational Scenario	SSAIAF Test Setup
Berthing Stage Configuration 1	
Attaching ITS S2 to S3	
Attaching Propulsion module	

Fig. 7   SSAIAF phase I pathfinder tasks and preliminary setup.

speed control; 9) automatic and semiautomatic camera focus and white control; 10) payload retention latching assembly (PRLA) control; 11) PRLA status talk backs; 12) rotation/translation hand controllers; and 13) multifunction cathode ray tube (CRT) display systems (MCDS) and MCDS keyboard.

The preceding interfaces are design requirements for the crew station mockup, and their functions will be incorporated into the user interface. The visual environment on the Space Shuttle for Space Station assembly will have two primary sources for visual feedback: 1) CCTV monitors which display views from Shuttle cameras, and 2) out the window views. Similarly, the SSAIAF CCTV monitors will be driven from both from cameras mounted on the actual SSAIAF hardware and simulated views generated with computer graphics (see Fig. 8). Whether the view will be real or simulated will be determined by a computer algorithm, and will depend on 1) if the real camera view can be provided given the physical limitations at the test facility, and 2) if the view is significant enough to carry out the operation (e.g., if during propulsion module installation, it will not do any good to have real camera views from Shuttle payload bay cameras A, B, C, or D, even if it is physically possible to provide them). The views that will help the operator complete the task are from special cameras near the interface between the propulsion module and the attachment system. Real views will be used when the camera view is representative of what would actually be seen on orbit. This will occur, for example, when the camera is focused on a close-up view of a test article or of a trunion being inserted into a V guide. Simulated views can be used for viewing parts of the environment that are not physically part of SSAIAF hardware. An example of this would be the Orbiter payload bay as viewed from an overhead camera view. Camera control will be controlled through computer simulation as well as operator inputs in some cases.

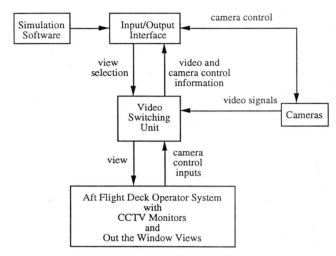

Fig. 8   Visual feedback control loop.

For out-the-window views through the aft flight deck windows, again a combination of real camera views and computer simulated views will be used. Fixed cameras will be used to capture images that would be seen through the aft flight deck windows wherever possible and necessary. For the overhead window views, which are located directly above the human operator's head, it is expected that a computer generated image will be used wherever the real views are not possible and necessary.

A major problem with representing out the window views through a camera view is that the resolution of the view must look like a direct view. Therefore high-resolution broadcast cameras will be used to provide realistic viewing conditions for out the window views. These same cameras will also be used to provide CCTV monitor views. However, when providing CCTV views, the camera output will be downgraded to reflect the lower resolution that would be provided through the CCTV monitors.

Another problem is how to provide the operator with depth cues for the out-the-window views. For actual out-the-window views that would occur in space, the operator would be able to receive depth cues to perceive the three-dimensional images of a direct visual view. To provide these realistic depth cues, wide angle collimator (WAC) windows will be used. The high-resolution monitor in the WAC assembly projects the image onto a beam splitter which then projects the image onto a parabolic mirror. The reflected image from the mirror provides the operator with what appears to be a three-dimensional image. The optical principles of WAC window are shown in Fig. 9.

As can be seen in Fig. 9, the WAC to be used for the SSAIAF is a reflective display system. It uses a spherical mirror to collimate or project the displayed scene from the CRT to optical infinity. This allows the production of a three-dimensional scene for the operator with depth cues as would occur with actual direct views.

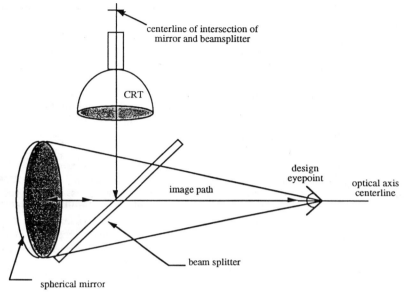

Fig. 9   WAC optics.

Mobile Remote Manipulator Development Facility

The MRMDF will provide a facility for Space Station crew members and ground support personnel to perform mobile servicing center (MSC) training, operational procedures development, CCTV operation, and payload design. As shown in Fig. 3, the MRMDF is located between the MDF and the 6-DOF table in building 9N at NASA/JSC.

The MRMDF is intended to replicate the appearance and functionality of the Canadian Space Agency's contribution to Station, the MSC. The MRMDF will provide MSC training on a ground-based system which provides similar flight crew interfaces and replicates MSC orbital functionality and performance characteristics as closely as possible. It will provide Station personnel with training for handling various payload elements for Station assembly operations and for handling orbital replacement units (ORUs) for maintenance operations. The MRMDF will assist operations personnel in developing procedures for the handling of payloads, ORUs, and Station elements. This includes moving payloads and ORUs along the truss, installing and removing payloads, handoff procedures, and Station construction.

The mobile closed circuit television (MCCTV) training system will be an operational replication of the television capability provided on the MSC. It will be used to allow developers to test and evaluate various replacements of CCTV equipment. It will provide the primary CCTV training interface for the crew using the MSC. Flight-similar replication of the flight system will give developers the ability to test and evaluate payloads and payload interfaces and evaluate sizes and shapes of potential payloads. This will allow crew training of simulated and actual payloads.

As shown in Fig. 10, the MSC includes the SSRMS, a manipulator base system (MBS) which supports the SSRMS at four different locations via power and data grapple fixtures, and the mobile transporter. An early concession to simplification of the MRMDF was to delete the mobility aspect of the MSC and fix the manipulator to a permanent base. The visual appearance of the base and transporter are provided by mockups.

The functional core is the mobile remote manipulator (MRM) which is a hydraulically actuated, computer-controlled, 7-DOF robotics manipulator used primarily to train operators in the use of the SSRMS. The MRM will simulate the SSRMS operation by tracking joint rates, angular limits, and physical size by providing booms and joints which match the SSRMS as closely as possible.

Attached to the wrist roll joint of the MRM is a replica of the SSRMS latching end effector (LEE). It will visually and mechanically duplicate the LEE functions. It also supports the power and data interfaces to the payloads. The MRM is mounted on a load-bearing truss which is anchored to the building floor. The MSC transporter and base are simulated with high-fidelity mockups to provide the proper visual cues and mechanical envelopes for positive training. The Station flight trusses are simulated with portable truss segments which provide the correct work site environment for maintenance training. The primary training workstation replicates the controls and displays of the station command and control workstation

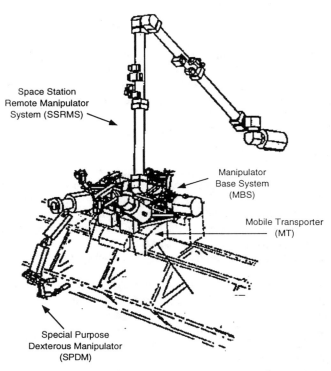

Space Station
Remote Manipulator
System (SSRMS)

Manipulator
Base System
(MBS)

Mobile Transporter
(MT)

Special Purpose
Dexterous Manipulator
(SPDM)

Fig. 10   Station MSC.

located in the cupola of the node. The training workstation consists of a computer workstation with three monitors, keyboard and trackball, and MRM rotational and translational hand controllers. The workstation will contain CCTV controls to operate the CCTV system. The workstation will be housed in a cupola mockup and located on the building floor in approximately the same relationship to the truss as it will be on orbit.

The operations complex contains the control computers and other avionics systems required to operate the MRMDF. It houses the test director's console which provides facility operations control as well as basic functions in MRM control.

One of the challenging design aspects of the MRMDF is the design of the manipulator arm. A prime design requirement is to replicate the physical features and functional operations of the flight arm as much as possible with a 500-lb payload. This requires not only matching the flight joint performance as shown in Table 2, but also the geometry. The design to meet these requirements has the end effector, wrist joints, and upper boom matching the flight design; however, the elbow joint and shoulder joints are substantially larger to withstand the higher moments resulting from the gravity effects that the SSRMS does not have to contend with. The lower boom is somewhat larger than the flight arm. Figures 11 and 12 show comparisons of the resulting geometries. The SSRMS weighs 3300 lb, the MRM exceeds 16,000 lb through the shoulder joints.

For the MRM to behave and respond like the SSRMS it must match the SSRMS control system bandwidth of 0.25 Hz maximum. This requires structural stiffness of the MRM of 0.75 Hz minimum. This stiffness requirement, combined with the geometry constraints for visual replication, arm mass distribution, and accommodations for joint flexibility resulted in ultrahigh modulus (UHM) graphite epoxy being chosen for the boom material. The approximately 700 lb of UHM fiber is the largest usage of this material to date in the United States. Each joint is driven by servo valve controlled high-speed hydraulic motors, one in the wrist roll joint and

Table 2   Comparison for SSRMS and MRMDF capabilities

	SSRMS		MRMDF	
Requirement	Min.	Max.	Min.	Max.
EE tip translation velocity (in./s)				
Unloaded	N/A	14.6	0.35	14
Loaded	N/A	0.79	0.35	14
EE tip translation velocity (deg/s)				
Unloaded	N/A	4.0	0.08	5.0
Loaded	N/A	0.15	0.08	5.0
Tip position accuracy (in.)	2.6	N/A	2.6	N/A
Tip attitude accuracy (deg)	0.71	N/A	0.71	N/A
Starting/stopping distance (in.)				
Unloaded	N/A	24	10	50
Loaded	N/A	50	10	50

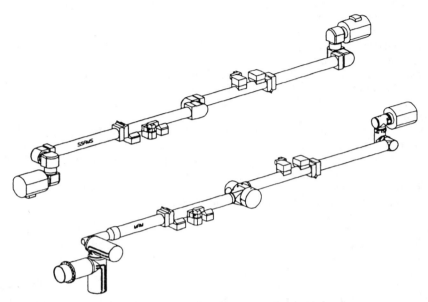

Fig. 11   Comparison of SSRMS and MRMDF configurations—isometric view.

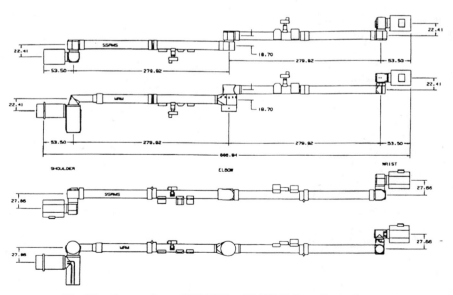

Fig. 12   Comparison of SSRMS and MRMDF configurations.

up to four in the shoulder joints, through planetary gear sets. High-speed motors were required to avoid low-speed ripple torques that would impact control system performance.

Summary

The first phase of the SSAIAF is now operational, and two subsequent phases to upgrade the facility are planned. The MRMDF-TS design phase will be complete in summer 1994 and after manufacturing is completed it will be ready for operations in late 1996 to support initialization of SSRMS operations in 1998. Both of these projects push the edges of simulation technology. To provide realistic contact dynamics and visual cues, the SSAIAF computer systems have to close dynamics loops with very high frame rates (e.g., milliseconds). Designed to emulate he SSRMS at the end effector despite the effects of its Earth gravity environment, the MRMDF-TS has a cutting-edge control system and specialized hardware and electronic designs. Both of these systems will help provide assurance that no surprises occur on orbit during the assembly operations phase of the Station program.

Author Index

PROGRESS IN ASTRONAUTICS AND AERONAUTICS
SERIES VOLUMES

*1. **Solid Propellant
Rocket Research** (1960)
Martin Summerfield
Princeton University

*2. **Liquid Rockets
and Propellants** (1960)
Loren E. Bollinger
Ohio State University
Martin Goldsmith
The Rand Corp.
Alexis W. Lemmon Jr.
Battelle Memorial Institute

*3. **Energy Conversion
for Space Power** (1961)
Nathan W. Snyder
*Institute for Defense
Analyses*

*4. **Space Power
Systems** (1961)
Nathan W. Snyder
*Institute for Defense
Analyses*

*5. **Electrostatic
Propulsion** (1961)
David B. Langmuir
*Space Technology
Laboratories, Inc.*
Ernst Stuhlinger
*NASA George C. Marshall
Space Flight Center*
J.M. Sellen Jr.
*Space Technology
Laboratories, Inc.*

*6. **Detonation and
Two-Phase Flow** (1962)
S.S. Penner
*California Institute
of Technology*
F.A. Williams
Harvard University

---

*Out of print.

*7. **Hypersonic Flow
Research** (1962)
Frederick R. Riddell
AVCO Corp.

*8. **Guidance and
Control** (1962)
Robert E. Roberson,
Consultant
James S. Farrior
*Lockheed Missiles
and Space Co.*

*9. **Electric Propulsion
Development** (1963)
Ernst Stuhlinger
*NASA George C. Marshall
Space Flight Center*

*10. **Technology of
Lunar Exploration** (1963)
Clifford I. Cummings
Harold R. Lawrence
Jet Propulsion Laboratory

*11. **Power Systems
for Space Flight** (1963)
Morris A. Zipkin
Russell N. Edwards
General Electric Co.

*12. **Ionization in High-
Temperature Gases** (1963)
Kurt E. Shuler, Editor
*National Bureau of
Standards*
John B. Fenn,
Associate Editor
Princeton University

*13. **Guidance and
Control—II** (1964)
Robert C. Langford
General Precision Inc.
Charles J. Mundo
Institute of Naval Studies

*14. **Celestial Mechanics
and Astrodynamics** (1964)
Victor G. Szebehely
*Yale University
Observatory*

*15. **Heterogeneous
Combustion** (1964)
Hans G. Wolfhard
*Institute for Defense
Analyses*
Irvin Glassman
Princeton University
Leon Green Jr.
*Air Force Systems
Command*

*16. **Space Power
Systems Engineering** (1966)
George C. Szego
*Institute for Defense
Analyses*
J. Edward Taylor
TRW Inc.

*17. **Methods in
Astrodynamics and
Celestial Mechanics** (1966)
Raynor L. Duncombe
U.S. Naval Observatory
Victor G. Szebehely
*Yale University
Observatory*

*18. **Thermophysics and
Temperature Control
of Spacecraft and
Entry Vehicles** (1966)
Gerhard B. Heller
*NASA George C. Marshall
Space Flight Center*

73. Combustion Experiments in a Zero-Gravity Laboratory (1981)
Thomas H. Cochran
NASA Lewis Research Center
ISBN 0-915928-48-5

74. Rarefied Gas Dynamics, Parts I and II (two volumes) (1981)
Sam S. Fisher
University of Virginia
ISBN 0-915928-51-5

75. Gasdynamics of Detonations and Explosions (1981)
J.R. Bowen
University of Wisconsin at Madison
N. Manson
Université de Poitiers
A.K. Oppenheim
University of California at Berkeley
R.I. Soloukhin
Institute of Heat and Mass Transfer, BSSR Academy of Sciences
ISBN 0-915928-46-9

76. Combustion in Reactive Systems (1981)
J.R. Bowen
University of Wisconsin at Madison
N. Manson
Université de Poitiers
A.K. Oppenheim
University of California at Berkeley
R.I. Soloukhin
Institute of Heat and Mass Transfer, BSSR Academy of Sciences
ISBN 0-915928-47-7

*77. Aerothermodynamics and Planetary Entry (1981)
A.L. Crosbie
University of Missouri-Rolla
ISBN 0-915928-52-3

78. Heat Transfer and Thermal Control (1981)
A.L. Crosbie
University of Missouri-Rolla
ISBN 0-915928-53-1

*79. Electric Propulsion and Its Applications to Space Missions (1981)
Robert C. Finke
NASA Lewis Research Center
ISBN 0-915928-55-8

*80. Aero-Optical Phenomena (1982)
Keith G. Gilbert
Leonard J. Otten
Air Force Weapons Laboratory
ISBN 0-915928-60-4

81. Transonic Aerodynamics (1982)
David Nixon
Nielsen Engineering & Research, Inc.
ISBN 0-915928-65-5

82. Thermophysics of Atmospheric Entry (1982)
T.E. Horton
University of Mississippi
ISBN 0-915928-66-3

83. Spacecraft Radiative Transfer and Temperature Control (1982)
T.E. Horton
University of Mississippi
ISBN 0-915928-67-1

84. Liquid-Metal Flows and Magnetohydrodynamics (1983)
H. Branover
Ben-Gurion University of the Negev
P.S. Lykoudis
Purdue University
A. Yakhot
Ben-Gurion University of the Negev
ISBN 0-915928-70-1

85. Entry Vehicle Heating and Thermal Protection Systems: Space Shuttle, Solar Starprobe, Jupiter Galileo Probe (1983)
Paul E. Bauer
McDonnell Douglas Astronautics Co.
Howard E. Collicott
The Boeing Co.
ISBN 0-915928-74-4

*86. Spacecraft Thermal Control, Design, and Operation (1983)
Howard E. Collicott
The Boeing Co.
Paul E. Bauer
McDonnell Douglas Astronautics Co.
ISBN 0-915928-75-2

87. Shock Waves, Explosions, and Detonations (1983)
J.R. Bowen
University of Washington
N. Manson
Université de Poitiers
A.K. Oppenheim
University of California at Berkeley
R.I. Soloukhin
Institute of Heat and Mass Transfer, BSSR Academy of Sciences
ISBN 0-915928-76-0

88. Flames, Lasers, and Reactive Systems (1983)
J.R. Bowen
University of Washington
N. Manson
Université de Poitiers
A.K. Oppenheim
University of California at Berkeley
R.I. Soloukhin
Institute of Heat and Mass Transfer, BSSR Academy of Sciences
ISBN 0-915928-77-9

*89. Orbit-Raising and
Maneuvering Propulsion:
Research Status and Needs
(1984)
Leonard H. Caveny
*Air Force Office of
Scientific Research*
ISBN 0-915928-82-5

90. Fundamentals
of Solid-Propellant
Combustion (1984)
Kenneth K. Kuo
*Pennsylvania State
University*
Martin Summerfield
*Princeton Combustion
Research Laboratories,
Inc.*
ISBN 0-915928-84-1

91. Spacecraft
Contamination: Sources
and Prevention (1984)
J.A. Roux
University of Mississippi
T.D. McCay
*NASA Marshall Space
Flight Center*
ISBN 0-915928-85-X

92. Combustion
Diagnostics
by Nonintrusive
Methods (1984)
T.D. McCay
*NASA Marshall Space
Flight Center*
J.A. Roux
University of Mississippi
ISBN 0-915928-86-8

93. The INTELSAT
Global Satellite
System (1984)
Joel Alper
COMSAT Corp.
Joseph Pelton
INTELSAT
ISBN 0-915928-90-6

94. Dynamics of Shock
Waves, Explosions, and
Detonations (1984)
J.R. Bowen
University of Washington
N. Manson
Université de Poitiers
A.K. Oppenheim
*University of California
at Berkely*
R.I. Soloukhin
*Institute of Heat and Mass
Transfer, BSSR Academy
of Sciences*
ISBN 0-915928-91-4

95. Dynamics of Flames
and Reactive Systems
(1984)
J.R. Bowen
University of Washington
N. Manson
Université de Poitiers
A.K. Oppenheim
*University of California
at Bereley*
R.I. Soloukhin
*Institute of Heat and Mass
Transfer, BSSR Academy
of Sciences*
ISBN 0-915928-92-2

96. Thermal Design
of Aeroassisted Orbital
Transfer Vehicles (1985)
H.F. Nelson
*University of Missouri-
Rolla*
ISBN 0-915928-94-9

97. Monitoring Earth's
Ocean, Land, and
Atmosphere from Space —
Sensors, Systems, and
Applications (1985)
Abraham Schnapf
*Aerospace Systems
Engineering*
ISBN 0-915928-98-1

98. Thrust and Drag:
Its Prediction and
Verification (1985)
Eugene E. Covert
*Massachusetts Institute
of Technology*
C.R. James
Vought Corp.
William F. Kimzey
*Sverdrup Technology
AEDC Group*
George K. Richey
U.S. Air Force
Eugene C. Rooney
*U.S. Navy Department
of Defense*
ISBN 0-930403-00-2

99. Space Stations
and Space Platforms —
Concepts, Design,
Infrastructure,
and Uses (1985)
Ivan Bekey
Daniel Herman
NASA Headquarters
ISBN 0-930403-01-0

100. Single- and Multi-
Phase Flows in an
Electromagnetic Field:
Energy, Metallurgical, and
Solar Applications (1985)
Herman Branover
*Ben-Gurion University
of the Negev*
Paul S. Lykoudis
Purdue University
Michael Mond
*Ben-Gurion University
of the Negev*
ISBN 0-930403-04-5

101. MHD Energy
Conversion:
Physiotechnical
Problems (1986)
V.A. Kirillin
A.E. Sheyndlin
*Soviet Academy
of Sciences*
ISBN 0-930403-05-3